T0324789

WINE TASTING

THIRD EDITION

WINE TASTING

A Professional Handbook

THIRD EDITION

RONALD S. JACKSON

Cool Climate Oenology and Viticulture Institute, Brock University, St. Catharines, Ontario, Canada

AMSTERDAM • BOSTON • HEIDELBERG • LONDON
NEW YORK • OXFORD • PARIS • SAN DIEGO
SAN FRANCISCO • SINGAPORE • SYDNEY • TOKYO
Academic Press is an imprint of Elsevier

British Library Cataloguing-in-Publication Data
A catalogue record for this book is available from the British Library

Library of Congress Cataloging-in-Publication Data
A catalog record for this book is available from the Library of Congress

ISBN: 978-0-12-801813-2

For Information on all Academic Press publications
visit our website at https://www.elsevier.com

Working together
to grow libraries in
developing countries

www.elsevier.com • www.bookaid.org

Publisher: Nikki Levy
Acquisition Editor: Nancy Maragioglio
Editorial Project Manager: Billie Jean Fernandez
Production Project Manager: Caroline Johnson
Designer: Ines Cruz

Typeset by MPS Limited, Chennai, India

Dedication

To the memory of inspiration, Suzanne Ouellet

Contents

Preface

The text has been primarily been written as an information conduit for enologists, winemakers, sensory scientists, and students in those endeavors. Nonetheless, material has been specifically added for those involved in wine appreciation, wine societies, sommeliers, and the serious wine lover. For example, the relevance of psychophysiology to wine tasting has been emphasized. As such, the text deals almost exclusively with a wine's intrinsic (sensory) qualities, relegating extrinsic aspects, such as the "best" vintages, producers, regions, grape varieties to others. There is more than an ample supply of this material in books, magazines, and articles on these subjective aspects of wine. It is what the wine itself communicates to our senses, and how the brain integrates and interprets this information that is covered, i.e., material which is within the purview of the scientist.

Winemakers approach tasting heuristically, either to direct production of particular types of wine, or to check for sensory problems before they become serious or intractable. In a similar manner, wholesalers and retailers use sampling to assist in their selection of wines that hopefully will pique the interest of their clients. Thus, both need to know the characteristics deemed important to their customers. For most connoisseurs, tasting wine is intended to enhance appreciation, either by itself, or in combination with food. These are the individuals toward whom most boutique or premium wine producers aim their products. For others, wine consumption is a social statement, acts as an affirmation of cultural heritage, or is just savored as is. For even more, especially in Europe, wine is consumed as the daily food beverage, tasting with discernment being relegated to special occasions.

In addition to the differences in how people view wine tasting, even the term *taste* has multiple meanings. To the sensory scientist, taste refers to a select group of compounds (sweet, sour, salty, bitter, umami) detected by modified epithelial cells, primarily located in taste buds on the tongue. In common usage, taste incorporates the sensations of mouth-feel and retronasal odor, generating the multimodal perception, flavor. As a verb, tasting refers to the process of sampling beverages and foods, usually in a deliberate, assessment mode. The terms organoleptic and degustation refer specifically to aspects of this process, but have been espoused by neither the scientific community nor popular literature.

The complexities surrounding the terms *taste* and *tasting* underscore a reality. Although the various sensations that a wine may express are initially analyzed in distinct areas of the brain, they are subsequently integrated in the orbitofrontal cortex. Here, they are also merged with previously generated vinous memories, creating our conscious image of the wine. Thus, any wine is interpreted through the lens of unconscious mental models constructed throughout a lifetime of tasting experiences. These constructs may be as potent as those that generate the Ames room illusion, where people are interpreted to grow or shrink as they walk from one corner of a room to the other (https://www.youtube.com/watch?v=EOeo8zMBfTA). Hopefully, training and conscious effort can free tasters from being prisoners of their own past, but this has not been established. Examples of taste illusions are the "sweet" sensation of many fruity odors in wines, the correlation between the mint flavor of gum and its sugar content, the nutty aspect of cracked-wheat bread, and the distortion of flavor by color.

With wine being viewed in a multiplicity of ways, it is not surprising that people construe wine and its tasting variously. For much of the text, wine tasting refers to the critical analysis of wine, either attempting to rank it within some concept of quality, cultivar, or appellation prototype, delineate its sensory diversity, or investigate the origins of its sensory characteristics. To do so, the analysis attempts to differentiate between sensation and perception. It is searching for human *reality*. Because people vary considerably in general as well as specific sensitivity to sensory inputs, panels of tasters are required to obtain data that may be both statistically significant and of human relevance (not always synonymous). Both training and experience are usually required for panel members to separate subjective responses from consistent, defined, objective evaluation.

Although training and experience can increase specific acuity, without proper application when used by consumers, they can lead to focusing on inessentials. There is even evidence that concentrating on odor descriptors can disrupt the ability to recall odor memories associated with distinguishing a wine's varietal, stylistic, or regional uniqueness. Nonetheless, most consumers seem content to appreciate, and appropriately so, wine for what it is—a salubrious, savory beverage—rather than to pursue a vain attempt to name every evanescent sensory nuance.

The techniques involved in critical sensory analysis ideally will assist grape growers and winemakers to improve the quality of their wines, and produce them in a diversity of styles to match the heterogeneity of their purchasers. In a perfect future, any wine worthy of the name should be of such quality that consumers need no advice from "experts" to find superb wine anywhere on the supermarket shelf.

Wine appreciation for many consumers reflects its supposed quality, which is the property that sells most expensive wines. Thus, most consumer-oriented wine tastings are conducted under conditions designed to favor such an impression. This usually involves extrinsic factors, such as the wine's price, the repute of the producer or estate, or accolades about the vintage. However, for most people, the principal value of wine is the pleasure it can provide. Thus, in the end, it behooves the vintner, the retailer, and the researcher to realize that more emphasis needs to be placed on understanding what attributes are the most likely to provide that pleasure. It is only for enophiles that initially unpleasant wines have an appeal. They know, or hope, that with proper storage, their patience will be rewarded with special and unique sensory joys, justifying the earlier outlay of (usually) considerable funds.

It is in the potential to provide a better understanding of the underpinnings of wine quality that sensory analysis has its greatest value. However, these clues will be of little long-term use until the chemical origins of quality are clear. This is a prerequisite for studying the conditions that influence the subtleties of grape, yeast, and bacterial secondary metabolism, i.e., those distinctive aspects that ultimately impact our senses. Chemistry is also essential to understanding the origins of the sensory uniqueness of varietal, stylistic, and regional wines, and how they can be manipulated in a rational fashion to meet the challenges of climate change. Winemakers can do wonders with blending, to optimize the qualities of the wines at their disposal, but the process would be speeded, if not also improved, if the chemical and psychophysiological nature of the benefits of blending were better known.

What also limits research efforts in improving wine quality is the absence of a clear, objective, generally accepted definition of what delineates quality (possibly because none is possible). Quality is a mental construct, ideally based on assiduous vinous sampling, versus extraneous factors that may bear little relation to sensory quality. Because wine is considered a sophisticated beverage, it is often viewed as an art form, and thus subject to the arbiters of taste, at least for those searching guidance. Most aficionados are sufficiently knowledgeable to be little influenced by the popular but valueless numerical scores. They realize that great wine is produced under the tutelage of vinous artisans, and no more amenable to ranking on a scale of 100 than paintings in an art gallery, classical music, or fine literature. For most consumers, ease of purchase, price, and the absence of unpleasant tastes are the principal predictors of acceptability (quality). Nonetheless, popular wine writers continue to construct their commentaries as if people cared, believed, or remembered the florid web of fantasies their fertile minds weave. For those desperate to achieve the impression of instant expertise, verbal effervescence has great appeal. Were it not for the disservice such logorrhea has on the profits of truly fine wine producers, it would not matter. This is particularly so with "label drinkers," who seem more interested in possessing the latest wines of repute (or high score), than assessing for themselves the wine's sensory worth. When I began studying wine, I too was beguiled by supposed experts, until I accidentally tasted some astounding wines, at about 10% the bloated price of the austere wines they flogged (possessing the sensory delights of unripened persimmons). Their recommendations gave no impression of wine being a "gift of the gods."

It is always useful to realize that sensory perception is *processed*, not *transmitted*, information. This has become particularly evident with individuals having highly localized strokes or specific sensory deficiencies. For example, affected people may distinguish objects but not color, trace objects perfectly but not recognize them, or be unable to recognize faces but readily identify other objects. Our sensory system evolved primarily to rapidly detect, recognize, and respond to discontinuities in time, space, or attribute (those with potential biological significance), avoiding the time and processing demands required to analyze slow, gradual changes. Our brains predate computer compression algorithms, ignoring uniform regions while concentrating on zones of transformation.

In critical tasting, it is especially important to attempt to negate the effect of ingrained bias on perception. These develop largely unawares, associated with personal experience, or are inculcated from external sources. People may enunciate concurrence with such sources, but whether this is sincere or due to servility, self-induced delusion, or polite silence is uncertain. Equally, one must be cautious about interpreting sensory data at face value. What seems obvious is not always true, and vice versa. For millennia, people thought the sun rotated around the earth—so much for apparent truth. Because the mind inherently constructs experience-based models to interpret sensory input, we must be constantly on guard so that these percepts do not dominate our thoughts, making it impossible to see its noumenon.

Tasting techniques are always personal, but the procedure described herein is an amalgam designed to maximize sensory detection. As such, it is particularly apropos for those involved in sensory analysis or wine-making. Although analytic, the process is also useful, possibly in simplified form, to anyone desirous of detecting the sensory pleasures a wine can bestow. Nonetheless, it is important that no one be under any illusion that it is easy to distinguish (at least consistently) varietal or regional styles. Many factors must be concurrent for clear expression. The wine must be made from grapes inherently possessing a unique varietal aroma, and grown under conditions conducive to its development, and the wine must be fermented and aged under conditions that permit the varietal and/or regional characteristics to be fully expressed. Although desirable and potentially enhancing the wine's sensory intrigue, such attributes are not essential for enjoyment. Wine is, after all, hopefully consumed for the pleasure it provides. Wine may periodically be "dissected" or revered, but not all the time.

Small wineries are unlikely to employ the staff necessary to conduct the detailed sensory analyses outlined in the text. Their wines are typically produced on the palate of the winemaker/cellar master, being viewed as creative, artisanal products. The wines sell because sufficient customers accept their perceptions. It also helps that most consumers are not always, or unduly, discriminating or demanding. Customers are often persuaded by or accept the opinions of others. This is not intended as a disparaging remark, just an affirmation of the truth. Some small wine estates produce absolutely superior wines, and their staff is exceptionally skilled. The same procedures, in less competent hands, can produce mediocre to eminently forgettable wines. The situation is quite different for large wineries. Their wines are sold internationally, produced in million-liter quantities, and sold continents away from direct contact with winery staff. Successful brands are created using blending techniques that require some of the most sensitive and critical sensory evaluation procedures available. Millions of dollars, and shareholder profits, can ride on the decisions made by blenders and evaluation panels. There is little margin for error. Quality control is critical.

The text commences by guiding the reader through the process of sampling wine, leading into a discussion of the psychophysiology of sensory perception. This is followed by detailing the optimal conditions for wine assessment and evaluation, the selection and training of tasting panels, the performance of various sensory procedures, and the analysis of its significance. Subsequently, wine classification and the origins of wine quality are covered with a disquisition on what can confidently be said about wine and food pairing.

Although major strides in understanding sensory perception have occurred in the past few years, it is also becoming clearer that perception is relative. What a person perceives depends not only on their sensory acuity and training, but also on their upbringing and life experiences, current emotional and physical health, and the tasting context. Within limits, the latter can be the most important aspect affecting a wine's perceived quality. Dogmatism concerning wine quality is as outmoded as the Model-T Ford.

Because of the historical and cultural connections between food and wine, and its popularity, the topic has more general interest than probably any other chapter in the book. Despite this, the best wines usually express their finest qualities when savored alone. For example, the development and finish of a wine are seldom detectable when consumed with food. To facilitate their detection, cognoscenti often analyze their wines prior to eating, or consume the wine slowly and conscientiously with simply prepared food with minimal flavoring. Equally, aperitif and dessert wines are more amenable to full appreciation when, as usual, they are taken by themselves.

As with the appreciation of other ethereal aspects of life, the status and well-being associated with sampling fine wines is of primal concern for many an aficionado. This can be further enhanced within the refined ambiance of an excellent dining establishment. This is one of the advantages of surplus income, where the pleasure of eating can be divorced from simply abating hunger. In addition, pairing wine with food also facilitates avoiding potential gustatory monotony at the table. Except where expert opinion is followed servilely, just the act of selecting a wine from the list can give anticipation a tantalizing gratification.

Wine can act as a wonderful accompaniment to a meal, helping to cleanse the palate, while providing a distinct and gratifying sensation, and can elevate mealtime to a sublime celebration of being alive. In turn, food freshens the palate to receive anew the flavors of the wine. Compatibility rests primarily in their differences, not similarities. In this regard, wine can be considered as a food condiment. Correspondingly, a central tenet of food and wine combination is that the attributes of the wine should neither clash with the food, nor be excessively mild in comparison with the predominant food flavors. This is, of course, based on the flavor sensitivities of the individual, and acceptance of the concept of balance as desirable. Although an interesting and endless topic of conversation, undue concern about pairing can overshadow what should be a pleasurable and relaxed occasion to nourish the body as well as the soul.

Hopefully, the information contained herein will allow the reader to strip away some of life's experience-based biases that too often afflict the appreciation of wine. As often as not, you should question your own perceptions as much as the views of others. Investigations have clearly demonstrated how powerful expectation or suggestion can be, not only on the higher cognitive centers of the brain, but also the responsiveness of our sensory receptors. Knowing how the brain can potentially "deceive" us may give us power over our pragmatic cerebral models of life. It is usually preferable to be in the driver's seat than be driven.

My final wish is that you will take the time to relax and contemplate the full sensory complement that food and wine can impart. Wines can embellish our all-to-short sojourn on this small but beautiful speck of the universe. But, for this to have its most hallowed effect, do this with a significant other. For me, all my vinous epiphanies were with Suzanne, as we savored leisurely meals, often with a wine about which we knew little (the element of surprise). Although I was the one academically involved with wine, we both merely loved assessing wine together, sharing our opinions, and being fascinated with what each other sensed.

I leave you with how André Simon, *bon vivant extraordinaire*, defined a connoisseur:

> "One who knows good wine from bad, and is able to appreciate the different merits of different wine."

Acknowledgments

Without the dedication of innumerable researchers, the complexities of human sensory acuity and perception would remain mysteries and this book would have been impossible.

Many thanks go to my students, participants of sensory panel tests, MLCC External Tasting Panel, the Manitoba Liquor Control Commission, and Elsevier for providing the opportunity to gain access to both the practical and theoretical sides of wine assessment.

Finally, but certainly not least, I must express my thanks to the assistance provided by the staff at Elsevier, notably Nancy Maragioglio. Their help and encouragement have been critical in bringing this book to fruition.

CHAPTER

1

Introduction

INTRODUCTION

As befits one of life's finest pleasures, wine merits serious attention. Nevertheless, no tasting procedure has achieved universal adoption. Most experienced tasters have their own preferred method. Although essential for critical tasting, detailed sensory analysis is too elaborate for most consumers. The difference is somewhat analogous to analyzing a musical score versus attending its performance. Critical tasting compares one or several wines against real or memory-derived archetypical standards. In contrast, wine with a meal is designed to be savored as its liquid accompaniment. Critical wine assessment is also ill designed for the dining room, due to its social and epicurean distractions. Nonetheless, even here, periodic concentration on a wine's attributes can reward the diner with enhanced consciousness.

TASTING PROCESS

The procedure outlined in Fig. 1.1 is a synthesis of views going back centuries, as well as experience gained from assessing tasters. The first known recorded description of studious wine tasting is noted in Francese Eiximenis (1384). Interestingly, the commentary is disparaging, comparing the procedure to how physicians of the time analyzed their patients' urine. Although no procedure is ideal for everyone, or under all situations, Fig. 1.1 provides a reasonable starting point. Probably the most essential requirement is the willingness, desire, and ability to focus attention on the intrinsic attributes of the wine itself, with a minimum of extrinsic information that might skew perception. Tasting with black International Standards Organization (ISO) glasses (Plate 6.1) is ideal for hiding visual details that can distort perception (e.g., features suggesting varietal or geographic origin, or wine age). This does delay appreciation of the wine's countenance, but it forces the taster to concentrate on the wine's essential sensory properties.

Peynaud (1987) advocated rinsing the mouth with a sample of the wines to be tasted before embarking on a dedicated assessment. Where the wines are unfamiliar, this could familiarize the tasters' to the wines' basic attributes. However, under most circumstances such a practice seems ill advised, as it is preferable to sample each wine unfettered by expectations. Peynaud also cautions against rinsing the palate between samples. He feels that this could alter sensitivity and complicate assessment. In this recommendation, he is at variance with essentially all other authorities. Only when the palate seems fatigued does he support palate cleansing. Peynaud's view assumes that tasters can perceive accurately when their senses have been or are becoming adapted, a dubious supposition. Most data would suggest it is preferable to encourage tasters to cleanse their palate between samples, to avoid, as much as possible, adaptation altering perception. Ideally, this should permit each wine to be assessed under uniform conditions. Nonetheless, when sampling very complex wines (e.g., vintage ports), olfactory adaptation can be of value. It can result in the "unmasking" of aromatic compounds whose presence was undetectable before adaptation to more potent flavorants (Goyert et al., 2007). Tasting over a prolonged period offers the chance to detect both quantitative and qualitative changes in fragrance throughout a tasting, as the composition of the headspace gases above the wine fluctuates dynamically. This can occur both in-glass and in-mouth. The understanding of this fascinating phenomenon is still in its infancy (Brossard et al., 2007; Baker and Ross, 2014).

Each sample should be poured into identical, clear, tulip-shaped, wine glasses. They should each be filled (1/4 to 1/3 full) with the same volume of wine.

I. Appearance

1 – View each sample at a 30° to 45° angle against a bright, white background.
2 – Record separately the wine's:
    clarity (absence of haze)
    color hue (shade or tint) and depth (intensity or amount of pigment)
    viscosity (resistance to flow)
    effervescence (notably sparkling wines)

II. Odor "in-glass"

1 – Sniff each sample at the mouth of the glass before swirling.
2 – Study and record the nature and intensity of the fragrance* (see Figs 1.3 and 1.4)
3 – Swirl the glass to promote the release of aromatic constituents from the wine.
4 – Smell the wine, initially at the mouth and then deeper in the bowl.
5 – Study and record the nature and intensity of the fragrance.
6 – Proceed to other samples.
7 – Progress to tasting the wines (III)

III. "In-mouth" sensations

(a) Taste and mouth-feel

1 – Take a small (6 to 10 ml) sample into the mouth.
2 – Move the wine in the mouth to coat all surfaces of the tongue, cheeks and palate.
3 – For the various taste sensations (sweet, acid, bitter) note where they are percieved, when first detected, how long they last, and how they change in perception and intensity.
4 – Concentrate on the tactile (mouth-feel) sensations of astringency, prickling, body, temperature, and "heat".
5 – Record these perceptions and how they combine with one another.

(b) Odor

1 – Note the fragrance of the wine at the warmer temperatures of the mouth.
2 – Aspirate the wine by drawing air through the wine to enhance the release of its aromatic constituents.
3 – Concentrate on the nature, development and duration of the fragrance. Note and record any differences between the "in-mouth" and "in-glass" aspects of the fragrance.

(c) Aftersmell

1 – Draw air into the lungs that has been aspirated through the wine for 15 to 30 s.
2 – Swallow the wine (or spit it into a cuspidor).
3 – Breath out the warmed vapors through the nose.
4 – Any odor detected in this manner is termed aftersmell; it is usually found only in the finest or most aromatic wines.

* Although fragrance is technically divided into the *aroma* (derived from the grapes) and *bouquet* (derived from fermentation, processing and aging), descriptive terms are more informative.

IV. Finish

1 – Concentrate on the olfactory and gustatory sensations that linger in the mouth.
2 – Compare these sensations with those previously detected.
3 – Note their character and duration.

V. Repetition of assessment

1 – Reevaluate the aromatic and sapid sensations of the wines, beginning at II.3—ideally several times over a period of 30 min.
2 – Study the duration and development (change in intensity and quality) of each sample.

Finally, make an overall assessment of the pleasurableness, complexity, subtlety, elegance, power, balance, and memorableness of the wine. With experience, you can begin to make evaluations of its *potential*— the likelihood of the wine improving in its character with additional aging.

FIGURE 1.1   Sequence of wine tasting.

Under most critical assessment conditions, wines are sampled using clear, tulip-shaped goblets, such as the ISO wine-tasting glass (Fig. 1.2; Plate 5.13 left). The primary exception involves sparkling wines, where flutes permit detailed analysis of the wine's effervescence (Plate 1.1), or simply scintillate the senses as these sprightly sparkles shimmer their way to the surface. However, because of the flute's narrowness, and a tendency to fill the glass to near the rim, effective swirling of the wine and concentration of its fragrance above the wine is prevented. However, these features are partially compensated by bubbles bursting as they reach the surface (Plate 1.2). They propel thousands of minuscule wine droplets into the air, and thereby enhance volatilization of the wine's aromatics.

All glasses in comparative tastings should be identical, made of crystal-clear glass, and filled to the same height (about one-quarter to one-third full). This facilitates each wine being sampled under equivalent conditions. Between 30 and 50 ml is adequate for most assessments. Not only are small volumes economical, but they also facilitate holding the glass at a steep angle (for viewing color and clarity) and permit vigorous swirling (to enhance the release of aromatics). Only under conditions where color differences between the wines are sufficiently marked as to potentially prejudice the perception of the wine's fragrance and flavor should the wines be served in black glasses (or under low-intensity, color-distorting, red illumination.)

Appearance

As noted above, except in situations where appearance might unduly bias the assessment, the wine's visual attributes are the first evaluated. To improve light transmission, the glass is usually tilted against a bright, white background (35° to 45° angle). This produces a stretched arch along the far side of the glass and varying depths of wine through which its visual attributes can be viewed.

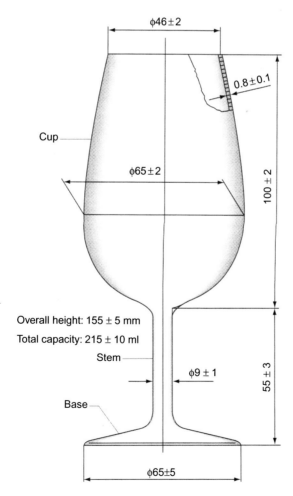

FIGURE 1.2 International Standards Organization (ISO) wine tasting glass. Dimensions are in mm. *Source: Courtesy of International Standards Organization, Geneva, Switzerland.*

PLATE 1.1   Illustration of the effervescence attributes of a sparkling wine in a flute. Notice the cordon de mousse around the edge of the glass. *Source: Photo courtesy of R. S. Jackson.*

PLATE 1.2   The surface discharge resulting from the bursting of bubbles from a flute containing champagne. *Source: Photo courtesy Collection CIVC Copyright Alain Cornu.*

A wine's countenance is typically studied simply as a source of pleasure, but can also provide clues as to other sensations to follow. Although a harbinger, particular colors do not always correlate with experience-based expectations. Thus, color can be a source of bias, often prejudicing a taster's perceptions, depending on their experience. Nonetheless, certain aspects of a wine's appearance may foretell the presence of particular off-flavors. Albeit an indicator, color should not unduly prejudge a wine. Assessment should be based on a full and honest evaluation of all of a wine's sensory characteristics.

Clarity

All wine should be brilliantly clear, the only exception being barrel samples tasted in a winery cellar. Some turbidity can be expected in a still-maturing wine. Cloudiness in bottled wine is always considered a fault, despite it seldom affecting the wine's taste or aromatic character. Because most sources of cloudiness are understood and controllable, crystal-clear clarity is now the norm. Even sediment in well-aged red wines is now relatively rare. Its disturbance and resuspension can by avoided by careful transfer of the bottle to the table, followed by slow and careful decanting.

Color

The two most significant features of a wine's color are its hue and depth. Hue denotes its shade or tint, whereas depth refers to intensity. Both aspects can provide clues relating to features such as grape maturity, duration of skin contact, oak cooperage exposure, and wine age. Immature white grapes yield almost colorless wines, whereas fully mature to overmature clusters can generate yellowish wines. Extended maturation on the vine may, but not consistently, enhance the potential color intensity of the grapes, and therefore both white and red wines. The extent to which these tendencies are realized largely depends on the duration of maceration (skin contact) before or during fermentation. Maturation in oak cooperage favors age-related color changes, as well as an initial enhancement in color depth. During aging, golden tints in white wines increase, whereas red wines lose color density. Eventually, brownish shades develop in all wines.

Because so many factors can influence color expression, it is difficult to be dogmatic about the significance of any particular shade. Only if the wine's origin, style, and age are known, may color suggest "correctness." An atypical color can be a sign of several faults, but not diagnostic of any by itself. The less known about a particular sample, the less value color becomes in assessing quality.

Tilting the glass has the advantage of creating a gradation in wine depths. Viewed against a bright background, the variation in depth provides a range of hues and density attributes. Pridmore et al. (2005) give a detailed discussion of these phenomena. Tilting can also be used in spectroscopically analyzing a wine's color (Hernández et al., 2009).

The rim of the wine generated on tilting provides one of the better measures of a wine's approximate age. A purplish-to-mauve hue is an indicator of a red wine's youth. By contrast, a brickish tint in the same zone is often the first indicator of aging. The best measure of color depth is obtained by looking down from the top of the glass.

The most difficult task in assessing wine color is expressing these impressions meaningfully. There is no accepted terminology for wine colors. The Munsell Color System exists, but is neither readily available nor particularly applicable. CIELAB color coordinates are scientifically useful in comparisons, but are essentially meaningless in expressing color in a manner that is applicable to tasting wine. Thus, color terms are seldom noted in a consistent or effective manner. Some tasters place a drop of the wine on the tasting sheet. Although this is an attempt at being objective, it does not even temporarily preserve an accurate record of the wine's color.

Until an effective, practical, wine-color standard is developed and widely accepted, the use of a few simple terms is probably the best that can be achieved. Terms such as purple, ruby, red, brick, and tawny for red wines, and straw, yellow, gold, and amber for white wines tend to have commonly accepted meaning. Combining these hue terms with qualifiers for color depth, such as pale, light, medium, and dark, should provide a suitable range of expressions. The terms are fairly self-explanatory and provide the potential for effective communication.

Viscosity

Viscosity refers to the resistance to flow of a liquid. Factors such as the sugar, glycerol, and alcohol contents can all affect wine viscosity. Typically, though, humanly observable differences are detectable only in dessert or highly alcoholic wines. Because these differences are minor, and of diverse origin, they are of little diagnostic value. Viscosity is ignored by most professional tasters.

Effervescence

Bubbles occasionally form along the sides and bottom of glasses of still table wines, occasionally also inducing a slight prickling in the mouth. Typically this is of no significance, being only the consequence of early bottling, before the excess carbon dioxide sorbed during fermentation has had a chance to escape. Infrequently, slight bubbling can also result from in-bottle malolactic fermentation. In the past, slight petillance could have been caused by the activity of spoilage microbes, but currently this is exceedingly rare. Active and continuous effervescence is usually found only in sparkling (or artificially carbonated) wines. In this case, the size, number, and duration of the bubbles can be important quality features.

Tears

Tears (alternatively called rivulets or legs) often develop and flow down the sides of the glass following swirling of the wine. They are little more than a crude indicator of a wine's alcohol content. Other than for the intrigue or visual amusement they can provide, tears are sensory trivia. Nonetheless, the film of wine formed during swirling favors the loss of aromatics from the thin veil, enhancing the wine's fragrance.

ODOR

When assessing wine fragrance, its qualitative, quantitative, and temporal aspects are assessed. Odor quality refers to the unique characteristics of the sensation, usually denoted in terms of resemblance to some particular object (e.g., roses, apples, truffles), category (e.g., flowers, fruit, vegetables), personal experiences (e.g., barnyard, hayfield, East Indian store), or emotional/esthetic perceptions (e.g., elegant, subtle, refined, complex, perfumed). Quantitative aspects refer to the intensity of the perception, be it that of a particular odor quality or the overall sensation. Temporal aspects refer to how the fragrance fluctuates in quality and intensity, both in-glass and in-mouth during repeat sampling.

ORTHONASAL (IN-GLASS) ODOR

The fragrance of a wine is best sampled successively, usually starting just above the mouth of the glass and prior to swirling. This permits initial assessment of the wine's most volatile aromatics. In comparing several wines, it is often easier to position the glasses close together and near the edge of the table or counter, quickly assessing the fragrance of each sample at the rim, rather than raise each glass in sequence up to one's nose. An alternative procedure preferred by some assessors is to sweep aromatics toward the nose by waving a hand over the rim of the glass. Some judges subsequently insert their nose deeply into the bowl of each glass, to compare the sensation with that taken at the rim of the glass, before swirling the glass.

Effective swirling, although simple, usually takes some practice. For those unfamiliar with the practice, it is safer to begin by slowly rotating the base of the glass on a level surface. The action involves cyclical shoulder movements, while the wrist remains motionless. Holding the glass by the stem assures a good grip and permits progressively more vigorous swirling. Once comfortable with this action, tasters should commence swirling with wrist action, and slowly lift the glass off the surface. Occasionally, some tasters hold the glass by the edge of the base, between the thumb and side of the curved index finger. While this works, its awkwardness seems designed as an affectation and overly pretentious. More secure is the normal habit of holding the glass jointly by its stem and base.

Because aromatics are released only at the wine's air interface, swirling enhances volatilization. This occurs primarily via the thin film of wine that coats the inner surfaces of the glass. In addition, swirling mixes the wine, replenishing the surface with aromatics. Diffusion of aromatics to the surface is slow without swirling, and only slightly improved by convection currents (Tsachaki et al., 2009). This is particularly important for highly volatile compounds, where the surface becomes rapidly depleted in these constituents. The now-popular "wine aerators" are a poor substitute for swirling. Even worse is the idea of pouring wine back and forth between decanters. Although it does facilitate volatilization, it obviates any opportunity to observe the fragrance "opening" (development). Why rush to some supposed optimum moment? Only about 0.5 percent of a wine's constituents are aromatic, and most impact compounds occur in minuscule amounts! Accentuating aroma loss before sampling is definitely ill advised and much potential sensory pleasure is missed.

The in-curved sides of tulip-shaped ISO glasses help not only concentrate released volatiles in the headspace above the wine, but also permit vigorous swirling. Other factors influencing the relative release of different aromatics with time are the equilibrium between dissolved and weakly bound odorants in the wine, and the wine's surface tension.

The different concentrations of aromatics detected during successive samplings at the rim and in the bowl can generate distinct sensations. With considerable attention, involving both inductive and deductive reasoning, it may be possible to recognize varietal, stylistic, vintage, age-related, and regional attributes. Where possible, this usually requires several attempts, extensive experience, and that most illusive of all skills, intuition. As the primary source of a wine's unique character, analyzing the wine's fragrance merits all the attention it requires. Murphy et al. (1977) consider that as much as 80 percent of the sensory significant information about what we drink or eat comes from olfaction.

When tasting under laboratory conditions, covers are often placed over the mouths of the glasses. These may be tightly fitting plastic Petri dish covers (see Plate 5.1), small pieces of plexiglass (see Plate 7.8), or watch glasses; even coffee-cup lids work. The covers serve several purposes. With highly fragrant wines, the cover limits aromatic contamination of the tasting room, especially those with inadequate ventilation. Such contamination can complicate the assessment of mildly aromatic wines. If the wines are poured in advance of the tasting, the covers guard against fragrance loss during the interval between pouring and sampling. In addition, the cover permits vigorous swirling (if the lid is held on tightly, usually with the index finger). This can be particularly useful when the wines are only delicately fragrant.

No special inhalation technique is deemed required for effective odor sleuthing (Laing, 1983). A single sniff is often adequate for olfactory identification (Laing, 1986), at least with simple aromatic solutions. Typical whiffs tend to last about 1.6 s, have an inhalation velocity of 27 L/min, and involve approximately 500 cm^3 of air (Laing, 1983). The duration and vigor are usually instinctive, and inversely correlated to odor intensity, unpleasantness, and ease of identification (Frank et al., 2006). Thus, although smelling for more than half a second rarely improves odor identification, at least for single compounds under laboratory conditions (Laing, 1982), extending the duration the sniff may be helpful with aromatically neutral wines. However, prolonged inhalation runs the risk of adaptation and loss of sensitivity to some aromatics. Thus, a combination of short sniffs and more prolonged inhalations might be optimal. Experimentation with each wine is the only way to tell.

The action of purposeful smelling itself activates the brain's olfactory centers (Sobel et al., 1998). This is similar to the activation of the gustatory cortex by the act of tasting (Veldhuizen et al., 2007). In addition, the intensity of a sniff affects the efficiency with which various odorants are deposited along the olfactory mucosa (Kent et al., 1996; Buettner and Beauchamp, 2010). Also, odorants adsorb onto, and diffuse through, the overlaying mucous layer at different rates. Thus, this may be another reason to vary the intensity of smelling during assessment (Mainland and Sobel, 2006). Longer inhalations appear to equalize odorant detection on both sides of the nasal septum (Sobel et al., 2000), negating any potential effects of the typical differences in flow rate between nostrils (Zhao et al., 2004).

Although extended inhalation induces adaptation to most aromatic compounds, adaptation can be informative with some aromatically complex wines, notably vintage ports. As certain olfactory receptors become adapted, detection of otherwise unnoticed compounds, or their combined influences, may become detectable.

When beginning concerted odor assessment of a series of wines, each sampling should be separated by about 30–60 s. Most olfactory receptors appear to take about this long to reestablish their intrinsic sensitivity. In addition, measurements of the rate of wine volatilization suggest that the headspace takes a minimum of 15 s to replenish itself (Fischer et al., 1996). Thus, assessment is best conducted at a leisurely pace. This may be another example of where "haste makes waste."

In comparative tastings, the wines' attributes should be assessed both sequentially and repeatedly over 20–30 min. Although this may seem excessive, considerable effort is required in making a detailed assessment and recording of the wines' sensory features. In addition, aspects such as development and duration of the fragrance take time to show themselves. Development and extended duration are highly regarded attributes, characteristics that should be expressed by any premium wine. The higher cost of these wines is justifiable only if accompanied by exceptional sensory endowments. Sequentially concentrating on the fragrance, taste, and flavor of a series of wines also diminishes the likelihood of sensory fatigue developing during the tasting session.

Regardless of how the assessment is conducted, it is important to record olfactory impressions as clearly and precisely as possible. This is easier said than done, and is difficult for everyone. This may be because we are not trained from an early age to develop verbal–olfactory associations. The difficulty even applies to naming common odorants in the absence of their associated visual clues. The situation has been aptly dubbed the "the-tip-of-the-nose" phenomenon (Lawless and Engen, 1977).

One of the advantages of recording olfactory perceptions is to focus attention on the central aromatic features that distinguish wines. Except in technical sensory evaluations, the actual terms used are less important than their relevance to the user. Dissecting a wine's sensory attributes is essential for the winemaker or sensory scientist, but does not necessarily enhance sensory appreciation. Wine enjoyment is not the equivalent of the sum of its sensory attributes, any more than the value of a poem is denoted by the number of similes, alliterations, metaphors, or analogies it possesses, or its adherence to a particular rhyming pattern. Most complex flavor perceptions are cerebral

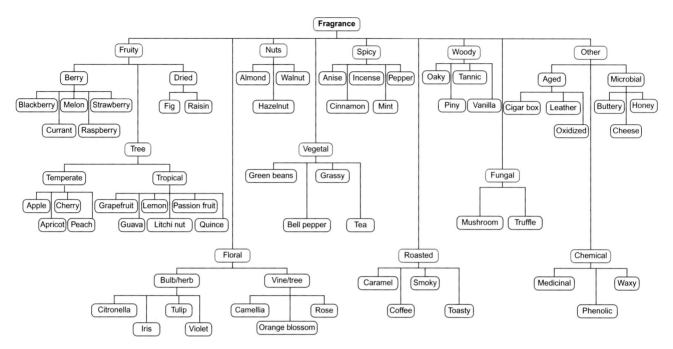

FIGURE 1.3   Wine fragrance chart. *Source: From Jackson, R.S., 2000. Wine Science: Principles, Practice, Perception, second ed. Academic Press, San Diego, CA (Jackson, 2000), reproduced by permission).*

creations, just as are our perceptions of colors. Odor perception starts with the activation of particular neurons, but soon involves interactions with other related neurons, often in several separate centers in the brain. These connect with established odor memory centers, the totality of which generates the constructs we perceive consciously. Thus, any recognizable fragrance, be it of wine, coffee, lilacs, or fried bacon, is a unique combination of multiple neuronal interactions, coordinated in the frontal regions of the brain (Figs. 3.11 and 3.38).

For detailed sensory analysis, tasters are usually trained using samples specifically designed for particular research projects. Reference samples for the various terms are commonly provided during tastings (Appendices 5.1 and 5.2). Fragrance and off-odor charts (Figs. 1.3 and 1.4) can assist in developing and maintaining a common, panel, wine lexicon. Terms help codify the uniquely distinctive aromatic attributes of particular groups of wines, similar to the characteristic note and harmonic patterns that embody the music of specific composers. However, without directed and extensive training, precise use of the most detailed tier of descriptors (e.g., violet, blackcurrant, truffle) is difficult. In general, middle-level terms (floral, berry, vegetal) seem more applicable, and are more effectively used by the majority of people. At the same time, it is important that consumers realize that odor analogy, as illustrated in Figs. 1.3 and 1.4, should not be construed as accurately describing a wine's aromatic features. At best, they only suggest some of a wine's more distinctive flavor characteristics. This is analogous to the difference between the architectural drawings of a house and the house itself. At worst, obsessive concern with descriptors can convince some consumers that, because they cannot personally recognize features supposedly present, that they are inherently incapable of appreciating wine. Analysis may enhance appreciation, like recognizing the names of plants in a garden, but it is not essential to enjoying being in a garden.

Thus, undue emphasis on descriptive terms by consumers can be counterproductive, especially in wine appreciation courses. Charts should be used only to encourage focusing on a wine's fragrance. Once the importance of concentrating on the wine's olfactory traits is ingrained, description in terms of specific fruits, flowers, vegetables, etc., is of limited or negative value. Inventing fanciful terms, in a vain attempt to be informative, is superfluous as well as disingenuous. This potential is aggravated by the legitimate difficulty people have in verbalizing olfactory sensations. It is generally more advantageous for consumers to concentrate on learning to recognize the sensory differences that exemplify varietal aromas, production styles, an aged bouquet and other features than articulate these in words. No one expects people to verbally describe the facial features that distinguish their friends, but that does not inhibit them from instantaneously recognizing them. Except for research purposes, collections of descriptive terms are best left for the purposes for which they were initially developed, descriptive sensory analysis.

In recording olfactory sensations, both positive and negative aspects should be noted. For this, use of an appropriate tasting sheet can be invaluable. Fig. 1.5 provides an example of a general tasting sheet for wine appreciation

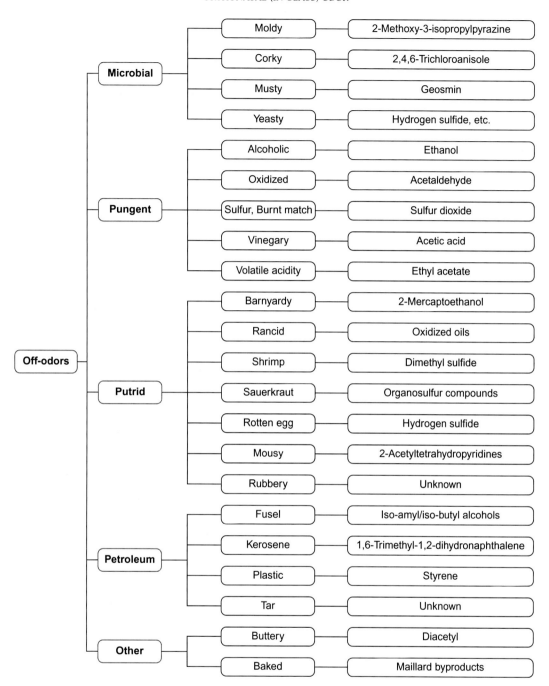

FIGURE 1.4   Wine off-odor chart (column at right notes examples of causal chemicals). *Source: From Jackson, R.S., 2000. Wine Science: Principles, Practice, Perception, second ed. Academic Press, San Diego, CA (Jackson, 2000), reproduced by permission.*

courses. Designed for reproduction on 11 × 17-inch sheets, the circles indicate the placement of six wine glasses. Where desired, photocopies of wine labels can be inserted, as illustrated in Fig. 6.3. Alternatively, a simple hedonic tasting sheet, such as that illustrated in Fig. 1.6, may be appropriate. Tasting sheets are discussed in greater depth in Chapter 5, Quantitative (Technical) Wine Assessment, and Chapter 6, Qualitative Wine Assessment. In addition to static verbal descriptions, similar to a composite photo, a line drawn on a rough time–intensity scale can effectively illustrate the dynamic nature of fluctuations in a wine's characteristics, equivalent to a montage (Fig. 1.7). Vandyke Price (1975) seems to have been the first to demonstrate this graphic method of recording both a wine's quantitative and qualitative attributes over time. The process easily and succinctly expresses the most noticeable and varying

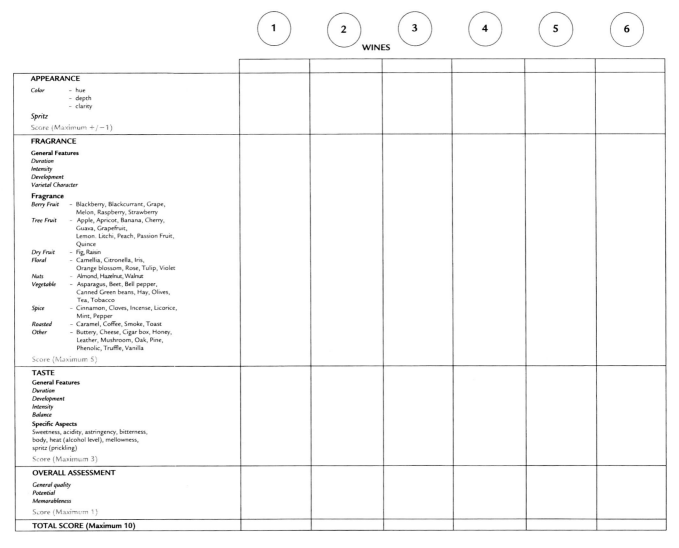

FIGURE 1.5   General wine tasting sheet (usually enlarged to 11 × 17-inch paper).

impressions of the taster. It is the forerunner of a more codified and precise sensory procedure termed temporal dominance of sensations (TDS) (Pineau et al., 2009) (see Chapter 5, Quantitative (Technical) Wine Assessment).

IN-MOUTH SENSATIONS

Taste and Mouth-Feel

After an initial assessment of fragrance, attention turns to taste and mouth-feel, before returning to reassess the wine's fragrance. As with odor, several attributes are evaluated. These include the quality, intensity, duration, and spatial distribution of each modality. Quality refers to the various expressions of a particular modality (e.g., the different expressions of astringency). Intensity refers to the perception of their relative strengths. The temporal characteristics relate to how the quality and intensity of each modality changes while the sample is in the mouth. The spatial pattern concerns the site(s) (tongue, cheeks, palate, or throat) where each modality is detected. The time–intensity and spatial pattern can be useful in differentiating between each modal expression.

Tasting commences with sipping about 6–10 mL of the sample. As far as feasible, the volume of each sample should be kept equivalent, to permit valid comparison among the wines. Active churning ("chewing") or aspirating the sample (see below) brings the wine into contact with all oral surfaces.

Sample Number: ___	Wine Category: ___	Exceptional	Very Good	Above Average	Average	Below Average	Poor	Faulty	Comments
Visual	Clarity								
	Intensity*								
Odor (orthonasal)	Duration**								
	Quality***								
Flavor (taste, mouth-feel, retronasal odor)	Intensity								
	Duration								
	Quality								
Finish (aftertaste and lingering flavor)	Duration								
	Quality								
Conclusion									

* *Intensity*: the perceived relative strength of the sensation—too weak or too strong are equally undesirable.
** *Duration*: the interval over which the wine develops or maintains its sensory impact; long duration is usually a positive feature if not too intense.
*** *Quality*: the degree to which the feature reflects appropriate and desirable varietal, regional, or stylistic features of the wine, plus the pleasure these features give the taster.

FIGURE 1.6   Hedonic wine tasting sheet for quality assessment. *Source: From Jackson, R.S., 2014. Wine Science: Principles and Applications, third ed. Academic Press, San Diego, CA (Jackson, 2014), reproduced by permission.*

The first taste modality typically detected, if present, is sweetness, followed by acidity (sourness). The sensation of sweetness is most evident on the tip of the tongue. By contrast, sourness is more evident along the sides of the tongue and insides of the cheeks (as astringency), depending on the individual. The sharp aspect of acidity typically lingers considerably longer than that of sweetness. Because bitterness begins to be detected slightly later, its increasing intensity often coincides with a decline in perceived sweetness, except in dessert or sweet fortified wines. In fortified wines, sweetness helps to diminish any bitterness the wine may show (e.g., ports). This is equivalent to the mollifying effect of sugar on the bitterness of coffee or tea. Bitterness can take upwards of 15 s before peak intensity is reached. Thus, to fully assess this attribute, the sample should remain in the mouth for at least 15 s. Bitterness is usually detected over the central, posterior portion of the tongue. Subsequently, or during this period, the taster should also focus on mouth-feel sensations, notably the dry, chalky, rough, dust-in-the-mouth, or velvety qualities of astringency. Additional mouth-feel sensations that may be present include the burning sensation of alcohol and the prickling/pain aspect of carbon dioxide. These and other tactile sensations are dispersed throughout the mouth, without specific localization.

The sequence, location, and temporal dynamics of each modality can be useful in distinguishing between these sensations and their identification (Kuznicki and Turner, 1986; Marshall et al., 2005; Laing et al., 2002). This capacity is, however, partially dependent on the taster's approach (Prescott et al., 2004). That is, analysis (conscious attention to each modality) versus synthesis (holistic integration of all in-mouth sensations). In contrast, the duration of

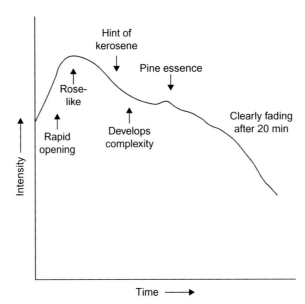

FIGURE 1.7   Example of a graphic representation of the development of a wine's fragrance during a tastings. Specific observations can be applied directly to the point on the graph where the perception was detected.

each sensation is not particularly diagnostic. Persistence reflects more the concentration and maximum perceived intensity of the tastant than its category (Robichaud and Noble, 1990).

Although significant in some critical tastings, the purpose of noting and recognizing individual sapid sensations is less important than how they integrate to form holistic perceptions, such as balance and body, and combine with retronasal odor to generate flavor. An example of integration is the creaminess of dairy products, being dependent not only on mouth-feel and fat-particle size, but also aroma (Kilcast and Clegg, 2002).

Sensory integration develops instinctively, as a consequence of experience, without the need for conscious participation (Hollowood et al., 2002). Wine-related examples of such unconscious integration include the illusion of sweetness often considered to be present in fruity, but dry, white wines, and the apparent greater flavor intensity of deeply colored wines. Nonetheless, this integration seems reversible. Actively concentrating on a complex sensation can often separate its component sensory aspects (van der Klaauw and Frank, 1996; Prescott, 1999). Thus, it may be up to the individual to decide whether the more inherently natural (integrated/holistic) approach, or a more analytic (dissective) approach, is desired. Perceived reality often depends as much on experience as on the context in which the sensation is detected.

Because wine tasting means different things to different people, there are differing opinions on whether taste and mouth-feel should be assessed during the first or subsequent samplings. Tannins react with saliva proteins in the mouth, diminishing their initial bitter and astringent perception. Although tannins induce enhanced saliva production, it is insufficiently rapid to compensate for dilution and precipitation during sampling. Thus, subsequent samplings of red wines can appear more astringent and bitter than during the first sip. If the purpose of the tasting were to involve food compatibility, then the first taste would provide a better equivalent. Otherwise, data from the second and subsequent samplings are likely to provide more comparable data with a series of wines.

In an attempt to minimize carryover effects from one wine to another, cleansing the palate between samples is recommended. Due to the importance of palate cleansers in food and wine sensory analysis, the effectiveness of different palate cleansers has been repeatedly investigated. For example, Colonna et al. (2004) found that a weak solution of pectin (1 g/L) was more effective than several traditional palate cleansers. Pectins (Hayashi et al., 2005) and ionic carbohydrates, such as xanthans and gum arabic (Mateus et al., 2004), appear to have their "cleansing" effect by limiting tannin–protein polymerization, much in the same manner as grape polysaccharides (rhamnogalacturonans) in mollifying wine astringency (Carvalho et al., 2006). This effect is most marked with galloylated flavonols (Hayashi et al., 2005), but becomes less effective as polymer size increases (Mateus et al., 2004). In a study by Brannan et al. (2001), the use of 0.55% carboxymethyl cellulose was recommended, due to its low residual effect in the mouth. In another study, crackers were assessed to be most effective in reducing the residual effects of red wine (Ross et al., 2007). Water, a commonly provided palate cleanser, has universally been found to be ineffective.

RETRONASAL (MOUTH-DERIVED) ODOR

As with in-glass odor, its relative intensity and qualitative aspects should be noted repeatedly over the full course of a tasting. Transfer of aromatics from the mouth and back of the throat into the nasal cavities is facilitated by aspirating the wine. Two interrelated events affect retronasal detection (Normand et al., 2004), namely, punctuated surges of air from the throat (immediately following swallowing), and the more tranquil outflow of air during breathing (Fig. 1.8). The pattern of flow can be regulated by adjusting swallowing (Fig. 1.8C), varying the breathing cycle, and tilting the head forward. These actions affect the movement of the velum (soft palate), epiglottis, and uvula. During swallowing, the velum and epiglottis block passage to the lungs, and the uvula forms a tight seal at the back of the nasopharynx, preventing passage to the nasal cavities (Buettner et al., 2001). After swallowing, the velum, epiglottis, and uvula return to their normal positions, allowing volatiles to again enter the nasal passages from the mouth, and airflow to and from the lungs. Intentional focus on deliberate expiration apparently enhances retronasal identification (Pierce and Halpern, 1996).

Because swallowing samples is discouraged during professional tastings (due to its reduction in taster accuracy), the importance of swallowing in retronasal odor detection, and thereby the perception of wine flavor, is of concern. Without swallowing, perceived aroma complexity can be both reduced and modified (Déléris et al., 2014). Thus, aspiration and focus during expiration take on increased importance during research-based assessments.

Wine aspiration can involve tightening the jaws, pulling the lips slightly ajar, and slowly drawing air through the wine. Alternatively, the lips can be pursed, before drawing air through the wine. Either procedure favors volatilization by increasing contact between the wine and air (analogous to swirling wine in the glass), as well as atomizing some of the wine. Vigorous agitation (mastication or chewing) of the wine in the mouth has a somewhat similar effect (de Wijk et al., 2003), but is less effective. Aspirating wine in public, unless done discretely, can be interpreted by those unfamiliar with the process as distinctly uncouth.

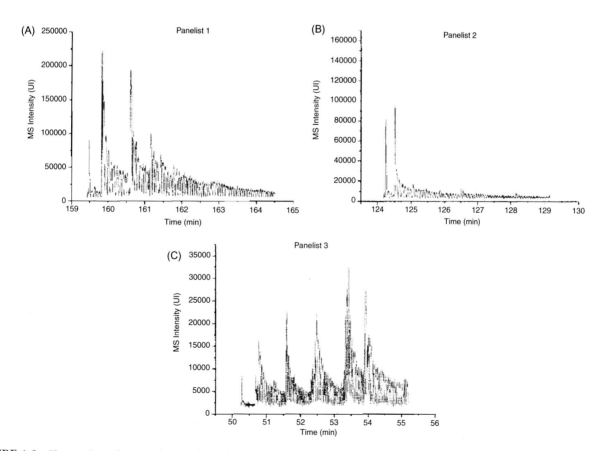

FIGURE 1.8 Flavor release from a solution of menthol sampled by three different subjects: (A) Panelist 1, (B) Panelist 2, and (C) Panelist 3. Each large peak corresponds to a swallow, smaller peaks to breathing. Panelist 3 was a flavorist who made many small swallows. This behavior clearly prolongs flavor release. *Source: Reproduced from Normand, V., Avison, S., and Parker, A., 2004. Modeling the kinetics of flavour release during drinking. Chem. Senses 29, 235–245, by permission of Oxford University Press.*

Wine fragrance perceived retronasally can be qualitatively different from that detected orthonasally (Negoias et al., 2008). This distinction is well known relative to the difference often detected between the smell and flavor of cheeses. This phenomenon probably has several origins. The concentration of aromatics reaching the olfactory patches via the oral cavity is considerably less than via the nose—the volume of airflow coming from the oral cavity being much less. As a result, some constituents detected retronasally may be below their detection thresholds. Additional factors may involve the increased temperature in the mouth (selectively modifying the relative volatility of compounds) and the action of both salivary and microbial enzymes. These could either degrade, generate, or facilitate the liberation of volatile compounds in the mouth. In addition, the perceived quality of a compound may be affected by the direction of airflow (Small et al., 2005). This may relate to the spatial positioning of different receptors along the olfactory patches, and, correspondingly, the temporal sequence with which they respond to the directional flow of the compound. This may be analogous to playing a short segment of music backward, or the interpretation of faces viewed upside down (Murray, 2004).

Although retronasal olfaction is of importance in and of itself, it is in its integration with taste and mouth-feel that retronasal smell has its greatest (and most underappreciated) significance: flavor. Foods and beverages lose most of their identifiable attributes when the retronasal component is lacking, an experience well known to anyone who has experienced a head cold. Pinching the nose has a similar effect.

At the end of oral sampling, tasters may complete their assessment with a prolonged inward aspiration. After swallowing the sample (if permitted) or expectoration, vapors in the lungs are slowly, but forcefully, exhaled. Any perception so derived is referred to as the after-smell. Whether the aromatics that induce this impression come from the lungs or from the oral cavity (a component of the wine's finish) is unclear. While occasionally informative, the procedure is typically of value only with complexly aromatic wines.

In contrast to wine sensory analysis and competitions, samples are typically consumed in wine appreciation courses, at wine society tastings, and the like. Because the number of wines being tasted is often small, and the assessments only for personal edification, consumption is unlikely to negatively affect tasting skill or results. However, when 20 or more wines are sampled, as in wine competitions or technical tastings, consumption must be avoided. Expectoration limits any significant increase in blood-alcohol content (Scholten, 1987). Nonetheless, sufficient tannic material may be consumed to cause a headache. This often can be avoided by taking a prostaglandin-synthesis inhibitor (e.g., acetylsalicylic acid, acetaminophen, or ibuprofen) about an hour before tasting. This and other "occupational hazards" of wine tasting are discussed at the end of Chapter 5, Quantitative (Technical) Wine Assessment.

FINISH

After assessing the wine's fragrance and flavor, concentration switches to the finish. Finish refers to the lingering flavor sensations in the mouth. It most likely arises from the thin film of wine that coats the mouth and throat, as well as compounds that have sorbed onto the mucous-layer lining of the throat (Bücking, 2000) and nasal passages. In addition, only those compounds that persist in, and subsequently escape from, the saliva and/or mucus are likely to be detected. Although tending to be subtle and fleeting (like a sunset), a delicate lingering finish is considered by many enophiles as a sine qua non of any quality wine. It may last only a few seconds, but it can be longer (Buettner, 2004). The French have coined a term for its duration, *caudalie*, each unit of which is equivalent to one second.

Although most table wines possess a relatively short finish, fortified wines, possessing more intense flavors, typically exhibit a much longer finish. Although a prolonged finish is usually viewed as desirable, features such as a persistent metallic sensation, the presence of off-odors, or excessively acidic, bitter, and astringent sensations are clearly not interpreted positively.

OVERALL QUALITY

After individually assessing a wine's sensory attributes, attention shifts to overall quality. As noted by Amerine and Roessler (1983), it is far easier to detect a wine's quality than define it. Wine quality is analogous to grammar; what is considered appropriate evolves out of local conditions and subsequently codified by authorities. This involves aspects of conformity with, and distinctiveness within, regional, varietal, and stylistic norms; aroma and bouquet development, duration, and complexity; the duration and character of the finish; the uniqueness of the tasting experience; and personal preference.

Many of the terms used for quality have been borrowed from the world of art criticism. Relative to wine, the term complexity refers to the presence of many, distinctive, aromatic elements, rather than one or a few easily recognizable odors. Balance (harmony) denotes the perceptive equilibrium of all olfactory and sapid sensations, where individual perceptions do not dominate. Balance without character is bland, but when dynamic and refined, it donates an almost otherworldly feeling of elegance. The complex interaction of sensory perceptions involved in this percept is evident in the reduced fruitiness of red wines possessing excessive astringency, or the hollowness of a sweet wine lacking sufficient fragrance and acidity. Balance often seems to be easier to achieve in white wines, due to their lower phenolic content. However, their reduced aromatic complexity often leaves white wines less inspiring than their red counterparts. Thus, refined balance is no easier to achieve in a white wine than a red. Occasionally, individual aspects may be sufficiently intense to give the impression that balance is on the brink of collapse. In this situation, balance has an element of nervousness that can be particularly fascinating. Development expresses changes in the aromatic character throughout sampling. These changes generate an aspect of intrigue than draws attention back to the latest transformation. Adequate duration denotes that the fragrance retains a unique, and usually evolving, character over the sampling period. Interest is the combined influence of all factors that sustain a taster's attention to detail. Implied, but often not specifically stated, is the requirement for both power and finesse. Without these attributes, attractiveness is short-lived. If the overall sensation is sufficiently remarkable, the experience becomes unforgettable, an attribute Amerine and Roessler (1983) term memorableness. This is the quintessential feature of a great wine. Knowing in advance the supposed "prestigious" nature of a wine almost guarantees that it will not generate an "Aha!" moment. A vinous apotheosis demands surprise.

The importance of amazement to a pleasurable experience has been demonstrated with functional magnetic resonance imaging (fMRI) (Berns et al., 2001). Unpredictability greatly enhances the reward stimulus associated with activating the cerebral orbitofrontal cortex. Distinctiveness, although less stunning than memorableness, is a component of it, and is particularly important in training tasters' odor memories and developing realistic expectations (Mojet and Köster, 2005).

Such quality terms lend themselves well to wine appreciation, but are too subjective and imprecise to have significance in analytic wine assessment, similar to the new buzzword descriptor "minerality." Its popularity seems strange, since it implies a sensation resembling the licking of a steel blade or stone, something totally estranged from wine tasting. In addition, there seems to be no consistency in what tasters mean by the term (Ballester et al., 2014; Heymann et al., 2014).

Most European authorities feel that tastings should be restricted to within regional appellations, counseling against tasting among regions or grape varieties. Although these limitations make wine assessments simpler, they negate much of their value, thus encouraging quality improvement. When tasting assesses general hedonic quality, rather than stylistic purity, comparative tasting can be especially revealing. Such tastings are more popular in England and the New World, where artistic merit tends to be considered more highly than maintenance of regional (appellation) styles.

POSTSCRIPT

The full assessment of a wine's sensory attributes may seem complicated and time consuming. However, it soon becomes almost automatic, and with accumulating experience (and expansion of odor memory), the only part that takes time is assessing the length of time the fragrance lingers and the wine's sensory changes during that period. Thus, the attributes of development and duration are often assessed only occasionally, missing out on two of the most wondrous attributes wines of quality can possess.

Despite the clear advantages of a systematic approach to wine assessment, permitted by the technique described above, few professional tasters appear to utilize such a disciplined approach (Brochet and Dubourdieu, 2001). This is clear from even a cursory look at most tasting notes. This paradox stems from most wine writers (and consumers) having no need (or inclination) to analyze wines critically, as tasting is conducted holistically, concentrating on the wine's general qualities or fanciful descriptions. Notes primarily highlight the most prominent of a wine's attributes, and the subjective reactions of the taster. Rarely do the terms used have verifiable or easily definable meanings. In addition, each taster tends to possess his or her own somewhat idiosyncratic lexicon. It presumably has relevance to the person who is using it, but little to others. In addition, term use often reflects more the life experiences of the taster than the wine's actual sensory attributes. At best, descriptors represent resemblances to past olfactory experiences, or reiterate traditionally used terms to describe regional, stylistic, or varietal wines. Ranking typically concentrates on how well, or poorly, wines express features desired by, or considered important to, the taster.

Holistic assessments appear to arise from selective activation of the brain's right hemisphere (Dade et al., 1998, 2002, Herz et al., 1999). This region deals principally with archetypic, creative, and emotive aspects of expression. This contrasts with the left hemisphere, where verbal aspects of language tend to be centralized; in left-handed individuals, this hemispheric specialization is usually reversed (Deppe et al., 2000, Knecht et al., 2000). Whether the tendency to express wine attributes in holistic terms reflects a lack of training in youth, or the lack of genetic "hardwiring" in the brain, is unknown. For example, reading and writing are solely learned skills, unlike our propensity to speak and walk. Either way, selective activation of the right hemisphere, and the small area of the brain set aside for processing olfactory information, may explain why humans have such a pauperized odor lexicon. Thus, people typically enunciate their sensory responses in terms of objects or events experienced in their daily life, or the emotions they engender. Thus, as noted earlier, the terms used by an individual may tell more about the taster than the wine (Brochet and Dubourdieu, 2001).

Regrettably, assessing wine holistically often robs the taster of the opportunity of detecting some of the wine's finest attributes. Thus, short-duration assessments, typical of commercial tastings or the dinner table, can exaggerate the apparent qualities of wines with forthright aromas, but which lack the ability to retain taster interest and being one-dimensional. The development and retention of exquisite sensory characteristics, and the ability to age, are traditional dual pillars of wine quality. Wine of the finest quality deserves to be fully appreciated. Nonetheless, it must be admitted that most wines are not so endowed. Regrettably, serious investigation may only demonstrate that they are instantaneously pleasant but mundane. In general, the older the wine, the greater the value of a full, detailed evaluation, assuming the wine has not been produced, as is now common, to be consumed when young.

Wine drinking is basically an aesthetic experience, or at least it is to the people who write wine books. Lehrer (1975)

References

Amerine, M.A., Roessler, E.B., 1983. Wines, Their Sensory Evaluation, 2nd. ed. Freeman, San Francisco, CA.

Baker, A.K., Ross, C.F., 2014. Sensory evaluation of impact of wine matrix on red wine finish: a preliminary study. J. Sens. Stud. 29, 139–148.

Ballester, J., Mihnea, M., Peyron, D., Valentin, D., 2014. Perceived minerality in wine: a sensory reality? Wine Vitic. J. 29 (4), 30–33.

Berns, G.S., McClure, S.M., Pagnoni, G., Montague, P.R., 2001. Predictability modulates human brain response to reward. J. Neurosci. 21, 2793–2798.

Brannan, G.D., Setser, C.S., Kemp, K.E., 2001. Effectiveness of rinses in alleviating bitterness and astringency residuals in model solutions. J. Sens. Stud. 16, 261–275.

Brochet, F., Dubourdieu, D., 2001. Wine descriptive language supports cognitive specificity of chemical senses. Brain Lang. 77, 187–196.

Brossard, C., Rousseau, F., Dumont, J.-P., 2007. Perceptual interactions between characteristic notes smelled above aqueous solutions of odorant mixtures. Chem. Senses 32, 319–327.

Bücking, M. (2000). Freisetzung von Aromastoffen in Gegenwart retardierender Substanzen aus dem Kaffeegetränk. Ph.D. Thesis, University of Hamburg, Germany, reported in Prinz, J. F., and de Wijk, R. (2004). The role of oral processing in flavour perception. In *Flavor Perception* (A. J. Taylor and D. D. Roberts, eds.). Blackwell Publishing, Oxford, UK.

Buettner, A., 2004. Investigation of potent odorants and afterodor development in two Chardonnay wines using the buccal odor screening system (BOSS). J. Agric. Food Chem. 52, 2339–2346.

Buettner, A., Beauchamp, J., 2010. Chemical input – sensory output: Diverse modes of physiology-flavor interaction. Food Qual. Pref. 21, 915–924.

Buettner, A., Beer, A., Hannig, C., Settles, M., 2001. Observation of the swallowing process by application of videofluoroscopy and real time magnetic resonance imaging – consequences for retronasal aroma stimulation. Chem. Senses 26, 1211–1219.

Carvalho, E., Mateus, N., Plet, B., Pianet, I., Dufourc, E., De Freitas, V., 2006. Influence of wine pectic polysaccharides on the interactions between condensed tannins and salivary proteins. J. Agric. Food Chem. 54, 8936–8944.

Colonna, A.E., Adams, D.O., Noble, A.C., 2004. Comparison of procedures for reducing astringency carry-over effects in evaluation of red wines. Aust. J. Grape Wine Res. 10, 26–31.

Dade, L.A., Jones-Gotman, M., Zatorre, R.J., Evans, A.C., 1998. Human brain function during odor encoding and recognition: a PET activation study. Ann. NY Acad. Sci. 855, 572–574.

Dade, L.A., Zatorre, R.J., Jones-Gotman, M., 2002. Olfactory learning: convergent findings from lesion and brain imaging studies in humans. Brain 125, 86–101.

Déléris, I., Saint-Eve, A., Lieben, P., Cypriani, M.-L., Jacquet, N., Brunerie, P., et al., 2014. Impact of swallowing on the dynamics of aroma release and perception during the consumption of alcoholic beverages. In: Ferreira, V., Lopez, R. (Eds.), *Flavour Science*. Proceedings from XIII Weurman Flavour Research Symposium. Academic Press, London, UK, pp. 533–537.

Deppe, M., Knecht, S., Lohmann, H., Fleischer, H., Heindel, W., Ringelstein, E.B., et al., 2000. Assessment of hemispheric language lateralization: a comparison between fMRI and fTCD. J. Cereb. Blood Flow Metab. 20, 263–268.

de Wijk, R.A., Engelen, L., Prinz, J.F., 2003. The role of intra-oral manipulation in the perception of sensory attributes. Appetite 40, 1–7.

Eiximenis, F. (1384) Terc del Crestis. Tome 3, Rules and Regulation for Drinking wine, cited in H. Johnson, 1989. *Vintage: The Story of Wine*. Simon and Schuster, New York, NY. p. 127.

Fischer, C., Fischer, U., Jakob, L., 1996. Impact of matrix variables, ethanol, sugar, glycerol, pH and temperature on the partition coefficients of aroma compounds in wine and their kinetics of volatization. In: Henick-Kling, T., Wolf, T.E., Harkness, E.M. (Eds.), *Proc. 4th Int. Symp. Cool Climate Vitic. Enol.*, Rochester, NY, July 16–20, 1996. NY State Agricultural Experimental Station, Geneva, New York. pp. VII 42–46.

Frank, R.A., Gesteland, R.C., Bailie, J., Rybalsky, K., Seiden, A., Dulay, M.F., 2006. Characterization of the sniff magnitude test. Arch. Otolaryngol. – Head Neck Surg. 132, 532–536.

Goyert, H., Frank, M.E., Gent, J.F., Hettinger, T.P., 2007. Characteristic component odors emerge from mixtures after selective adaptation. Brain Res. Bull. 72, 1–9.

Hayashi, N., Ujihara, T., Kohata, K., 2005. Reduction of catechin astringency by the complexation of gallate-type catechins with pectin. Biosci. Biotechnol. Biochem. 69, 1306–1310.

Hernández, B., Sáenz, C., Fernández de la Hoz, J., Alberdi, C., Alfonso, S., Diñeiro, J.M., 2009. Assessing the color of red wine like a taster's eye. Col. Res. Appl 34, 153–162.

Herz, R.S., McCall, C., Cahill, L., 1999. Hemispheric lateralization in the processing of odor pleasantness versus odor names. Chem. Senses 24, 691–695.

Heymann, H., Hopfer, H., Bershaw, D., 2014. An exploration of the perception of minerality in white wines by projective mapping and descriptive analysis. J. Sens. Stud. 29, 1–13.

Hollowood, T.A., Linforth, R.S.T., Taylor, A.J., 2002. The effect of viscosity on the perception of flavour. Chem. Senses 28, 11–23.

Jackson, R.S., 2000. *Wine Science: Principles, Practice, Perception*, second ed. Academic Press, San Diego, CA.

Jackson, R.S., 2014. *Wine Science: Principles and Applications*, third ed. Academic Press, San Diego, CA.

Kent, P.F., Mozell, M.M., Murphy, S.J., Hornung, D.E., 1996. The interaction of imposed and inherent olfactory mucosal activity patterns and their composite representation in a mammalian species using voltage-sensitive dyes. J. Neurosci. 16, 345–353.

Kilcast, D., Clegg, S., 2002. Sensory perception of creaminess and its relationship with food structure. Food Qual. Pref. 13, 609–623.

Knecht, S., Drager, B., Deppe, M., Bobe, L., Lohmann, H., Floel, A., et al., 2000. Handness and hemispheric language dominance in healthy humans. Brain. 123, 2512–2518.

Kuznicki, J.T., Turner, L.S., 1986. Reaction time in the perceptual processing of taste quality. Chem. Senses 11, 183–201.

Laing, D.G., 1982. Characterization of human behaviour during odour perception. Perception 11, 221–230.

Laing, D.G., 1983. Natural sniffing gives optimum odour perception for humans. Perception 12, 99–117.

Laing, D.G., 1986. Identification of single dissimilar odors is achieved by humans with a single sniff. Physiol. Behav. 37, 163–170.

Laing, D.G., Link, C., Jinks, A.L., Hutchinson, I., 2002. The limited capacity of humans to identify the components of taste mixtures and taste-odour mixtures. Perception 31, 617–635.

Lawless, H.T., Engen, T., 1977. Associations of odors, interference, mnemonics and verbal labeling. J. Expt. Psychol. Human Learn. Mem 3, 52–59.

Lehrer, A., 1975. Talking about wine. Language 51, 901–923.

Liger-Belair, G., Beaumont, F., Vialatte, M.-A., Jérou, S., Jeandet, P., Polidori, G., 2008. Kinetics and stability of the mixing flow patterns found in champagne glasses as determined by laser tomography techniques: likely impact on champagne tasting. Anal. Chim. Acta 621, 30–37.

Mainland, J., Sobel, N., 2006. The sniff is part of the olfactory percept. Chem Senses 31, 181–196.

Marshall, K., Laing, D.G., Jinks, A.J., Effendy, J., Hutchinson, I., 2005. Perception of temporal order and the identification of components in taste mixtures. Physiol. Behav. 83, 673–681.

Mateus, N., Carvalho, E., Luís, C., de Freitas, V., 2004. Influence of the tannin structure on the disruption effect of carbohydrates on protein–tannin aggregates. Anal. Chim. Acta 513, 135–140.

Mojet, J., Köster, E.P., 2005. Sensory memory and food texture. Food Qual. Pref. 16, 251–266.

Murphy, C., Cain, W.S., Bartoshuk, L.M., 1977. Mutual action of taste and olfaction. Sens. Process 1, 204–211.

Murray, J.E., 2004. The ups and downs of face perception: evidence for holistic encoding of upright and inverted faces. Perception 33, 387–398.

Negoias, S., Visschers, R., Boelrijk, A., Hummel, T., 2008. New ways to understand aroma perception. Food Chem. 108, 1247–1254.

Normand, V., Avison, S., Parker, A., 2004. Modeling the kinetics of flavour release during drinking. Chem. Senses 29, 235–245.

Peynaud, E. (Trans. by M. Schuster) (1987). *The Taste of Wine. The Art and Science of Wine Appreciation.* Macdonald & Co., London.

Pierce, J., Halpern, B.P., 1996. Orthonasal and retronasal odorant identification based upon vapor phase input from common substances. Chem. Senses 21, 529–543.

Pineau, N., Schlich, P., Cordelle, S., Mathonnière, C., Issanchou, S., Imbert, A., et al., 2009. Temporal dominance of sensations: construction of the TDS curves and comparison with time-intensity. Food Qual. Pref. 20, 450–455.

Prescott, J., 1999. Flavor as a psychological construct: implications for perceiving and measuring the sensory qualities of foods. Food Qual. Pref. 10, 349–356.

Prescott, J., Johnstone, V., Francis, J., 2004. Odor-taste interactions: effects of attentional strategies during exposure. Chem. Senses 29, 331–340.

Pridmore, R.W., Huertas, R., Melgosa, M., Negueruela, A.I., 2005. Discussion on perceived and measured wine color. Color Res. Appl. 30, 146–152.

Robichaud, J.L., Noble, A.C., 1990. Astringency and bitterness of selected phenolics in wine. J. Sci. Food Agric 53, 343–353.

Ross, C.F., Hinken, C., Weller, K., 2007. Efficacy of palate cleansers for reduction of astringency carryover during repeated ingestions of red wine. J. Sens. Stud. 22, 293–312.

Scholten, P., 1987. How much do judges absorb? Wines Vines 69 (3), 23–24.

Small, D.M., Gerber, J.C., Mak, Y.E., Hummel, T., 2005. Differential neural responses evoked by orthonasal versus retronasal odorant perception in humans. Neuron 47, 593–605.

Sobel, N., Prabhakaran, V., Desmond, J.E., Glover, G.H., Goode, R.L., Sullivan, E.V., et al., 1998. Sniffing and smelling: separate subsystems in the human olfactory cortex. Nature 392, 282–286.

Sobel, N., Khan, R.H., Hartley, C.A., Sullivan, E.V., Gabrieli, J.D.E., 2000. Sniffing longer rather than stronger to maintain olfactory detection threshold. Chem. Senses 25, 1–8.

Tsachaki, M., Linforth, R.S.T., Taylor, A.J., 2009. Aroma release from wines under dynamic conditions. J. Agric. Food Chem. 57, 6976–6981.

van der Klaauw, N.J., Frank, R.A., 1996. Scaling component intensities of complex stimuli: the influence of response alternatives. Environ. Int. 22, 21–31.

Vandyke Price, P.J., 1975. The Taste of Wine. Random House, New York, NY.

Veldhuizen, M.G., Bender, G., Constable, R.T., Small, D.M., 2007. Trying to detect taste in a tasteless solution: modulation of early gustatory cortex by attention to taste. Chem. Senses 32, 569–581.

Zhao, K., Scherer, P.W., Hajiloo, S.A., Dalton, P., 2004. Effect of anatomy on human nasal air flow and odorant transport patters: implications for olfaction. Chem. Senses 29, 365–379.

2

Visual Perceptions

As noted in Chapter 1, Introduction, wine appearance can provide indicators of quality, style, and varietal origin. Unfortunately, visual clues can also prejudice assessment. In this chapter, the nature, origin, and relevance of a wine's countenance are more fully investigated.

COLOR

Color Perception and Measurement

The perceived color of an object depends on its properties to selectively absorb, transmit, and reflect visible radiation; how several photoreceptors in the eye respond to light; and finally how the brain interprets the impulses the eyes transmit. For wines, this begins with how its pigments reflect and transmit light, generating the properties of hue and brilliance. These properties depend on the amount and chemical nature of the pigments present, and correspondingly the intensity and spectral quality of the visible radiation reflected and transmitted. Color purity depends on the relative absorptive properties of the pigments across the visible spectrum. The more uniform the spectrum, the less specifically the wine's color will represent a particular portion of the visible spectrum. As noted in Fig. 2.1, the wine becomes less red as it ages, denoted by the higher tint value (estimated by absorbance at 420 mm divided by absorbance at 520 mm). Fig. 2.2 provides an example of the different reflectance values characteristic of young white, rosé, and red wines.

The spectral characteristics of a wine can be most precisely measured with a spectrophotometer (Hernández et al., 2009), but its relevance to human color perception is far from direct. Spectrophotometric measurements assess the intensity of individual wavelengths, whereas human color perception results from a cerebral construct based partially on impulses derived from three types of receptors (cones) in the eye. Each cone contains multiple copies of a single type of photosensitive pigment: P424 (S, blue), P530 (M, green), or P560 (L, red). As noted in parentheses, the pigments may alternatively be named after the range of the visual spectrum absorbed (S, short; M, medium; L, long) or the dominant color of the principal zone to which they respond. Each pigment is variably responsive to a range of visible radiation, showing considerable overlap, notably P530 and P560 (Plate 2.1). The retina contains about 6–7 million cones, densely packed into a central region called the fovea. Their relative number and distribution differ, with those most sensitive to blue light being the least common, whereas those most responsive to middle and long wavelengths are about equally represented. Color hue is generated from subtracting responses from select cones at the boundary where color changes, and relative to surrounding colors. This probably explains why the color of objects appears to stay constant throughout the day, despite diurnal changes in sunlight spectral quality (Brou et al., 1986). In contrast, the brightness of a color is based on summing cone responses. The various attributes of color and vision are processed in different sites in the visual cortex at the back of the brain before being integrated into the conscious perception of color.

Because the brain can adjust remarkably to the varying spectral quality of most light sources, there is no need to assess wine color under north-reflected light, as occasionally asserted. For example, a sheet of white paper looks the same color under sunlight or incandescent light (which contains a high proportion of radiation in the yellow portion of the visible spectrum). In addition, it is rarely possible to conduct tastings during the day, or in a room with north-facing windows. However, the brain's ability to adjust its interpretation to the light source has a limit.

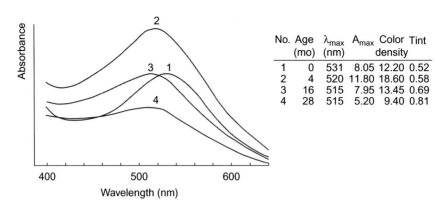

FIGURE 2.1   Absorbance (A) scans of a single cultivar port (Touriga Nacional, 1981) across the visible spectrum at different ages (*A*, absorbance, *λ*, wavelength in nm). *From Bakker, J. Timberlake, C.F., 1986. The mechanism of color changes in aging port wine. Am. J. Enol. Vitic. 37, 288–292, reproduced by permission.*

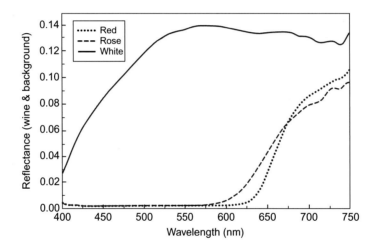

FIGURE 2.2   Example of the reflectance spectra of a white, rosé, and red wine, taken at the center of a wine glass. *From Huertas, R., Yebra, Y., Pérez, M.M., Melgosa, M., Negueruela, A.I., 2003. Color variability for a wine sample poured into a standard glass wine sampler. Color Res. Appl. 28, 473–479, reproduced by permission of John Wiley and Sons, copyright 2003.*

This is purposely used when there is a desire to distort the wine's color, in an attempt to avoid color biasing tasters' assessments. A typical example is the use of low-intensity red light.

That color perception is a construct (Kaiser and Boynton, 1996; Squire et al., 2008) helps explain the difference between spectral and nonspectral colors. Spectral colors correspond to specific named ranges in the visible spectrum (e.g., violet, blue, green, yellow, and red, or regions adjacent to the black line of Plate 2.2). Although, even with some of these terms there is no consistency in their application. In contrast, nonspectral colors arise from specific combinations of light wavelengths, and have no equivalent in the solar spectrum (e.g., purple, orange, brown, and, of course, white being a mixture of all visible wavelengths) (central portion of colored portion of Plate 2.2). Thus, the purplish color of a young red wine is the combination of blue and red wavelengths (those not absorbed, and thereby reflected or transmitted by anthocyanins).

As noted in Chapter 1, Introduction, there is no generally accepted classification of wine colors, presumably because they are difficult to adequately represent on paper. People can differentiate thousands of color gradations by direct comparison. However, they identify comparatively few consistently by name (Fig. 2.3), using related terms interchangeably (Chapanis, 1965). Therefore, it seems best to use only a few simple terms, such as those illustrated in Table 2.1, so that consistent and effective communication is possible.

Color terms should include aspects relating to hue (wavelength purity), saturation (grayness), and brightness (capacity to reflect or transmit light). Fig. 2.4 illustrates these color attributes. However, in practice, it is not easy to

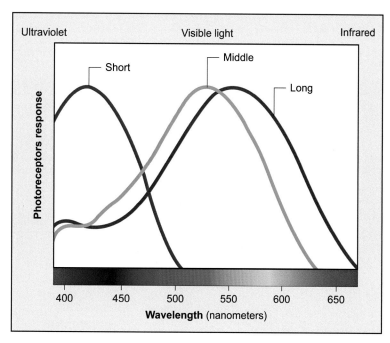

PLATE 2.1 Illustration of the absorption curves of the three human opsin-based photoreceptors in the retina. Even though photons across the expanse of visible radiation possess different energy values, photoreceptor response is dependent on their ability to absorb the radiation not energy level. Thus, cone cell response does not encode the wavelength of the absorbed light. For the brain to distinguish color, it compares signals from cones with different the different photosensitive pigments.

differentiate between these aspects clearly. For example, brown is commonly described as a hue, but technically is an impure yellow-red (combining yellow-red and blue spectral elements of low brilliance). Equally, moderate pink is a partially saturated maroon red.

The availability of a standard for wine colors would increase the value of color in sensory evaluation. The Munsell color notation (Munsell, 1980) has a long history of use in the food industry and scientific investigation. However, the Munsell notation does not fully represent the range of human color vision, which itself shows considerable individual (genetic) variability. Lack of full color vision is imprecisely called color blindness. While inability to detect color does exist (absence of any cones), it is rare. More common are various distortions of color vision, based on the absence of one or two of the three cones, or where the color peaks of the pigments are shifted. Color perception also changes with age. Yellow pigment accumulates in the lens and retina, resulting in a slow loss in blue sensitivity. Nonetheless, this is seldom noticed. The brain adapts its interpretation to the changes.

On a more popular level, Bouchard Aîné et Fils (Beaune, France) has prepared several large, attractive posters, one of which represents various wine colors (http://www.bouchard-aine.fr/en/tours-and-tastings.r-16/the-wine-shop.r-106/our-wine-posters.r-109/?valid_legal=1). Each representation is intended to illustrate the color of a particular named wine. As such, it is not applicable as a color classification scheme, but it could be an appealing visual aid in a wine course.

While difficulties remain in correlating spectral absorbency and perceived color (Kuehni, 2002), simple techniques can often yield useful data. For example, the color of red wine is often estimated by its absorbency at 420 nm and 520 nm (Somers and Evans, 1977). The sum of these values is a measure of color density (intensity, saturation), while their ratio estimates tint (hue). Greater absorbency at 420 nm provides an indicator of a brownish cast, whereas greater absorbency at 520 nm reflects redder shades. As red wines age, the level of yellowish polymeric pigments increases, whereas the impact of monomeric red anthocyanins decreases. Young red wines often have E_{420}/E_{520} ratios of 0.4 to 0.5, whereas old reds frequently show values between 0.8 and 0.9 (Fig. 2.1). An alternate method incorporates absorbency at 620 nm (Glories, 1984). Problems associated with turbidity may be taken into consideration by measurements taken at 700 nm (Mazza et al., 1999), or following sample centrifugation (Birse, 2007).

Additional information may be derived from estimates of the proportion of colored (ionized) monomeric anthocyanins, total anthocyanin, and phenolic contents. The proportion of anthocyanins complexed to various phenolic polymers can be derived from acidification with hydrochloric acid, decolorization with metabisulfite, and

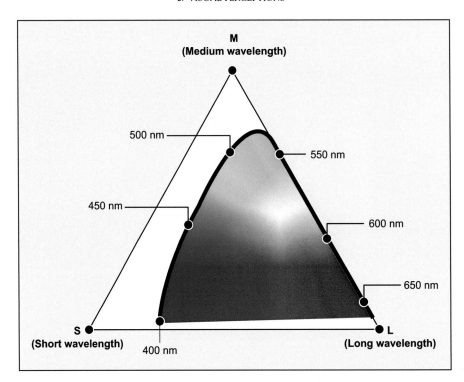

PLATE 2.2   Human color vision mapped as a triangle. The colors perceived by humans that refer to specific spectral wavelengths plot along the black curve. Nonspectral colors, represented below the curve are constructs generated in the brain by combining the responses from several cones, changes at the edges of colors, and other factors.

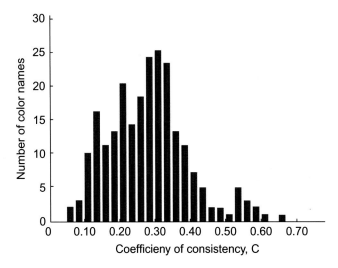

FIGURE 2.3   Coefficients of consistency for the selections made to 233 color names. *From Chapanis, A., 1965. Color names for color space. Am. Scientist 53, 327–345.*

subsequent recoloration with acetaldehyde (Somers and Evans, 1977). Because the proportion of complexed anthocyanins increases with age, it has been described as the wine's "chemical age." Several studies have shown a strong correlation between the amount of colored (ionized) anthocyanins and the perceived quality of young red wines (Somers and Evans, 1974; Somers, 1998). Regrettably, the measurements do not account for the presence of important pigments such as pyranoanthocyanins. In addition, besides the resulting ferruginous shift in color, aging is associated with pigment loss and reduction in color intensity.

TABLE 2.1 Proposed Set of Color Terms for Wines. Shaded Areas Are Term Combinations Inconsistent With Colors Found in Wines

Wine type	Color depth				
Red wines	Pale	Light	Medium	Dark	Intense
Purple red	░				
Ruby red					
Red					
Brick red				░	░
Tawny red	░			░	░

White wines	Pale	Light	Medium	Dark	Intense
Greenish yellow		░	░	░	░
Straw yellow			░	░	░
Yellow				░	░
Yellow gold	░				░
Golden brown	░	░			░

Rosé wines	Pale	Light	Medium	Dark	Intense
Pink			░	░	░
Rosé			░	░	░
Orangish rosé			░	░	░

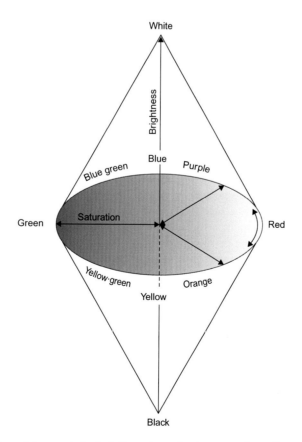

FIGURE 2.4 Schematic representation of the psychological dimensions of color space. *From Chapanis, A., 1965. Color names for color space. Am. Scientist 53, 327–345.*

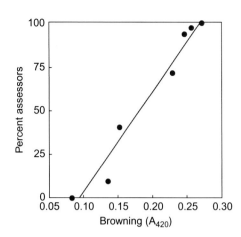

FIGURE 2.5   Percentage of 31 assessors rating wines with various levels of browning (measured by absorbance at A_{420}) as visually unaccept-able. *From Peng, Z., Duncan, B., Pocock, K.F., Sefton, M.A., 1999. The influence of ascorbic acid on oxidation of white wines: Diminishing the long-term antibrowning effect of SO2. Aust Grapegrower Winemaker 426a, 67–69, 71–73, reproduced by permission.*

For white wines, absorbency at 420 nm is used as an indicator of browning (Fig. 2.5). To simplify assessing this attribute in bottled wines (as an indicator of commercial acceptability), Skouroumounis et al. (2003) have developed correlations between measurements taken in cuvettes (10-mm path length) with those of several bottle-glass colors. As with any color estimate, the values derived may miss important subtleties (Skouroumounis et al., 2005).

Tristimulus colorimetry is another technique often used in assessing wine color. The procedure approximates the response of the human eye. Tristimulus colorimetry involves taking three, separate, spectrophotometer measure-ments, using red, green, and blue filters. Mathematical transformations convert the measurement into tristimulus colorimetry values. In contrast, tristimulus colorimeters directly translate readings into tristimulus measurements. Estimates of brightness (light/dark), saturation or chroma (degree of grayness), and hue (basic color) are generated.

However, the most internationally accepted standard for assessing color involves absorbency measurements over the full range of the visible spectrum. The CIELAB system, designed by La Commission Internationale de l'Eclairage, uses these measurements to derive values for L^* (relative lightness), a^* (relative redness, on a red–green axis), and b^* (relative yellowness, on a yellow–blue axis). The values are typically calculated using software packages that accompany the spectrophotometer. The data are combined with the absorption curves of the three retinal color pig-ments to approximate human color vision. Brightness is calculated by combining the response patterns generated by the red (long wavelength) and green (medium wavelength) pigments, whereas hue is derived from comparing the response patterns of the red and green pigments (red–green axis), and a comparison of the pattern derived from the blue and both red and green responses (blue–yellow axis) (Gegenfurtner and Kiper, 2003).

Although CIELAB measurements are often used to measure wine color, Negueruela et al. (1995) and Ayala et al. (1997) have proposed changes to make the values more applicable to wine. Such data can be used to design the blending of several wines to a predetermined color (Negueruela et al., 1990). To simplify determining CIELAB values in wineries, without access to the equipment and software normally required, Pérez-Magariño and González-San José (2002) have proposed a set of absorbance measurements.

Use of color measurements in normal winery practice has been hindered by marked differences in how assess-ments have been made. In spectrophotometric measurements, wines (often diluted) are placed in cuvettes (usually 1 or 2 mm for undiluted red wine, or 10 mm if diluted). In contrast, direct visual assessment involves wine in a glass, under poorly defined light conditions. In the latter situation, there can be considerable light scattering and marked differences in color noted. These differences have recently been described analytically by Huertas et al. (2003). Finally, chemical interpretation can be complicated as different combinations of pigments, in various states of ionization, oxidation, and polymerization, can produce the same subjective color impression. The best, but still imperfect, correlation between taster and CIELAB color assessments were obtained with hue values assessed at the rim of the wine (Hernández et al., 2009).

An alternative color assessment technique uses digital photography. With computerized, pixel-by-pixel analy-sis of the image, good correlation with human perception has been reported (Pointer et al., 2002; Brosnan and Sun, 2004; Cheung et al., 2005). Digital photography is also less expensive than the purchase and operation of a spectrophotometer, is easier to use, assays a larger surface area (useful where the color is nonhomogeneous), is rapid, and the data can easily be transferred to a computer for analysis (Yam and Papadakis, 2004; León et al., 2006). An example of digital camera analysis used to evaluate wine color is given by Martin et al. (2007).

Significance in Tasting

Color can affect the perception of a wine's quality (Figs. 2.6 and 2.7), as well as its taste and odor perception (Pangborn et al., 1963; Maga, 1974; Clydesdale et al., 1992). Correspondingly, wine assessors must guard against its potential to distort perceptions surreptitiously. For example, the perceived flavor intensity (and quality) of red wines has been correlated with color intensity (density, saturation) (Iland and Marquis, 1993) and hue (the proportion of red "ionized" anthocyanins) (Somers and Evans, 1974; Bucelli and Gigliotti, 1993). Although there is some justification for this association—most grape flavorants being primarily located in the skins, and likely to be extracted under the conditions that promote pigment extraction—this is not always the case. Color intensity is also considered an indicator of aging potential. Where color-induced bias is considered likely to unduly affect assessments (e.g., comparing wines of markedly different hue or color density), the wines should be sampled in black wine glasses, or, where such glasses are unavailable, under low-intensity red light.

The degree to which perception is based on color association is exemplified by the observation that people who can correctly identify cola-flavored beverages when dark brown frequently misidentify the solution as orange or tea when colored orange (Sakai et al., 2004, 2005). Conversely, an orange-flavored solution was often misidentified when colored dark brown. In another experiment, cherry-flavored beverages were frequently misidentified as lime when colored green (DuBose et al., 1980). With wine, the influence of color was clearly demonstrated by Morrot et al. (2001). The addition of tasteless anthocyanin (red) pigments to a white wine induced participants to describe the wine in terms typical of a red wine (descriptors possessing red or somber colors). Admittedly, the red-colored white wine (a Sauvignon blanc) does have aromatic similarities to Cabernet Sauvignon. Both wines were ones

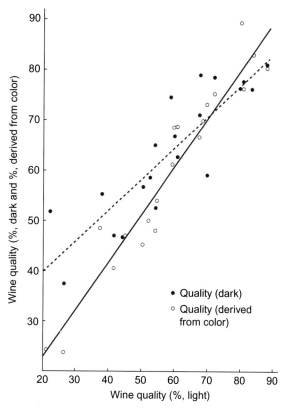

FIGURE 2.6   Illustration of the relationship between wine quality assessed by smell and taste (blind) and by sight alone (color). *From Tromp, A., van Wyk, C.J., 1977. The influence of colour on the assessment of red wine quality. In: Proc. S. Afr. Soc. Enol. Vitic. pp. 107–117, reproduced by permission.*

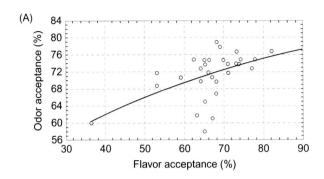

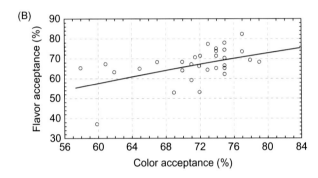

FIGURE 2.7   Relation between flavor and odor acceptance (A) and color and flavor acceptance (B). *From Pokorný, J., Filipům M., Pudil, F., 1998. Prediction of odour and flavour acceptancies of white wines on the basis of their colour. Nahrung 42, 412–415, reproduced by permission.*

the enology students had sampled frequently in previous tasting sessions. Color-induced distortion of perception may be less pronounced with professional tasters, at least when the color aberration seriously conflicts with other sensory clues as to origin (Parr et al., 2003). When tasters are presented with a single varietal wine, where color variance was limited, color appeared to play little role in quality assessment (Valentin et al., 2016). In another experimental variation, white, rosé, and red wines were assessed in black glasses (Ballester et al., 2009). Both wine experts and novices could identify whether the wines were white or red (by smell alone), but had difficulty with rosé wines. This applied equally to whether the participants were asked to categorize the wines as a white, rosé, or red wine, or whether they were asked to describe their fragrance verbally. In the second situation, white wines were described using terms for objects with light yellow or orange colors, whereas the red wines were described in terms of dark-colored objects. Rosés were more commonly described in terms appropriate for white wines than red wines.

Sampling wine under any illumination other than white is usually limited to laboratory conditions, where low-intensity red light is considered to distort color perception sufficiently to exclude wine color from affecting perception. Nonetheless, sampling wine under colored light of similar luminance had less effect on quality perception; blue and red apparently enhanced the appreciation of a white wine (Rheingau Riesling) (Oberfeld et al., 2009). In another experiment, green light enhanced the perceived freshness of a red wine (tasted in black glasses), whereas red illumination favored liking (Spence et al., 2014).

The effect of modified color intensity on experienced tasters has not been performed using identical wines. However, anecdotal evidence suggests that darker wines are inherently more highly rated. Under laboratory test conditions (Zellner and Whitten, 1999), increasing color intensity only marginally affected perceived odor intensity, but significantly affected the perceived appropriateness of the colored solution. Similar results were detected by Kemp and Gilbert (1997), with odor intensity being associated with darker colors. Not surprisingly, there was no influence when the participants were blindfolded (Koza et al., 2005). The influence of a solution's color on perceived odor intensity has even been observed at the neuronal level (Österbauer et al., 2005). Presentation of samples with the color typically associated with a particular fragrance (e.g., red with strawberry) enhances the orbitofrontal complex response. This is the cerebral region most associated with the integration of sensory

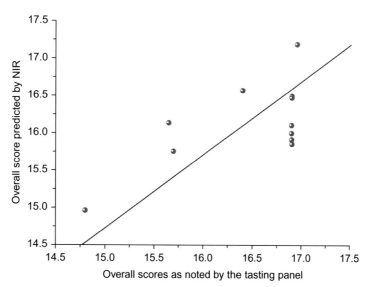

FIGURE 2.8 Near-infrared (NIR) measurement-predicted scores versus reference value for commercial Shiraz wines. *Reprinted from Cozzolino, D., Cowey, G., Lattey, K. A., Godden, P., Cynkar, W. U., Dambergs, R. G., Janik, L., and Gishen, M., 2008. Relationship between wine scores and visible-near infrared spectra of Australian red wines. Anal. Bioanal. Chem. 391, 975–981., with kind permission from Springer Science + Business.*

impulses. The inverse occurs with inappropriate colors (e.g., blue with strawberry). That this influence is not just a laboratory phenomenon is suggested by the effect of colors on wine quality (Fig. 2.6) and inappropriate color on acceptability (Figs. 2.5 and 2.7).

Although color can affect wine assessment, not all tasters are equally influenced (Williams et al., 1984b). This may relate to the manner in which tasters do their assessment. From taste–taste and taste–odor interaction studies, the influences are most marked when assessments are requested to be holistic, and least evident when conducted attribute by attribute. Directions to ignore color in assessment may result in tasters disregarding color in their evaluations (Williams et al., 1984a). How effective this is has not been confirmed.

The influence of color on wine perception appears to be based on experience associating particular colors with certain wines. For example, young dry white wines generally range from nearly colorless to pale straw. A more obvious yellow tint could suggest more mature grapes, extended skin contact (maceration), or maturation in oak cooperage. More golden colors are also associated with prolonged aging or may indicate a sweet botrytized wine. Sherries vary from pale straw to dark golden brown, depending on the style (*finos* the lightest, *olorosos* the darkest). Rosé wines are expected to be pale to light pink to raspberry colored, without shades of blue. Red wines vary from deep purple to tawny red. Initially, most red wines possess a purplish-red hue, especially noticeable when viewed on an angle at the wine–glass interface. With prolonged aging, the wine loses much of its color intensity and increasingly takes on a brickish to ferruginous hue. Red ports, depending on the style, may be deep red, ruby, or tawny colored.

Because all wines eventually take on brownish hues, browning is often used as an indicator of age. The change in red wines is often indicated by a lowering of the E_{420}/E_{520} ratio (Somers and Evans, 1977). However, a brownish cast may equally indicate premature oxidation, or excessive heat exposure. Therefore, wine age, type, and style must be known before interpreting the meaning and significance of a brownish hue (or any wine color). Brown shades are acceptable only if associated with the development of a desirable processing or aged bouquet. The heating of madeira, which gives the wine its brown coloration and baked bouquet, is an example of process-produced browning. Because most wines fail to develop a desirable aged bouquet, brown casts often mean that the wine is oxidized, or in the vernacular, "over the hill."

Another, more modern variant of objectively assessing red wine quality is based on visible (VIS) and near-infrared (NIR) spectra (Cozzolino et al., 2008). The wine scores of a large number of Australian red wines were correlated with measurements between 400 and 2,500nm). Although inapplicable as a substitute for human assessment, the correlation between wine scores and spectroscopic measurements may warrant its use in screening large numbers of wines before assessment in a competition (Fig. 2.8). It could reduce the number of samples meriting full assessment.

Origin and Characteristics

Red Wines

Anthocyanins are the primary color determinants in red grapes and wine. In grapes, anthocyanins occur predominantly as glucosides—conjugates with one or more glucose molecules. The bonding increases both chemical stability and water solubility. The glucosides may also be complexed with acetic, coumaric, or caffeic moieties.

Five main classes of anthocyanins occur in grapes and wine: cyanins, delphinins, malvins, peonins, and petunins. They differ based on the number and position of hydroxyl and methyl groups on the B-ring of the anthocyanidin molecule (Table 2.2). The content and relative amounts of each class vary considerably among cultivars and with growing conditions (Wenzel et al., 1987). The anthocyanins' hydroxylation/methylation pattern influences color hue and stability, with free hydroxyl groups enhancing blueness, whereas methylation augments redness. In addition, the presence of hydroxyl groups adjacent to each other (o-diphenols) on the B-ring markedly enhances potential oxidation. Thus, wine with a high proportion of malvin or peonin, neither of which possess o-diphenol arrangements, significantly enhances color stability. Resistance to oxidation is also a function of conjugation of the anthocyanin with sugar(s) and other moieties (Robinson et al., 1966). In most red grapes, malvin is the predominant anthocyanin. Because malvin is the reddest of anthocyanins, its concentration is the principal contributor to a young red wine's hue.

An additional source of color variation results from the proportion of anthocyanins in one of five molecular states. Four are free-forms, whereas one is bound to sulfur dioxide (Fig. 2.9). Most of these states are colorless within the pH range of wine. Those molecules in the flavylium state generate a red hue, whereas those in the quinoidal state donate a bluish tint. The proportion in each state depends primarily on the pH and sulfur dioxide content of the wine. Low pH enhances redness, by favoring the flavylium state, whereas high pH generates a blue-mauve cast, by favoring the quinoidal state. Color density is also affected. The bonding of sulfur dioxide to an anthocyanin can diminish (bleach) a wine's color. As a wine ages, progressive bonding of anthocyanins with tannins and other wine constituents causes further changes to the color absorbency of the molecule.

In grapes, anthocyanins exist primarily in stacked conglomerates. These occur primarily as hydrophobic interactions between individual anthocyanins (self-association), or between anthocyanins and other phenolic compounds (copigments). Both complexes increase light absorbency and, thereby, color intensity. During vinification and maturation, these conglomerates begin to disassociate. Anthocyanin molecules freed into the acidic environment of wine lose their bluish color. In addition, disassociation results in reduced light absorption and loss in color. Typical losses in color density can vary from two- to five-fold, depending on the pH, ethanol, and tannin contents of the wine.

TABLE 2.2   Anthocyanins Occurring in Wine[a]

Specific name	R_3	R_4	R_5
Cyanidin	OH	OH	
Peonidin	OCH_3	OH	
Delphinidin	OH	OH	OH
Petunidin	OCH_3	OH	OH
Malvidin	OCH_3	OH	OCH_3

Derivatives	Structure
Monoglucoside	R_1 = glucose (bound at the glucose 1-position)
Diglucoside	R_1 and R_2 = glucose (bound at the glucose 1-position)

[a]After Methods for Analysis of Musts and Wines, MA Amerine and CS Ough, Copyright 1980 John Wiley and Sons, Inc. Reprinted by permission of John Wiley & Sons, Inc.

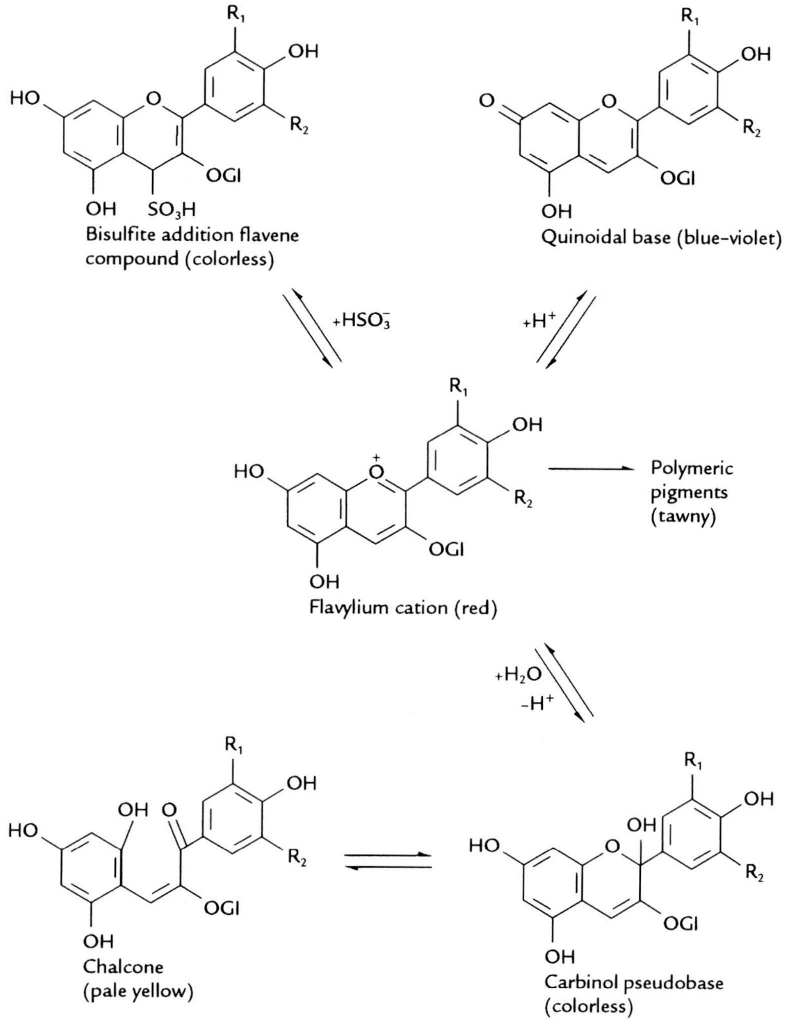

FIGURE 2.9 Equilibria among the major forms of anthocyanins in wine (*Gl*, glucose). *From Jackson, R.S., 2014. Wine Science: Principles and Applications, 4th ed. Academic Press, San Diego, CA, reproduced by permission.*

Nevertheless, sufficient copigmentation complexes may survive, or form during fermentation, to contribute to the purple tint characteristic of most young red wines. These color changes may occur without modification in the wine's absolute anthocyanin content.

During wine maturation, not only do anthocyanin aggregates disassociate, but anthocyanins also tend to lose their sugar and acyl (acetate, caffeate, or coumarate) moieties. This makes them both more susceptible to irreversible oxidation (browning) and conversion from colored flavylium states to colorless hemiacetals. To limit these events, it is important that the wine contain significant quantities of flavonoids: catechins, proanthocyanidins (dimers, trimmers, and tetramers of catechins), and their polymers (condensed tannins). These combine with free anthocyanins to form more color-stable polymers. The polymers also extend light absorption into the blue region. This partially explains the ferruginous shift that occurs during aging.

Correspondingly, the concurrent extraction of various flavonoids with anthocyanins during fermentation is crucial to long-term color stability in red wines. These compounds begin to polymerize with free anthocyanins almost immediately. By the end of fermentation, some 25 percent of the anthocyanins may be polymerized with catechins and their polymers. Anthocyanin polymerization can rise to about 40 percent within 1 year in oak cooperage (Somers, 1982). Subsequent polymerization continues more slowly, and may approach 100 percent within several years (Fig. 2.10). Thus, red color reflects the initial amount, nature, and states of the anthocyanins in the wine, as well as the types and amounts of flavonoids extracted and retained during and after vinification, and their

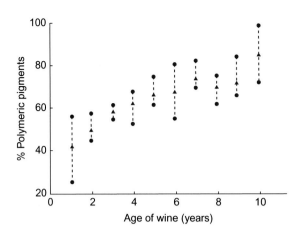

FIGURE 2.10   Increase in the contribution of polymeric pigments to wine-color density during the aging of Shiraz wines (▲, mean values; ●, extremes). *From Somers, T.C., 1982. Pigment phenomena – From grapes to wine. In: Webb, A.D., (Ed.), Grape Wine Cent. Symp. Proc. pp. 254–257. University of California, Davis, reproduced by permission.*

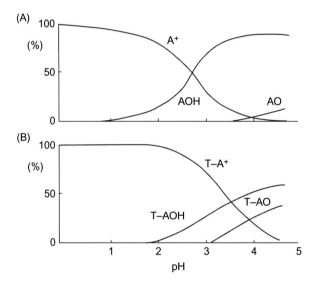

FIGURE 2.11   Equilibria among different forms of (A) free anthocyanins and (B) combined anthocyanins (T-A) extracted from wine ($^+$, red flavylium cation; O, blue-violet quinoidal base; OH, colorless carbinol pseudobase). *From Ribéreau-Gayon, P., Glories, Y., 1987. Phenolics in grapes and wines. In: Lee, T. (Ed.), Proc. 6th Aust. Wine Ind. Tech. Conf. Australian Industrial Publ., Adelaide, Australia, pp. 247–256, reproduced by permission.*

complex polymerization. The poor color stability of most red muscadine wines appears to be due to the absence of appropriate flavonoids and acylated anthocyanins in muscadine grapes (Sims and Morris, 1986). A somewhat similar situation may explain why most Pinot noir wines are poorly colored.

Polymerization helps protect the anthocyanin molecule not only from oxidation but also from other chemical modifications. Polymerization also tends to increase solubility, reducing both flavonoid (tannin) and pigment loss. Finally, polymerization augments the number of anthocyanin molecules in colored (flavylium and quinoidal) states. For example, about 60 percent of anthocyanin–tannin polymers are colored at pH 3.4, whereas 20 percent of equivalent free anthocyanins are colored (Fig. 2.11). The yellow-brown flavylium and quinoidal anthocyanin–tannin polymers are thought to generate most of the age-related brickish shades of red wines.

Anthocyanin polymerization with flavonoids occurs slowly in the absence of oxygen. Nevertheless, it is accelerated in association with oxygen inadvertently absorbed during racking or other cellar activities. This has induced some winemakers to use microoxygenation to regulate the rate and degree of polymerization, especially when using inert cooperage. The small amounts of peroxide generated, as oxygen reacts with phenolics in the wine, oxidizes ethanol to acetaldehyde. Subsequent reactions between acetaldehyde and anthocyanins further favor their polymerization with flavonoids. The initially small anthocyanin–acetaldehyde–flavonoid polymers are thought to enhance the violet shift so typical of young red wines (Dallas et al., 1996). Acetaldehyde also reacts with sulfur dioxide, favoring its disengagement from anthocyanins. This not only reverses any bleaching action of sulfur dioxide added before fermentation, but also liberates the anthocyanin for polymerization with catechins and their polymers.

Other mechanisms involved in color stabilization include various yeast metabolites, notably pyruvic acid. It can react with anthocyanins (Fulcrand et al., 1998), generating a tawny red pigment. Monoglucosides and coumaroyl monoglucosides of malvin, the predominant anthocyanin in grapes, can also complex with 4-vinylphenol, generating red-orange pigments (Fulcrand et al., 1997). These and related compounds are termed pyranoanthocyanins. For details see Rentzche et al. (2007) and Jackson (2014). In addition, internal rearrangements of anthocyanins and other flavonoids produce yellow-orange xanthylium products. Colorless flavonoids also generate ferruginous products upon oxidation. Thus, although the color of red wines starts out primarily produced by anthocyanins, the aged color of red wines is a complex of, among others, anthocyanin–tannin polymers, oxidized tannins, pyranoanthocyanins, and xanthylium products. Its complexity is such that even more sophisticated analytical instruments than currently exist may be needed to discover the critical and decisive events involved in the evolution of a red wine's color. Table 2.3 highlights some of the various forms producing yellow, yellow-red, yellow-brown, red, and violet shades. The reduction in color density that accompanies aging results from oxidation, structural changes in anthocyanin–tannin polymers, and their precipitation with tartrate salts and soluble proteins.

Red wines vary from deep red-purple to pale tawny-red. As noted, the purplish-red hue of young red wines partially reflects the continuing presence of anthocyanin complexes and anthocyanin–acetaldehyde–flavonoid oligomers, but can also be an indicator of the wine's pH. A light color can indicate grape immaturity or poor winemaking practices. However, certain cultivars, such as Gamay and Pinot noir, seldom yield wines with deep colors. Spätburgunder (Pinot noir) wines from Germany are typically so pale as to often resemble a rosé. Cool climatic conditions are not conducive to the production of dark-colored red wines. In these situations, the varietal origin of the wine should be known to avoid unduly penalizing the wine (unless color bias is avoided by sampling in black glasses).

More intensely pigmented varieties, such as Shiraz and Cabernet Sauvignon, may remain deep red for decades. Dark shades often correlate with rich flavors, coextracted along with anthocyanins from the skins. Because older vinification procedures favored the uptake of high levels of tannins, they generated bitter/astringent sensations that could take decades to soften. This can be reduced with modern techniques, such as the use of rotary fermenters. These favor the early extraction of berry flavors and intense coloration, before tannin uptake reaches high levels. Even with standard fermenters, shorter skin-contact time during fermentation can generate milder but still highly flavored, deeply colored wines. What may be compromised is the long aging potential of the wine.

TABLE 2.3 Color and Molecular Weight of Some Wine Phenols[a]

Name[b]	Color	Molecular weight (kDa)
A⁺	Red	
AOH	Noncolored	500
AO	Violet	
AHSO₃	Noncolored	
P	Noncolored	600
T	Yellow	1000–2000
T-A⁺	Red	
T-AOH	Noncolored	1000–2000
T-AO	Violet	
T-AHSO₃	Noncolored	
TC	Yellow-red	2000–3000
TtC	Yellow-brown	3000–5000
TP	Yellow	5000

[a]From Ribéreau-Gayon and Glories (1987), reproduced by permission.
[b]A, Anthocyanin; HSO₃, bisulfite addition compound; O, quinoidal base; OH, carbinol pseudobase; P, proanthocyanidin; T, tannin; TC, condensed tannin; TtC, very condensed tannin; TP, tannin condensed with polysaccharides.

Most red wines begin to take on a noticeable brickish cast within a few years, especially when long-aged in oak cooperage (Fig. 2.1). Ferruginous or tawny-red colors are acceptable only if associated with the development of a favorable aged bouquet. In most standard red wines, these hues indicate only that the wine has lost the fruitiness of its youth. In young red wines, brick shades may suggest overheating (as in a warehouse where a baked odor might also be present), or a faulty closure (and associated with an oxidized phenol odor).

Rosé Wines

Rosé wines are expected to be pale pink, sour cherry, or raspberry colored, without shades of blue. The actual shade depends on the amount and types of anthocyanins found in the cultivar(s) used. An orangish cast is generally undesirable, but can be characteristic of rosés made from Grenache. Otherwise, hints of orange usually suggest oxidation. Pale bluish hints often signify that the wine is too high in pH and may taste flat.

White Wines

Less is known about the chemical nature and development of color in white wines. The minor phenolic content of most white wines consists primarily of hydroxycinnamates, such as caftaric acid and related derivatives. On crushing, these readily oxidize and form S-glutathionyl complexes. These generally do not turn brown. Thus, it is believed that most of the yellowish pigmentation in young white wines is derived from the extraction and oxidation of flavonols, such as quercetin and kaempferol. Constituents extracted from oak cooperage, if used, can add to the color of white wines. The deepening yellow-gold of older white wines probably comes from the oxidation of phenols or galacturonic acid (a breakdown product of grape-derived pectins). However, gold shades may also develop following the slow formation of melanoidin compounds by Maillard reactions, and sugar caramelization. Occasionally, a pinkish cast is detectable in some white wines. For example, the occasional pinking of Sauvignon blanc wines is attributed to the oxidation of dehydrated leucoanthocyanins (flavan-3,4-diols). However, the pinkish coloration of some Gewürztraminer wines comes directly from anthocyanins extracted from pinkish-red clones of the cultivar. Some so-called white (blush) wines, such as white Zinfandels, come from red grapes pressed early to minimize color extraction. They are rosés by another name. In fortified sweet wines, much of the color comes from either oxidation of wine phenolics, melanoid pigments formed during baking (*estufagem*) in the production of madeira, or the concentration of grape juice used for sweetening (e.g., *surdo*, *mistela*).

Typically, young, dry, white wines range from nearly colorless to pale straw. A more obvious yellow tint may be considered suspicious, unless associated with overmaturation of the grapes, extended skin contact (maceration), or maturation in oak cooperage. Deeper hues develop with aging. If associated with the development of an appreciated aged bouquet, it is desirable. If associated with accidental oxidation ("browning"), and the presence of a degraded odor, it is a fault. In contrast, unusually pale colors may suggest the use of unripe grapes (absence of typical coloration, high acidity, little varietal character), removal of the juice from the skins without maceration (extraction of few phenolics and reduced varietal flavor), or the excessive use of sulfur dioxide (which has a bleaching action). Sweet white wines generally are more intensely colored, being straw-yellow to yellow-gold. This may result from in-berry oxidation of grape constituents during overripening.

Sherries vary from pale straw to golden brown, depending on the particular style (*fino* to *oloroso*), and the method and degree to which the wines are sweetened. Madeiras are typically amber colored (unless decolorized), due to the heat processing they undergo (younger versions made with Tinta negra cultivar retain a reddish color). Although white wines typically darken with age, some fortified white wines may lighten (e.g., Marsala). This results from the precipitation of melanoid pigments.

CLARITY

In contrast to the complexity of interpreting the significance of color, haziness is always considered a fault. With modern techniques of clarification, consumers have come to expect a crystal-clear product. This does demand considerable effort to achieve though.

Crystals

Young wines typically are supersaturated with tartrate salts after fermentation. This results as the increasing alcohol content generated during fermentation reduces their solubility. Solubility also declines due to salt isomerization during wine maturation. Given sufficient time, these crystals precipitate spontaneously. In northern regions, the low temperatures found in unheated cellars can induce adequately rapid precipitation. Where spontaneous precipitation is inadequate, refrigeration often achieves rapid and satisfactory bitartrate stability.

Crusty, flake- or needle-like crystals are usually potassium bitartrate, whereas fine crystals are typically calcium tartrate. Illustration of tartrate and other salts potentially found in wine is given in Lüthi and Vetoch (1981) and Edwards (2006).

Bitartrate crystallization is based on free-salt concentration. Association with protective colloids can mask positively charged sites on bitartrate crystals, retarding their crystallization (Lubbers et al., 1993). Interaction between tannins, tartrate, and potassium ions also delays crystallization. Thus, cold treatment may be insufficient to permanently stabilize some wines, due to subsequent disassociation of the aforementioned complexes.

Although comparatively rare now, the accumulation of tartrate salt crystals in a wine should not be a cause for consumer rejection. The crystals are tasteless (but crunchy), and usually remain in the bottle with any sediment that may have developed during aging. Alternatively, tartrate crystals may form on the underside of the cork. Because white wines are transparent and typically sold in pale to colorless bottles, any crystal formation is likely to be more obvious in a white wine than a red wine. In addition, white wines are typically chilled or stored cool, accentuating crystal formation. Some producers mention their possible occurrence by the euphemistic term "wine diamonds." Unfortunately, some consumers still unwittingly mistake tartrate crystals for glass slivers.

Another type of crystal occasionally found in wine is formed from calcium oxalate. Oxalic acid is a minor constituent of grapes, especially if they contain raphide or druse crystals, but higher than normal amounts in a wine are usually associated with "contamination" of the grape crush with grape leaves. Crystal formation occurs primarily in older wines due to the slow oxidation of any ferrous or ferric oxalate. Because ferric oxalate is unstable, dissociation liberates free oxalic acid. It can subsequently associate with calcium, leading to the formation of calcium oxalate crystals. Corks are another potential source of oxalate.

Other potential troublesome sources of crystals are saccharic and mucic acids. Both are produced by, and extracted from Botrytis-infected grapes, subsequently forming insoluble calcium salts in bottled wine. Calcium mucate crystals are usually the source of yellowish particles occasionally found in bottles of sauternes.

Sediment

Resuspension of sediment was probably the most frequent source of clouding in older red wine. Sediment typically consisted of a complex of polymerized anthocyanins, tannins, proteins, and tartrate crystals. Depending on its composition (Quinsland, 1978), sediment had a bitter or chalky taste. To some wine aficionados, the presence of sediment is considered a sign of quality. This view was based on exaggerated concern that clarification and fining removed critical flavorants. However, avoidance of these procedures does not guarantee a finer or more flavorful wine. Currently, most red wines are fined sufficiently to rarely if ever produce a noticeable sediment. Vintage ports are the one major exception, being bottled early in their development.

Proteinaceous Haze

Although an uncommon source of wine rejection, protein haze can still cause considerable economic loss in bottle returns. Proteinaceous haze results from the clumping of dissolved proteins into light-dispersing particles. Its formation is enhanced by exposure to heat, and the presence of sulfites, tannins, and trace amounts of metal ions. Proanthocyanins (oligomers of catechin flavonoids) bind well with proteins containing proline (Siebert, 2006). However, these proteins typically precipitate during fermentation or maturation, or are removed during fining. Thus, they are not the problem they can be in beer (Asano et al., 1984). The proteins that appear to be the principal source of problems in bottled wine are pathogenesis-related (PR) proteins (notably chitinase and acid-stable thaumatin-like proteins) (Waters et al., 1996b; Dambrouck et al., 2003). They are produced in response to infection by pathogens or other stresses, including damage during harvest (Pocock and Waters, 1998). Although yeast mannoproteins and grape arabinogalactan–protein complexes may promote heat-induced protein hazes, specific members can also reduce their formation (Pellerin et al., 1994; Dupin et al., 2000).

Phenolic Haze

Excessive use of oak chips during wine maturation, or the accidental incorporation of leaf material during grape crushing (Somers and Ziemelis, 1985), can occasionally induce rare forms of phenolic haze. The first situation results from the excessive extraction of ellagic acid, leading to the formation of fine, off-white to fawn-colored crystals.

Fine yellow quercetin crystals, extracted from leaf material during crushing, can induce a flavonol haze in white wines. This is particularly likely if the wine has been bottled before the crystals have had a chance to precipitate fully (Somers and Ziemelis, 1985). The use of sulfur dioxide in the production of red wines (presumably to counter the action of laccases released from diseased grapes) has also been associated with cases of phenolic haze.

Casse

Several insoluble metal salts can generate haziness in bottled wine (*casse*). Of these, the most important are induced by iron (Fe^{3+} and Fe^{2+}) and copper (Cu^{2+} and Cu^{+}) ions. The primary source of troublesome concentrations of metallic ions is corroded winery equipment, or the use of copper-based fungicides too close to harvest.

Two forms of ferric *casse* are recognized. White wines may be affected by an off-white haze that forms as soluble ferrous phosphate oxidizes to insoluble ferric phosphate. The haze results either from particles of ferric phosphate alone, or a complex formed between ferric phosphate and soluble proteins. In red wine, oxidation of ferrous to ferric ions can generate a blue *casse*. In this instance, ferric ions form insoluble particles with anthocyanins and tannins.

In contrast to iron-induced haziness, copper *casse* forms only under reduced (anaerobic) conditions. The *casse* develops as a fine, reddish-brown deposit when the redox potential of bottled wine falls during aging. Exposure to light speeds the reaction. The particles consist of cupric and cuprous sulfides, or their complexes with proteins. Copper *casse* is primarily a problem in white wines, but can also occur in rosé wines.

Deposits on Bottle Surfaces

Occasionally sediment may adhere tightly to the inner surfaces of bottles. This generally occurs as an elongated, elliptical deposit on the lower side of the bottle. Less frequently, a lacquer-like deposit may form in bottles of red wine. It consists of a film-like tannin–anthocyanin–protein complex (Waters et al., 1996a). Champagne may also develop a thin, film-like deposit on the inner surfaces of the bottle. This phenomenon, called *masque*, results from the deposition of a complex between albumin (used as a fining agent) and fatty acids (probably derived from auto-lyzing yeast cells). It develops after the second, in-bottle fermentation (Maujean et al., 1978).

Microbial Spoilage

Haze can also result from the action of spoilage organisms (yeasts and bacteria). The most important spoilage yeasts in bottled wine are species of *Zygosaccharomyces* and *Brettanomyces*. Three groups of bacteria may also induce spoilage: lactic acid bacteria, acetic acid bacteria, and on rare occasions, *Bacillus* species.

Zygosaccharomyces bailii can generate both flocculent and granular deposits (Rankine and Pilone, 1973), notably in white and rosé wines. Contamination usually originates from yeast colonization of improperly cleaned and steril-ized bottling equipment. In contrast, *Brettanomyces* spp. induce a distinct haziness. It has been reported to become evident at less than 10^2 cells/ml (Edelényi, 1966). More commonly, though, noticeable cloudiness becomes visible only at concentrations above 10^5 cells/ml. Yeast-induced haziness may occur without spoilage, but frequently is associated with vinegary (*Z. bailii*) or mousy, barnyardy (*Brettanomyces*) off-odors. Other fungi can cause clouding or pellicle formation, but only under aerobic conditions (a situation not found in properly sealed bottled wine).

Certain lactic acid bacteria generate a cloudy, viscous formation in red wines, a situation called *tourne*. The affected wine also develops a sauerkrauty or mousy taint, carbon dioxide accumulation, and a dull ferruginous color. Other bacterial strains may synthesize profuse amounts of mucilaginous polysaccharides (β-1,3-glucans). They hold the bacteria together in long silky chains. The filaments often appear as floating threads, generating a situation termed *ropiness*. When dispersed, the polysaccharides give the wine an oily look and viscous texture. Although visually unappealing, ropiness is not consistently associated with off-odors. These spoilage conditions principally develop at pH values ≥ 3.7.

Acetic acid bacteria have long been associated with wine spoilage. For years, these bacteria were thought to be obligately aerobic (requiring molecular oxygen for growth). It is now known that quinones (oxidized phenolics) can supply the oxygen needed (Aldercreutz, 1986). Thus, acetic acid bacteria may grow in bottled wine, if acceptable electron acceptors are present. In addition, even traces of oxygen, incorporated in procedures such as racking or bâttonage (associated with *sur lies* maturation) can activate growth. If they multiply in wine, they can form a hazi-ness associated with the development of vinegary odors and tastes.

VISCOSITY

Although more a mouth-feel than a visual attribute, if detectable, it is noticeable as a slight sluggishness in the wine's flow (fluidity). The few instances where detectable increases in viscosity occur involve glycerol contents >25 g/liter (as in highly botrytized wines), or in the presence of bacterial mucopolysaccharides (wines showing *ropiness* or *tourne*).

Sugar contents (e.g., 15 g fructose and 5 g glucose) can generate viscosity values equivalent to 25 g glycerol (Nurgel and Pickering, 2005), about 1.5 cP (mPa). At or above this value, viscosity differences begin to be humanly perceptible. Besides affecting fluidity, viscosity at or above 1.5 cP reduces the perception of astringency and sourness (Smith and Noble, 1998). It may also reduce the perceived intensity of aromatic compounds (Cook et al., 2003). However, the latter phenomenon appears to occur at viscosity values more typical of food and desserts than wine.

PETILLANCE/EFFERVESCENCE

Still wines are supersaturated with carbon dioxide after fermentation. During maturation, the excess CO_2 usually dissipates as equilibration develops between the carbon dioxide in the wine and the ullage of the cooperage. However, if the wine is bottled early, before this has come to completion, bubbles may form along the sides and bottom of the glass, and be detectable in the mouth as a slight petillance. This is most likely to be observed in bottles of Beaujolais nouveau, or young wines not left on their sides for about 24 h after cork insertion. This does not occur in wines aged in-bottle for several years as any excess carbon dioxide eventually escapes via the cork, or between the cork and the neck of the bottle.

Another source of mild effervescence can be as a byproduct of postbottling malolactic fermentation. If this is the cause, it typically is associated with the formation of a fine, haze-producing sediment (bacterial cells). Light petillance can also be associated with *tourne*.

Usually, marked and continuous effervescence is intentional. In sparkling wines, bubble size, association in chains, and duration are considered important quality features. Slow, continuous effervescence is favored by prolonged contact between the wine and its lees, following the second fermentation. Yeast autolysis (self-digestion) releases colloidal mannoproteins (cell-wall constituents) into the wine. The weak association formed between carbon dioxide and these proteins is thought essential to the production of a steady stream of bubbles following opening. Whether the sound of bursting bubbles (or the hiss/pop associated with cork removal) has sensory significance to the perceived quality of sparking wines, as in carbonated beverages (Zampini and Spence, 2005), has yet to be assessed.

Many factors affect the solubility of carbon dioxide, notably the temperature and the wine's sugar and ethanol contents. Increasing any of these factors decreases solubility. Once the wine is poured, atmospheric pressure becomes the critical factor promoting bubble formation. Pressure on the wine drops from 6 ATM (in the bottle) to 1 ATM (ambient). This reduces carbon dioxide solubility from about 14 g/liter to 2 g/liter. This initiates the release of 5–6 liters of gas (from a 750-ml bottle). In the absence of agitation, however, there is insufficient free energy for the CO_2 to escape immediately. It enters a metastable state, from which it slowly escapes.

Carbon dioxide escapes from the wine via several mechanisms (Fig. 2.12). The initial foaming associated with pouring is activated by the free energy connected with pouring (homogeneous nucleation). However, the continuous chains of bubbles so desired in a sparkling wine are produced by heterogeneous nucleation. This begins when carbon dioxide diffuses into microscopic gas pockets, usually in particulate matter adhering to the sides of the glass or floating in the wine. This material usually consists of cellulosic fragments left behind during drying, or that have fallen out of the air (Liger-Belair et al., 2002). Nonetheless, some bubbles can initiate from suspended crystals of potassium bitartrate. Nucleation sites probably originate during the initial burst of homogenous bubbling, as wine is poured into the glass. As carbon dioxide diffuses into these nucleation sites, nascent bubbles sequentially enlarge, bud off, and begin their ascent (Fig. 2.13). During their ascent, bubbles enlarge as they continue to absorb carbon dioxide from the wine. This increases their rate of ascent and causes their separation within the chain (Fig. 2.14). The slow, progressive release of bubbles is the principal, and desirable, means by which CO_2 escapes from a sparkling wine.

In contrast, gushing, the sudden to explosive release of carbon dioxide, results from a number of separate processes. Mechanical shock waves produced by agitation (or shaking), or during pouring, provides sufficient energy to weaken the bonds between water and carbon dioxide. As bubbles reach a critical size, they incorporate more CO_2

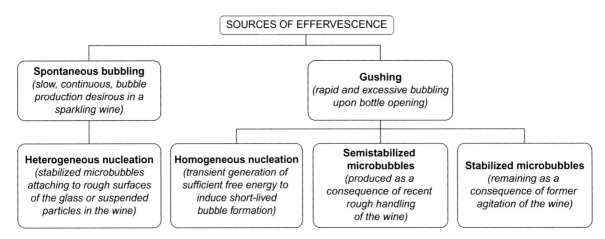

FIGURE 2.12   Mechanisms of effervescence (CO₂ escape) from sparkling wine. *From Jackson, R.S., 2014. Wine Science: Principles and Applications, 4th ed. Academic Press, San Diego, CA, reproduced by permission.*

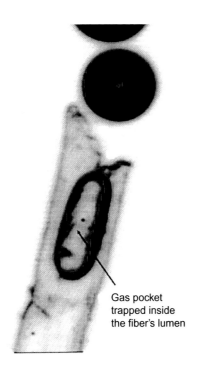

FIGURE 2.13   Closeup of a cellulose fiber acting as a bubble nucleation site. *From Liger-Belair, G., 2005. The physics and chemistry behind the bubbling properties of champagne and sparkling wines: A state-of-the-art review. J. Agric. Food Chem. 53, 2788–2802, reproduced by permission. Copyright 2005 American Chemical Society.*

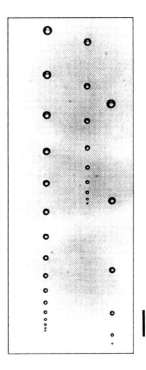

FIGURE 2.14   Simultaneous formation of distinct bubble chains from a collection of differently sized nucleation sites on the side of a champagne flute (*dark line*=1 mm). *From Liger-Belair, G., 2005. The physics and chemistry behind the bubbling properties of champagne and sparkling wines: A state-of-the-art review. J. Agric. Food Chem. 53, 2788–2802, reproduced by permission. Copyright 2005 American Chemical Society.*

than they lose, enlarging as they rise. In addition, semistabilized to stabilized microbubbles, previously generated in the bottle, are ready to enlarge explosively if sufficient free energy is provided.

Another feature in assessing the quality of sparkling wines is the accumulation of bubbles (*mousse*) on the surface of the wine in the center of the glass, and a ring of bubbles (*cordon de mousse*) along the edge of the glass (Plate 1.1). Prolonged duration of the mousse, as in the head on beer, is undesirable. The durability of these formations is largely dependent on the nature of the wine's surfactants (such as soluble proteins, polyphenols, and polysaccharides). They decrease surface tension, easing bubble fusion. This results as gravity removes fluid from between the bubbles, inducing them to fuse and take on angular shapes. As their size increases, so does their tendency to rupture.

Natural surfactants originate as degradation products of yeast autolysis. Their concentration in the wine increases two to three times within the first year of maturation, following the second, CO_2-producing fermentation. Contaminant surfactants, notably soap or detergent residues, left after glass washing, can effectively dampen bubble formation. These residues coat the particles that are essential for the initiation (nucleation) step in bubble formation. This can be readily demonstrated with detergent-washed, but poorly rinsed, glasses—preferably with a soft drink rather than a fine champagne.

The duration of effervescence can be extended by pouring the wine slowly down the sides of a flute, at about a 45°angle (Liger-Belair et al., 2010). After a small amount of the wine has been poured, pouring should stop until the initial accumulation of bubbles subsides. Pouring can subsequently recommence without the undue loss of carbon dioxide.

TEARS

Droplets that form and slide down the sides of a glass after swirling (Fig. 2.15) have been noticed for centuries. These have been variously termed tears, legs, or rivulets. They have attracted considerable interest from physiochemists (Walker, 1983; Neogi, 1985; Fournier and Cazabat, 1992; Vuilleumier et al., 1995). Tears form after swirling as alcohol evaporates more rapidly than water from the thin film of wine that coats the surface of the glass. This results because the surface tension (γ) of an alcohol–water mixture increases from a value of about 0.43 at 15% alcohol, to 0.48 (at 10%), to 0.56 (at 5%), and 0.73 for pure water (at 20°C). Corresponding density changes are 0.986, 0.982, 0.989, and 0.998 (kg/L). In addition, a temperature differential develops across the film (Venerus and Simavilla, 2015). These factors results in wine being drawn up the sides of the glass and the water molecules pulling together more tightly. As the density of the liquid film is higher at its rim than elsewhere, droplets start to sag, producing arches. Finally, drops begin slide down, forming the tears. On reaching the surface of the wine, fluid is lost, density decreases, and the drop may pull back slightly.

After swirling, the rim of wine adhering to the glass begins to descend, occasionally so quickly that tears do not form. The descent may be slightly counteracted by upward flow, but eventually only a rim of wine remains about 1 mm above the surface of the wine (the meniscus). This is generated by the film's surface-tension gradient (Gugliotti and Silverstein, 2004), which tends to draw wine up the glass more than capillary action alone. However, the demonstration the authors provide hardly represents typical conditions (1 ml of a 50:50 alcohol solution in a watch glass). The temperature gradient generated by the differential cooling produced by alcohol evaporation also generates convection currents across and up the glass, facilitating the flow of warmer wine to the surface (the Marangoni effect). This can be observed with a light surface-dusting of nonwettable powder (such as *Lycopodium*

FIGURE 2.15　Illustration of the flow of wine up a wine glass (*lower arrows*) and the formation of tears. These begin to flow down the sides as ethanol escapes from the thin film of wine adhering to the glass (*upper arrows*), which results in an increase of surface tension of the remaining water.

or talcum powder). Adding a drop of food coloring is more evident, but has the disadvantage that its constituents (such as propylene glycol) may modify the wine's surface physicochemistry.

The duration of tear formation depends on factors affecting the density of the wine film, notably its temperature, alcohol content and liquid/air interface, as well as the wettability and slope of the sides of glass (and thereby the influence of gravity). Contrary to past belief, glycerol neither significantly affects nor is required for tear formation.

Postscript

For the aficionado, the countenance, and especially the color, are possibly wine's most sensual pleasure. The sight of sunlight shimmering through a wine can be ecstasy. That the appearance can also foretell sensory delights to follow can contribute to the viewer's anticipation.

For the winemaker, analysis of a wine's visual aspects acknowledges its significance to consumer appreciation. It reinforces the value of the effort taken to provide wine with an appealing countenance. A splendid *robe* preconditions the taster to sample the wine in a positive frame of mind, just as an intriguing book cover entices the reader to look inside.

More specifically, color can provide clues as to flavor intensity, varietal, and stylistic origins, as well as age. Although potentially prejudicing the consumer's opinion in advance of tasting, in most instances this is a plus. Clarity is also a feature perceived positively, further encouraging an auspicious expectation. Although first impressions are not always fulfilled, it is clearly better to commence confidently. Conversely, poor first impressions precondition the taster to expect disappointment, if not worse. Even if negative anticipations are not subsequently realized, it is unlikely the overall impression will generate a desirable impression, and encourage repeat sales.

Suggested Readings

Foster, D.H., 2011. Color Constancy. Vision Res. 51, 674–700.

Livingstone, M., 2002. Vision and Art: The Biology of Seeing. Abrams, New York, NY.

Pridmore, R.W., Huertas, R., Melgosa, M., Negueruela, A.I., 2005. Discussion on perceived and measured wine color. Color Res. Appl. 30, 146–152.

Waterhouse, A.L., Kennedy, J.A. (Eds.), 2004. *Red Wine Color. Revealing the Mysteries*. ACS Symposium Series, No. 886, American Chemical Society Publication, Washington, DC.

Zellner, D.A., 2013. Color-odor interactions: A review and model. Chem. Percept. 6, 155–169.

References

Aldercreutz, P., 1986. Oxygen supply to immobilized cells. 5. Theoretical calculations and experimental data for the oxidation of glycerol by immobilized *Gluconobacter oxydans* cells with oxygen or *p*-benzoquinone as electron acceptor. Biotechnol. Bioeng. 28, 223–232.

Amerine, M.A., Ough, C.S., 1980. Methods for Analysis of Musts and Wines. John Wiley, New York.

Amerine, M.A., Berg, H.W., Kunkee, R.E., Ough, C.S., Singleton, V.L., Webb, A.D., 1980. The Technology of Wine Making, 4th ed AVI Publ. Co, Westport, CN.

Asano, K., Ohtsu, K., Shinagawa, K., Hashimoto, N., 1984. Affinity of proanthocyanidins and their oxidation products for haze-forming proteins of beer and the formation of chill haze. Agric. Biol. Chem. 48, 1139–1146.

Ayala, F., Echávarri, J.F., Negueruela, A.I., 1997. A new simplified method for measuring the color of wines. II. White wines and brandies. Am. J. Enol. Vitic. 48, 364–369.

Bakker, J., Timberlake, C.F., 1986. The mechanism of color changes in aging port wine. Am. J. Enol. Vitic. 37, 288–292.

Ballester, J., Abdi, H., Langlois, J., Peyron, D., Valentin, D., 2009. The odors of colors: Can wine experts and novices distinguish the odors of white, red, and rosé wines? Chem. Percept. 2, 203–213.

Birse, M.J., 2007. The Color of Red Wine. PhD Thesis, School of Agriculture, Food and Wine. University of Adelaide, Australia.

Brosnan, T., Sun, D.W., 2004. Improving quality inspection of food products by computer vision. J. Food Engin 61, 3–16.

Brou, P., Sciascia, T.R., Linden, L., Lettvin, J.Y., 1986. The colors of things. Sci. Amer. 255 (3), 84–91.

Bucelli, P., Gigliotti, A., 1993. Importanza di alcuni parametri analatici nella valutazione dell'attitudine all'invecchiamento dei vini. Enotecnico 29 (5), 75–84.

Chapanis, A., 1965. Color names for color space. Am. Scientist 53, 327–345.

Cheung, V., Westland, S., Li, C., Hardeberg, J., Connah, D., 2005. Characterization of trichromatic color cameras by using a new multispectral imaging technique. J. Optic. Soc. Am. A 22, 1231–1240.

Clydesdale, F.M., Gover, R., Philipsen, D.H., Fugardi, C., 1992. The effect of color on thirst quenching, sweetness, acceptability and flavor intensity in fruit punch flavored beverages. J. Food Quality 15, 19–38.

Cook, D.J., Hollowood, T.A., Linforth, R.S.T., Taylor, A.J., 2003. Oral shear stress products flavour perception in viscous solutions. Chem. Senses 28, 11–23.

Cozzolino, D., Cowey, G., Lattey, K.A., Godden, P., Cynkar, W.U., Dambergs, R.G., et al., 2008. Relationship between wine scores and visible-near infrared spectra of Australian red wines. Anal. Bioanal. Chem. 391, 975–981.

Dallas, C., Ricardo-da-Silva, J.M., Laureano, O., 1996. Products formed in model wine solutions involving anthocyanins, procyanidin B_2, and acetaldehyde. J. Agric. Food Chem. 44, 2402–2407.

Dambrouck, T., Narchal, R., Marchal-Delahaut, L., Parmentier, M., Maujean, A., Jeandet, P., 2003. Immunodetection of protein from grapes and yeast in a white wine. J. Agric. Food Chem. 51, 2727–2732.

DuBose, C.V., Cardello, A.V., Maller, O., 1980. Effects of colorants and flavorants on identification, perceived flavor intensity, and hedonic quality of fruit-flavoured beverages and cake. J. Food Sci. 45, 1393–1399.

Dupin, V.S., McKinnon, B.M., Ryan, C., Boulay, M., Markides, A.J., Jones, G.P., et al., 2000. *Saccharomyces cerevisiae* mannoproteins that protect wine from protein haze: Their release during fermentation and lees contact and a proposal for their mechanism of action. J. Agric. Food Chem. 48, 3098–3105.

Edelényi, M., 1966. Study on the stabilization of sparkling wines (in Hungarian). Borgazdaság 12, 30–32. (reported in Amerine *et al.*, 1980).

Edwards, C.G., 2006. Illustrated Guide to Microbes and Sediments in Wine, Beer, and Juice. WineBugs LLC, Pullman, WA.

Fournier, J.B., Cazabat, A.M., 1992. Tears of wine. Europhys. Lett. 20, 517–522.

Fulcrand, H., Cheynier, V., Oszmianski, J., Moutounet, M., 1997. The oxidized tartaric acid residue as a new bridge potentially competing with acetaldehyde in flavan-3-ol condensation. Phytochemistry 46, 223–227.

Fulcrand, H., Benabdeljalil, C., Rigaud, J., Chenyier, V., Moutounet, M., 1998. A new class of wine pigments generated by reaction between pyruvic acid and grape anthocyanins. Phytochemistry 47, 1401–1407.

Gegenfurtner, K.R., Kiper, D.C., 2003. Color vision. Annu. Rev. Neurosci. 26, 181–206.

Glories, Y., 1984. La couleur des vins rouge. 2e Partie. Mesure, origin et interpretation. Conn. Vigne Vin 18, 253–271.

Gugliotti, M., Silverstein, T., 2004. Tears of wine. J. Chem. Educ. 81, 67–68.

Hernández, B., Sáenz, C., Fernández de la Hoz, J., Alberdi, C., Alfonso, S., Diñeiro, J.M., 2009. Assessing the color of red wine like a taster's eye. Col. Res. Appl. 34, 153–162.

Huertas, R., Yebra, Y., Pérez, M.M., Melgosa, M., Negueruela, A.I., 2003. Color variability for a wine sample poured into a standard glass wine sampler. Color Res. Appl. 28, 473–479.

Iland, P.G., Marquis, N., 1993. Pinot noir – Viticultural directions for improving fruit quality. In: Williams, P.J., Davidson, D.M., Lee, T.H. (Eds.), Proc. 8th Aust. Wine Ind. Tech. Conf. Adelaide, 13-17 August, 1992. Winetitles, Adelaide, Australia, pp. 98–100.

Jackson, R.S., 2014. Wine Science: Principles and Applications, 4th ed. Academic Press, San Diego, CA.

Kaiser, P.K., Boynton, R.M., 1996. Human Color Vision, 2nd ed. Optical Society of America, Washington, DC.

Kemp, S.E., Gilbert, A.N., 1997. Odor intensity and color lightness are correlated sensory dimensions. Am. J. Psychol. 110, 35–46.

Koza, B.J., Cilmi, A., Dolese, M., Zellner, D.A., 2005. Color enhances orthonasal olfactory intensity and reduces retronasal olfactory intensity. Chem. Senses 30, 643–649.

Kuehni, R.G., 2002. CIEDE2000, milestone or final answer? Col. Res. Appl. 27, 126–127.

León, K., Mery, D., Pedreschi, F., León, J., 2006. Color measurement in L*a*b* units from RGB digital images. Food Res. Int. 39, 1084–1091.

Liger-Belair, G., 2005. The physics and chemistry behind the bubbling properties of champagne and sparkling wines: A state-of-the-art review. J. Agric. Food Chem. 53, 2788–2802.

Liger-Belair, G., Marchal, R., Jeandet, P., 2002. Close-up on bubble nucleation in a glass of champagne. Am. J. Enol. Vitic. 53, 151–153.

Liger-Belair, G., Bourqet, M., Villaume, S., Jeandet, P., Pron, H., Polidori, G., 2010. On the losses of dissolved CO_2 during champagne serving. J. Agric. Food Chem. 58, 8768–8775.

Lubbers, S., Leger, B., Charpentier, C., Feuillat, M., 1993. Effet colloide protecteur d'extraits de parois de levures sur la stabilité tartrique d'une solution hydroalcoolique model. J. Int. Sci. Vigne Vin 27, 13–22.

Lüthi, H., Vetoch, U., 1981. Mikroskopische Beurteilung von Weinen und Fruchtsäften in der Praxis. Heller Chemie-und Verwaltsingsgesellschaft mbH. Schwäbisch Hall, Germany.

Maga, J.A., 1974. Influence of color on taste thresholds. Chem. Senses Flavor 1, 115–119.

Martin, M.L.G.-M., Ji, W., Luo, R., Hutchings, J., Heredia, F.J., 2007. Measuring colour appearance of red wines. Food Qual. Pref. 18, 862–871.

Maujean, A., Haye, B., Bureau, G., 1978. Étude sur un phénomène de masque observé en Champagne. Vigneron Champenois 99, 308–313.

Mazza, G., Fukomoto, L., Delaquis, P., Girard, B., Ewert, B., 1999. Anthocyanins, phenolics and color in wine from Cabernet Franc, Merlot and Pinot noir in British Columbia. J. Agric. Food Chem. 47, 4009–4017.

Morrot, G., Brochet, F., Dubourdieu, D., 2001. The color of odors. Brain Lang. 79, 309–320.

Munsell, A.H., 1980. Munsell Book of Color – Glossy Finish. Munsell Color Corporation, Baltimore, MD.

Negueruela, A.I., Echávarri, J.F., Pérez, M.M., 1995. A study of correlation between enological colorimetric indexes and CIE colorimetric parameters in red wines. Am. J. Enol. Vitic. 46, 353–356.

Negueruela, A.I., Echávarri, J.F., Los Arcos, M.L., Lopez de Castro, M.P., 1990. Study of color of quaternary mixtures of wines by means of the Scheffé design. Am. J. Enol. Vitic. 41, 232–240.

Neogi, P., 1985. Tears-of-wine and related phenomena. J. Colloid Interface Sci. 105, 94–101.

Nurgel, C., Pickering, G., 2005. Contribution of glycerol, ethanol and sugar to the perception of viscosity and density elicited by model white wines. J. Texture Studies 36, 303–323.

Oberfeld, D., Hecht, H., Allendorf, U., Wickelmaier, F., 2009. Ambient lighting modifies the flavor of wine. J. Sens. Stud. 24, 797–832.

Österbauer, R.A., Matthews, P.M., Jenkinson, M., Beckmann, C.F., Hansen, P.C., Calvert, G.A., 2005. Color of scents: Chromatic stimuli modulate odor responses in the human brain. J. Neurophysiol. 93, 3434–3441.

Pangborn, R.M., Berg, H.W., Hansen, B., 1963. The influence of colour on discrimination of sweetness in dry table wines. Am. J. Psyc. 76, 492–495.

Parr, W.V., White, K.G., Heatherbell, D., 2003. The nose knows: Influence of colour on perception of wine aroma. J. Wine Res. 14, 99–121.

Pellerin, P., Waters, E., Brillouet, J.-M., Moutounet, M., 1994. Effet de polysaccharides sur la formation de trouble proteique dans un vin blanc. J. Int. Sci. Vigne Vin 28, 213–225.

Peng, Z., Duncan, B., Pocock, K.F., Sefton, M.A., 1999. The influence of ascorbic acid on oxidation of white wines: Diminishing the long-term antibrowning effect of SO_2. Aust. Grapegrower Winemaker 426a, 67–69. 71–73.

Pérez-Magariño, S., González-San José, M.L., 2002. Prediction of red and rosé wine CIELab parameters from simple absorbance measurements. J. Sci. Food Agric. 82, 1319–1324.

Pocock, K.F., Waters, E.J., 1998. The effect of mechanical harvesting and transport of grapes, and juice oxidation, on the protein stability of wines. Aust. J. Grape Wine Res. 4, 136–139.

Pointer, M.R., Attridge, G.G., Jacobson, R.E., 2002. Food colour appearance judged using images on a computer display. Imaging Sci. J. 50, 25–37.

Pokorný, J., Filipům, M., Pudil, F., 1998. Prediction of odour and flavour acceptancies of white wines on the basis of their colour. Nahrung 42, 412–415.

Quinsland, D., 1978. Identification of common sediments in wine. Am. J. Enol. Vitic. 29, 70–71.

Rankine, B.C., Pilone, D.A., 1973. *Saccharomyces bailii*, a resistant yeast causing serious spoilage of bottled table wine. Am. J. Enol. Vitic. 24, 55–58.

Rentzche, M., Schwarz, M., Winterhalter, P., 2007. Pyranoanthocyanins – An overview on structures, occurrence, and pathways of formation. Trends Food Sci. Technol. 18, 526–534.

Ribéreau-Gayon, P., Glories, Y., 1987. Phenolics in grapes and wines. In: Lee, T. (Ed.), Proc. 6th Aust. Wine Ind. Tech. Conf. Australian Industrial Publ., Adelaide, Australia, pp. 247–256.

Robinson, W.B., Weirs, L.D., Bertino, J.J., Mattick, L.R., 1966. The relation of anthocyanin composition to color stability of New York State wines. Am. J. Enol. Vitic. 17, 178–184.

Sakai, N., Kobayakawa, T., Saito, S., 2004. Effect of description of odor on perception and adaptation of the odor. J. Jpn. Assoc. Odor Environ. 35, 22–25. [in Japanese].

Sakai, N., Imada, S., Saito, S., Kobayakawa, T., Deguchi, Y., 2005. The effect of visual images on perception of odors. Chem. Senses 30 (suppl 1), i244–i245.

Siebert, K.J., 2006. Haze formation in beverages. LWT 39, 987–994.

Sims, C.A., Morris, J.R., 1986. Effects of acetaldehyde and tannins on the color and chemical age of red Muscadine (*Vitis rotundifolia*) wine. Am. J. Enol. Vitic. 37, 163–165.

Skouroumounis, G.K., Kwiatkowski, M., Sefton, M.A., Gawel, R., Waters, E.J., 2003. *In situ* measurement of white wine absorbance in clear and in coloured bottles using a modified laboratory spectrophotometer. Aust. J. Grape Wine Res. 9, 138–148.

Skouroumounis, G.K., Kwiatkowski, M.J., Francis, I.L., Oakey, H., Capone, D.L., Peng, Z., et al., 2005. The influence of ascorbic acid on the composition, colour and flavour properties of a Riesling and a wooded Chardonnay wine during five years' storage. Aust. J. Grape Wine Res. 11, 355–368.

Smith, A.K., Noble, A.C., 1998. Effects of increased viscosity on the sourness and astringency of aluminum sulfate and citric acid. Food Qual. Pref. 9, 139–144.

Somers, T.C., 1982. Pigment phenomena – From grapes to wine. In: Webb, A.D. (Ed.), Grape Wine Cent. Symp. Proc.. University of California, Davis, pp. 254–257.

Somers, T.C., 1998. The Wine Spectrum. Winetitles, Adelaide, Australia.

Somers, T.C., Evans, M.E., 1974. Wine quality: Correlations with colour density and anthocyanin equilibria in a group of young red wine. J. Sci. Food. Agric. 25, 1369–1379.

Somers, T.C., Evans, M.E., 1977. Spectral evaluation of young red wines: Anthocyanin equilibria, total phenolics, free and molecular SO_2 "chemical age.". J. Sci. Food. Agric. 28, 279–287.

Somers, T.C., Ziemelis, G., 1985. Flavonol haze in white wines. Vitis 24, 43–50.

Spence, C., Velasco, C., Knoeferle, K., 2014. A large sample study on the influence of the multisensory environment on the wine drinking experience. Flavor 3, 8. (12 pp).

Squire, L.R., Bloom, F.E., Spitzer, N.C. (Eds.), 2008. Fundamental Neuroscience, 3rd ed. Academic Press, San Diego, CA.

Tromp, A., van Wyk, C.J., 1977. The influence of colour on the assessment of red wine quality. Proc. S. Afr. Soc. Enol. Vitic, 107–117.

Valentin, D., Parr, W.V., Peyron, D., Grose, C., Ballester, J., 2016. Colour as a driver of Pinot noir wine quality judgements: An investigation involving French and New Zealand wine professionals. Food Qual. Pref. 48, 251–261.

Venerus, D.C., Simavilla, D.N., 2015. Tears of wine: New insights on an old phenomenon. Sci. Rep. 5, 16162.

Vuilleumier, R., Ego, V., Neltner, L., Cazabat, A.M., 1995. Tears of wine: The stationary state. Langmuir 11, 4117–4121.

Walker, J., 1983. What causes the "tears" that form on the inside of a glass of wine? Sci. Amer. 248, 162–169.

Waters, E.J., Peng, Z., Pocock, K.F., Williams, P.J., 1996a. Lacquer-like bottle deposits in red wine. In: Stockley, S. (Ed.), Proc. 9th Aust. Wine Ind. Tech. Conf. Winetitles, Adelaide, Australia, pp. 30–32.

Waters, E.J., Shirley, N.J., Williams, P.J., 1996b. Nuisance proteins of wine are grape pathogenesis-related proteins. J. Agric. Food Chem. 44, 3–5.

Wenzel, K., Dittrich, H.H., Heimfarth, M., 1987. Die Zusammensetzung der Anthocyane in den Beeren verschiedener Rebsorten. Vitis 26, 65–78.

Williams, A.A., Langron, S.P., Noble, A.C., 1984a. Influence of appearance of the assessment of aroma in Bordeaux wines by trained assessors. J. Inst. Brew. 90, 250–253.

Williams, A.A., Langron, S.P., Timberlake, C.F., Bakker, J., 1984b. Effect of color on the assessment of ports. J. Food Technol. 19, 659–671.

Yam, K.L., Papadakis, S.E., 2004. A simple digital imaging method for measuring and analyzing color of food surfaces. J. Food Engin. 61, 137–142.

Zampini, Z., Spence, C., 2005. Modifying the multisensory perception of a carbonated beverage using auditory cues. Food Qual. Pref. 16, 632–641.

Zellner, D.A., Whitten, L.A., 1999. The effect of color intensity and appropriateness on color-induced odor enhancement. Am. J. Psychol. 112, 585–604.

3

Olfactory Sensations

OLFACTORY SYSTEM

Nasal Passages

Our ability to sense odor is dependent on two, small, seemingly insignificant patches of tissue in the upper recesses of our nasal passages (Fig. 3.1A). Aromatic compounds can reach these patches either directly, via the nostrils (orthonasally), or indirectly, via the back of the throat (retronasally). The direct nasal route generates what is termed smell, whereas the "back-door" route generates most of what is called flavor (the integration of olfactory, gustatory, and chemesthesis sensations in the mouth, as well as influences by visual and sound inputs).

Anatomically, the nasal passages are divided bilaterally by a central septum into right and left halves. The epithelial olfactory patches occupy a small portion of the nasal cavity, directly below the cribriform plate (and directly across from the ears). The cribriform plate is a unique, perforated portion of the skill. It is through this perforated region that receptor cells of the olfactory epithelium directly connect to the olfactory bulb at the base of the brain (Fig. 3.1B).

Each nasal passage is also incompletely subdivided by three transverse outgrowths, i.e., the turbinate bones. They increase contact between the nasal epithelium with incoming air, cleaning, warming, and moistening it before it enters the throat and lungs. The baffle-like turbinate bones also limit airflow past the olfactory regions (located above the superior turbinate). Consequently, only about 5–10% of the air taken in by the nose passes over the olfactory patches during ordinary breathing (Hahn et al., 1993; Zhao et al., 2004). Even with intense sniffing, this value may reach only 20%. Higher flow rates may enhance odor perception (Sobel et al., 2000), but the duration (Fig. 3.2), number, or interval between sniffs does not appreciably enhance perceived odor intensity (Fig. 3.3). Nonetheless, the action of sniffing activates the olfactory system for odorant response (Mainland and Sobel, 2006). The traditional recommendation to take short, swift, sniffs probably relates to limiting odor adaptation, although this is debatable (Lee and Halpern, 2013).

Our limited abilities to recognize even familiar odors without visual or other cues may be an indirect consequence of our primate ancestors' shift from a nocturnal to a diurnal habit. Therefore, increased emphasis was placed on visual over olfactory acuity (Gilad et al., 2004). This view possesses logic since primates are the only mammals to possess tricolor vision (through duplication and mutation of the P560 photoreceptor gene), presumably to better recognize ripe from unripe fruit (possible with tricolor vision).

Of the aromatic molecules reaching the olfactory patches, only a portion is absorbed into the mucus that coats the epithelium. Of those absorbed, only a fraction probably reaches reactive sites on receptor cells. In some animals, high concentrations of cytochrome-dependent oxygenases accumulate in the olfactory mucus (Dahl, 1988). These enzymes catalyze a wide range of reactions. Some may increase the hydrophilic properties of odorants, aiding their diffusion through, or escape from, the mucus. Enzymes may also interact with olfactory UDP-glucuronosyl transferase, catalyzing reactions that terminate receptor activation (Lazard et al., 1991). The mucous layer is thought to be recycled about every 10 min.

Olfactory Epithelium, Receptor Neurons, and Cerebral Connections

The olfactory patches consist of a thin layer of epithelium and underlying tissue covering an area of about $2.5\,cm^2$, within which about 10 million receptor neurons, and associated supporting and basal cells occur (Fig. 3.4). Receptor

41

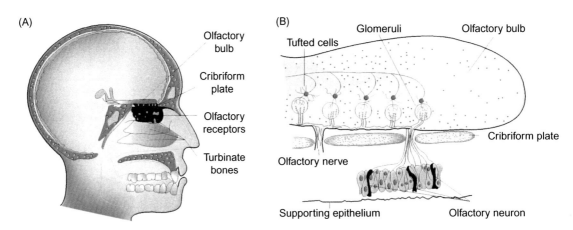

FIGURE 3.1   Location of the olfactory region in the nasal cavity (A) and an enlarged section (not to scale) showing the olfactory neurons (receptors) and their connections to the olfactory bulb (B). *From Jackson, 2014, reproduced by permission.*

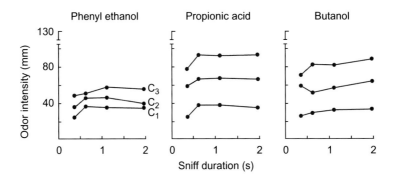

FIGURE 3.2   The perceived intensities of three concentrations (C_1, C_2, and C_3) of three odorants are plotted as a function of the duration of a sniff. *From Laing, 1986, reproduced by permission.*

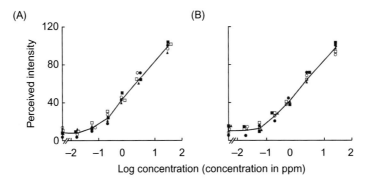

FIGURE 3.3   Arithmetic means of estimates of the perceived odor intensity of different concentration of pentyl acetate. These were obtained by having subjects use (A) their natural sniffing technique (●) or one (■), three (□), five (▲), or seven (○) natural sniffs; (B) natural sniffing (●) or three natural sniffs separated by intervals of 0.25 s (■), 0.5 s (□), 1.0 s (○), and 2.0 s (▲). *From Laing, 1983, Natural sniffing gives optimum odor perception for humans. Perception 12, 99–117, Pion Limited, London, reproduced by permission.*

neurons are modified epithelial cells. They produce dendritic extensions up and just into the mucous layer. The bulbous end generates thin, hair-like, 1 to 2-μm-long, cell-membrane extensions termed cilia (about 20 per cell) (Fig. 3.5). They greatly expand odorant exposure to receptor proteins, which are embedded in the cilial membrane. Supporting cells (and the glands underlying the epithelium) produce several classes of odorant-binding proteins as well as the special mucus that coats the olfactory epithelium (Hérent et al., 1995; Garibotti et al., 1997). A major constituent also produced by supporting cells is olfactomedin. It is thought to act as a neurotrophic factor, activating the growth and differentiation of olfactory epithelial cells into receptor neurons.

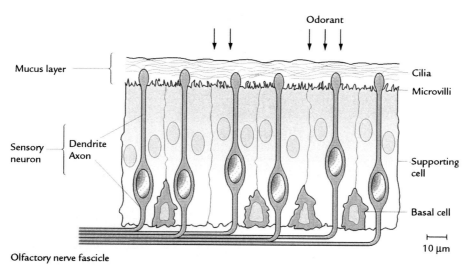

FIGURE 3.4 Drawing of the olfactory neuroepithelial layer (from Lancet, D. (1986). Vertebrate olfactory reception. *Reproduced with permission from the Annual Review of Neuroscience, Volume 9. Copyright 1986 by Annual Reviews, www.AnnualReviews.org.)*

FIGURE 3.5 Scanning electron micrographs of the human olfactory mucosal surface (A) and olfactory dendritic knobs and cilia (B). *Photo courtesy of Drs. Richard M. Costanzo and Edward E. Morrison, Virginia Commonwealth University.*

Odorant-binding proteins (OBP) are a class of small mucous proteins. They reversibly bind to odorants (Briand et al., 2002), and appear to prolong the odor signal (Yabuki et al., 2011). This may partially explain the concentration of odorant molecules being 10^3 to 10^4-fold higher in the mucus than in the inflowing air stream (Senf et al., 1980). OBPs also have been proposed to assist in transporting odorants across the mucous layer (Nespoulous et al., 2004), transferring them to olfactory receptors (Taylor et al., 2008). OBPs may also limit adaptation by degrading odorants. Various mucous enzymes, such as P450, also appear to metabolize aromatic compounds, presumably functioning to limit the duration of adaptation, but also modulate odor quality. Although humans appear to produce only one type of OBP, other mammals produce several, notably those mammals depending predominantly on their sense of smell to secure food and for survival (e.g., rats, rabbits, and pigs produce three, and porcupines produce eight). Their enzymatic activity may also limit bacterial and viral activity. This is particularly valuable as there is direct access to the brain via the cribriform plate. Subsets of epithelial cells containing receptors to bitter tastants are another component protecting against microbial infection. These cells respond to microbial constituents (e.g., acyl-homoserine lactones and bitter tastants), inducing the respective cells to release nitric oxide or defensins that are toxic to bacteria (Lee and Cohen, 2015).

Basal cells differentiate into receptor neurons and replace them as they degenerate. Receptor neurons remain active for an indifferent period, possibly as long as one year, but averaging about 60 days. As basal cells differentiate, they produce both dendritic and axonal extensions. Axons grow upward, through the cribriform plate perforations, and connect with the olfactory bulb at the base of the skull (Key and St Johns, 2002). In humans, olfactory and gustatory neurons are the only nerve cells known to regenerate regularly (excepting neurons in the hippocampus associated with short-term memory). Sudden jarring of the neck and head can result in olfactory loss due to severing of receptor axons. However, receptor neuron regeneration eventually reestablishes their connection with the olfactory bulb, and the sense of smell returns.

The nonmyelinated axons of receptor neurons associate in bundles as they pass through the cribriform plate. Supporting cells electrically isolate adjacent receptor cells, and are thought to maintain normal function.

Odor quality, the unique perceived characteristics of an odorant (or mixture thereof), is not associated with any obvious anatomical differentiation of receptor neurons. Differentiation is based on the type of olfactory receptor (OR) proteins embedded in the cilial membrane. The receptive part of the OR protein extends out into the mucous layer. Depending on the physicochemical properties of the aromatic compound, it may bind to one or more different OR proteins (Bautze et al., 2012). Binding activates adenylate cyclase, liberating cyclic adenosine monophosphate (AMP) that opens ion channels in the cell membrane. The result is in an influx of Na^+ and Ca^{2+} ions (Murrell and Hunter, 1999). The resultant membrane depolarization initiates a self-propagating impulse along the axon, eventually activating various olfactory centers in the brain.

Although OR proteins were first discovered and associated with OR cells, they have recently been found in the membrane of blood leukocytes (Geithe et al., 2015). What function they might have in that location is anyone's guess.

The distinctive character (quality) of an odorant, or aromatic object, may arise from the differential sensitivity of multiple receptor neurons, and the associated temporospatial pattern of activation generated across the olfactory patches. Neurons receptive to specific ligands (the part of a molecule that fits into the OR active site) tend to be organized spatially into distinct, nonoverlapping regions in the olfactory epithelium (Johnson and Leon, 2007). The pattern of responses they generate may be modified in the olfactory bulb, and finally integrated with other sensations in higher centers of the brain, generating what Shepard (2006) calls "odor images." Because the temporal response pattern generated by orthonasal detection may, in some instances, be the reverse of its retronasal pattern, the same compound (or mixtures thereof) may be recognized as having distinctly different qualities, depending on the mode of presentation (nose versus mouth). The markedly divergent response to the smell and flavor of several cheeses, notably Limberger, Époisses, and Olomouc, or fruit such as durian (*Durio zibethinus*), is almost legendary.

Detection of any compound depends on the presence of one or more reactive groups. If these ligands differ, they may react with various OR proteins, each associated with a separate receptor neuron (Buck and Axel, 1991; Olender and Lancet, 2012). Each receptor neuron possesses multiple copies of a single OR protein, encoded by one of about 340–400 odorant receptor genes (Malnic et al., 2004). There are also about 600 OR pseudogenes, which are inactive gene sequences, having similarity to and probably derived from active equivalents. The OR gene cluster constitutes the largest gene family in mammals (Fuchs et al., 2001), and may constitute about 1–2 percent of the human genome (Buck, 1996). The genes are related to those that encode for rhodopsin (the photoreceptor found in the rods of the eye), as well as those that encode for the receptors to sweet and bitter tastants.

Each OR protein possesses seven domains that span and extend beyond the cell membrane. These domains may associate with particular ligand groupings, e.g., a hydroxyl, methyl, or a more complex structure. That individual receptor proteins possess several domains means that each receptor neuron can potentially be activated by several distinct odorants if they possess the same ligand. Equally, aromatic compounds possessing several distinct ligands may activate ORs on more than one type of receptor neuron (Figs. 3.6 and 3.7). Upon interaction between a ligand and an OR protein, a closely associated G-protein initiates impulse generation.

Odorant activation of OR receptor sites is thought to be analogous to the lock-and-key mechanism involved in the recognition of the active sites of enzymes or neuroreceptors with their respective substrates or neurotransmitters. However, in the case of odor recognition, it may involve the pattern generated by the simultaneous or sequential activation of a series of ORs. Thus, the pattern generated may be akin to simultaneously playing the notes of a chord, or in sequence, such as an arpeggio. Recently, an alternative or supplemental hypothesis for OR activation has been proposed, dubbed the swipe-card model (Brookes et al., 2012). Although often incorporating both structural and quantum vibrational attributes of the ligand, it is highly controversial (Gane et al., 2013; Block et al., 2015; Turin et al., 2015).

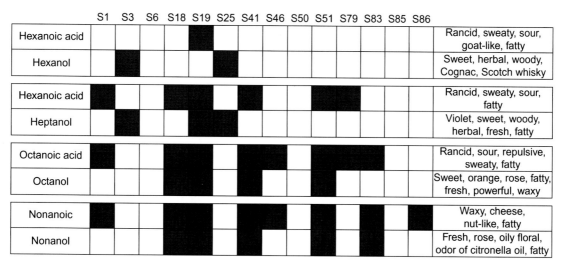

FIGURE 3.6 Comparison of the receptor codes for odorants that have similar structures but different odors. Aliphatic acids and alcohols with the same carbon chains were recognized by different combinations of olfactory receptors, thus providing a potential explanation for why they are perceived as having strikingly different odors. Perceived odor qualities shown on the right were obtained from Arctander (1994), The Good Scents Company (http://www.thegoodscentscompany.com/), and The Chemfinder Web Server (http://chenbiofinder.cambridgesoft.com). *From Malnic et al., 1999, reproduced by permission.*

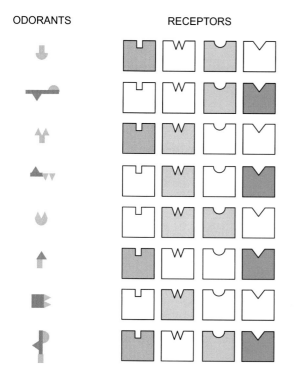

ODORANTS   RECEPTORS

FIGURE 3.7 Model for the combinatorial receptor codes for odorants. The receptors shown in color are those that recognize the odorant on the left. The identities of different odorants are encoded by different combinations of receptors. However, each olfactory receptor can serve as one component of the combinatorial receptor codes for many odorants. Given the immense number of possible combinations of olfactory receptors, this scheme could allow for the discrimination of an almost unlimited number and variety of different odorants. *From Malnic et al., 1999, reproduced by permission.*

Although most studies of olfaction have investigated single aromatic compounds, most natural odors are mixtures, with a few impact constituents being the principal activating components (Lin et al., 2006). In addition, the odor quality of a compound often varies, depending on its concentration (e.g., low concentrations potentially activating only a single type of OR, whereas higher concentrations may activate several ORs as well as trigeminal receptors) (Gross-Isseroff and Lancet, 1988). Thus, there appears to be no simple correlation between chemical structure and odor quality.

Most patterns of olfactory receptor activity appear to become associated (memorized) with the experiences with which they occurred. Odor memories may develop in association with a single odorant (e.g., sulfur dioxide, hydrogen sulfide, 2,4,6-trichloroanisole), or more frequently with a mixture (e.g., derived from a particular fruit, flower, or more generically, with fruits or flowers).

It appears that structurally complex aromatic compounds (and possibly mixtures) generate more odor "notes," and tend to be perceived as more pleasant than structurally simpler compounds (Fig. 3.8; Plate 3.1). This is likely based on bond-type variation, diversity in elemental composition, and symmetrical complexity activating more ORs, and, correspondingly, more neural networks (Sezille et al., 2015). Encoding such patterns in memory is often greatly

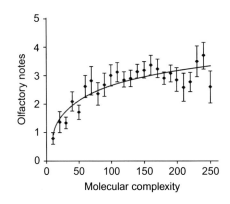

FIGURE 3.8　Molecular complexity of monomolecular odorants influences the number of perceived olfactory (but not trigeminal) notes. A significant logarithmic relationship is observed between molecular complexity and the number of olfactory notes. *Reprinted by permission from Macmillan Publishers Ltd. from Kermen, F., Chakirian, A., Sezille, C., Joussain, P., Le Goff, J., Ziessel, A., Chastrette, M., Mandairon, N., Didier, A., Rouby, C., and Bensafi, M. (2011) Molecular complexity determines the number of olfactory notes and the pleasantness of smells. Sci. Reports* **1**, *206., Copyright 2011.*

(A)

Cinnamon	Apricot	Woody	Hay	Floral
			Sweet	Leafy
	Apple	Peppery	Spicy	Balsamic
Smokey	Fruity	Dry	Herbaceous	Earthy
Spicy	Sour	Spicy	Nut	Fruity
Furan (3)	Allyl propionate (4)	Eugenol (4)	Tobacco	Spicy
			Coumarin (6)	Sweet
				Aceteugenol (7)

Number of olfactory notes

(B)

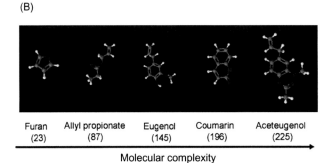

| Furan (23) | Allyl propionate (87) | Eugenol (145) | Coumarin (196) | Aceteugenol (225) |

Molecular complexity

PLATE 3.1　Examples of odorants described by olfactory notes (A) and displaying various degrees of molecular complexity (B). (A) Odorants can be described by few or many olfactory notes. The number of olfactory notes evoked by each odorant is shown in brackets. (B) At the molecular level, odorant molecules display various degrees of complexity. The molecular complexity value for each odorant is shown in brackets. All odorants were selected from the Arctander (1994), which provides data regarding olfactory notes. Molecular complexity values were obtained from the PubChem database and 3-dimensional molecular drawings were obtained from http://www.thegoodscentscompany.com/. *Reprinted by permission from Macmillan Publishers Ltd. from Kermen, F., Chakirian, A., Sezille, C., Joussain, P., Le Goff, J., Ziessel, A., Chastrette, M., Mandairon, N., Didier, A., Rouby, C., and Bensafi, M. (2011) Molecular complexity determines the number of olfactory notes and the pleasantness of smells. Sci. Reports* **1**, *206., Copyright 2011.*

enhanced if the olfactory experience coincides with a distinctive emotional response. Direct connections with the amygdala probably explain the strong and immediate emotional responses that certain odors can beget. In contrast, the qualitative aspects of odor memories appear to reside in the piriform cortex. Other components of odor memory, such as verbal encoding and retrieval occur in other regions of the brain (Fletcher et al., 1995). Figs. 3.9 and 3.38A illustrate the known connections relating to odor detection and identification. Thus, odor memories resemble other memories, in that different aspects are sequestered in different parts of the brain.

With there being up to 400 different functional OR genes potentially expressed, with each receptor cell containing two copies of its particular OR gene, each of which can potentially occur in several phenotypically different forms (alleles), it is not surprising that the number of recognizable odor qualities has been estimated in the trillions (Bushdid et al., 2014). As yet, though, ligands have been identified for only 48 of the 350–400 expressed human ORs. In addition, the quality of an odorant can vary with concentration, as noted above, but also in association with other aromatic compounds. Furthermore, sensitivity can range over several orders of magnitude (Suprenant and Butzke, 1996; Tempere et al., 2011). Thus, human ability to discriminate between pairs of odors is fantastic. It is when it comes to recognizing such differences in isolation, or identifying them by name, that our capacities seem to fail us. This is equivalent to our ability to distinguish between millions of color hues and sounds (on direct comparison), but difficulty in being able to identify, let alone name, these differences when they are presented individually. In practice, the primary controlling factors determining the number of recognizable odorants and aromatic objects appear to be individual significance (Li et al., 2008), their emotional impact (Kass et al., 2013), and probably intensity (Distel et al., 1999). In addition, the subjective response to an odor is not "fixed," often being affected by the term applied to it. For example, the response to being supplied with a sample of menthol was markedly different when presented as an example of "breath mint" versus "chest medicine" (Herz and von Clef, 2001). Repeat exposure improves discrimination, but the improvement is not universal, applying only to the compounds sampled (Li et al., 2006). Improved discrimination has even been detected in the brain with functional magnetic resonance imaging (fMRI), as well as odor memories detected down to the level of individual cortical neurons (Rolls et al., 1996).

Increased sensitivity to particular odorants, such as androstenone (Wysocki et al., 1989), upon repeat exposure may also result from the selective reproduction of receptor neurons. This also appears to apply to related odorants, but not to unrelated compounds (Stevens and O'Connell, 1995). This phenomenon may also relate to plasticity in higher olfactory centers in the brain (Li et al., 2006). Plasticity is most characteristic of women (during their childbearing years), but relatively uncommon in men (Fig. 3.10).

On stimulation, receptor neurons send impulses up to the olfactory bulb. Receptors of the same type appear to terminate together (Tozaki et al., 2004; Johnson and Leon, 2007), in specific regions termed glomeruli (Fig. 3.1B). Nonetheless, these are interconnected by several types of nerve cells. They are thought to be involved in feedback inhibition of receptor neurons, as well as being receptive to feedback signals from higher centers in the brain. This

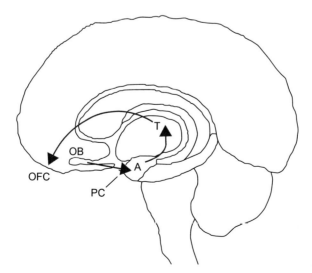

FIGURE 3.9 Schematic drawing of some relevant olfactory pathways. *A*, amygdala; *OB*, olfactory bulb; *OFC*, orbitofrontal cortex; *PC*, piriform cortex; *T*, thalamus (medial dorsal nucleus). From the amygdala, diffuse projections emerge to the limbic system. *Reprinted from González, J., Barros-Loscertales, A., Pulvermuller, F., Meseguer, V., Sanjuán, A., Belloch, V., and Ávila, C. (2006) Reading cinnamon activates olfactory brain regions. NeuroImage* **32**, *906–912, with permission from Elsevier.*

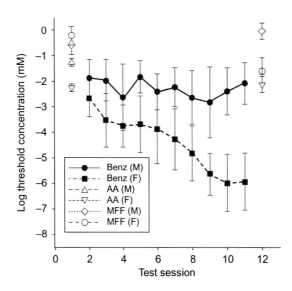

FIGURE 3.10 Gender effects of repeated test exposures to benzaldehyde (*Benz*) on mean benzaldehyde and control thresholds. Although initial thresholds for naive and experienced volunteers differed, the changes over trials did not, and thus data from the two groups were combined. Differences in sensitivity to two other olfactants, to which the subjects were not exposed repeatedly [amyl acetate (AA) and 5-methylfurfural (MMF)], did not change. Two-way, repeated-measure analyses of variance (ANOVAs) with gender as the between-group factor showed a significant interaction between group and time ($F (9,90) = 3.29; p < 0.001$). *Reprinted by permission from Macmillan Publishers Ltd: Nature Neurosci., Dalton, P., Doolittle, N., and Breslin, P. A. S. (2002). Gender-specific induction of enhanced sensitivity to odors. 5, 199–200.*

modulation (Christensen et al., 2000) may partially explain why the perceived odor quality of an odorant mixture (such as a wine) seldom resembles the fragrance qualities of its components. In addition, neurons in the piriform cortex can be activated by mixtures of odorants not activated by their individual constituents (Zou and Buck, 2006). This may further help to explain why the odor qualities of individual wine aromatics are difficult to isolate and recognize.

Neurons in the piriform or olfactory cortex are hypothesized to be associated with particular odorant memory (Wilson et al., 2004). Intriguingly, the lifespan of these neurons may be dependent on odorant stimulation (Lledo and Gheusi, 2003). This could explain why varietal aroma memory needs continual practice. In other words, tasters cannot rest on their laurels.

Although frequent exposure may benefit odor memory, in the short term, extended exposure can result in adaptation (loss of detection). Adaptation can involve direct suppression of receptor activation, such as with benzyl acetate, 2-phenylethanol, or geraniol (Teleuchi et al., 2009), or by activation of trigeminal receptors (e.g., TRPA1) (Richards et al., 2010). Adaptation can be observed within a few seconds as reduced activity in the higher, sensory-perceptive centers in the brain (Li et al., 2006). Although seemingly undesirable during tasting, it may permit detection of other odorants masked by dominant flavorants. On a broader scale, adaptation has benefits by reducing cognizance of persistent (background) odors (Best and Wilson, 2004), while facilitating detection of changes in the olfactory environment (Kadohisa and Wilson, 2006).

The orbitofrontal cortex is not only the site of conscious sensory perception, but also where taste, mouth-feel, odor, and visual impulses converge and interact (Fig. 3.11). This is the origin of multisensory perceptions such as flavor. When olfactory patterns are recognized, they activate experience-associated memories. Examples include the frequent association of a vanilla odor with sweetness, the aroma of Gewürztraminer with litchi nuts, the aroma of thyme with turkey stuffing, or pine oil with Christmas trees.

Predictably, odorants with the strongest association with sweetness have the greatest influence in enhancing the perceived sweetness of sugars (Stevenson et al., 1999; Fig. 3.12). They also have the greatest influence in reducing perceived sourness. Conversely, sweetness and other taste modalities can influence the perceived fruitiness of certain aromatics (Cook et al., 2003). The general stability of such affiliations (Stevenson et al., 2003) probably explains why identification can be poor when components (usually visual) of the association are missing (Laing et al., 2002). However, without repeat enforcement, or dissociation upon future exposures, learned linkages can weaken.

The integration of sensory modalities into holistic percepts such as flavor may explain why "taste" attributes are associated with particular fragrances (such as the perceived sweetness in dry, fruity wines), and conversely, why aromatic intensity can be affected by particular tastants (Dalton et al., 2000; Labbe et al., 2007).

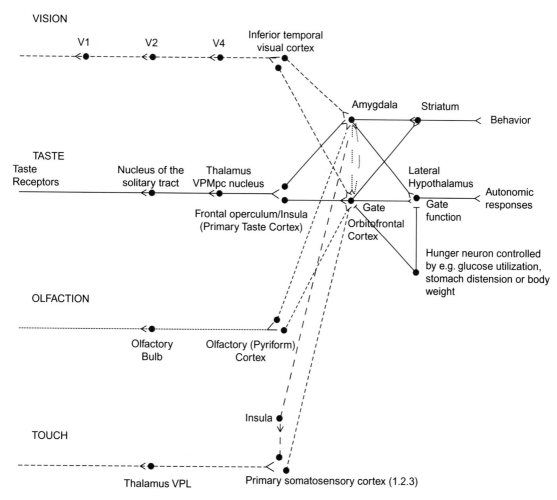

FIGURE 3.11 Schematic diagram of the taste and olfactory pathways in primates (including humans) showing that they converge with each other and with visual pathways. The gate functions shown refer to the finding that the responses of taste neurons in the orbitofrontal cortex and the lateral hypothalamus are modulated by hunger: *V1, V2, V4*, visual cortical areas; *VPL*, ventral posterolateral nucleus; *VPMpc*, ventral posteromedial thalamic nucleus. *Reproduced from Rolls, E. T. (2005). Taste, olfactory, and food texture processing in the brain, and the control of food intake. Physiol. Behav.* **85**, *45–56, with permission from Elsevier.*

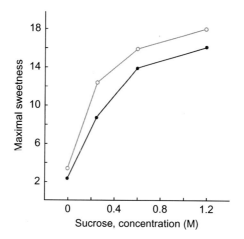

FIGURE 3.12 Perceived sweetness as affected by the presence (○) or absence (●) of strawberry fragrance. *From Frank and Bryam, 1988.*

ODORANTS AND OLFACTORY STIMULATION

There is no precise definition of what constitutes an olfactory compound. For air-breathing animals, though, odorants must be volatile at normal temperatures. This places an upper limit on molecular size ($\leq$300 daltons). However, low molecular weight implies neither volatility nor fragrance. Most aromatic compounds are partially fat and water soluble, have low polarity, bond weakly with cellular constituents, and dissociate readily.

As noted, odor quality can be associated with individual compounds or mixtures thereof. Interactions between compounds can be additive, synergistic, or suppressive. Although wines possess hundreds of aromatic compounds, less than 50 may occur at concentrations sufficient to directly influence their fragrance. Because interactions are often complex, wine odor memories are normally stored as holistic, integrated patterns (their odor gestalt). Each is based as much on direct sensory stimulation as on the context of the experience, present and past. Memory development is also affected by the attention paid to the wine, and any expectations the taster may have (Wilson and Stevenson, 2003). Major exceptions to this generality are off-odors, where the offending compound may be sufficiently intense to be identified as such.

Typically a wine's varietal character is dependent on multiple aromatic constituents (e.g., Guth, 1998). In addition, the wine's color can also influence how a wine's aroma is perceived. Equally, knowing in advance the varietal nature of a wine can induce a search for particular traits, possibly distorting their apparent presence, or even invent sensory illusions. More commonly, professional tasters use odor memories, developed and refined over their career, to confirm a wine's suspected varietal, regional, or stylistic origins. The degree to which a taster can recognize a wine's origin depends on the potency and veracity of these odor archetypes. Odor memories need not be associated with a specific descriptor. Typically, descriptors are imperfect representations of some aromatic attribute of a wine, but can act as a focus for creating an archetype. The nebulous character of most descriptors is suggested by the frequent use of suffixes, such as "-like" or "-y" (e.g., rose-like, apple-like, grassy, barnyardy). They evoke resemblances, not identities. This may be equivalent to mentally constructing images from dots (Fig. 3.13), or objects imagined in ink blots (Rorschach test) or cloud formations. Our ability to interpret identical data differently (Figs. 3.14 and 6.4) may also help to explain the various fragrances found in wine by different people, or that may appear at different stages during a tasting.

FIGURE 3.13   Example of Gestalt perception. The image may initially appear as a random set of black patches on a white background. With continued observation the appearance of a Dalmatian dog sniffing the ground with a tree trunk in the background appears. This sudden "observation" is dependent on one possessing memory of the basic shape of a dog. (Despite repeated efforts, we have been unsuccessful in our attempts to locate the photographer Ron James or the original source of this image. Please contact the publisher if you have questions regarding the use of this image).

FIGURE 3.14   Example of a perception set. The viewer can visualize an illustration of an old lady or a young woman in sequence, but never at the same time. It was first used as an example of perceptual set in the 1930s by Edwin G. Boring, and subsequently by Leeper (1935). It seemingly was first published by the British cartoonist W. E. Hill in 1915, but may be based on a earlier French version produced in 1888.

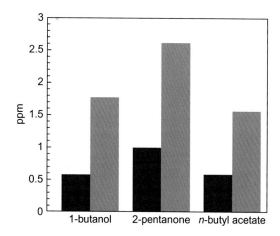

FIGURE 3.15   Mean thresholds of detection for 1-butanol, 2-pentanone, and *n*-butyl acetate alone (pale color), and in a mixture (deep color) of all three compounds (mix). Data from 40 subjects. *From Patterson et al., 1993, reproduced by permission.*

The association of OR proteins with an odorant involves one or more of its physicochemical properties, notably electrostatic attractions, hydrophobic bonds, van der Waals forces, hydrogen bonding, dipole–dipole interactions, and seemingly quantum vibration. Even minor atomic alignment within a molecule, such as that found in stereoisomers, can markedly affect the relative intensity and perceived quality of an odorant. For example, D- and L-carvone stereoisomers possess spearmint-like and caraway-like qualities, respectively.

Aromatic compounds belonging to the same chemical group may show competitive inhibition, despite possessing markedly different odor qualities (Pierce et al., 1995). This phenomenon, called cross-adaptation, results in suppressed detection of one odorant associated with prior (or simultaneous) exposure to a related compound, but dissipates with extended exposure (Lawless, 1984a). This may explain some of the apparent "unmasking" of compound(s) as a tasting progresses. However, unmasking could also be associated with the changing composition of aromatics in the headspace of the glass during tasting.

When aromatic compounds are combined, they often lose their individual recognizable qualities, or are detected at diminished intensities. Such suppression can occur at the level of olfactory reception in the nose, where members of diverse chemical groups antagonize activation of specific OR proteins (Sanz et al., 2006). However, at low concentrations, synergistic effects can occur (Fig. 3.15). This could benefit the detection of desirable aromatics, but also enhance the perception of off-odors (Laing et al., 1994). An expression of the complexity of such reactions is given by Piggott and Findlay (1984). They found both synergistic and suppressive influences with different pairs of esters, with opposite effects being expressed at different concentrations. Synergistic effects can even occur across modalities, such that the combination of subthreshold concentrations of aromatic and taste compounds result in the detection of both (Dalton et al., 2000). This phenomenon appears to occur only with sensations that have already been integrated through experience (Breslin et al., 2001). These reactions, combined with human variability in sensory acuity, partially explain why people differ so frequently in their responses to wines. Individual variation in perceptive abilities may also rise from subthreshold concentrations of compounds, where detection may occur at a subconscious level but not consciously. This possibility is suggested from brain-imaging studies (Lorig, 2012).

CHEMICAL COMPOUNDS INVOLVED

The major chemical constituents in wine (alcohols, acids, phenolics) tend to generate its gustatory sensations. In contrast, the minor or trace constituents donate a wine's distinctive aromatic character. For example, the most common wine phenolics (e.g., tannins) elicit taste and mouth-feel sensations, whereas trace phenolics (e.g., vinyl phenols, syringaldehyde) possess aromatic properties The primary exception is ethanol. Although principally inducing mouth-feel sensations, ethanol also possesses a mild but distinctive odor.

Volatility, an essential attribute of an odorant, is influenced by many factors (Goubet et al., 1998). In wine, one of the more significant involves other constituents—termed the matrix (Munõz-González et al., 2015). For example, glucose can increase volatility, whereas ethanol can suppress it (Fig. 3.16). With many wines now possessing higher alcohol contents than traditional, suppression of fruit, and enhanced herbaceous, attributes (Table 3.1) could undo the intended value of leaving grapes to overmature or partially dehydrate.

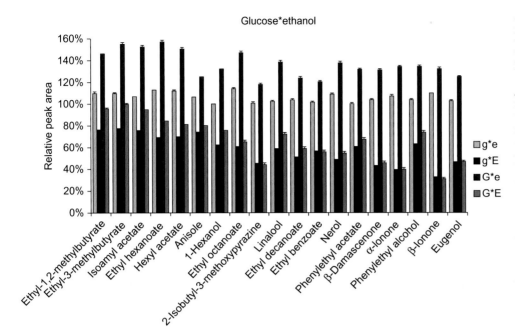

FIGURE 3.16 The effect on headspace partitioning of several compounds in a solution in the presence of matrix constituents (*e*, ethanol, at 14 percent vol/vol; *g*, glucose, at 240 g/L). Data points represent the mean peak area, relative to the mean peak area observed in the water. Capital letters denote the presence of the matrix component, whereas lowercase letters denote their absence. *Reprinted with permission from Robinson, A. L., Ebeler, S. E., Heymann, H., Boss, P. K., Solomon, P. S., and Trengove, R. D. (2009) Interactions between wine volatile compounds and grape and wine matrix components influence aroma compound headspace partitioning. J. Agric. Food Chem.* **57**, *10313–10322. Copyright 2009 American Chemical Society.*

TABLE 3.1   Mean Aroma Attributes of 23 Malbec Wines Within Two Ethanol Ranges

Attribute	Mean aroma attribute from ethanol range (%) ± SEM	
	10.0–12.0	14.5–17.2
Fruity	2.6 ± 0.2	1.8 ± 0.2[a]
Citrus	1.5 ± 0.1	1.8 ± 0.2
Strawberry	3.2 ± 0.3	2.1 ± 0.1[a]
Plum	3.2 ± 0.2	2.4 ± 0.2[b]
Raisin	2.6 ± 0.4	1.8 ± 0.2
Spicy	2.6 ± 0.3	2.6 ± 0.2
Cooked fruit	2.9 ± 0.3	1.9 ± 0.3[b]
Floral	2.2 ± 0.2	2.3 ± 0.3
Honey	2.5 ± 0.3	1.6 ± 0.2[b]
Herby	2.0 ± 0.3	3.2 ± 0.3[b]
Sweet pepper	2.5 ± 0.3	2.2 ± 0.2

From Goldner et al., 2009, J. Sens Stud., reproduce with permission of John Wiley and Sons, copyright.
SEM, standard error of the mean.
[a]$P < 0.01.$
[b]$P < 0.05.$

The matrix effect can be very marked. For example, the threshold of β-damascenone is 1000-fold higher in red wine than in a hydroalcoholic solution (Pineau et al., 2007). Additive effects have also been noted, as with berry flavors and total ester content (Escudero et al., 2007). In addition, mannoproteins can retard the volatility of some flavorants, such as β-ionone, ethyl hexanoate, and octanal, but enhance the liberation of others, such as ethyl octanoate and ethyl decanoate (Lubbers et al., 1994). Many of these interactions are of particular concern as they occur at concentrations typically found in wine (Chalier et al., 2007). Other wine constituents that can affect volatility include

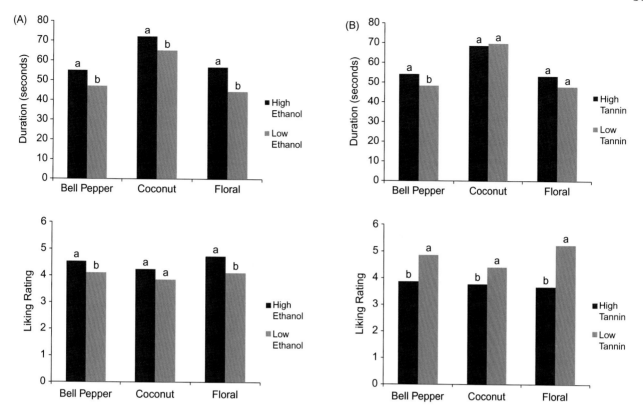

FIGURE 3.17 Effect of composition on the duration (top) and liking (bottom) of the wine's finish at two concentrations of ethanol (A) (low: 9 percent; high, 14 percent w/v) and (B) tannin contents (Low: ≤140 mg/L, High ≤1400 mg/L catechin equivalents) of a dealcoholized Syrah wine. Flavor compounds were added to the wine to generate final concentrations for floral (20.4 mg/L 2-phenylethanol), bell pepper (20 ng/L 3-isobutyl-2-methoxypyrazine), and coconut (23.8 mg/L oak lactone) attributes. *From Baker and Ross, 2014, J. Sens. Stud., reproduced by permission of John Wiley and Sons, copyright.*

glycerol (Robinson et al., 2009), polysaccharides, proteins (Voilley et al., 1991), and polyphenolics (Aronson and Ebeler, 2004; Lund et al., 2009). These effects can even modify basic perceptions. For example, a wine's nonvolatile matrix can shift the typical yellow, citrus, and tropical notes frequently ascribed to a white wine to the black, red, and dry fruit attributes usually associated with red wine, and vice versa (Sáenz-Navajas et al., 2010). A wine's matrix not surprisingly also affects the wine's finish, but in ways that might not have been predicted. For example, duration and flavors were enhanced in wine with higher alcohol but liking reduced at higher tannin contents (Fig. 3.17). Adaptation occurred during the finish, as expected, but timing was compound specific (Fig. 3.18). How these results relate to dilution of the alcohol content in the mouth was not assessed.

Acids

Most volatile acids found in wine are the byproducts of microbial metabolism. Of these, acetic acid is the most common. Other examples include formic, butyric, and propionic acids, but they seldom occur at above threshold values. All have marked odors, with acetic acid being vinegary, formic acid having a strong pungent odor, propionic acid possessing a fatty smell, butyric acid resembling rancid butter, and those that are six to ten carbons long possessing goaty odors. Correspondingly, volatile acids are typically associated with off-odors. A partial exception is acetic acid. At up to its recognition threshold, acetic acid may add complexity to a wine's bouquet, but above this value, it becomes a fault.

In contrast, the major organic acids found in grapes (tartaric and malic) are nonvolatile. Lactic acid, primarily the byproduct of malolactic fermentation, is relatively nonvolatile, possessing a mild, insignificant odor.

Alcohols

Although ethanol has a mild fragrance, the most significant aromatic alcohols are the higher (fusel) alcohols, those with carbon chains three to six carbons long. Examples such as 1-propanol, 2-methyl-1-propanol (isobutyl alcohol),

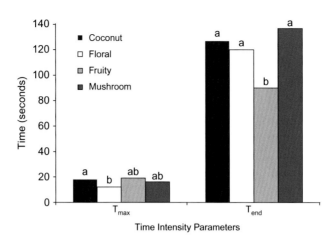

FIGURE 3.18 Aspects of the finish of a model white wine (time to maximum intensity [T_{max}] and duration [T_{end}]) spiked with single flavors: coconut (47.6 mg *cis-/trans*-oak lactone), floral (43.1 mg/L linalool), fruity (65.2 mg/L ethyl hexanoate), and mushroom (41.5 mg/L 1-octen-3-ol). *Reprinted from Goodstein, E. S., Bohlscheid, J., Evans, M., and Ross, C. F. (2014) Perception of flavor finish in model white wine: A time-intensity study. Food Qual. Pref.* **36**, *50–60, with permission from Elsevier.*

2-methyl-1-butanol, and 3-methyl-1-butanol (isoamyl alcohol) tend to have fusel odors, whereas hexanols possess a herbaceous scent. The major phenol-derived alcohol, 2-phenylethanol (phenethyl alcohol), has a rose-like scent.

At low concentrations (~0.3 g/L or less), higher alcohols may contribute complexity to the bouquet. At higher levels, these alcohols increasingly dominate the fragrance. In distilled beverages, such as brandies and whiskeys, these alcohols give the beverage much of its distinctive fragrance. It is only in port that a slight fusel character is considered a positive quality attribute. This property comes from the unrectified fortifying spirits added during port's production. Although most higher alcohols are yeast by-products, various fungi (grape pathogens or contaminants in cork, oak, barrels, etc.) can donate potentially important alcohols, notably 1-octen-3-ol, possessing a mushroomy odor.

Aldehydes and Ketones

Acetaldehyde is the major vinous aldehyde. It often constitutes more than 90% of a wine's aldehyde content. Above threshold values in table wines, acetaldehyde is considered an off-odor. Combined with other oxidized compounds, it contributes to the traditional bouquet of sherries and other oxidized wines. Furfural and 5-(hydroxymethyl)-2-furaldehyde are other aldehydes having sensory impact. Their caramel-like aspects are most evident in baked wines.

Phenolic aldehydes, such as cinnamaldehyde and vanillin, may accumulate to sensorially significant concentrations in wines matured in oak. Other phenolic aldehydes, such as benzaldehyde, may have diverse origins. Benzaldehyde's almond fragrance has occasionally been considered characteristic of certain wines, for example, those produced from Gamay grapes. In addition, some aldehydes, such as phenylacetaldehyde and methional (a sulfur-containing aldehyde) can suppress the fruitiness of red wines (San-Juan et al., 2011).

Although not having a direct sensory effect, hydroxypropanedial (triose reductone) characteristically occurs in botrytized wines (Guillou et al., 1997). It exists in tautomeric equilibrium between 3-hydroxy-2-oxopropanal and 3-hydroxy-2-hydroxypro-2-enal. Reductones, such as hydroxypropanedial, can play a role in preserving a wine's fragrance by retarding aromatic volatilization.

Many ketones are produced during fermentation, but few appear to have sensory significance. The major exception is diacetyl (biacetyl, or 2,3-butanedione). At low concentrations, diacetyl can donate a buttery, nutty, or toasty flavor. However, at much above its sensory threshold, diacetyl can generate an overly noticeable buttery, lactic off-odor. This commonly occurs in association with spoilage induced by certain strains of lactic acid bacteria.

Acetals

Acetals form when an aldehyde (or ketone) reacts with the hydroxyl group of an alcohol. They generally possess vegetal odors. Because they form primarily during oxidative aging and distillation, they tend to occur only in significant amounts in sherries and brandies.

Esters

More than 160 esters have been isolated from wine. Because most esters occur in trace amounts, and have either low volatility or mild scents, their importance to wine fragrance is negligible. However, the more common esters may occur at or above their sensory thresholds. Because some esters have a fruity fragrance, they can contribute significantly to the bouquet of young wines. Their influence tends to be comparatively short-lived (Fig. 8.28) as they hydrolyze back to their acid and alcoholic moieties during aging.

There are three main categories of vinous esters: those formed between ethanol and short-chain fatty acids, acetic acid and various short-chain alcohols, and nonvolatile acids and ethanol. Several other ester groupings have been isolated from wine, but appear to have little sensory significance.

Ethyl acetate, formed between ethanol and acetic acid, is the most significant wine ester. Typically, it occurs at concentrations below 50–100 mg/L. At low levels (<50 mg/L), it may contribute to a wine's aromatic complexity. However, at above 150 mg/L, ethyl acetate generates an acetone-like off-odor (Amerine and Roessler, 1983). Ethyl acetate can be either a byproduct of microbial activity (yeast or bacterial), or form abiotically by a reaction between acetic acid and ethanol.

Other than ethyl acetate, the major ethanol-based esters are those formed with higher alcohols, such as isoamyl $[(CH_3)_2CH_2OH]$ and isobutyl $[(CH_3)_2CHCH_2OH]$ alcohols. These lower-molecular-weight esters are often termed fruit esters, because of their fruit-like fragrances. Isoamyl acetate (3-methylbutyl acetate) has a banana-like scent, whereas benzyl acetate has an apple-like aspect. They play a significant role in the bouquet of young wines, notably white versions (Vernin et al., 1986). As the length of the acid's hydrocarbon chain increases, the ester's odor shifts from being fruity to soap-like and, finally, lard-like with C_{16} and C_{18} fatty acids. The presence of certain of these esters—for example, hexyl acetate and ethyl octanoate—has occasionally been considered an indicator of red wine quality (Marais et al., 1979). A poster or downloadable graphic color illustration of the fragrance characteristic of many esters can be obtained from Kennedy (2013).

Esters of the major nonvolatile wine acids (tartaric, malic and lactic) form slowly during aging. Nevertheless, because of their weak odors, they are seldom of sensory significance. In contrast, the methanolic and ethanolic esters of succinic acid appear to contribute to the aroma of muscadine wines (Lamikanra et al., 1996).

Occasionally, other esters of significance may be found in grapes. The phenolic ester, methyl anthranilate, is a prime example. It contributes to the grapy essence of some *V. labrusca* varietal wines. Another is ethyl 9-hydroxynonanoate, synthesized by *B. cinerea*. It may contribute to the distinctive aroma of botrytized wines (Masuda et al., 1984).

Hydrogen Sulfide and Organosulfur Compounds

Hydrogen sulfide and sulfur-containing organics generally occur only in trace amounts in bottled wine—thankfully. They usually have unpleasant to nauseating odors. However, because their sensory thresholds are typically low (often in parts per trillion), they can occasionally generate off-odors.

Hydrogen sulfide (H_2S) is a byproduct of yeast sulfur metabolism. At near-threshold levels, hydrogen sulfide is part of the yeasty odor of newly fermented wines. Above threshold levels, it generates a putrid, rotten-egg odor.

The simplest organosulfur compounds found in wine are the mercaptans. A significant member is ethanethiol (ethyl mercaptan). It possesses a rotten onion, burnt-rubber odor even at threshold levels. At higher concentrations, it has a skunky, fecal odor. Related thiols, such as 2-mercaptoethanol, methanethiol, and ethanedithiol, have barnyard, rotten cabbage, and sulfur–rubber off-odors, respectively. Additional, retronasally detected sulfurous odors can arise from the presence of sulfur-containing amino acids (e.g., cysteine) in the mouth (Hettinger et al., 1990).

Exposure to light can stimulate the reductive synthesis of organosulfur compounds in bottled wine. An example is the cooked-cabbage–shrimp-like odor generated by dimethyl sulfide in the *goût de lumière* taint of champagne (Charpentier and Maujean, 1981).

Although organosulfur compounds generate several repulsive odors, an increasing number of thiols are being implicated in the varietal aroma of grape cultivars. For example, several thiols contribute to the aroma of Sauvignon blanc wines. These include 4-mercapto-4-methylpentan-2-ol, 3-mercaptohexan-1-ol, 4-mercapto-4-methylpentan-2-one, 3-mercaptohexyl acetate, and 3-mercapto-3-methylbutan-1-ol (Tominaga et al., 1996, 1998). In addition, 4-mercapto-4-methylpentan-2-ol plays a central role in the characteristic aroma of Scheurebe (Guth, 1997b) and Macabeo (Escudero et al., 2004) wines. Often these constituents occur as nonvolatile complexes in grapes. It is the enzymic action of yeasts that liberates their volatile ingredients. Another example of the significance of aromatic organosulfur compound is 3-mercaptohexan-1-ol. It contributes to the fruity bouquet of many rosé wines (Murat, 2005).

Hydrocarbon Derivatives

Several grape-derived hydrocarbons are the progenitors of several important aromatics. Examples are β-damascenone (floral-like), α- and β-ionone (violet-like), vitispirane (eucalyptus–camphor-like), (E)-1-(2, 3, 6-trimethylphenyl)buta-1,3-diene (cut grass), and 1,1,6-trimethyl-1,2-dihydronaphthalene (TDN). Some can also mask the bell pepper odor of isobutyl methoxypyrazine (IBMP), for example β-damascenone, at least in hydroalcoholic solutions (Pineau et al., 2007). Many of the compounds noted above are the hydrolytic breakdown products of carotenoids.

Possibly the most significant hydrocarbon derivative is the norisoprenoid, TDN. After several years in-bottle, the concentration of TDN in some white wines can rise to 40 ppb or above (Rapp and Güntert, 1986). At above 20 ppb TDN donates a smoky, kerosene, bottle-aged fragrance (Simpson, 1978).

A cyclic hydrocarbon occasionally found in wine is styrene. It has been detected as a taint in wine stored in some plastic cooperage or transport containers (Hamatschek, 1982). Additional hydrocarbon taints may come from methyl tetrahydronaphthalene, implicated in some types of corky off-odors (Dubois and Rigaud, 1981).

Lactones and Other Oxygen Heterocycles

Lactones are cyclic esters formed by internal esterification between carboxyl and hydroxyl groups. Lactones coming from grapes seldom contribute varietal odors. One exception is 2-vinyl-2-methyltetrahydrofuran-5-one. It has been proposed as contributing to the distinctive aroma of Riesling and Muscat varieties (Schreier and Drawert, 1974). Because lactone formation is enhanced during heating, some of the raisined character of sun-dried grapes may come from lactones, such as 2-pentenoic acid-γ-lactone. Sotolon (4,5-dimethyl-tetrahydro-2,3-furandione) is characteristic of botrytized wine (Masuda et al., 1984), as well as sherries (Martin et al., 1992). It has attributes described as nutty, sweet, and burnt. Sotolon also tends to occur along with its ethyl analog, abhexon (5-ethyl-3-hydroxy-4-methyl-2(5H)furanone) (Bailly et al., 2009). Another lactone appearing in sauternes is 2-nonen-4-olide (Stamatopoulos et al., 2014). It is described as possessing an overripe orange fragrance.

Although lactones can form during fermentation and aging, the most commonly found lactones are those extracted from oak. Of these, the most important are termed oak lactones (isomers of β-methyl-γ-octalactone). Yeasts can also synthesize these lactones in small amounts. They possess oaky, mildly coconut-like attributes.

Among other heterocyclic compounds, vitispirane appears to be the most significant (Etiévant, 1991). Vitispirane forms slowly during aging, reaching concentrations of 20–100 ppb. Its two isomers have different odor qualities. The cis-isomer has a chrysanthemum flower–fruity fragrance, whereas trans-vitispirane has a heavier, exotic, fruit-like scent.

Terpenes and Oxygenated Derivatives

Terpenes donate the characteristic fragrance to many flowers, fruits, seeds, leaves, woods, and roots. Chemically, terpenes are composed of two or more, basic, five-carbon, isoprene units. Unlike many other wine aromatics, terpenes are primarily derived from grapes (Strauss et al., 1986). Only those terpenes that occur free (unbound to sugars) contribute to a wine's fragrance.

Terpenes contribute to the varietal character of several important grape varieties, most notably members of the Muscat and Riesling families (see Rapp, 1998). A sesquiterpene, rotundone, has been identified as the source of the peppery aroma of Shiraz wines (Wood et al., 2008). Other cultivars produce terpenes, but they appear to play little role in their varietal distinctiveness (Strauss et al., 1987b).

Although terpenes are unaffected by fermentation, grape infection by B. cinerea both reduces and modifies their terpene content. This undoubtedly plays a major role in the minimal varietal character of most botrytized wines (Bock et al., 1988).

During aging, the types and proportions of terpenes change significantly (Rapp and Güntert, 1986). Although some increase in sensory impact, as they are liberated from their glycosidic bondage, losses due to oxidation are more common. In the latter reactions, most monoterpene alcohols are converted to terpene oxides. These have sensory thresholds approximately 10 times higher than their precursors. In addition, these changes affect odor quality. For example, the muscaty, iris-like odor of linalool is progressively replaced by the musty, pine-like scent of α-terpineol.

During aging, additional changes can modify the structure of wine terpenes. Some terpenes become cyclic and form lactones, for example, 2-vinyl-2-methyltetrahydrofuran-5-one (from linalool oxides). Other terpenes may transform into ketones, such as α- and β-ionone, or spiroethers such as vitispirane. Recently, the monoterpene ketone, piperitone, has be associated with the mint nuance detected in aged red Bordeaux wines (Picard et al., 2016).

Although most terpenes have pleasant odors, some generate off-odors. The musky-smelling sesquiterpenes produced by *Penicillium roquefortii* in cork (Heimann et al., 1983) are a prime example. *Streptomyces* species may also synthesize earthy-smelling sesquiterpenes in cork or cooperage wood.

Phenolics

The most distinctive, grape-derived, volatile phenolic is methyl anthranilate (Robinson et al., 1949). It is a major aroma component in some *Vitis labrusca* varieties. Another significant volatile phenolic is 2-phenylethanol. It produces the rose-like fragrance typical of some *V. rotundifolia* cultivars (Lamikanra et al., 1996). Volatile phenols are also a part of the complex of aromatic compounds found in some *vinifera* wines (Moio and Etiévant, 1995; Rapp and Versini, 1996).

Although volatile phenolics may contribute to the varietal aroma of some cultivars, the most significant sources of volatile phenolics are hydroxycinnamic esters, generated during fermentation or derived from oak cooperage. They can be metabolized to unpleasant smelling compounds by spoilage microbes (Chatonnet et al., 1997). Their derivatives, vinylphenols (4-vinylguaiacol and 4-vinylphenol) and ethylphenols (4-ethylphenol and 4-ethylguaiacol) can donate spicy, pharmaceutical, clove-like odors, and smoky, phenolic, animal, stable-like notes, respectively. Off-odor detection frequently occurs when ethylphenol contents exceed 400 µg/L, or 725 µg/L for vinylphenols. Eugenol, another clove-like volatile phenol, can also occur in wine. At usual concentrations, eugenol adds only a general spicy note. However, guaiacol can generate a sweet-like, smoky off-odor (Simpson, 1990).

Oak cooperage is the source of several volatile phenolic acids and aldehydes. Benzaldehyde is particularly prominent and possesses an almond-like odor. Its occurrence in sherries may participate in their nut-like bouquet. Other significant phenolic aldehydes are vanillin and syringaldehyde, both of which possess vanilla-like fragrances. They form during the breakdown of wood lignins. The toasting of oak staves during barrel construction is another source of volatile cyclic aldehydes, notably furfural and related compounds.

Pyrazines and Other Nitrogen Heterocyclics

Pyrazines are cyclic nitrogen-containing compounds that contribute significantly to the flavor of many whole and baked foods. They are also important to the varietal aroma of several grape cultivars. 2-Methoxy-3-isobutylpyrazine plays a major role in the bell (green) pepper odor often detectable in Cabernet Sauvignon and related cultivars, such as Sauvignon blanc and Merlot. At concentrations of about 8–20 ng/L, methoxybutylpyrazine may be desirable, but, at above these values, it starts to generate an overpowering vegetal, herbaceous aroma. Related pyrazines are present, but generally occur at concentrations at or below their detection thresholds (Allen et al., 1996).

Pyridines are another group of aromatic cyclic nitrogen compounds periodically isolated from wine. Thus far, their involvement in wine flavor appears to be restricted to the production of mousy off-odors. This attribute has been associated with 2-acetyltetrahydropyridines, 2-ethyl-tetrahydropyridine, and 2-acetyl-1-pyrroline (Heresztyn, 1986; Grbin et al., 1996). They are produced by some strains of the spoilage yeast, *Brettanomyces*.

SENSATIONS OF CHEMESTHESIS (THE COMMON CHEMICAL SENSE)

Although it is a much more common sensation detected in the oral cavity (see Chapter 4, Oral Sensations (Taste and Mouth-Feel)), free nerve endings of trigeminal nerves also occur throughout the nasal epithelium (excepting the olfactory patches). The location of these receptors throughout the nasal passages is poorly characterized, but appears to be random (Frasnelli et al., 2004). Some of these receptors respond to low concentrations of certain volatile compounds, generating perceptions referred to as pungent, putrid, or irritant. Wine-derived examples can include hydrogen sulfide, sulfur dioxide, and carbon dioxide. However, nasal trigeminal nerve endings can respond to almost any volatile compound at high enough concentrations (in the range of 3–5 magnitudes greater that their olfactory thresholds) (Cometto-Muñiz and Abraham, 2016). Vanillin is an apparent exception, not activating trigeminal receptors (Savic et al., 2002).

Trigeminal receptors can elicit sensations such as burning, cooling, stinging, tingling, and pain. Thus, odorants at high concentrations can provoke nasal irritation, especially since trigeminal receptors show little tendency toward adaptation (Cain, 1976). An example of how the quality of an aromatic can change with concentration is indole. At low concentrations it is perceived as jasmine-like, but at higher levels fecal-like (Kleene, 1986), presumably due to activation of certain trigeminal receptors. A similar situation may explain why hydrogen sulfide can contribute to a

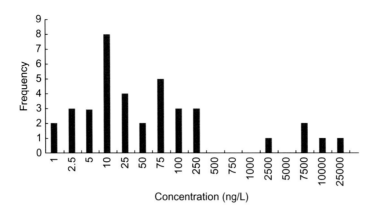

FIGURE 3.19   Frequency distribution for individual 2,4,6-trichloroanisole thresholds in Sauvignon blanc wine. *From Suprenant and Butzke, 1996, reproduced by permission.*

yeasty bouquet in young wines at low concentrations (~1 µg/L) (MacRostie, 1974), but at slightly higher concentrations can generate a putrid, rotten-egg odor.

Trigeminal stimulants can also suppress the perception of olfactory compounds (Richards et al., 2010). For example, carbon dioxide can suppress the detection of vanillin (Kobal and Hummel, 1988), and sulfur dioxide, at sufficiently high concentrations, could mask the subtleties of the wine's fragrance.

ODOR PERCEPTION

Differences in odor perception among individuals have long been known (Pangborn, 1981). Variation can affect the ability to detect, differentiate, or identify odorants, assess intensity, or react emotionally or hedonically to their presence.

Some of these differences have been codified, with the lowest concentration inducing a particular type of conscious response being defined as its threshold. For example, the detection threshold is the concentration at which the presence of a substance becomes noticeable, usually calculated as the value at which just over 50 percent of test subjects detect the compound above chance. Thresholds for different compounds can range over 10 orders of magnitude, from about 2×10^{-2} M for ethane to between 10^{-10} and 10^{-12} M for mercaptans. Even for related compounds, such as pyrazines, thresholds can vary markedly (Siefert et al., 1970). The range in sensitivity among individuals also can show marked variation, some patterns resembling a bell-shaped curve, others skewed (e.g., Poisson distribution), or occasionally expressed as bimodal (e.g., McRae et al., 2013) (Figs. 3.19 and 5.19). Typical detection thresholds for several important wine aromatics are noted in Table 3.2.

Although giving the impression of precision, threshold values are far from absolute values, being denoted as the mean or mode of a range of responses. In addition, threshold values are significantly influenced by the solvent and matrix constituents. Increasing ethanol content changes the proportion of ethanol clusters, reducing hydrophobic hydration (D'Angelo et al., 1994), augmenting nonpolar solubility, and reducing volatility (Aznar et al., 2004). For example, the threshold of ethyl isobutanoate can change from 0.03 ng/L in air (Grosch, 2001) to 15 µg in 10 percent ethanol (Guth, 1997b) (clearly affecting the potential for detection of this ester "in-glass" versus "in-mouth"). Conversely, the fruity and floral aspects of esters increase when the wine's alcohol content decreases (Guth, 1998). This may partially explain why many German Riesling wines are so floral in nature (often having alcohol contents significantly below the 11–13 percent typical of most table wines). Such effects are often highly specific (Table 3.3), presumably involving modification of partition constants. In addition, because the volatility of esters is most marked at low alcohol contents (Fig. 3.20), their sensory significance may be most noticeable in the finish. This effect may be equally significant when assessing a wine's fragrance in the glass, as ethanol evaporates from the film of wine coating the glass following swirling. The headspace concentration of a range of aromatics was quickly replenished by alcohol-induced convection currents at low ethanol concentrations (5 percent) (Tsachaki et al., 2009).

Guth and Sies (2002) consider that alcohol's effect on fruitiness resulted not so much through volatilization but on perception. However, Escalona et al. (1999) found that increasing alcohol content progressively decreased the volatility of several higher alcohol and aldehydes, as did Conner et al. (1998) for some ethyl esters. The potential

TABLE 3.2 Detection thresholds of some aromatic compounds found in wine[a]

Compound	Threshold (μg/L)	Reference
ACIDS		
Acetic acid	200,000	Guth (1997b)
Butyric acid	1732	Ferreira et al. (2000)
Decanoic acid	1000; 15,000	Ferreira et al. (2000); Guth (1997b)
Hexanoic acid	420; 3000	Ferreira et al. (2000); Guth (1997b)
Isobutyric acid	2,300; 10,000	Ferreira et al. (2000); Guth (1997b)
Isovaleric acid	33	Ferreira et al. (2000)
Octanoic acid	500	Ferreira et al. (2000)
Propanoic acid	420	Ferreira et al. (2000)
ALCOHOLS		
Ethanol	2.8 to 90 × 10^9	Meilgaard and Reid (1979)
1-Hexanol	8000	Guth (1997b)
3-Hexenol	400	Guth (1997b)
Isobutanol	40,000	Guth (1997b)
Isoamyl alcohol	30,000	Guth (1997b)
Methionol	1000	Ferreira et al. (2000)
ALDEHYDES & KETONES		
Acetaldehyde	500	Ferreira et al. (2000); Guth (1997b)
Acetoin	150,000	Ferreira et al. (2000)
Diacetyl	100	Guth (1997b)
ESTERS		
Ethyl acetate	12,200; 7500	Etiévant (1991); Guth (1997b)
Ethyl butyrate	14; 20	Ferreira et al. (2000); Guth (1997b)
Ethyl decanoate	200	Ferreira et al. (2000)
Ethyl hexanoate	14; 5	Ferreira et al. (2000); Guth (1997b)
Ethyl isobutyrate	15	Guth (1997b)
Ethyl isovalerate	3	Ferreira et al. (2000)
Ethyl 2-methylbutyrate	18; 1	Ferreira et al. (2000); Guth (1997b)
Ethyl octanoate	5; 2	Ferreira et al. (2000); Guth (1997b)
Isoamyl acetate	30	Guth (1997b)
3-Methylbutyl acetate	30	Guth (1997b)
LACTONES		
γ-Decalactone	88; 0.7	Etiévant (1991); Ferreira et al. (2004)
γ-Dodecalactone	7	Ferreira et al. (2004)
Z-6-Dodecenoic acid γ-lactone	0.1	Guth (1997b)
4-Hydroxy-2, 5-dimethy1-3-(2*H*)-furanone	5	Guth (1997b)
γ-Nonalactone	25	Ferreira et al. (2004)
cis-Oak lactone	67	Etiévant (1991)
Sotolon	5	Guth (1997b)

(*Continued*)

TABLE 3.2   (Continued)

Compound	Threshold (µg/L)	Reference
NORISOPRENOIDS		
β-Damascenone	0.05	Guth (1997b)
β-Ionone	0.09	Ferreira et al. (2000)
PHENOLS AND PHENYL DERIVATIVES		
Ethyl dihydrocinnamate	1.6	Ferreira et al. (2000)
trans-Ethyl cinnamate	1	Ferreira et al. (2000)
4-Ethylguaiacol	33	Ferreira et al. (2000)
4-Ethyl phenol	440	Boidron et al. (1988)
Eugenol	5	Guth (1997b)
Guaiacol	10	Guth (1997b)
2-Phenethyl acetate	250	Guth (1997b)
Phenylacetaldehyde	5	Ferreira et al. (2000)
2-Phenylethanol	14,000; 10,000	Ferreira et al. (2000); Guth (1997b)
Vanillin	200	Guth (1997b)
4-Vinylguaiacol	10; 40	Ferreira et al. (2000); Culleré et al. (2004)
PYRAZINES		
2-methoxy-3-isobutylpyrazine	0.002; 0.015	Ferreira et al. (2000); Roujou de Boubée et al., 2002
3-Isobutyl-2-methoxypyrazine	0.002	Ferreira et al. (2000)
SULFUR COMPOUNDS		
Benzenemethanethiol	0.0003	Tominaga et al. (2003)
Dimethyl sulfide (DMS)	25; 60	Goniak and Noble (1987); De Mora et al. (1987)
Dimethyl disulfide (DMDS)	29	Goniak and Noble (1987)
Ethyl mercaptan (ethanethiol)	1.1	Goniak and Noble (1987)
Hydrogen sulfide	5 to 80	Wenzel et al. (1980)
3-Mercaptohexyl acetate	0.004	Tominaga et al. (1988)
4-Mercapto-4-methylpentan-2-one	0.0006	Guth (1997b)
3-Mercaptohexan-1-ol	0.06	Tominaga et al. (1998)
2-Methyl-3-furanthiol	0.005	Ferreira et al. (2002)
Methyl mercaptan (methanethiol)	2	Fors (1983)
3-(Methylthio)-1-propanol	1000; 500	Ferreira et al. (2000); Guth (1997b)
TERPENES		
Geraniol	30	Guth (1997b)
Linalool	25; 15	Ferreira et al. (2000); Guth (1997)
cis-Rose oxide	0.2	Guth (1997b)
Wine lactone	0.01	Guth (1997b)

[a]Odor thresholds often vary considerable depending on the procedure used, solvent involved, type of wine, presence of other aromatic compounds, and the acuity of the taster(s). Thus, threshold values are primarily useful in comparing the potential flavor impact of various compounds.

TABLE 3.3   Effect of ethanol on the odor threshold of some wine aromatics in air (ethanol in the gas phase 55.6 mg/L)

Compound	Odor threshold (ng/L)		
	Without ethanol (*a*)	With ethanol (*b*)	Factor b/a
Ethyl isobutanoate	0.3	38	127
Ethyl butanoate	2.5	200	80
Ethyl hexanoate	9	90	10
Methylpropanol	640	200,000	312
3-Methylbutanol	125	6300	50

*Source: Reproduced from Grosch, W. (2001). Evaluation of the key odorants of foods by dilution experiments, aroma models and omission. Chem. Senses **26**, 533–545, based on data from Guth, 1997a, by permission from Oxford University Press.*

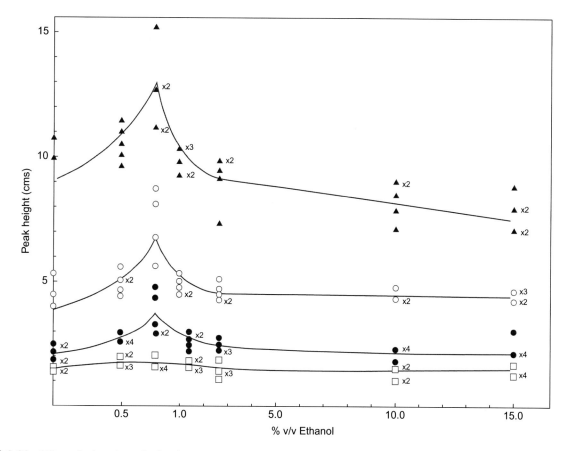

FIGURE 3.20   Effect of ethanol on the headspace composition of a synthetic mixture. Ethyl acetate (●), ethyl butrate (○), 3-methylbutyl acetate (▲), 2-methylbutanol (□). *From A.A. Williams and P.R. Rosser (1981). Aroma enhancing effect of ethanol. Chem. Senses **6**, 149–153. Reproduced by permission of Oxford University Press.*

complex influence of alcohol on wine flavor is further illustrated by synergistic and suppressive effects on β-methyl-γ-octalactone (oak lactone) and isoamyl acetate, respectively (Le Berre et al., 2007). Further complicating the issue, experimental conditions can markedly effect results (Tsachaki et al., 2005). Most studies are conducted after equilibrium has been established between aromatics in the wine and the headspace above the glass. However, under typical tasting conditions, odorants are still actively escaping from the wine into the headspace. Fig. 3.21 illustrates results under such dynamic conditions. Although differences in alcohol content of as little as 0.1 percent have been suggested to have detectable influences on sensory perception (Wollan, 2005; Caporn, 2011), adjusting the alcohol

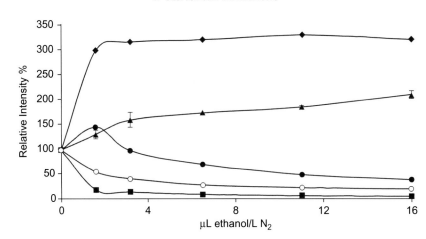

FIGURE 3.21   Intensity of octanal (●), diacetyl (■), linalool (▲), ethyl butyrate (◆), and limonene (○) as ethanol concentration of the ethanol vapors increased. *Reproduced from Aznar, M., Tsachaki, M., Linforth, R. S. T., Ferreira, V., and Taylor, A. J. (2004). Headspace analysis of volatile organic compounds from ethanolic systems by direct APCIMS. Int. J. Mass Spectrometry* **239**, *17–25, with permission from Elsevier.*

content of a wine to reach such a "sweet spot" has not been confirmed experimentally (King and Heymann, 2014). Under their conditions, the minimum detectable difference occurred in the range of 0.4 percent alcohol.

Numerous wine constituents other than ethanol can influence volatility. For example, grape (rhamnogalacturonans and arabinogalactan protein) and yeast (mannoprotein) polysaccharides influence the liberation of esters, higher alcohols, and diacetyl (Dufour and Bayonove, 1999a); flavonoids affect the release of various esters and aldehydes (Dufour and Bayonove, 1999b); and anthocyanins can associate with volatile phenolics, such as vanillin (Dufour and Sauvaitre, 2000). Thus, the detection threshold of compounds can vary considerably, depending on the wine in which they are assessed. For example, the detection threshold of diacetyl varied from 0.2 mg/L in a Chardonnay, to 0.9 mg/L in a Pinot noir, and to 2.8 mg/L in a Cabernet Sauvignon (Martineau et al., 1995). Even near- or subthreshold concentrations of nonvolatile constituents can influence volatility (Dalton et al., 2000; Friel et al., 2000; Pfeiffer et al., 2005). For example, low sugar concentrations (1 percent) enhanced the volatility of ethyl acetate and ethanol (Nawar, 1971), but decreased that of acetaldehyde (Maier, 1970).

Such matrix effects probably result from a combination of the wine's physicochemistry, as well as the multisensory nature of odor cognition and memory. Thus, wines tend to be more easily identified when vision and taste are involved than when only smell is involved (Ballester et al., 2005); this is why sucrose is crucial to the continuance of a mint flavor in gum (Cook et al., 2003). Such cross-modal interactions can be highly individualistic (Hort and Hollowood, 2004), reflecting innate differences in sensitivity and experience.

Saliva and oral microbiology are additional factors that can affect the release and perception of aromatics. Saliva, by diluting wine constituents, can affect volatility, whereas salivary and microbial enzymes can change a wine's chemical makeup. Typically, saliva appears to reduce volatility (as measured by headspace concentration), depending on the wine matrix (Muñoz-González et al., 2014). In addition, salivary enzymes have been shown to degrade esters and thiols (Buettner, 2002a), as well as reduce aldehydes to their corresponding alcohols (Buettner, 2002b). In contrast, the concentrations of 2-phenylethanol and furfural, responsible for rose and toasted almond fragrances, increased, while those of vitispirane and TDN decreased in the mouth (Genovese et al., 2009). The smoky attribute of wines produced from grapes exposed to smoke from brush fires may partially originate from salivary enzymes releasing smoke-derived phenols from glycoconjugates (Kennison et al., 2008). In contrast, the protein-binding (astringent) ability of polyphenols was reduced. Although the action of salivary enzymes may be minimal during sampling (the duration of wine in the mouth typically being about 15 s), their effects could affect expression of a wine's finish (Fig. 3.22), where the contact period is longer. Because saliva chemistry varies between individuals and diurnally, this is likely another source of the idiosyncratic responses people have to wine.

When detection thresholds are markedly below normal, the condition is termed anosmia. Anosmia can be general or specific, affecting only a select range of compounds (Amoore, 1977). The occurrence of specific anosmias varies widely. For example, about 3 percent of humans are anosmic to isovalerate (sweaty), whereas 47 percent are anosmic to 5α-androst-16-en-3-one (urinous). Differences in the allelic nature and copy number of OR genes in an individual probably explains most specificity in detecting aromatic compounds (Nozawa et al., 2007, Young et al., 2008). For

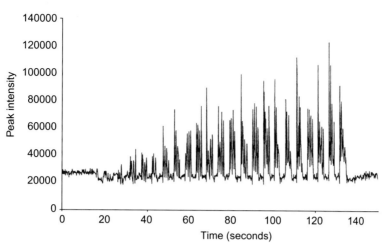

FIGURE 3.22 Assessment of hexanol released from hexyl-β-D-glucopyranoside in the mouth of the subject. The sample was introduced at 15s. *Reproduced from Hemingway, K. M., Alston, M. J., Chappell, C. G., and Taylor, A. J. (1999). Carbohydrate flavor conjugates in wine. Carbohydrate Polymers 38, 283–286, with permission from Elsevier.*

example, the ability to detect *cis*-3-hexen-1-ol (possessing a grassy odor) has been correlated with production of a specific OR protein, i.e., OR2J3 (McRae et al., 2012). Nonetheless, absence of the principal responsive OR protein does not necessarily lead to complete anosmia, as other ORs may weakly respond to the odorant.

Hyperosmia, the detection of odors at abnormally low concentrations, is poorly understood. One of the most intriguing accounts of hyperosmia relates to a 3-week experience of a medical student, suddenly able to recognize people and objects by their odors (Sachs, 1985). Nonetheless, this may only be a more pronounced incidence of a more common, but little realized human attribute (Wells and Hepper, 2000; Lundström et al., 2009). People are often capable of recognizing kin, at rates significantly above chance, by odor alone. This may be a function of olfactory receptors responding to major histocompatibility (MHC)-induced odor factors (Spehr et al., 2006).

Hyperosmia may also occur in Parkinson's patients taking L-dopa (a synthetic neurotransmitter), as well as some individuals afflicted with Tourette's syndrome. Reduced mucus production in the nasal passages appears to be associated with increased odor sensitivity and may explain some age-related differences in odor sensitivity (Cain and Gent, 1991). Even more intriguing are instances of synesthesia, where odors are associated with particular colors or other sensory inputs (Stevenson and Tomiczek, 2007).

The origin of the limited odor identification (naming) skills of humans is unknown. It may arise from the small size of the human olfactory epithelium and olfactory bulbs. For example, dogs possess an olfactory epithelium up to 15 cm² in surface area (containing about 200 million receptor cells), whereas the olfactory patches in humans cover about 2–5 cm² (Moran et al., 1982), and possess an estimated 6–10 million receptors (Dobbs, 1989). Considering the effective olfactory area (the cilia), dogs may have 7 m² (Moulton, 1977), in comparison to the about 22 cm² of humans (Doty, 1998). Comparative measurements of the odor thresholds of dogs and humans suggest that dogs are at least 100 times more sensitive than humans (Moulton et al., 1960). However, this does not apply to all odorants. For example, humans are as sensitive as, or slightly better than, dogs with some aliphatic aldehydes (Laska et al., 2000). As far as functional olfactory (OR) genes, dogs possess about 970 (Olender et al., 2004), and mice about 913 (Godfrey et al., 2004), whereas humans possess about 340 (Malnic et al., 2004). These factors may account for some of our limited identification ability.

Additional reasons may involve repression of odor sensitivity during human evolution, to minimize social conflict and favor nuclear families (Stoddart, 1986). However, the process of inactivation of olfactory genes (65 percent in humans) appears to have started earlier, correlating with the development of tricolor vision in our primate ancestors (Gilad et al., 2004), and continuing in association with movement of the eyes forward, to facilitate stereoscopic vision. Thus, our limited olfactory competence may be based as much or more on increased reliance on vision (reduced need for olfactory acuity) than a need to diminish odor-based territoriality in social groupings.

Although human evolution has been associated with a loss of inherent olfactory abilities, our cerebral processing abilities have been significantly enriched. In addition, our responses to odors are largely learned, not hardwired. Thus, combined with our remarkable capacity for memory and language, our discriminatory powers concerning odors are still remarkable. What is lacking is concentration on naming at the same time as we associate particular

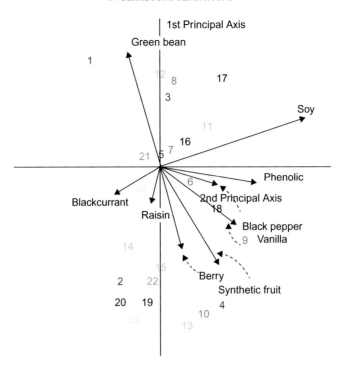

FIGURE 3.23 Plot of principal component scores of 24 Bordeaux wines based on 9 aroma attributes: 1–5, St. Estèphe; 6–10, St. Julien; 11–15, Margaux; 16–20, St. Émilion; 21,22, Haut Médoc; 23, Médoc; 24, Bordeaux. Note the considerable dispersion in sensory attributes of these regional wines. *Williams et al., 1984, reproduced with permission.*

odor qualities with the objects they characterize, especially during our formative years. This may be the reason why, in some cultures, where the ability to describe odors in abstract (musty) versus concrete (banana) or evaluative (pleasing) terms has significance, people can name scents as well as colors (Majid and Burenhult, 2014). Timing of learning to name odors could be decisive, if there is a critical period in odor learning, similar to that for language acquisition.

Two additional olfactory thresholds may be designated: differentiation and recognition. The differentiation threshold is the concentration difference between samples required for sensory discrimination. The recognition threshold refers to the minimum concentration at which an aromatic compound can be correctly identified. The recognition threshold is typically higher than the detection threshold and demands more processing power than differentiation. It is the equivalent of differentiating shades of color, side by side, from naming these shades when presented separately, especially sometime later.

As already noted, people have considerable difficulty identifying odors, especially without visual or contextual clues with which the odor memories originated (Engen, 1987). Nevertheless, it is often thought that expert tasters and perfumers have superior identification skills. For example, Jones (1968) estimated that perfumers were able to identify from 100 to 200 odorants. Although sensory skills obviously differ, whether wine professionals, such as sommeliers and Masters of Wine, possess abilities as acute is unsubstantiated (Noble et al., 1984; Morrot, 2004). In addition, even if such superior skills were to exist (equivalent to superior athletic skill), of what practical relevance is it to the majority of us mere mortals? Even winemakers are known to frequently fail to recognize their own wines in blind tastings, and experienced wine professionals frequently misidentify the varietal and geographical origin of wines (Morrot, 2004). In addition, variability in sensory acuity among tasters is remarkable (Fig. 5.19). Although not conforming with the aura generally presented in the popular press, it is easy to comprehend the difficulty experienced even by "experts," confronting the subtle and ephemeral aromatic differences that exist between similar wines, and the marked variability within regional and varietal wines (Fig. 3.23). What is amazing, relative to the difficulties involved, is the skill some tasters can develop. For example, experienced tasters at Davis, California were able to correctly identify the varietal origin of more than 40 percent of experimental wines (Winton et al., 1975). As expected, the most easily recognized were those from familiar cultivars with pronounced aromas (e.g., Cabernet Sauvignon, Zinfandel, Muscat, Gewürztraminer, and Riesling). The features that may distinguish superior tasters include a well-developed olfactory memory (associated with extensive experience), combined with a serious interest (motivation), and inherently better sensory acuity (see Ross, 2006).

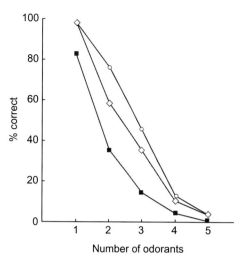

FIGURE 3.24 Percent correct identification of the constituents of mixtures containing up to five odorous chemicals by untrained (■), trained (◇), and expert (○) subjects. *From Laing, 1995, reproduced by permission.*

The issue of odor identification becomes increasingly intractable when they occur in mixtures (Jinks and Laing, 2001). Odor recognition appears to be primarily based on selective pattern detection, similar to facial recognition (Murray, 2004). Thus, the relative stimulation of several ORs (some of which may be the same) by a mixture of aromatic compounds increasingly complicates identification of its individual components (Laing, 1994). Odor mixtures are often perceived as being qualitatively distinct and different from those generated by their individual components (Sinding et al., 2014). However, with training, this does not necessarily prevent identification of some, if not all of the components (Jinks and Laing, 2001). A similar situation exists with wines possessing multiple aromatics below, at, or above their individual thresholds. Correspondingly, identification of varietal, stylistic, or regional differences is generally thought to be based on a holistic response to the pattern generated by the wine's multiple flavorants, and less commonly on individual constituents. Since tasters differ in sensitivity and expertise, it is not surprising that odor memories vary widely, and the terms used to describe them reflect those differences and limitations (Brochet and Dubourdieu, 2001). For example, wine critics often use quality-related expressions, whereas winemakers are more likely to use chemical or enologically based terms. Nevertheless, with training, most tasters can come to agree on a common set of descriptive terms. Those that remain idiosyncratic or inconsistent in term use are typically excluded from a tasting panel. Training tends to bring olfactory term use down to a common denominator, facilitating interpretation of the assessments generated (Case et al., 2004). Although this probably eliminates much human sensory variability, tasting panels are not designed to represent consumers. Instead, they are substitute analytic instruments, where standardization is important in obtaining statistically valid (but possibly consumer irrelevant) data.

In typical studies on odor recall, the sample contains a single compound or, if a mixture, the task is simple: name the source (e.g., a particular fruit). As noted, identifying the individual components of an odor mixture rapidly decreases as the number of aromatics increases (Fig. 3.24). This may explain why identifying wine off-odors often varies with the wine (its matrix) (Martineau et al., 1995; Mazzoleni and Maggi, 2007; Fig. 3.25). Extensive training improves identification ability, but it tends to be limited to four components (Livermore and Laing, 1998).

Typically, the odor quality of binary or simple mixtures reflects the constituent possessing the highest perceived intensity, even where the differences are slight. The situation is less straightforward with complex mixtures. The constituents may blend, generating a distinctly different odor quality, express the quality of the most intense constituent(s) (such as with some wine off-odors), express to varying degrees the odor qualities of its constituents, or enhance the fragrance of other constituents (Thomas-Danguin et al., 2014). A famous example of the latter is the use of the aliphatic aldehyde, 2-methylundecanal, to enhance the fragrance of the floral aspects of Chanel N° 5. A wine example is probably the activity of the thiol, 3-mercapto-3-methylbutan-1-ol, in accentuating the fruity–floral aspect of several white wines (Tominaga et al., 2000).

Perfumers appear to have superior identification skills, but this may relate as much to assessments being based on paper wicks dipped into samples as experience. As individual components tend to evaporate at different rates from wicks, the process of identification may resemble crude gas chromatography–olfactometry (see below), liberating constituents somewhat sequentially. Identification also improves when the participants in a test are first

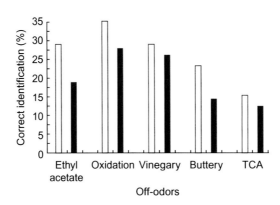

FIGURE 3.25   Detection of off-odors in different wines: □ white wine; ■ red wine; ethyl acetate (60 mg/L); oxidation (acetaldehyde, 67 mg/L); vinegary (acetic acid, 0.5 g/L); buttery (diacetyl, mg/L); TCA (2,4,6-trichloroanisole 15 μg/L). Based on results from 42 subjects.

allowed to become adapted to specific constituents prior to testing (Goyert et al., 2007). This could partially explain the "appearance" and "disappearance" of particular fragrances throughout a tasting.

Most odors are nominally grouped in categories based on origin (e.g., fruity, floral, vegetal, smoky), or associated with specific events or locations (e.g., Christmas, barbecuing, bonfires, hospitals). Thus, odor terms typically are concrete, referring to a specific or group of objects or experiences, and not to the olfactory perception itself. Usually, the more significant the event, the more intense and stable the memory. Engen (1987) views this memory pattern as equivalent to how young children associate words, categorizing objects and events functionally, rather than conceptually. For example, a hat is something a person might wear on their head, rather than a representative of an article of clothing. This may partially clarify why it is so difficult to use unfamiliar terms (e.g., chemical names) for familiar odors. The difficulty of correctly naming odors may also arise from the localization of language and concepts in different hemispheres of the brain, typically left versus right, respectively (Herz et al., 1999). People also have difficulty envisaging odor sensations, in contrast to the ease of conjuring up visual and auditory memories. Verbal or contextual clues improve identification (de Wijk and Cain, 1994a), in the same way that suggestion can influence what can be perceived in cloud formations. Pronouncing the name of a cultivar or an odor often induces its apparent detection, even in its absence. Suggestion seems to organize (or reorganize) odor perception in a manner similar to the way hints reshapes vision or memory (Murphy, 1995). In addition, experience can induce the perception of objects, for example, in a painting, that are not actually there, but could be considered appropriate to be there (Livingstone, 2002). Although supplying verbal, visual, or contextual clues often facilitates odor identification, odors learned in the absence of visual clues (e.g., the smell of toast or frying bacon) are easily recognized by odor alone.

Odor terms derived from personal experience are generally more easily recalled than those generated by others (Lehrner et al., 1999a). Thus, term retention should be better when the participants in sensory evaluation are involved in developing the lexicon (Gardiner et al., 1996; Herz and Engen, 1996). In contrast, the terms consumers and wine critics use tend to express more their personal holistic or emotional response to the wine than its sensory attributes (Lehrer, 1975; Dürr, 1985).

As with other attributes, perceived odor intensity, relative to concentration, often varies considerably. For example, a 3-fold increase in perceived intensity was correlated with a 25-fold increase in the concentration of propanol, but a 100-fold increase in the concentration of amyl butyrate (Cain, 1978). Aromatics that typically activate both olfactory and trigeminal receptors at low concentrations, such as hydrogen sulfide and mercaptans, often appear intense even at their recognition threshold.

SOURCES OF VARIATION IN OLFACTORY PERCEPTION

Small sex-related differences have been detected in olfactory acuity. Women are generally more sensitive to, and more skilled at, identifying odors than men (Fig. 3.26; Choudhury et al., 2003). This may relate to women's acuity increasing more than men on repeat exposure, by up to 5 orders of magnitude with some odorants (Dalton et al., 2002). In addition, the cerebral activity of women on exposure to odors is considerably more marked than in men (Yousem et al., 1999). The types of odors identified may also show sex- (or probably experience-) related differences.

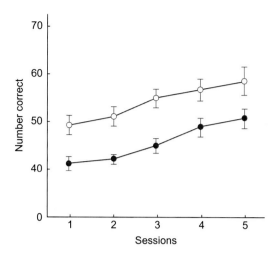

FIGURE 3.26   Mean number of correct odor identifications (±1SEM) by males (●) and females (○) over the course of five sessions. *From Cain, 1982, reproduced by permission.*

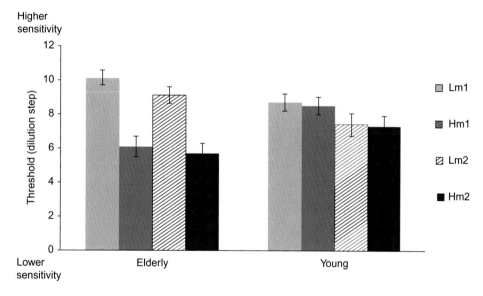

FIGURE 3.27   Effect of age on threshold values of comparatively light and heavy aromatic compounds. Threshold values (given in dilution steps) of the single molecules light (Lm1, *cis*-3-hexenol) and heavy (Hm1, β-ionone), and of the binary mixtures composed of light (Lm2, γ-valerolactone and γ-heptalactone) or heavy molecules (Hm2, γ-decalactone and γ-dodecalactone) in the groups of older people (50–70 years old) and of young adults (18–30 years old). The highest value of the *y*-axis (12) is the highest dilution step (lower threshold). *Reprinted from Sinding, C., Puschmann, L., and Hummel, T. (2014) Is the age-related loss in olfactory sensibility similar for light and heavy molecules. Chem. Senses* **39**, *383–390., by permission of Oxford University Press.*

Women often identify floral and food odors better than men, whereas men tend to do better at identifying petroleum odors. In addition, women experience modulation in olfactory discrimination, correlated with cyclical hormonal changes (see Doty, 1986). However, the most significant factor differentiating people's sensory skill may be the remarkable genetic diversity among individuals in the olfactory receptor (OR) genes they express (Menashe et al., 2003; Nozawa et al., 2008). Each person tested had a distinctive OR gene pattern. The results also indicate that ethnic groups may differ slightly in olfactory sensitivity, with acceptance/rejection of particular odors partially dependent on particular OR gene clusters (Eriksson et al., 2012; Jaeger et al., 2013). Nonetheless, twin studies indicate that differences in experience affect perceived odor intensity and pleasantness more than genetic factors (Knaapila et al., 2008).

Age also influences olfactory acuity. Identification ability appears to be optimal in young adults (Cain and Gent, 1991; de Wijk and Cain, 1994b). Acuity loss typically expresses itself as an increase in all three odor thresholds (detection, identification, and discrimination) (Lehrner et al., 1999a). Acuity loss may be progressive, but appears to affect structurally complex odorants more than simpler compounds (Fig. 3.27). Identification of even common foods can

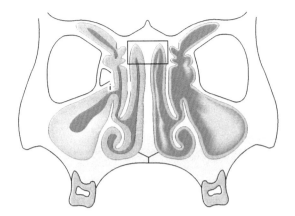

FIGURE 3.28   Schematic drawing of the nasal cavity and paranasal sinuses. The left side is typical of chronic rhinitis, illustrating restricted airflow to the olfactory epithelium caused by congestion of the ostiomeatal complex. The right side resembles typical airflow. *From Smith and Duncan, 1992, reproduced by permission.*

surprisingly different. For example, the identification rates for banana, broccoli, carrot, and cabbage in the absence of textural differences (in a purée) were 41, 30, 63, and 7 percent for a group of students, but only 24, 0, 7, and 4 percent for elderly subjects (Schemper et al., 1981). Compound-specific loss in acuity (detection and identification thresholds) also occurs (Seow et al., 2016). For example I once could detect small traces of Gewürztraminer in a blend; now I can only recognize a Gewürztraminer wine by the label. Even fresh litchis do not smell as I remember them. However, when cooked, I detect that distinctive and delectable odor exactly as before. Thus, the receptor(s) required are still there, but the threshold for activation has gone up significantly.

Age-related reduction in short-term odor memory has been reported (Lehrner et al., 1999b). This could result from reduced activity in the hypothalamus, a reduction in the regeneration olfactory neurons (Doty and Snow, 1988), their reduced sensitivity, or extenuating factors such as medications, smoking, or a history of nasal problems (Mackay-Sim et al., 2006). In addition, neuronal losses in the olfactory bulb and connections to the olfactory cortex cannot be ruled out. Olfactory regions frequently experience earlier degeneration than other parts of the brain (Schiffman et al., 1979), explaining why smell is often viewed as the first of the chemical senses to show age-related loss. In addition, it is clear that cognitive functions tend to decline with age. This is often particularly marked in terms of memory, such as learning odor names (Davis, 1977; Cain et al., 1995). Although sensory ability tends to decline with age, it does not seem to be inevitable (Nordin et al., 2012), at least to all compounds. Experience and mental concentration seems to compensate for some sensory loss. In a study of centenarians, cognitive status and health significantly influenced olfactory function (Elsner, 2001).

As might be expected, the effective nasal volume associated with the olfactory epithelium influences acuity (Damm et al., 2002). This is well known to anyone having experienced nasal congestion. A thick mucus coating limits airflow to and over the olfactory patches, as well as retards the diffusion of aromatics to the receptive neuronal layer (Fig. 3.28). In addition, infection may also accelerate certain degenerative changes, producing long-term acuity loss. Differences also exist between individuals in the shape, pattern, and size of their olfactory patches (Read, 1908).

Olfactory loss can also involve nerve destruction in association with several bacterial and viral infections, notably meningitis, osteomyelitis, and polio. In addition, some genetic diseases are associated with generalized anosmia, notably Kallmann syndrome (a deficiency in gonadotropin-releasing hormone). Many medications and drugs also can temporarily and adversely affect smell (Doty and Bromley, 2004), for example some antibiotics, antidepressants, antihistamines, bronchodilators, and cardiac medications, as well as recreational and illicit drugs, such as cocaine.

Smoking produces both short- and progressive long-term impairment of olfactory skills (Fig. 3.29). Recovery, after cessation, may occur but is often slow. Although smoking has not prevented some winemakers and cellar masters from becoming highly skilled at their craft (e.g., André Tchelistcheff), there is no evidence that smoking helped.

It is commonly believed that hunger moderately enhances olfactory acuity, whereas satiation lowers it. This view is supported by a report that both hunger and thirst increase the general reactiveness of the olfactory bulb and cerebral cortex (Freeman, 1991). Ghrelin, liberated in association with hunger, seems to be the activator (Tong et al., 2011).

Odor adaptation is another source of short-term, altered olfactory perception (Fig. 3.30). Adaptation can result from temporary loss in receptor excitability, reduced sensitivity in the brain, or both (Zufall and Leinders-Zufall, 2000). Generally, the more intense the odor, the slower adaptation becomes evident, but the longer it persists.

Olfactory response occurs quickly, followed by a slower period of adaptation (Fig. 3.31). Correspondingly, wine tasters have usually been counseled to take short sniffs, but this seems ill advised (Fig. 3.32); slower regular breathing

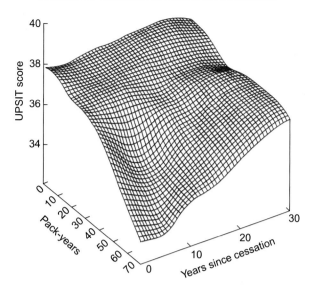

FIGURE 3.29 Effect of cumulative smoking dose and years since cessation from smoking for previous smokers. The individual data were fit to a distance-weighted least-squares regression to derive the surface plot. Although a few subjects evidenced a smoking dose greater than 70 pack-years, the pack-year scales was limited for clarity of surface illustration. UPSIT (University of Pennsylvania Smell Identification Test). *From Frye et al., 1990, reproduced by permission.*

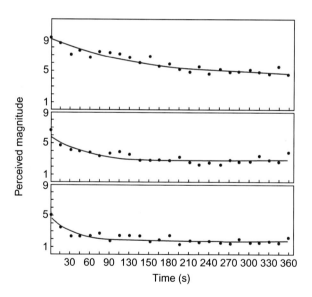

FIGURE 3.30 Adaptation in perceived magnitude of *n*-butyl acetate at 0.8 mg/L (bottom), 2.7 mg/L (middle), and 18.6 mg/L (top). *From Cain, 1974, reproduced by permission.*

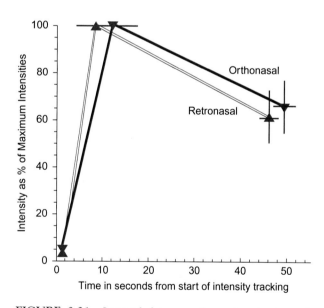

FIGURE 3.31 Summed data on orthonasal and retronasal response intensity to five commercial flavor extracts (anise, coffee, orange, peppermint, and strawberry), measured at 100 msec intervals. *From Lee and Halpern, 2013, reproduced by permission.*

is more effective in supplying odorants to the olfactory patches. Improved retronasal odorant transfer may be obtained by training to keep the velum–tongue closure open during sampling (Buettner et al., 2003).

The onset of adaptation appears to be compound dependent (Yoder et al., 2012), and partially based on its interaction with other volatiles. The dynamics of adaptation may also depend on whether, and to what degree, the aromatic compound stimulates trigeminal nerve receptors, as well as the effectiveness with which the odorant is absorbed by the mucus lining the epithelium, and subsequent degradation and/or release. It is estimated that the nasal epithelium can remove up to 75 percent of aromatics in the air in a single pass, depending on the properties

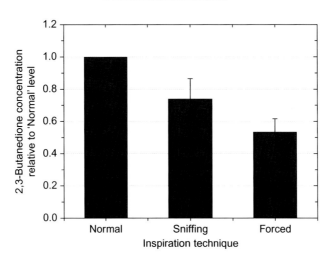

FIGURE 3.32   Mean maximum intranasal odorant concentration of 2,3-butadedione directly at the olfactory epithelium, as a function of inhalation technique (normal breathing, sniffing with repetitive irregular inhalation, and forced inhalation with deep breathing under high airflow. Error bars represent the standard error on the mean. Data taken from Beauchamp et al., 2009. *Reprinted from Buettner, A., and Beauchamp, J. (2010). Chemical input – Sensory output: Diverse modes of physiology-flavor interaction. Food Qual. Pref.* **21**, *915–924, with permission from Elsevier.*

of the compound (Keyhani et al., 1997). Little is known about the dynamics of a return from adaptation to baseline sensitivity. However, with H_2S, loss of olfactory adaptation appears to follow a curve that is the inverse of its adaptation (Ekman et al., 1967). Anecdotal evidence suggests that selective adaptation can reveal different components of a wine's fragrance. This view is supported by one of the few studies on adaptation in odorant mixtures (de Wijk, 1989). With intricate dynamics of detection and adaptation, there is additional rationale for assessing a wine's fragrance over an extended period, especially aromatically complex wines, such as vintage ports.

Synergy among aromatics can be of particular significance in wine, where hundreds of compounds occur at or below their respective detection thresholds (Ryan et al., 2008). For example, β-damascenone, carboxylic acids, dimethyl sulfur and esters can enhance the fruitiness of wines at subthreshold concentrations. In addition, subliminal concentrations of off-odors can induce odor masking (Guth, 1998; Czerny and Grosch, 2000). For example, ethylphenols, the source of a *Brett* character, slightly suppressed the fruity (but not floral) character in a Merlot wine at subliminal concentrations (Tempere et al., 2016). The effect of higher concentration was much more pronounced. In some instances masking may result from the activation of trigeminal nerves, or by disrupting the odor pattern generated by a milder aromatic compound. In the case of trichloroanisole (TCA), masking involves a generalized depression of olfactory sensitivity (Takeuchi et al., 2013).

In addition to wine-induced matrix effects, food undoubtedly has similar influences. Although little studied, combinations of wine and cheese indicate a mutual reduction in flavor intensity (Nygren et al., 2003a, b). Subjectively, this may be perceived as beneficial to their respective appreciations (see Chapter 9, Wine and Food Combination).

Multisensory interactions such as those noted above can depend on how they are assessed. For example, interactions are less marked or absent if taste and olfactory sensations are assessed individually, rather than holistically (Frank et al., 1993). Thus, how odor and flavor memories are accessed may be as important as the sensations themselves.

The origin of flavor preferences is poorly understood, but appears to begin even before we are born (Mennella et al., 2001), based on what the mother eats during pregnancy. Likes and dislikes continue to develop during childhood, but apparently establish themselves most noticeably between the ages of 6 and 12 (Garb and Stunkard, 1974). Subsequent changes can be significant, depending largely on cultural influences and social pressures (Moncrieff, 1967). Equivalent effects affecting wine partiality are also well known (Williams et al., 1982). Of factors involved, threshold intensity appears not to be particularly important (Trant and Pangborn, 1983), but the tasting environment and personal sensitivity often play a major role.

Associated factors such as taste (Zampini et al., 2008), color intensity (Iland and Marquis, 1993; Zellner and Whitten, 1999), and other visual clues (Sakai et al., 2005) can modify perception of a wine's fragrance and flavor. A classic example is the degree to which the addition of tasteless anthocyanins to a white wine could distort its perception (Fig. 3.33). A Sauvignon blanc, colored red, and tasted along with its uncolored version, was described with terms typically ascribed to red wines, whereas the white version was described in terms typical for a white wine. Although striking, the enology students were in the habit of sampling both Sauvignon blanc and Cabernet

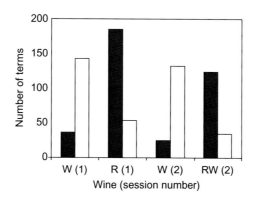

FIGURE 3.33 Frequency distribution of olfactory terms representative of red/dark objects (*solid bars*) and yellow/light objects (*open bars*) used to describe two paired tastings of a white and a red wine (*R*, Cabernet Sauvignon; *RW*, the Sémillon/Sauvignon blanc wine colored red with anthocyanins; *W*, Sémillon/Sauvignon blanc). Tasting 1 paired the unmodified white and red wines, whereas in tasting 2 the unmodified white wine was sampled (unknowingly) with its red-colored version. *Reproduced from Morrot, G., Brochet, F., and Dubourdieu, D. (2001). The color of odors. Brain Lang.* **79**, *309–320, with permission from Elsevier.*

Sauvignon. Because both cultivars possess some related aroma profiles, this probably explains part of the reason color could have such a markedly influence on term use. In a somewhat similar investigation, experienced tasters were less influenced by discordant color and flavor combinations (Parr et al., 2003). Additionally, revealing studies on the influence of extrinsic information on perception have been presented in Brochet and Morrot (1999), Lange et al. (2002), and Hall et al. (2010). Contextual effects can also arise by comparison (e.g., the appreciation of a good wine being enhanced when preceded by poorer wines, but appearing less appealing when tasted with better wines). These influences are not caused by adaptation as they occur independent of their position in a tasting.

Comments from authorities (Aqueveque, 2015), or even other tasters, have the potential to skew perception. For example, suggesting the presence of a hazelnut odor often engenders confirmation from other tasters. Whether this is due to its actual perception, illusion, or peer pressure is a moot point. Familiarity with a particular odor can increase its apparent intensity, perceived pleasantness (Distel and Hudson, 2001), and identification (Degel et al., 2001). Underlying connotations, occasionally associated with particular terms, can also influence how the odor is perceived (Herz and von Clef, 2001). For example, common descriptors (cat's piss versus passion fruit) for one of the varietal aroma characteristics of Sauvignon blanc wines conjure up quite different olfactory illusions. In addition, wine memories often incorporate episodic experiences with the odor. When the ambiance is not the same as that which generated the memory (e.g., on vacation), the impression of the wine may be very different (e.g., at home). Thus, odor memories may reflect as much, or more, the occasion than the wine.

Taster training ideally generates archetypal sensory memories sufficiently precise and inclusive to represent variations in wine varietal, stylistic, and regional expression, as well as the influences of aging. These prototypes provide models against which new samples can be compared and judged (Hughson and Boakes, 2002). In their absence, novice tasters are left to base judgment on general sensory attributes, such as sweetness, or personal hedonic impressions (Solomon, 1988).

The development and use of wine memories by professionals are supported by fMRI. For example, sommeliers show greater activity in the left insula and orbitofrontal complex than novice tasters. The principal areas activated in novice tasters were the primary gustatory cortex and regions associated with emotional processing (Castriota-Scanderbeg et al., 2005). Further evidence of preconditioning has been presented by McClure et al. (2004) concerning cola drinks. Without brand knowledge, response was similar. However, when brand identification was supplied, the response pattern changed considerably, affecting not only sensory regions, but also emotional, reward, and memory centers of the brain. Thus, brand marketing, label knowledge (Brochet and Morrot, 1999), and price (Fig. 3.34) can change perception, even when sensory clues are insufficient to permit discrimination, or opposed to previously noted preferences. The influence of extrinsic factors on sensory perception is well known. Surprisingly, the most expensive (and presumably desirable) red wines in a recent study were characterized by higher concentrations of compounds often ascribed to oxidized or faulty wines (phenylacetaldehyde, ethyl phenols, methional, (*E*)-2-alkenals, and acetic acid), whereas less expensive wines showed the highest concentration of fruity, floral, and fragrant aromas (San-Juan et al., 2014). If a common phenomenon, it does beg the question why people purchase expensive wines, as it is seemingly not for the sensory pleasure they provide. Admittedly, there are people who attest to savoring the burning pain of chili peppers (capsaicin) and horseradish (allyl isothiocyanate).

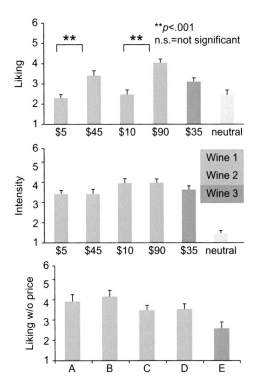

FIGURE 3.34   Response of subjects to the same wines sampled unknowingly twice, but each time noted as costing different prices (upper and middle) and again without price cues (lower). The neutral sample was a electrolyte solution similar to that of saliva. Reported pleasantness (upper) and taste intensity (middle) for the wines with monetary value denoted and reported pleasantness (lower) for the wines sampled again during a postexperimental session without price cues *(from Plassmann, H., O'Doherty, J., Shiv, B., and Rangel, A. (2008) Marketing actions can modulate neural representations of experienced pleasantness. PNAS 105, 1050–1054. Reproduced with permission, Copyright (2008) National Academy of Sciences, U.S.A.).*

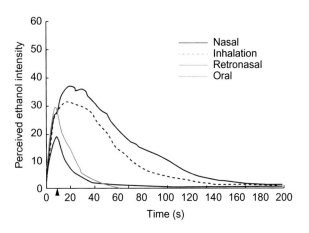

FIGURE 3.35   Average time-intensity curves for nasal (sniff), inhalation (inhale), retronasal (sip), and oral (sip, nose plugged) response for ethanol intensity of 10% v/v ethanol. *From Lee, 1989, reproduced by permission.*

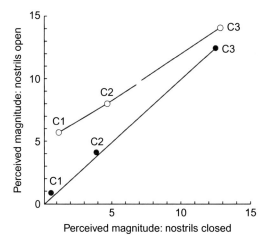

FIGURE 3.36   Perceived sweetness intensity of sodium saccharin $(0, 7.20 \times 10^{-4}, 1.68 \times 10^{-4}$ and $3.95 \times 10^{-4})$ in the presence of ethyl butyrate $(1.33 \times 10^{-3}\%$ by volume) when sipped with the nostrils either open (○) or closed (●). *Modified from Murphy et al., 1977, reproduced by permission.*

Although it is usual to think primarily of the nose in relation to olfaction, its retronasal component is equally significant (Figs. 3.35 and 3.36). However, clear differentiation between these two modes of olfactory perception can be complicated, especially when combined with gustatory and trigeminal sensations. In addition, orthonasal and retronasal responses to the same compound can differ in several aspects (Negoias et al., 2008). Not only is identification of odorants poorer retronasally, but also threshold values are generally higher (Fig. 3.37). Retronasal versus orthonasal qualities may also be distinct, due to different spatiotemporal patterns of receptor activation (Frasnelli et al., 2005), differential intensity of receptor activation, differential activation of the olfactory bulb (especially at threshold concentrations) (Furudono et al., 2013), and/or differential and localized activation in various higher centers in the brain (Small et al., 2005; Iannilli et al., 2014). For example, retronasal olfaction selectively activates

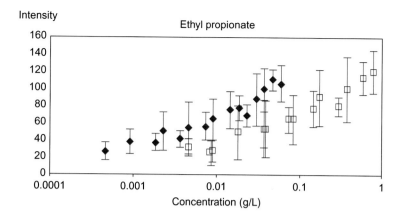

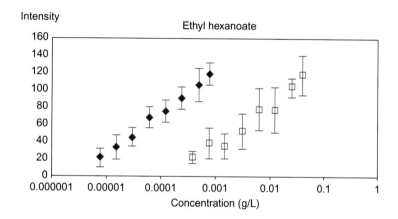

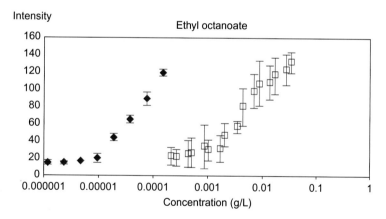

FIGURE 3.37 Comparison of orthonasal (♦) and retronasal (□) dose-response curves of a homologous series of esters. *From Diaz, 2004, Flavor Fragr. J., reproduced by permission John Wiley and Sons, copyright.*

the central sulcus, a site associated with impulses from the oral cavity. This may partially explain why the olfactory component of flavor is interpreted as coming from the mouth (in an absolute sense they do, they arise from the same site).

Other differences between orthonasal and retronasal perception, include a loss of orthonasal smell not necessarily resulting in an equivalent loss in retronasal smell (Landis et al., 2005), both seeming to be processed separately. In addition, while olfactory memory is often a combination of a compound's (or mixture's) quality, and the context of its detection, retronasal olfactory memory is usually associated with several other modalities (notably gustatory,

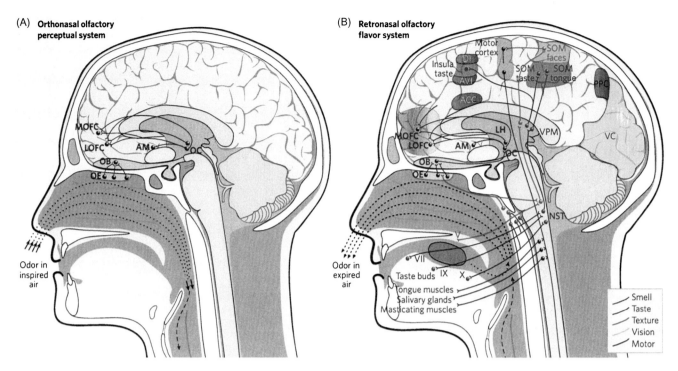

FIGURE 3.38 The dual olfactory system. (A) Brain systems involved in smell perception during orthonasal olfaction (sniffing in). (B) Brain systems involved in smell perception during retronasal olfaction (breathing out), with food in the oral cavity. Airflows indicated by dashed and dotted lines; dotted lines indicate air carrying odor molecules. *ACC*, nucleus accumbens; *AM*, amygdala; *AVI*, anterior ventral insular cortex; *DI*, dorsal insular cortex; *LH*, lateral hypothalamus; *LOFC*, lateral orbitofrontal cortex; *MOFC*, medial orbitofrontal cortex; *NST*, nucleus of the solitary tract; *OB*, olfactory bulb; *OC*, olfactory cortex; *OE*, olfactory epithelium; *PPC*, posterior parietal cortex; *SOM*, somatosensory cortex; *V, VII, IX, X*, cranial nerves; *VC*, primary visual cortex; *VPM*, ventral posteromedial thalamic nucleus. *From Shepard, 2006, reproduced by permission.*

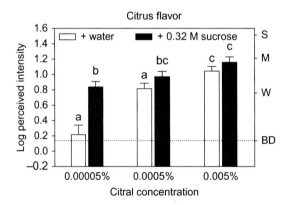

FIGURE 3.39 Influence of sucrose on the retronasal perception of citral (citrus flavor) at three concentrations: in the presence of water (white bars) or a sucrose solution (dark bars). Letters on the right *y*-axis represent semantic labels (*BD*, barely detectable; *W*, weak; *M*, moderate; *S*, strong. *Reprinted from Fujimaru, T., and Lim, J. (2013) Effects of stimulus intensity on odor enhancement by taste. Chem. Percept.* **6**, *1–7., with kind permission of Springer Science + Business.*

tactile, textural, and visual sensations, as well as context) (Fig 3.38). The various aspects of flavor may be assessed separately, with varying degrees of success, but more commonly are synthesized into a single, multisensory perception, termed flavor.

Thus, retronasal memory is as strongly allied with matrix factors and experience as is orthonasal memory. For example, sweetness can enhance retronasal detection (Green et al., 2012). This is particularly noticeable when odorant concentration is low, a situation typical with wines (Fig. 3.39). Retronasal olfaction can also modify taste perception (Murphy et al., 1977, Labbe et al., 2008). Nonetheless, color apparently affects perceived odor intensity differentially, depending on whether it is experienced nasally or orally (Koza et al., 2005).

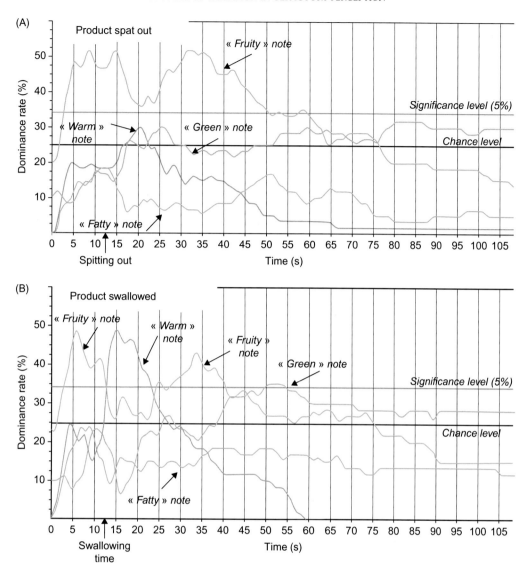

FIGURE 3.40 Temporal dominance of sensations (TDS) curves evaluated using the consumption of a diluted alcoholic beverage with: (A) spitting out or (B) swallowing the sample. The lower and upper horizontal lines indicate the chance and significance levels, respectively. Only TDS curves above the significance lever are considered to be significantly dominant. Time 0 corresponds to the moment at which the product was placed in the mouth. *Reprinted from Déléris, I., Saint-Eve, A., Guo, Y., Lieben, P., Cypriani, M.-L., Jacquet, N., Brunerie, P., and Souchon, I. (2011) Impact of swallowing on the dynamics of aroma release and perception during the consumption of alcoholic beverages. Chem. Senses **36**, 701–713., by permission of Oxford University Press.*

As might be expected, the concentration of aromatics escaping from the oral cavity via the nose is markedly lower than that emanating in the mouth from the same solution (Linforth et al., 2002). Nonetheless, dilution in the saliva did not appear to significantly affect volatility. The rate of airflow out the nose appeared to be the most important factor, followed by dilution up into the nasal passages, and adsorption by the nasal mucosa. However, decreased flow rate alone does not appear to explain the difference between orthonasal and retronasal detection (Heilmann and Hummel, 2004). Retronasal olfaction responded both more slowly and requiring higher concentrations. As already noted, this may relate to retronasal olfaction involving differential processing (Furudono et al., 2013).

Swallowing is a major factor in directing airflow from the oral cavity into the nasal passages and, therefore, the concentration of volatiles present for detection (Hodgson et al., 2005). How much this may distort the perception of a wine's aroma during critical assessment, where wine is not swallowed but expectorated, is unknown (Fig. 3.40), but is a source of concern. Exhalation generates necessary airflow, but only after partial loss via absorption in the lungs. With concentrations diminished, adaptation is presumably a less significant factor affecting retronasal detection (Cook et al., 2003).

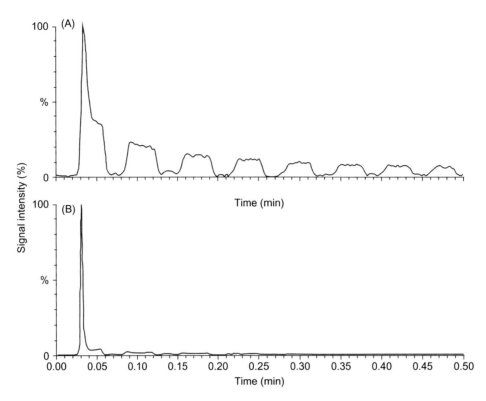

FIGURE 3.41　Breath-by-breath profiles after swallowing an aqueous solution of (A) anethole (anise-like) and (B) *p*-cymene (an essential oil found in cumin and thyme). Swallowing is indicated as 0 min, with the intensity for each compound followed at 0.5 min intervals. The high initial breath concentration falls over the course of the first exhalation to a plateau, which declines in successive breaths. *From Linforth and Taylor, 2002, reproduced by permission.*

Possibly the most significant contribution of retronasal olfaction to wine appreciation may be its involvement in a wine's finish. The vapor pressure differences between aromatics, their relative absorption in and/or destruction by saliva enzymes, rates of absorption in the pharynx and nasal mucosa, and destruction by mucosal enzymes all likely influence the perceived duration and intensity of both the wine's perceived flavor and finish. Aromatic compounds can remain in the mouth for several minutes (Wright et al., 2003). Examples of the periodicity of release and diminishing concentrations in the breath are illustrated in Figs. 3.41 and 3.42. Additional factors influencing residual odor presence include reversible associations with the wine matrix (Munõz-González et al., 2015). The physicochemical properties of nonvolatile complexes formed in the wine matrix are unknown, but may be similar to the incorporation of molecules in cyclodextrin micelles (Auzély-Velty et al., 2001). These possess hydrophobic interiors and hydrophilic exteriors. The multisensory nature of olfactory memories further complicates issues, affecting perceived intensities without changing concentrations in the nose (Hollowood et al., 2002).

ODOR ASSESSMENT IN WINE TASTING

The fragrance of a wine may be its most difficult attribute to assess, but it is also wine's most diverse, distinctive, interesting, and informative feature. This may be obscured in most tasting sheets and score cards, submerged under the term flavor. Nevertheless, it is a wine's aromatic constituents that generate most of its unique market appeal.

In this regard, most connoisseurs are fascinated with the positive, pleasure-giving aspects of a wine's fragrance, and justly so. Regrettably, little progress has been made in describing varietal, regional, stylistic, or aged attributes in a consumer-meaningful way. Fig. 1.3 provides a range of wine descriptive terms, with Appendix 5.1 listing methods for their preparation. Unfortunately, samples are a nuisance to prepare, maintain, and standardize. Even "pure" chemicals can contain contaminants that alter odor quality (Meilgaard et al., 1982). Microencapsulated

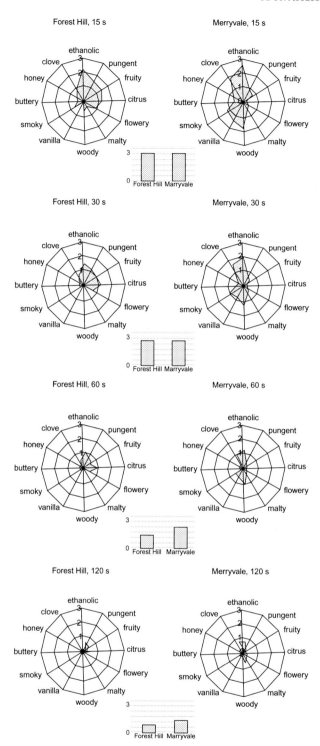

FIGURE 3.42 Time-resolved retronasal evaluation of the intensities of odor attributes and their overall odor intensities (middle graph) after intraoral application and expectoration of two Chardonnay wines. *From Buettner, 2004, reproduced by permission.*

("scratch-and-sniff") strips would be ideal and efficient, but are commercially unavailable. Wine aroma kits may be convenient in representing terms for wine appreciation courses, but too often may imply that the objective of appreciation is odor identification (or at least naming). This deflects attention from the legitimate goal of enjoying wine and developing sensory memories for various varietal and stylistic types of wines. These features are memorized as holistic entities, just as are facial types. Attempting to concentrate on individual, presumed odor resemblances may be counterproductive, inducing depreciation (Dijksterhuis et al., 2006). In critical sensory analyses, only a few descriptors are usually necessary to differentiate among samples of similar wines (Lawless, 1984b).

Two categories of wine fragrance are generally recognized, their differentiation being based on origin. Aroma traditionally refers to fragrances, or their precursors, derived from grapes. Although usually applied to compounds giving grape varieties their distinctive aromatic character, aroma can also include aromatics derived as a result of autofermentation (carbonic maceration), overripening, partial dehydration (*appassimento*), raisining, sunburning, and infection. Currently, there is no evidence supporting the belief that grapes derive specific flavors from the soil. However, the discovery of polysulfances, possessing flint-like odors (Starkenmann et al., 2016), indicates that aromatics with desirable *goût de terroir* and mineral attributes may still be found.

The other category, bouquet, refers to scents that develop during fermentation, processing, and aging. Fermentative bouquets include yeast (alcoholic) and bacterial (malolactic)-derived aromatic compounds. Processing bouquets refer to flavorants derived from procedures such as the addition of distilled wine (port), baking (madeira), *flor* action (sherry), yeast autolysis (sparkling wines), maturation, in either inert or oak cooperage, and *sur lies* maturation. Aging bouquets refer to fragrant elements that develop during in-bottle aging.

Although the use of aroma and bouquet is common, their distinction in practice is often impossible. Similar or identical compounds may be derived from grape, yeast, or bacterial metabolism, or be generated by abiotic reactions. Even where differences exist, it may take considerable experience to make the distinction.

OFF-ODORS

The chemical nature of wine quality may still be illusive, but this is not the case with most wine faults. It is the equivalent of recognizing a violin played off-key. In contrast, explaining precisely why a Stradivarius violin sounds sublime is not easy, maybe impossible. The following section summarizes the characteristics of several important wine off-odors, and Appendix 5.2 gives directions for preparing faulty samples—for training purposes.

Quick and accurate identification of off-odors is necessary for the winemaker and wine merchant alike. For the winemaker, early remedial action can often correct the problem, before the fault becomes more serious or intractable. For the wine merchant, avoiding losses associated with faulty wines improves the profit margin. Consumers should also know more about wine faults, but use the knowledge wisely and with discretion. Rejection should be based on genuine faults, rather than unfamiliarity, inappropriate expectations, the presence of bitartrate crystals, dryness (occasionally incorrectly termed "vinegary"), or worst of all, to suggest connoisseurship in a restaurant.

No precise definition of what constitutes a wine fault exists; human perception is too variable. Wine faults are like grammatical errors, that is, designated as such by consensus. Nevertheless, off-odors tend to have a common property. They tend to mask the wine's fragrance. Examples include the suppression induced by trichloroanisole (TCA) (Takeuchi et al., 2013) and *Brett* taints (Licker et al., 1999; Botha, 2010). Nonetheless, some compounds considered faults at above detection thresholds may be desirable at or near their detection thresholds. For example, dimethyl sulfide (DMS) may donate off-odors, and mask other fragrances at concentrations significantly above its threshold (Spedding et al., 1982; Segurel et al., 2004). By contrast, DMS may augment the fruitiness of red wines at close to its threshold, and is considered to contribute to truffle and black olive notes in an aged bouquet. Other wine aromatics known to potentially mask odors are benzaldehyde, benzyl acetate, β-damascone, geraniol, linalool, and nonenal (Takeuchi et al., 2009). One off-odor can even suppress the perception of another, e.g., acetic acid on ethyl phenol (Wedral, 2007). Thus, depending on relative concentration and personal sensitivity, a wine that can be appreciated by one person can be objectionable to another. In addition, when aromatically subtle, they may be an essential and a traditional attribute in certain wines. Examples are the oxidized bouquets of oloroso sherries and the fusel nuances in ports. In other instances, off-odors such as barnyardy (usually ascribed to ethyl phenols) may be considered part of the *goût de terroir* of certain wines, and as such, highly appreciated (apparently), or at low doses, the source of the leathery aspect of aged red wines (Mahaney et al., 1998). Even aspects considered normally desirable may be considered a fault to some, e.g., the so-called overoaked aspect in some wines. It is the equivalent of whether there is too much, too little, or just enough salt, pepper, or spice in a recipe. Prestige of the wine may also delude some connoisseurs into considering that a fault cannot exist (or be a quality feature), e.g., the noticeable ethyl acetate character in several sauternes, or a *Brett* character is some expensive wines.

Acetic Acid (Volatile Acidity)

Accumulation of acetic acid to detectable levels usually results from the action of acetic acid bacteria. This may occur in the vineyard (secondary infection of damaged or diseased grapes), in the winery, or after bottling (both due to excessive oxygen uptake). Wines that are actually "vinegary" are sharply acidic, with an irritating odor derived from the combined effects of acetic acid and ethyl acetate. Acetic acid concentrations in wine should not exceed 0.7 g per liter. People unfamiliar with wine may mistake the bright acidic note of some dry table wines as being vinegary.

Baked

Fortified wines, such as madeira, are purposely heated to over 45°C for several weeks. Under such conditions, the wine develops a distinctive, baked, caramelized odor. Although characteristic and expected in wines such as madeira, a baked aspect is a negative feature in table wines. If found, it usually indicates exposure to high temperatures during transit or storage. A baked essence can develop within a few weeks at above 35°C. It involves a series of thermal degradation reactions, often involving sugars (notably fructose) and amino acids. Reactions of this type generate what are termed Maillard products.

Buttery

Diacetyl (biacetyl, butanedione) is usually found in wine at low concentrations as a result of yeast and/or bacterial metabolism. It may also be an oxidative byproduct of maturation in oak. Although typically considered to possess a butter-like odor, diacetyl is a negative quality feature at significantly above its recognition threshold. For some individuals (the author included), other constituents accompanying diacetyl give it a vile, crushed-earthworm attribute. Despite diacetyl typically being considered to have a buttery character, wine possessing this attribute may have concentrations below its detection threshold (Bartowsky et al., 2002). Thus, other compounds, such as acetoin and 2,3-pentanedione, may play a supplemental role in contributing to the buttery attribute.

Corky/Moldy

Wines may show a range of corky, musty, moldy odors. One of the most common culprits is 2,4,6-trichloroanisole (TCA). It usually develops as a consequence of fungal growth on or in cork. The fungi are thought to metabolize a pesticide, pentachlorophenol (PCP), to TCA. The pesticide was formerly used as an insecticide on cork trees and other wood products. Alternatively, TCA can be derived from microbial metabolism of chlorine used to surface sterilize or whiten stoppers.

TCA produces a distinctive, musty, chlorophenol odor at a few parts per trillion, depending on the wine contaminated (Mazzoleni and Maggi, 2007). Possibly of more significance, currently, is the ability of TCA to suppress olfactory response at concentrations much lower than those that generate a recognizable musty odor (Takeuchi et al., 2013). It suppresses the activation of one of the ion channels (CHG) required for receptor neuron activation. Integration of TCA into the membrane of receptor neurons seems to explain their slow recovery. Thus, TCA may have more widespread negative (masking) effects on wine fragrance than previously thought. Fig. 3.43 compares the relative activity of several phenolics (including TCA) in suppressing olfactory receptor response.

Other corky/moldy off-odors may come from the presence of a related compound, 2,4,6-tribromoanisole (TBA) (Chatonnet et al., 2004), or 2-methoxy-3,5-dimethylpyrazine (Simpson et al., 2004). Musty off-odors in cork may also originate from sesquiterpenes produced by filamentous bacteria, such as *Streptomyces*. Additional moldy, earthy odors can develop from the production of guaiacol and geosmin by fungi, notably *Penicillium* and *Aspergillus* spp. growing on cork or moldy grapes. Although most moldy (corky) taints come from cork, oak cooperage can also be a source of similar off-odors, e.g., TBA.

Ethyl Acetate

Wines spoiled by the presence of ethyl acetate are apparently far less common than in the past (Sudraud, 1978). At concentrations below 50 mg/L, ethyl acetate may add a subtle, agreeable fragrance. However, above about 100 mg/L, it begins to have a negative influence, possibly due to suppressing of the fragrance of other aromatic compounds such as fruit esters (Piggott and Findlay, 1984). At above 150 mg/L, ethyl acetate generates an obvious acetone- (Cutex-) like off-odor. Threshold values vary with the type and intensity of the wine's fragrance. Correspondingly, it is more readily apparent in white than red wines. Spoilage microbes are the most common source of ethyl acetate off-odors. Nonetheless, it can accumulate abiotically from the esterification of ethanol and acetic acid.

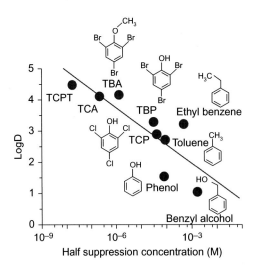

FIGURE 3.43   Comparison of the suppression olfactory receptor cell activation by several phenolic compounds: TCPT (trichlorophenetole, a synthesized TCA derivative), TCA (trichloroanisole), TBA (tribromoanisole), TCP (trichlorophenol), TBP (tribromophenol). *From Takeuchi, H., Kato, H., and Kurahashi, T. (2013) 2,4,6-Trichloroanisole is a potent suppressor of olfactory signal transduction. PNAS 110, 16235–16240, reproduced with permission of the National Academy of Sciences, USA.*

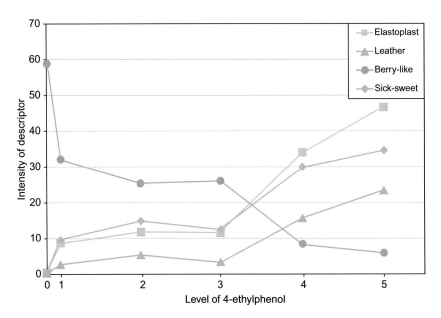

FIGURE 3.44   Change in sensory profile of Pinotage wine due to the addition of 4-ethylphenol. 0 (0 μg/L); 1 (82 μg/L); 2 (227 μg/L); 3 (623 μg/L); 4 (1711 μg/L); 5 (4695 μg/L). *From Botha, 2010, reproduced with permission.*

Ethyl Phenols

Ethyl phenols (4-ethyl phenol, 4-ethylguaiacol, and 4-ethylcatechol) can donate a variety of off-odors, depending primarily on their concentrations. These can vary from smoky, spicy, phenolic, to medicinal, at near-threshold values, to sweaty, leather, horse, barnyardy, or manure-like at higher concentrations. The expression of these attributes can be partially masked by other wine constituents, for example oak and pyrazines (Bramley et al., 2008), or acetic acid (Wedral, 2007). Conversely, 4-ethyl phenol can suppress the berry-like fragrances of wine and generate a sickly-sweet character (Fig. 3.44). Despite being considered to generate off-odors, there is considerable diversity (contention) over subjective responses to *Brett* odors. For example, some winemakers consider these compounds donate a desired distinctiveness to their wines, and seemingly their presence is acceptable to some consumers. Ethyl phenols have even been suggested to be the source of the leathery character of aged wines (presumably in small amounts).

In contrast, most professionals and consumers find affected wines vary from lacking in fruitiness to unacceptable (Lattey et al., 2010). This diversity of opinion may arise from a combination of differences in detection threshold and wine experience (Tempère et al., 2014), the action of ethyl phenols at (to most people) subliminal concentrations (Tempere et al., 2016), and the simultaneous production of additional unpleasant and masking volatiles (e.g., isovaleric acid and tetrahydropyridines).

Ethyl phenols are metabolic byproducts of the yeast *Brettanomyces bruxellensis*. Different strains of *Brettanomyces* vary in their ability to first oxidize hydroxycinnamic acids to vinyl phenols, and subsequently ethyl phenols. The yeast can contaminate wine from various sources, but most often from improperly cleansed oak cooperage. Because more red wines receive oak aging than white wines, red wines tend to be the wines most affected by *Brett* odors. Alternative sources of contamination include passage through contaminated transfer tubing and connections. The yeast can grow on even minute remnants of wine.

Fusel

During fermentation, yeasts produce limited amounts of higher (fusel) alcohols. At concentrations close to their detection thresholds, they may add to a wine's aromatic complexity. If these alcohols accumulate to levels greater than 300 mg/L, they are typically considered a negative quality factor. Nevertheless, the presence of a fusel note is characteristic and expected in ports (*porto*). It comes from the concentration of fusel alcohols during distillation in the production of the largely unrectified wine spirits used to fortify ports.

Geranium-Like

A geranium-like off-odor arose from the use of sorbate as an antimicrobial agent. Metabolism of sorbate by certain lactic acid bacteria converted it to 2-ethoxyhexa-3,5-diene. This compound has a sharp, penetrating odor resembling scented geranium. As a consequence, the commercial use of sorbate to control unwanted yeast growth has ceased.

Light-Struck

Light-struck (*goût de lumière*) refers to a series of off-odors that can develop in wine exposed to light. In champagne, it apparently results from the generation of dimethyl sulfide, methyl mercaptan, and dimethyl disulfide (Charpentier and Maujean, 1981). However, D'Auria et al. (2002) failed to confirm this, finding instead an increased presence of 2-methylpropanol (a fusel alcohol), and a significant modification and reduction in the presence of fruit esters.

Mousy

Several lactic acid bacteria and *Brettanomyces* yeasts can generate mousy taints. The odor is caused by the synthesis of several tetrahydropyridines and related compounds. Because they are not readily volatile at wine pH values, their presence is seldom detected on smelling the wine. Their odors become evident only on tasting. Winemakers often put a small amount of wine on their hands and use the "palm-and-sniff" technique to quickly detect a mousy taint. Detection of this taint can vary by 2 orders of magnitude (Grbin et al., 1996).

Oxidation

Presence of an obviously oxidized aspect is now comparatively rare in commercial table wine. Oxidation produces a "flat" sensation, a range of ill-defined odors, and a loss (or masking) of desirable flavors typical of the wine. Off-odor descriptors may include terms such as cardboard, cooked vegetables, boiled potato, hay, and woody. As per usual, especially with something potentially as complex as oxidation, its olfactory expression is diverse and attempts to describe it in words are difficult. For example, premature oxidation in red wines has been associated with a prune-like character, due to the production of several lactones, notably 3-methyl-2,4-nonanedione (Pons et al., 2008). Table 3.4 lists compounds that have been associated with oxidized wine, their detection thresholds, and some typical descriptors. These off-odors are often associated with browning in white wines, and the premature development of a brickish hue in red wines. Nevertheless, short-term exposure to air is not correlated with the detection or accumulation of acetaldehyde (Escudero et al., 2002). This apparent anomaly probably results from the

TABLE 3.4 Compounds Associated with Oxidized Odors, their Descriptors, and Olfactory Thresholds

Compound	Odor descriptor	Odor threshold (µg/L)
(E)-2-ALKENALS		
(E)-2-heptenal	Soapy, fatty	4.6[a,9]
(E)-2-hexenal	Green apple	4[a,9]
(E)-2-nonenal	Green, fatty, sawdust	0.17[a,14]
(E)-2-octenal	Fatty, nutty	3[a,9]
STRECKER ALDEHYDES		
Methional	Cooked potato-like	0.5[b,7]
2-methylpropanal	Malty	6[b,9]
3-methylbutanal	Malty	4.6[b,9]
2-phenylacetaldehyde	honey, floral	1[b,9]
FURANS		
Furaneol	Caramel	37[b,15]
Homofuraneol	Caramel	10[b,15]
Sotolon	Curry, seasoning	15[b,12]
ALDEHYES		
Benzaldehyde	Bitter almond-like	2000[a,16]
Furfural	Sweet, bread	14 100[b,17]
Hexanal	Grassy, green	20[a,16]
5-methylfurfural	Sweet, bitter almond	2000[a,16]
ALCOHOLS		
Maltol	Caramel	5000[b,18]
Methionol	Cooked potato-like	1000[b,17]
Eugenol	Clove-like	6[b,17]

Reprinted with permission from Mayr, C. M., Capone, D. L., Pardon, K. H., Black, C. A., Pomeroy, D., and Francis, I. L. (2015) Quantitative analysis by GC-MS/MS of 18 aroma compounds related to ocidative off-flavor in wines. J. Agric. Food Chem. 63, 3394–3401. Copyright 2015, American Chemical Society.
[a]*Odor detection threshold determined in water.*
[b]*Odor detection threshold determined in model wine (10 percent aqueous ethanol, 7 g/L glycerol, pH 3.2).*

low concentration of *o*-diphenols in white wines (acetaldehyde being an indirect byproduct of phenol oxidation), the rapidity with which acetaldehyde can form nonvolatile complexes, and the short duration of oxygen exposure in the experiment (1 week).

Sherries possess a complex oxidized odor, due primarily to the slow accumulation of acetaldehyde and branched (methyl) aldehydes (Culleré et al., 2007). The latter can possess dried fruit to orange-like flavors. In contrast, oxidized table wines possess few branched aldehydes, but significant amounts of unbranched aldehydes such as methional, 2-phenylacetaldehyde, 2-nonenal, and 2-octenal (Culleré et al., 2007). In other studies, concentrations of methional were distinctly associated with oxidized white wines (Escudero et al., 2000), as well as methional, 2-phenylacetaldehyde, 3-methylbutanal, and 2-hexenal (Mayr et al., 2015). Several of these compounds are known to be able to mask a wine's fragrance. By contrast, unoxidized aged wines contained both branched and unbranched aldehydes. Branched aldehydes appear to suppress the perception of the undesirable, unbranched forms (Culleré et al., 2007).

Premature oxidation in table wine appears to be correlated with the use of faulty corks, or their improper insertion (causing creases that impede a tight seal with the bottle neck). In addition, cork stoppers may contain sufficient oxygen to induce detectable oxidative browning in sensitive white wine (Caloghiris et al., 1997). Oxygen can also seep into wine through the cork, or probably more frequently between it and the neck. Because of the comparative oxygen permeability of older types of synthetic corks, they were often associated with oxidative changes. Rapid temperature changes (and associated changes in pressure in the bottle) can loosen the seal of the cork, favoring oxygen ingress. Leaving a bottle upright for an extended period can result in cork shrinkage (due to its drying), followed by oxygen gaining access to the wine. Nevertheless, even under seemingly proper storage conditions, white wines may begin to show signs of browning within 4–5 years.

The most well-known reaction associated with wine oxidation involves the production of hydrogen peroxide, from the interaction of oxygen and wine phenolics. Hydrogen peroxide can subsequently react with other wine constituents, such as alcohols, fatty acids, esters, and terpenes. Because ethanol is the reactant occurring in the highest concentration, the principal byproduct of wine oxidation is acetaldehyde. Because it rapidly binds with other wine constituents, the free (volatile) acetaldehyde content remains below its detection threshold. Only after extensive oxidation does acetaldehyde accumulate to the point where it could influence a wine's fragrance.

As noted, oxidative reactions with other wine constituents progressively degrade the wine's fragrance. Oxidation and polymerization of wine phenolics lead to the formation of brown pigments. Although browning occurs in both red and white wines, it becomes apparent earlier in white wines, due to their paler color. White wines, although possessing fewer phenolic compounds than red wines, are less likely to generate H_2O_2, thereby limiting other forms of oxidation. That browning can affect the acceptance ratings of white wines in laboratory tests is clear. Whether it is as important factor to consumers is less certain.

In contrast to the slow oxidative changes in bottled wine, wine packaged in bag-in-box containers often oxidize noticeably within a year. This appears to result from the infiltration of oxygen around or through the spigot. New advancements in spigot design and manufacture may limit this deficiency, in an otherwise consumer-friendly package.

Oxidation problems can occur at any stage in wine production. However, what happens shortly after bottle opening is technically not oxidation, although this is what it is called in the popular press. During the normal time frame over which wine is consumed, there appears to be no sensorially detectable wine oxidation (Singleton, 1987; Russell et al., 2005). Detectable chemical changes on exposure to oxygen take several days to express themselves. They may be detected as an accumulation of ethyl acetate (possessing a nail-polish-remover-like odor), and the loss of compounds, such as phenylethyl alcohol (rose), ethyl hexanoate (fruity), phenyl acetaldehyde (rose), and nonanoic acid (rancid) (Lee et al., 2011). Oxygen uptake by wine can be highly variable (Lee et al., 2011). Over extended periods, oxidation of aromatic compounds such as terpenes occurs, combined with the development of animal, dairy, and bitter attributes (Roussis et al., 2005).

Nonetheless, if the wine is not fully consumed within a few hours, it often begins to lose its original fragrance. What appears to be part of the cause is a progressive loss of the fruity character donated by ethyl and acetate esters (Roussis et al., 2005). In addition, there is the incremental escape of aromatics from the wine, significantly depauperating the wine's store of aromatic compounds. For example, the headspace/wine contact in a 750-ml bottle at opening is about $0.4 cm^2/100 ml$, becoming $7.5 cm^2/100 ml$ (about a 20-fold increase) after half the contents have been removed. As soon as the bottle is opened, the equilibrium between aromatics in the headspace, dissolved in the wine, and weakly associated with matrix constituents, shifts, favoring release. While initially desirable, and promoted by swirling wine in the glass, this eventually has undesirable consequences in a partially filled bottle. Because most aromatics occur in wine in trace amounts, their escape into the surrounding air impoverishes their presence in the wine. This likely explains why wine poured from a partially emptied bottle after several to many hours appears to have lost part of its original character. However, smelling the neck of the bottle shows that much of that original character (fragrance) is found in the headspace gases of the bottle. Because low temperatures slow this process, storing the wine at cold (refrigerator) temperatures can delay fragrance loss. Alternatively, any leftover wine (or wine predicted not to be consumed shortly after opening) can be decanted into small, easily sealed bottles. This is preferable to attempting to evacuate the air in a partially empty bottle. Volatiles can still escape into the headspace above the wine.

Although oxygen uptake begins slowly on contact with air, molecular oxygen is rarely directly involved in oxidizing wine. It is oxygen's conversion to peroxide or free radicals (notably superoxide and hydroxyl) whereby oxygen begins to react with and oxidize organic compounds. Peroxide is produced as a consequence of the oxidation of *o*-diphenols, whereas free radicals can be generated in the presence of metal catalysts (such as iron and copper), or riboflavin (or its protein complexes) on exposure to light.

Reduced-Sulfur Odors

Hydrogen sulfide and several reduced-sulfur organics may be produced during wine fermentation and maturation. The rotten-egg odor of hydrogen sulfide, if noticeable, usually dissipates quickly during swirling in the glass.

Unfortunately, this does not eliminate off-odors generated by most volatile organosulfur compounds. Mercaptans, which impart off-odors reminiscent of farmyard manure, rotten onions, etc., only oxidize slowly to disulfides. Disulfides also possess unpleasant, cooked-cabbage to shrimp-like odors, but at thresholds about 30 times higher than their corresponding mercaptans. Many other related compounds, such as 2-mercaptoethanol and 4-(methyl-thio)butanol, produce intense barnyardy and chive–garlic odors, respectively. These compounds may be produced by spoilage microbes, but more commonly form abiotically in lees under highly reducing conditions.

Sulfur Dioxide

Sulfur dioxide is typically added at one or more points during winemaking, maturation, or at bottling (white wines). It is also generated by yeast metabolism. Nevertheless, a reduction in sulfur dioxide use has meant that wines are unlikely to show a burnt-match odor, typically expressed at above threshold concentrations. Even at above threshold levels, its odor usually dissipates quickly, rapidly escaping into the air during swirling.

Atypical Aging Flavor (*Untypischen Alterungsnote*, UTA)

An atypical aging (ATA) [*untypischen Alterungsnote*] (UTA) flavor possesses naphthalene (mothball), furniture varnish, wet wool, or related odor(s). It has been proposed to be due to 2-aminoacetophenone (AAP) (Rapp et al., 1993), a byproduct of indole acetic acid (IAA) metabolism by yeasts. ATA tends to develop several months (to years) after bottling, typically in white wines, and appears to be correlated with the presence of unfavorable growing conditions, notably water deficit and/or a shortage of nitrogen. The fault is also characterized by a lack of varietal aroma, possibly due to the masking aspect of the off-odor. The appearance of a similar phenomenon in New York State has been correlated with the presence of 1,1,6-trimethyl-1,2-dihydronaphthalene (TDN), and the loss of volatile terpenes (Henick-Kling et al., 2005). The accumulation of TDN, possessing a kerosene-like odor, can also develop during white wine aging. As a "fault" on its own, TDN's rejection threshold can be 4–10 times that of its detection threshold (Ross et al., 2014). In some countries, ATA appears to be associated with exposure to high-intensity sun-light and warm growing conditions (Rapp, 1998). Whether these atypical aging phenomena are various expressions of the same physiological problem or distinct disorders remains to be clarified (Sponholz and Hühn, 1996). A full understanding of atypical aging is complicated by the poor correlation between 2-aminoacetophenone content and fault expression, and the incomplete development of sensory characteristics of ATA upon adding AAP (Schneider, 2014). An unrelated source of a naphthalene off-odor may come from naphthalene derived from winery equipment, or absorbed by corks during storage prior to use.

Vegetative Odors

Several herbaceous off-odors are recognized. The best understood is that associated with the presence of "leaf" (C_6) aldehydes and alcohols. Additional sources of vegetative to herbaceous odors come from the presence of several methoxypyrazines, notably those that characterize many Cabernet Sauvignon and Sauvignon blanc wines. They can donate an overpowering green bell pepper, green bean odor.

Other Off-Odors

Additional recognized off-odors include those produced by butanoic acid (spoiled butter) and propanoic acid (goaty), the raisined aspect derived from sun-dried grapes, the cooked flavor of wines fermented at high tempera-tures, the stemmy feature produced by the presence of green grape stems during fermentation, and the *rancio* char-acter of old oxidized red wines. *Rancio* is apparently associated with 3-methyl-2,4-nonanedione (Pons et al., 2013). Off-odors of unknown identities or origins are rubbery (likely a thiol), weedy (possibly a synonym for vegetal), and earthy (perhaps one or more sesquiterpenes).

Wines tainted with a smoke flavor in Australia gain this attribute from grapes exposed to volatile phenols from fires, notably guaiacol, 4-methylguaiacol, syringol, 4-methylsyringol *o-*, *m-*, *p*-cresol, and their glycoconjugates. Enzymes in the saliva release additional volatile versions in the mouth by hydrolyzing nonvolatile glycoconjugates of these compounds (Mayr et al., 2014). Another aromatic taint, derived from a vineyard environment is eucalyptol (1,8-cineole), from adjacent eucalyptus groves (Capone et al., 2011).

CHEMICAL NATURE OF VARIETAL AROMAS

The presence of a distinctive varietal aroma is one of a wine's premium quality features. Unfortunately, distinctive varietal attributes are not consistently expressed, even by famous cultivars (notably Pinot noir). Varietal expression can vary with the clone, cultivation, climatic conditions, as well as the production skills of the winemaker and subsequent storage conditions.

For several cultivars, specific aromatic attributes tend to epitomize their varietal character (Tables 7.2 and 7.3). In a few cases, these can be correlated with one or a few impact compounds. With other cultivars, the varietal aroma appears to be associated with the relative concentrations of several volatile compounds. In most instances, though, no particular compound, or group of compounds, appears to have singular importance. This may be because there are no unique impact compounds, or they have yet to be discovered. Based on the recent discovery of the importance of thiol compounds and sesquiterpenes in several cultivars, which occur in ng amounts, many crucial impact compounds likely await discovery.

Determining the chemical nature of a varietal aroma is fraught with difficulty. Not only must the compound(s) exist in forms unmodified by current extraction techniques, but they must also occur in concentrations that make isolation and identification possible. Isolation is easier when the varietal compound(s) exist in volatile forms in both fresh grapes and the wine. Unfortunately, aroma compounds often exist in grapes primarily as nonvolatile conjugates (Fig. 3.45). Their volatile constituents may be released only on crushing, through yeast activity, or during aging.

Even with the highly sophisticated analytical tools presently available, great difficulty can be encountered in detecting certain groups (e.g., aldehydes bound to sulfur dioxide). The situation is even more demanding when impact compounds are labile, or occur in trace amounts.

However, presence does not necessarily, or often, signify sensory significance. It is estimated that less than 5 percent of the volatile compounds in food are of sensory significance (Grosch, 2001). The situation with wine appears similar. Determining sensory impact requires detailed analysis, involving its concentration, how that relates to its threshold, and direct confirmation of its role in model wines, either by addition or omission. Because compounds can jointly activate similar receptors in the nose, their effects may be synergistic, or combine to produce a qualitative response distinct from that generated by its constituents.

For comparative purposes, volatile ingredients are often grouped into subjective categories, termed "impact," "contributing," or "insignificant." Impact compounds elicit distinctive fragrances, and are varietally or varietal-group distinctive, the equivalent of Cyrano's nose. Contributing compounds add complexity, but are not in themselves varietally unique. Examples are the acetate esters of higher alcohols that contribute to the fruity odor to young wines (Ferreira et al., 1995; Ferreira, 2010). Most of these are the byproducts of yeast and bacterial metabolism, or develop during maturation and/or aging. Thus, they are not inherently varietally distinctive, but their production can be influenced by a variety's distinctive nutrient composition. Insignificant compounds constitute the vast majority of volatile constituents. They occur at concentrations insufficient to significantly alter a wine's aroma or bouquet, directly or indirectly.

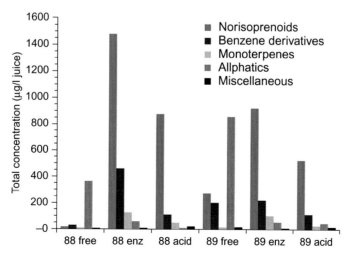

FIGURE 3.45 Concentration of five categories of volatiles, observed as free compounds (*free*) or after release by either glycosidase enzyme (*enz*) or acid hydrolysis (*acid*) of precursor fractions from the 1988 and 1989 Chardonnay juices. *From Sefton et al., 1993, reproduced by permission.*

Most grape varieties are not known to possess a distinctive varietal aroma. This may result from few varieties having been studied sufficiently to know whether they possess a distinctive aroma. For example, were the fame of Pinot noir not already established, few growers or winemakers would spend the time and effort to occasionally produce a wine that shows the cultivar's potential. Nevertheless, there is a growing list of cultivars in which impact compounds have been found. The following summarizes some of what is presently known.

Several *Vitis labrusca* cultivars show a foxy character. This has been variously ascribed to ethyl 3-mercaptopropionate (Kolor, 1983), N-(N-hydroxy-N-methyl-γ-aminobutyryl)glycine (Boison and Tomlinson, 1988), or 2-aminoacetophenone (Acree et al., 1990). Although aminoacetophenone has also been isolated from several *V. vinifera* wines, it can also be synthesized by certain strains of *S. cerevisiae* (Ciolfi et al., 1995), and various members of the grape epiphytic flora (Sponholz and Hühn, 1996). Other *V. labrusca* cultivars possess a strawberry odor, probably induced by furaneol (2,5-dimethyl-4-hydroxy-2,3-dihydro-3-furanone) and its methoxy derivative (Schreier and Paroschy, 1981), or methyl anthranilate and β-damascenone (Acree et al., 1981). Methyl anthranilate has become so associated with grape juice in North America that it is considered to epitomize the odor of grapes.

The bell-pepper (*Capsicum*) character of Cabernet Sauvignon (Boison and Tomlinson, 1990) and Sauvignon blanc (Lacey et al., 1991) wines is primarily due to the presence of 2-methoxy-3-isobutylpyrazine. Isopropyl and *sec*-butyl methoxypyrazines are also present, but at lower concentrations. Methoxypyrazines are not restricted to the Cabernet and related varieties, but often accumulate to sensory-significant amounts in those cultivars. The source of the desirable blackcurrant fragrance of some Cabernet Sauvignon wines is uncertain, but may be related to the presence of β-damascenone and eugenol, and/or the absence of fruit esters such as ethyl butanoate and ethyl octanoate (Guth and Sies, 2002). 4-Mercapto-4-methyl-pentan-2-one (4MMP) is also reported to have a blackcurrant, box-tree fragrance at low concentrations. It occurs in several French cultivars (Rigou et al., 2014) and at concentrations apparently sufficient to generate a blackcurrant aroma. Related thiols, 3-(mercapto)hexyl acetate (3MHA) and 3-mercapto-1-hexanol (3MH), may act as aroma enhancers. In wines rich in these thiols, the blackcurrant aroma may mask other aromatics in the wine.

Another source of a peppery aspect, but this time black pepper (*Piper nigrum*), is the sesquiterpene, rotundone (Wood et al., 2008). It donates the peppery aroma of Shiraz wines. Rotundone may also contribute to a peppery character of several other French (Mourvèdre and Durif), Italian (Schioppettino and Vespolina) and Austrian cultivars (Grüner Veltliner) (Mattivi et al., 2011). Rotundone can be detected at a few parts per trillion (ng/L). The association of rotundone with the aroma of black pepper is not unexpected, as it occurs along with piperine in the fruit of *Piper nigrum*, but at much higher concentrations. As usual, there is considerable variation in human sensitivity to rotundone, with about 20 percent of those tested being essentially anosmic to the compound.

Several complex thiols have been implicated as crucial impact compounds in several varietal aromas. The first group of cultivars in which their significance was discovered was the Carmenets (Cabernet Sauvignon, Merlot, Sauvignon blanc, etc.) (Bouchilloux et al., 1998; Tominaga et al., 1998, 2004). Examples are 4-mercapto-4-methylpentan-2-one, 3-mercaptohexyl acetate, 4-mercapto-4-methylpentan-2-ol, and 3-mercaptohexan-1-ol. The first two are considered to donate a box-tree odor, whereas the last two may give citrus zest–grapefruit and passion fruit–grapefruit essences, respectively. 4-Mercapto-4-methylpentan-2-one has also been found to be central to the aroma of Scheurebe (Guth, 1997b), Colombard (du Plessis and Augustyn, 1981), and Macabeo (Escudero et al., 2004), and the blackcurrant aroma noted above. These compounds tend to accumulate in grapes as nonvolatile, cysteinylated precursors, being released through the action of yeast enzymes during or after fermentation. Several of these thiols are also important in the flavor of several rosé wines (Murat, 2005).

The intensely aromatic character of Gewürztraminer has been associated with the presence of 4-vinyl guaiacol, along with several terpenes (Versini, 1985), and more recently with *cis*-rose oxide, several lactones, and esters (Guth, 1997b; Ong and Acree, 1999). Interestingly, the same compounds that give litchi nuts their distinctive aroma are found in Gewürztraminer wines, giving credence to the presence of a litchi-like fragrance.

Muscat varieties, which may also be designated as Moscato and Moscatel cultivars, are distinguished by the prominence of monoterpene alcohols and C_{13}-norisoprenoids in their varietal aromas. Similar monoterpene alcohols are important, but occur at lower concentrations in Riesling and related cultivars. The relative and absolute concentrations of these compounds, and their respective sensory thresholds, distinguish the varieties in each group (Rapp, 1998). Thiols may also contribute significantly to their varietal distinctiveness (Tominaga et al., 2000).

Occasionally, impact compounds are fermentation byproducts, e.g., 2-phenylethanol. It is also an aroma compound central to the distinctive fragrance of muscadine wines (Lamikanra et al., 1996). Another likely example is isoamyl acetate, a distinctive flavorant of Pinotage wines (van Wyk et al., 1979).

Several well-known, aromatically distinctive cultivars, such as Chardonnay (Lorrain et al., 2006) and Pinot noir (Fang and Qian, 2005), appear to possess no distinctive impact compounds. Their varietal distinctiveness may arise from quantitative, rather than qualitative, aromatic attributes (Le Fur et al., 1996; Ferreira et al., 1998). Examples are β-damascenone, a prominent component in the aroma profile of Chardonnay and Riesling wines (Simpson and Miller, 1984; Strauss et al., 1987a), β-ionone, typical of Muscat wines (Etiévant et al., 1983), and α-ionone and benzaldehyde, characteristic of Pinot noir and Gamay wines (Dubois, 1983). Particular combinations have been considered to generate varietal characteristics. For example, Moio and Etiévant (1995) suggest that four esters—ethyl anthranilate, ethyl cinnamate, 2,3-dihydrocinnamate, and methyl anthranilate—donate the central varietal flavor to Pinot noir wines, whereas particular combinations of terpenes, lactones and 1-octen-3-ol characterize sweet Fiano wine (Genovese et al., 2007).

Although grapes are the principal source of impact compounds, even if they may be released by yeast action or during maturation, there are other sources of important wine flavorants. Yeast-derived esters and volatile phenolics have already been mentioned. Another, but little-recognized, source of impact compounds are derived from oak cooperage. For example, several oak constituents are similar to grape flavorants. This may explain why the varietal expression of certain wines is enhanced by maturation in oak (Sefton et al., 1993). In contrast, oak lactones and quaiacol can suppress the fruitiness of some esters, their masking action being complex and nonlinear (Atanasova et al., 2004). Thus, the inadvisability to mature mild-flavored white wines in oak barrels.

Although not varietal, many esters are essential to the fruity character of both white and red wines. The diversity and concentration of these esters can contribute significantly to the expression of wine's varietal character. Many esters may not occur individually at above their threshold values, but can still act in a quantitative manner to influence the overall fruity character of a wine. An example of this synergy is given for the blackberry and fresh-fruit aspects of several red wines (Lytra et al., 2013).

Although hundreds of aromatic compounds have been isolated from wine, assessing their relative significance to a wine's aroma and bouquet is far from simple. The task has partially been simplified by estimating the odor activity value (OAV) of its aromatic constituents. OAV is determined by dividing the concentration of a compound by its detection threshold. A list of olfactory thresholds and concentration ranges is provided in Table 3.2. Additional information has been compiled by Francis and Newton (2005). Compounds with an OAV greater than unity have a much greater likelihood of influencing a wine's fragrance than those possessing OAVs less than unity. Unfortunately, there is no direct correlation between a compound's OAV and its sensory impact. The perceived intensity of various compounds at and above their threshold can differ markedly, as does the slope at which perceived intensity increases with concentration (Ferreira et al., 2003). In addition, interpretation of data is complicated by multimodal interactions among constituents, i.e., the norm. This can involve various associations within the wine itself (affecting the dynamics of volatility), interaction at the neuronal level in the olfactory patches, and subsequent processing in the olfactory bulb and higher centers of the brain. This can either reduce or augment a compound's perceived intensity as well as its perceived quality. Finally, threshold values are an average for a particular group of tasters, and the conditions under which the threshold was assessed (e.g., water, aqueous ethanol, a wine-like solution, or a specific wine). Nonetheless, OAVs are a good starting point in studying the potential sensory significance of a wine's chemical composition.

Several techniques have been developed to further clarify the sensory significance of particular compounds. Gas chromatography–olfactometry (GC-O) is frequently used (San-Juan et al., 2010), often in combination with other analytic techniques. These data may be used in aroma extract dilution analysis (AEDA) (Grosch, 2001; Ferreira et al., 2002). It involves adding compounds with OAVs greater than unity to model wines and assessing their character, relative to a particular varietal wine. It has already been used to demonstrate the importance of certain flavorants in varietal wines, such as Grenache (Ferreira et al., 1998), and the existence of a significant difference between the Cabernet Sauvignon and Merlot wines. The latter are apparently characterized by a distinguishing caramel odor. It is suspected to be due to a difference in the concentration of 4-hydroxy-2,5-dimethylfuran-3(2H)-one and 4-hydroxy-2(or 5)-ethyl-5(or 2)-methylfuran-3(2H)-one (Kotseridis et al., 2000). The technique has also supplied evidence for the probable methoxypyrazine origin of the pronounced stemmy character in Cabernet Sauvignon and Chardonnay wines fermented in contact with their stems (Hashizume and Samuta, 1997). In addition, the importance of several, previously presumed, sensory-significant compounds will likely need to be reassessed.

In spite of theoretical concerns about the applicability of OAV evaluations, their value has been demonstrated in addition, omission, and reconstruction experiments (Escudero et al., 2004; Grosch, 2004; Escudero et al., 2007). For example, the flavor of Gewürztraminer and Scheurebe wines could be reproduced in a synthetic wine with 29 and 42 flavorants that occurred at above-threshold values in the respective wines (Guth, 1998). Of these, cis-rose oxide and 4-mercapto-4-methylpentan-2-one were essential to the distinctive character of Gewürztraminer and Scheurebe wines, respectively. In contrast, elimination of compounds such as acetaldehyde, β-damascenone, and geraniol had little effect. Addition of 13 other aromatics, occurring at below their threshold value in the original wine, did not

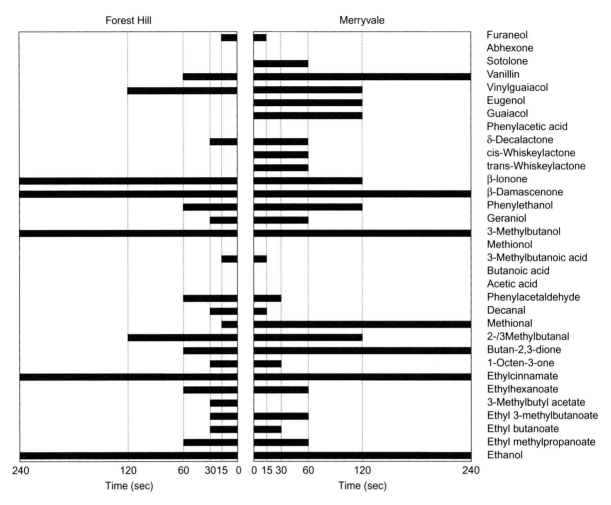

FIGURE 3.46 Comparative buccal odor screening system (BOSS) analysis illustrating how long various aromatics from two different Chardonnay wines remained in the oral cavity after expectoration. *From Buettner, reproduced by permission.*

affect the aroma characteristics of model wines. Reconstruction experiments have also expanded to involve both volatile and nonvolatile constituents, to determine their respective involvement in the flavor of wine, for example with a Dornfelder wine (Frank et al., 2011). This, and other studies, such as Sáenz-Navajas et al. (2010), have provided additional evidence for the importance of nonvolatile constituents to the flavor characteristics of wines.

Another technique, atmospheric pressure chemical ionization mass spectroscopy (APCI-MS, or MS-Nose), provides real-time, human analysis of volatile constituents (Aznar et al., 2004; Tsachaki et al., 2009). Although it has limited identification power, it has proven particularly useful in studying the dynamic change in aromatics following release from food (Taylor et al., 2000). It has been little used with wine due to disrupted ionization in solutions containing more than 4 percent ethanol. However, it presumably could be used in assessing the dynamics of aroma release associated with a wine's finish, or the influence of food on wine flavor. Nonetheless, the technique has already demonstrated that it is not the absolute content of aromatics in food that is critical, but changes in their concentration that tend to influence perceived odor intensity (Linforth and Taylor, 2000). It has also confirmed the continued liberation of aromatic compounds from the oral cavity, minutes after swallowing (Wright et al., 2003; Hodgson et al., 2005). The duration of a particular compound's presence is markedly affected by its partition coefficient (K_{al}). Those with low coefficients are liberated more slowly, but over a longer period. Using another technique, buccal odor screening system (BOSS) has indicated that the fruity character of Chardonnay wines disappears more rapidly from the oral cavity than more slowly volatilized oak flavorants (Fig. 3.46).

All these studies have a consistent message, namely, that wine quality is predominantly derived from its particular combination of aromatic constituents. Some contribute positively, some negatively, and most seemingly

TABLE 3.5 Compounds Associated with Positive and Negative Attributes of Spanish Red Wines

Vector	Regression coefficient	Compounds
Minor branched ethyl esters	0.083	Ethyl 4-methylpentanoate
		Ethyl 3-methylpentanoate
		Ethyl 2-methylpentanoate
		Ethyl ciclohexanoate
Major ethyl esters	0.006	Ethyl propanoate
		Ethyl butyrate
		Ethyl hexanoate
		Ethyl octanoate
		Ethyl decanate
		Ethyl 2-methylpropanoate
		Ethyl 2-methylbutyrate
		Ethyl 3-methylbutyrate
Norisoprenoids	0.038	β-damascenone
		β-ionone
Acids	0.598	Butyric acid
		Hexanoic acid
		Octanoic acid
		Decanoic acid
		2-methylpropanoic acid
		2-methylbutyric acid
		3-methylbutyric acid
Aging-related compounds	0.222	Eugenol
		E-isoeugenol
		E-whiskylactone
		Z-whiskylactone
Enolones	0.088	Furaneol
		Homofuraneol
Methional	−0.150	
Phenylacetaldehyde	−0.201	
4-ethylphenol	−0.327	
Acetic acid	−0.283	

Reprinted with permission from San-Juan, F., Ferreira, V., Cacho, J., and Escudero, A. (2011) Quality and aromatic sensory descriptors (mainly fresh and dry fruit character) of Spanish red wines can be produced from their aroma-active chemical composition. J. Agric. Food Chem. 59, 7916–7924, Copyright 2011 American Chemical Society.

play little role. An excellent example of this is illustrated in San-Juan et al. (2011). The authors studied a set of 25 high-quality Spanish red wines by aroma-active composition. The wine's positive attributes correlated well with the wines' contents in fruity ester, volatile fatty acids, norisoprenoids, enolones (e.g., furaneol), and oak-derived constituents, but negatively with attributes often associated with oxidized and other off-odors, notably 4-ethylphenol, acetic acid, phenylacetaldehyde, and methional (Table 3.5). Vegetal aspects were associated with dimethyl sulfide, methanethiol and 1-hexanol, but their expression appeared to be suppressed in the presence of acetaldehyde, linear fatty acids, and ethyl esters.

Postscript

One of the principal lessons that has emerged in studying olfaction and its associated memories is the gap separating sensory-detected and perceived reality. Perceived olfactory reality is a cerebral construct that is often multimodal in nature, often incorporating visual, gustatory, and somatosensory (trigeminal) inputs. Once a pattern has been learned and reinforced, it can skew perceptions to conform to expectations. Thus, although it is desirable for consumers, and necessary for wine assessors, to develop prototypic sensory memories of the attributes that characterize particular wine types and age effects, tasters must always guard against these constructs from warping what is being sensed. It is the essence of wine's uniqueness that should be encoded in memory, the commonality that underlies the complexity that is the purvey of any wine.

One of the more intriguing aspects of olfaction is its malleability, often being dependent on the context under which it is detected. By contrast, our other senses are measurable by clearly defined physical/chemical factors; hearing is equated to sound vibration and amplitude, temperature to the degree of molecular thermal vibration, equilibrium to orientation and acceleration, touch by pressure and movement, color by the spectral quality and the relative photon energy of a select range of electromagnetic radiation. Also, most tastes relate to the concentration of a small group of compounds. In contrast, odor detection is largely based on holistic pattern recognition, where (as yet) we cannot analyze aroma patterns as we can distinguish the letters in a word, or specifically isolate aspects of a face. In addition, until an aromatic compound becomes associated in memory with a particular olfactory response pattern, its olfactory quality is undefinable. If it is not pungent or putrid, or readily identifiable in a mixture, an odorant's quality is only part of the pattern associated with other compounds with which it was associated; it is a dynamic, kaleidoscopic perception, susceptible to change due to concentration and sensory adaptation. Thus, it is somewhat amazing that we are able at all, with experience, to isolate such evanescent diffuse perceptions with varietal aromas, stylistic characteristics, or regional expressions. Maybe it is the attribute of the brain to search for meaning, equivalent to our ability to isolate a set of sounds, from what to can inherently seem to be a jumble, and recognize them as words of a language we know.

In addition, assessing a wine's fragrance can seem analogous to listening to a symphony. Detection of the string section may be easy, but differentiation among the first, second and third violin sections is far less simple, at least for nonmusicians. Thus, lists of potentially odor-active compounds, and their respective qualitative (descriptor) terms does not provide useful information on how their aromatic sources interact with other components that make up the wine's matrix. A related problem arises when looking at spider diagrams of a wine's aroma. Spider diagrams prove useful visual representations for sensory analysis, but do not in themselves provide a means to imagine how the wine would be sensed in the glass. Thus, they are not the equivalent to a musical score for a musician.

Finally, the lack of a unique odor lexicon, associated with odor memory, may relate to its lack of social and evolutionary value. Detecting differences rather than identification may have been more important; that is, it may have been more important to quickly recognize the significance of an odor than it was to name it, in a manner somewhat equivalent to rapidly detecting the presence of a potential predator in the bush than naming it.

Suggested Readings

Breer, H., 2008. The sense of smell. Reception of flavors. Ann. N.Y. Acad. Sci. 1126, 1–6.

Cahill, J., 2014. The triumph of perception over reality. Wine Vitic. J. 29 (3), 8–9.

Ferreira, V., 2010. Volatile aroma compounds and wine sensory attributes. In: Reynolds, A.G. (Ed.), Managing Wine Quality. Vol. 1. Viticulture and Wine Quality, Woodhead Publishing Ltd, Cambridge, UK, pp. 3–28.

Firestein, S., Beauchamp, G.,K. (Eds.), 2008. The Senses: a Comprehensive Reference, Vol. 4, Olfaction and Taste, Academic Press, Elsevier, Oxford.

Francis, I.L., Newton, J.L., 2005. Determining wine aroma from compositional data. In: Blain, C.J., Francis, M.E., Pretorius, I.S. (Eds.) Advances in Wine Science, The Australian Wine Research Institute, Glen Osmond, SA, Australia, pp. 201–212.

Gottfried, J.A., 2008. Perceptual and neural plasticity of odor quality coding in the human brain. Chem. Percept. 1, 127–135.

Leland, J.V., Scheiberke, P., Buettner, A., Acree, T.E. (Eds.), 2004. Gas Chromatography-Olfactometry: The State of the Art. ACS Symposium Series, American Chemical Society, Washington, DC.

Linforth, R., Taylor, A., 2006. The process of flavour release. In: Voilley, A., Etiévant, P. (Eds.) Flavour in Food, Woodhouse Publ. Inc., Cambridge, UK, pp. 287–307.

Murthy, V.N., 2011. Olfactory maps in the brain. Annu. Rev. Neurosci. 34, 233–258.

Rolls, E.T., Critchley, H.D., Verhagen, J.V., Kadohisa, M., 2010. The representation of information about taste and odor in the orbitofrontal cortex. Chem. Percept. 3, 16–33.

Ryan, D., Prenzler, P.D., Saliba, A.J., Scollary, G.R., 2008. The significance of low impact odorants in global odor perception. Trends Food Sci. Technol. 19, 383–389.

Silva Teixeria, C.S., Cerqueira, N.M.F.S.A., Silva Ferreira, A.C., 2016. Unraveling the olfactory sense: From the gene to odor perception. Chem. Senses 41, 105–121.

Spence, C., 2013. Multisensory flavour perception. Curr. Biol. 23, R365–R369.

Spence, C., 2016. Oral referral: On the mislocalization of odours to the mouth. Food Qual. Pref. 50, 117–128.

Styger, G., Prior, B., Bauer, F.F., 2011. Wine flavor and aroma. J. Ind. Microbiol. Biotechnol. 38, 1145–1159.

Taylor, A.J., Cordell, R., Linforth, R.S.T., 2009. Dynamic flavour analyses suing direct MS. pp. 24–28. In: Märk, T.D., Holzner, B. (Eds.) 4th International Conference on Proton Transfer Reaction Mass Spectrometry and its Application, Innsbruck University Press, Innsbruck, Austria.

Thomas-Danguin, T., Sinding, C., Romagny, S., El Mountassir, F., Atanasova, B., Le Berre, E., Le Bon, A.-M., Coureaud, G., 2014. The perception of odor objects in everyday life: a review on the processing of odor mixtures. Front. Psychol. 5, 1–18.

Wilson, D.A., Stevenson, R.J.,2006.Learning to Smell: Olfactory Perception from Neurobiology to Behavior. Johns Hopkins University Press, Baltimore, MD.

Yeshurun, Y., Sobel, N., 2010. An odor is not worth a thousand words: Form multidimensional odors to unidimensional odor objects. Annu. Rev. Psychol. 61, 219–241.

References

Acree, T.E., Braell, P.A., Butts, R.M., 1981. The presence of damascenone in cultivars of Vitis vinifera (Linnaesus), rotundifilia (Michaux), and labruscana (Bailey). J. Agric. Food Chem. 29, 688–690.

Acree, T.E., Lavin, E.H., Nishida, R., Watanabe, S., 1990. o-Aminoacetophenone, the "foxy" smelling component of Labruscana grapes Wöhrmann Symposium. Wädenswil, Switzerland.49–52.

Allen, M.S., Lacey, M.J., and Boyd, S.J. (1996). Methoxypyrazines: New insights into their biosynthesis and occurrence. In Proc. 4th Int. Symp. Cool Climate Vitic. Enol., Rochester, NY, July 16–20, 1996 (T. Henick-Kling, T. E. Wolf, and E. M. Harkness, eds.), pp. V-36–39. NY State Agricultural Experimental Station, Geneva, New York.

Amerine, M.A., Roessler, E.B., 1983. Wines, Their Sensory Evaluation, 2nd ed. Freeman, San Francisco, CA.

Amoore, J.E., 1977. Specific anosmia and the concept of primary odors. Chem. Senses Flavours 2, 267–281.

Aqueveque, C., 2015. The influence of experts' positive word-of-mouth on a wine's perceived quality and value: the moderator role of consumers' expertise. J. Wine Res. 26, 181–191.

Arctander, S., 1994. Perfume and Flavor Chemicals (Aroma Chemicals). Allured Publ. Corp., Carol Stream, IL.

Aronson, J., Ebeler, S.E., 2004. Effect of polyphenol compounds on the headspace volatility of flavors. Am. J. Enol. Vitic. 55, 13–21.

Atanasova, B., Thomas-Danguin, T., Langlois, D., Nicklas, S., Etievant, P., 2004. Perceptual interactions between fruity and woody notes of wine. Flavour Fragr. J. 19, 476–482.

Auzély-Velty, R., Péan, C., Djedaïni-Pilard, F., Zemb, Th, Perly, B., 2001. Micellization of hydrophobically modified cyclodextrins.2. Inclusion of guest molecules. Langmuir 17, 504–510.

Aznar, M., Tsachaki, M., Linforth, R.S.T., Ferreira, V., Taylor, A.J., 2004. Headspace analysis of volatile organic compounds from ethanolic systems by direct APCI-MS. Int. J. Mass Spectrom. 239, 17–25.

Baker, A.K., Ross, C.F., 2014. Sensory evaluation of impact of wine matrix on red wine finish: a preliminary study. J. Sens. Stud. 29, 139–148.

Bailly, S., Jerkovic, V., Meurée, A., Timbermans, A., Collins, S., 2009. Fate of key odorants in Sauternes wines through aging. J. Agric. Food Chem. 57, 8557–8563.

Ballester, J., Dacremont, C., Le Fur, Y., Etiévant, P., 2005. The role of olfaction in the elaboration and use of the Chardonnay wine concept. Food Qual. Pref. 16, 351–359.

Bartowsky, E.J., Francis, I.L., Bellon, J.R., Henschke, P.A., 2002. Is buttery aroma perception in wines predictable from the diacetyl concentration? Aust. J. Grape Wine Res. 8, 180–185.

Bautze, V., Bär, R., Fissler, B., Trapp, M., Schmidt, D., Beifuss, U., et al., 2012. Mammalian-specific OR37 receptors are differentially activated by distinct odorous fatty aldehydes. Chem. Senses 37, 479–493.

Beauchamp, J., Scheibe, M., Hummel, T., Buettner, A., 2009. Characterisation of odorant pathways in olfactory dysfunction. International Conference on Breath and Breath Odor Research. Dortmund, Germany, http://breath2009.isas.de/.

Best, A.R., Wilson, D.A., 2004. Coordinate synaptic mechanisms contributing to olfactory cortical adaptation. J. Neurosci. 24, 652–660.

Block, E., Jang, S., Matsunami, H., Sekharan, S., Dethier, B., Ertem, M.Z., et al., 2015. Implausibility of the vibrational theory of olfaction. Proc. Natl. Acad. Sci. 112, E2766–E2774.

Bock, G., Benda, I., Schreier, P., 1988. Microbial transformation of geraniol and nerol by Botrytis cinerea. Appl. Microbial. Biotechnol 27, 351–357.

Boidron, J.N., Chatonnet, P., Pons, M., 1988. Influence du bois sur certaines substances odorantes des vins. Conn. Vigne Vin 22, 275–294.

Boison, J., Tomlinson, R.H., 1988. An investigation of the volatile composition of Vitis labrusca grape must and wines, II. The identification of N-(N-hydroxy-N-methyl-γ-aminobutyryl)glycin in native North American grape varieties. Can. J. Spectrosc. 33, 35–38.

Boison, J.O.K., Tomlinson, R.H., 1990. New sensitive method for the examination of the volatile flavor fraction of Cabernet Sauvignon wines. J. Chromatogr. 522, 315–328.

Botha, J.J., 2010. Sensory, chemical and consumer analysis of Brettanomyces spoilage in South African wines. Masters Thesis. Department of Food Science, Stellenbosch University, Stellenbosch, S. A.

Bouchilloux, P., Darriet, P., Henry, R., Lavigne-Cruège, V., Dubourdieu, D., 1998. Identification of volatile and powerful odorous thiols in Bordeaux red wine varieties. J. Agric. Food Sci. 46, 3095–3099.

Bramley, B., Curtin, C., Cowey, G., Holdenstock, M., Coulter, A., Kennedy, E., et al., 2008. Wine style alters the sensory impact of 'Brett' flavour compounds in red wines. In: Blair, R.J., Williams, P.J., Pretorius, I.S. (Eds.), 13th Australian Wine Industry Technical Conference. Australian Wine Industry Technical Conference, Inc., Adelaide, Australia, pp. 45.

Breslin, P.A., Doolittle, N., Dalton, P., 2001. Subthreshold integration of taste and smell: the role of experience in flavor integration. Chem. Senses 26, 1035.

Briand, L., Eloit, C., Nespoulous, C., Bezirard, V., Huet, J.-C., Henry, C., et al., 2002. Evidence of an odorant-binding protein in the human olfactory mucus: location, structural characterization, and odorant-binding properties. Biochemistry 41, 7241–7252.

Brochet, F., Dubourdieu, D., 2001. Wine descriptive language supports cognitive specificity of chemical senses. Brain Lang. 77, 187–196.

Brochet, F., Morrot, G., 1999. Influence du contexte sure la perception du vin. Implications cognitives et méthodologiques. J. Int. Sci. Vigne Vin 33, 187–192.

Brookes, J.C., Horsfield, A.P., Stoneham, A.M., 2012. The swipe card model of odorant recognition. Sensors 12, 15709–15749.

Buck, L., Axel, R., 1991. A novel multigene family may encode odorant receptors: A molecular basis for odor recognition. Cell 65, 175–187.

Buck, L.B., 1996. Information coding in the vertebrate olfactory system. Annu. Rev. Neurosci. 19, 517–544.

Buettner, A., 2002a. Influence of human salivary enzymes on odorant concentration changes occurring in vivo. 1. Esters and thiols. J. Agric. Food Chem. 50, 3283–3289.

Buettner, A., 2002b. Influence of human saliva on odorant concentrations.2. Aldehydes, alcohols, 3-alkyl-2-methoxypyrazines, methoxyphenols, and 3-hydroxy-4,5-dimethyl-2(5H)-fruanone. J. Agric. Food Chem. 50, 7105–7110.

Buettner, A. (2003). Physiology and chemistry behind retronasal aroma perception during winetasting. In A. Lonvaud-Funel, G. de Revel, & P. Darriet (Eds.), Proceedings of the VIIth International Symposium d'Oenology. (A. Lonvaud-Funel, G. de Revel, and P. Darriet, Eds.) Editions Tec & Doc., Paris.

Buettner, A., 2004. Investigation of potent odorants and after odor development in two Chardonnay wines using the buccal odor screening system (BOSS). J. Agric. Food Chem. 52, 2339–2346.

Buettner, A., Beauchamp, J., 2010. Chemical input – Sensory output: Diverse modes of physiology-flavor interaction. Food Qual. Pref. 21, 915–924.

Bushdid, C., Magnasco, M.O., Vosshall, L.B., Keller, A., 2014. Humans can discriminate more than 1 trillion olfactory stimuli. Science 343, 1370–1372.

Cain, W.S., 1974. Perception of odour intensity and the time-course of olfactory adaption. Trans. Am. Soc. Heating, Refrigeration Air-Conditioning Engin 80, 53–75.

Cain, W.S., 1976. Olfaction and the common chemical sense: Some psychophysical contrasts. Sens. Processes 1, 57.

Cain, W.S., 1978. The odoriferous environment and the application of olfactory research In: Carterette, E.C. Friedman, P.M. (Eds.), Handbook of Perception: Tasting and Smelling, Vol. 6A. Academic Press, New York, pp. 197–229.

Cain, W.S., 1982. Odor identification by males and females: Predictions and performance. Chem. Senses 7, 129–141.

Cain, W.S., Gent, J.F., 1991. Olfactory sensitivity: Reliability, generality, and association with aging. J. Expt. Psychol. Human Precept. Perform. 17, 382–391.

Cain, W.S., Stevens, J.C., Nicou, C.M., Giles, A., Johnston, I., Garcia-Medina, M.R., 1995. Life-span development of odor identification, learning, and olfactory sensitivity. Perception 24, 1457–1472.

Caloghiris, M., Waters, E.J., Williams, P.J., 1997. An industry trial provides further evidence for the role of corks in oxidative spoilage of bottled wines. Aust. J. Grape Wine Res. 3, 9–17.

Capone, D.L., van Leeuwen, K., Taylor, D.K., Jeffery, D.W., Pardon, K.H., Elsey, G.M., et al., 2011. Evolution and occurrence of 1,8-cineole (eucalyptol) in Australian wine. J. Agric. Food Chem. 59, 953–959.

Carpentier, N., Maujean, A., 1981. Light flavours in champagne wines. In: Schreier, P. (Ed.), Flavour '89: 3rd Weurman Symp. Proc. Int. Conf. de Gruyter, Berlin, pp. 609–615.

Case, T.I., Stevenson, R.J., Dempsey, R.A., 2004. Reduced discriminability following perceptual learning with odors. Perception 33, 113–119.

Castriota-Scanderbeg, A., Hagberg, G.E., Cerasa, A., Committeri, G., Galati, G., Patria, F., et al., 2005. The appreciation of wine by sommeliers: a functional magnetic resonance study of sensory integration. Neuroimage 25, 570–578.

Chalier, P., Angot, B., Delteil, D., Doco, T., Gunata, Z., 2007. Interactions between aroma compounds and whole mannoprotein isolated from Saccharomyces cerevisiae strains. Food Chem. 100, 22–30.

Charpentier, N., Maujean, A., 1981. Sunlight flavours in champagne wines. In: Schreier, P. (Ed.), Flavour ' 81. Proc. 3rd Weurman Symp. de Gruyter, Berlin, pp. 609–615.

Chatonnet, P., Viala, C., Dubourdieu, D., 1997. Influence of polyphenolic components of red wines on the microbial synthesis of volatile phenols. Am. J. Enol. Vitic. 48 443–338.

Chatonnet, P., Bonnet, S., Boutou, S., Labadie, M.-D., 2004. Identification and responsibility of 2,4,6-tribromoanisole in musty, corked odors in wine. J. Agric. Food Chem. 52, 1255–1262.

Choudhury, E.S., Moberg, P., Doty, R.L., 2003. Influences of age and sex on a microencapsulated odor memory test. Chem. Senses 28, 799–805.

Christensen, T.A., Pawlowski, V.M., Lei, H., Hildebrand, J.G., 2000. Multi-unit recordings reveal context-dependent modulation of synchrony in odor–specific neural ensembles. Nat. Neurosci. 3, 927–931.

Ciolfi, G., Garofolo, A., Di stefano, R., 1995. Identification of some o-aminophenones as secondary metabolites of Saccharomyces cerevisiae. Vitis 34, 195–196.

Cook, D.J., Davidson, J.M., Linforth, R.S.T., Taylor, A.J., 2003. Measuring the sensory impact of flavour mixtures using controlled delivery. In: Deibler, K.D., Delwicke, J. (Eds.), Handbook of Flavor Characterization: Sensory Analysis, Chemistry and Physiology. Marcel Dekker, New York, NY, pp. 135–150.

Cometto- Muñiz, J.E., Abraham, M.H., 2016. Dose-response functions for the olfactory, nasal trigeminal, and ocular trigeminal detectability of airborne chemicals by humans. Chem. Senses 41, 3–14.

Conner, J.M., Birkmyre, L., Paterson, A., Piggott, J.R., 1998. Headspace concentrations of ethyl esters at different alcoholic strengths. J. Sci. Food Agric. 77, 121–126.

Culleré, L., Escudero, A., Cacho, J., Ferreira, V., 2004. Gas chromatography-olfactometry and chemical quantitative study of the aroma of six premium quality Spanish aged red wines. J. Agric. Food Chem. 52, 1653–1660.

Culleré, L., Cacho, J., Ferreira, V., 2007. An assessment of the role played by some ozidation-related aldehydes in wine aroma. J. Agric. Food Chem. 55, 876–881.

Czerny, M., Grosch, W., 2000. Potent odorants of raw Arabica coffee. Their changes during roasting. J. Agric. Food Chem. 48, 868–872.

Dahl, A.R., 1988. The effect of cytochrome P-450 dependent metabolism and other enzyme activities on olfaction. In: Margolis, F.L., Getchell, T.V. (Eds.), Molecular Neurobiology of the Olfactory System. Plenum, New York, pp. 51–70.

Dalton, P., Doolittle, N., Nagata, H., Breslin, P.A.S., 2000. The merging of the senses: integration of subthreshold taste and smell. Nature Neurosci. 3, 431–432.

Dalton, P., Doolittle, N., Breslin, P.A.S., 2002. Gender-specific induction of enhanced sensitivity to odors. Nature Neurosci. 5, 199–200.

Damm, M., Vent, J., Schmidt, M., Theissen, P., Eckel, H.E., Lötsch, J., et al., 2002. Intranasal volume and olfactory function. Chem. Senses 27, 831–839.

D'Angelo, M., Onori, G., Santucci, A., 1994. Self-association of monohydric alcohols in water: compressibility and infrared absorption measurements. J. Chem. Phys. 100, 3107–3113.

d'Auria, M., Emanuele, L., Mauriello, G., Racioppi, R., 2002. On the origin of "goût de lumiere" in champagne. J. Photochem. Photobiol A: Chemistry 158, 21–26.

Davis, R.G., 1977. Acquisition and retention of verbal association to olfactory and abstract visual stimuli of varying similarity. J. Exp. Psychol. Learn. Me,. Cog 3, 37–51.

Degel, J., Piper, D., Köster, E.P., 2001. Implicit learning and implicit memory for odors: the influence of odor identification and retention time. Chem. Senses 26, 267–280.

Déléris, I., Saint-Eve, A., Guo, Y., Lieben, P., Cypriani, M.-L., Jacquet, N., et al., 2011. Impact of swallowing on the dynamics of aroma release and perception during the consumption of alcoholic beverages. Chem. Senses 36, 701–713.

De Mora, S.J., Knowles, S.J., Eschenbruch, R., Torrey, W.J., 1987. Dimethyl sulfide in some Australian red wine. Vitis 26, 79–84.

de Wijk, R.A. (1989). "Temporal Factors in Human Olfactory Perception." Doctoral thesis, University of Utrecht, The Netherlands. (cited in Cometto-Muñiz, J. E., and Cain, W. S. (1995). Olfactory adaptation. In Handbook of Olfaction and Gustation (C. L. Doty, ed.), pp. 257–281. Marcel Dekker, New York.)

de Wijk, R.A., Cain, W.S., 1994a. Odor quality: Discrimination versus free and cued identification. Percept. Psychophys. 56, 12–18.

de Wijk, R.A., Cain, W.S., 1994b. Odor identification by name and by edibility: Life-span development and safety. Hum. Factors 36, 182–187.

Diaz, M.E., 2004. Comparison between orthonasal and retronasal flavour perception at different concentrations. Flavour Fragr. J. 19, 499–504.

Dijksterhuis, A., Bos, M.W., Nordgren, L.F., van Baaren, R.B., 2006. On making the right choice: The deliberation-without-attention effect. Science 311, 1005–1007.

Distel, H., Ayabe-Kanamura, S., Martínez-Gómez, M., Achicker, I., Koyayakawa, T., Saito, S., et al., 1999. Perception of everyday odors–correlation between intensity, familiarity and strength of hedonic judgement. Chem. Senses 24, 191–199.

Distel, H., Hudson, R., 2001. Judgement of odor intensity is influenced by subject's knowledge of the odor source. Chem. Senses 26, 247–251.

Dobbs, E., 1989. The scents around us. The Sciences, 46–53. (Nov/Dec).

Doty, R.L., 1986. Reproductive endocrine influences upon olfactory perception, a current perspective. J. Chem. Ecol. 12, 497–511.

Doty, R.L., 1998. Cranial nerve I: olfaction. In: Goltz, C.G., Pappert, E.J. (Eds.), Textbook of Clinical Neurology. Saunders, Philadelphia, PA, pp. 90–101.

Doty, R.L., Bromley, S.M., 2004. Effects of drugs on olfaction and taste. Otolaryngol. Clin. N. Am. 37, 1229–1254.

Doty, R.L., Snow Jr., J.B., 1988. Age-related alterations in olfactory structure and function. In: Margolis, F.L., Getchell, T.V. (Eds.), Molecular Neurobiology of the Olfactory System. Plenum, New York, pp. 355–374.

Dubois, P., 1983. Volatile phenols in wines. In: Piggott, J.R. (Ed.), Flavour of Distilled Beverages. Ellis Horwood, Chichester, UK, pp. 110–119.

Dubois, P., Rigaud, J., 1981. Á propos de goût de bouchon. Vignes Vins 301, 48–49.

Dufour, C., Bayonove, C.L., 1999a. Influence of wine structurally different polysaccharides on the volatility of aroma substances in a model system. J. Agric. Food Chem. 47, 671–677.

Dufour, C., Bayonove, C.L., 1999b. Interactions between wine polyphenols and aroma substances. An insight at the molecular level. J. Agric. Food Chem. 47, 678–684.

Dufour, C., Sauvaitre, I., 2000. Interactions between anthocyanins and aroma substances in a model system. Effect on the flavor of grape-derived beverages. J. Agric. Food Chem. 48, 1784–1788.

du Plessis, C.S., Augustyn, O.P.H., 1981. Initial study on the guava aroma of Chenin blanc and Colombar wines. S. Afr. J. Enol. Vitic. 2, 101–103.

Dürr, P., 1985. Gedanken zur Weinsprache. Alimentia 6, 155–157.

Ekman, G., Berglund, B., Berglund, U., Lindvall, T., 1967. Perceived intensity of odor as a function of time of adaptation. Scand. J. Psychol. 8, 177–186.

Elsner, R.J.F., 2001. Odor threshold, recognition, discrimination and identification in centenarians. Arch. Gerontol Geriat. 33, 81–94.

Engen, T., 1987. Remembering odors and their names. Am. Sci. 75, 497–503.

Eriksson, N., Wu, S., Do, C.B., Kiefer, A.K., Tung, J.Y., Mountain, J.L., et al., 2012. A genetic variant near olfactory receptor genes influences cilantro preference. Flavour J. 1, 22. (1–7).

Escalona, H., Piggott, J.R., Conner, J.M., Paterson, A., 1999. Effects of ethanol strength on the volatility of higher alcohols and aldehydes. Ital. J. Food Sci. 11, 241–248.

Escudero, A., Hernandez-Orte, P., Cacho, J., Ferreira, V., 2000. Clues about the role of methional as character impact odorant of some oxidized wines. J. Agric. Food Chem. 48, 4268–4272.

Escudero, A., Asensio, E., Cacho, J., Ferreira, V., 2002. Sensory and chemical changes of young white wines stored under oxygen. An assessment of the role played by aldehydes and some other important odorants. Food Chem. 77, 325–331.

Escudero, A., Gogorza, B., Melús, M.A., Ortín, N., Cacho, J., Ferreira, V., 2004. Characterization of the aroma of a wine from Maccabeo. Key role played by compounds with low odor activity values. J. Agric. Food Chem. 52, 3516–3524.

Escudero, A., Campo, E., Farina, L., Cacho, J., Ferreira, V., 2007. Analytical characterization of the aroma of five premium red wines. Insights into the role of odor families and the concept of fruitiness of wines. J. Agric. Food Chem. 55, 4501–4510.

Etiévant, P.X., 1991. Wine. In: Maarse, H. (Ed.), "Volatile Compounds in Foods and Beverages". Marcel Dekker, New York, pp. 483–546.

Etiévant, P.X., Issanchou, S.N., Bayonove, C.L., 1983. The flavour of Muscat wine, the sensory contribution of some volatile compounds. J. Sci. Food Agric. 34, 497–504.

Fang, Y., Qian, M., 2005. Aroma compounds in Oregon Pinot Noir wine determined by aroma extract dilution analysis (AEDA). Flavour Fragr. J. 20, 22–29.

Ferreira, V., 2010. Volatile aroma compounds and wine sensory attributes. In: Reynolds, A.G. (Ed.), Managing Wine Quality. Vol. 1. Viticulture and Wine Quality. Woodhead Publishing Ltd, Cambridge, UK, pp. 3–28.

Ferreira, V., Fernández, P., Peña, C., Escudero, A., Cacho, J., 1995. Investigation on the role played by fermentation esters in the aroma of young Spanish wines by multivariate analysis. J. Sci. Food Agric. 67, 381–392.

Ferreira, V., López, R., Escudero, A., Cacho, J.F., 1998. The aroma of Grenache red wine: Hierarchy and nature of its main odorants. J. Sci. Food Agric. 77, 259–267.

Ferreira, V., López, R., Cacho, J.F., 2000. Quantitative determination of the odorants of young red wine from different grape varieties. J. Sci. Food Agric. 80, 1659–1667.

Ferreira, V., Ortin, N., Escudero, A., Cacho, J., 2002. Chemical characterization of the aroma of Grenache rosé wines: Aroma extract dilution analysis, quantitative determination, and sensory reconstitution studies. J. Agric. Food Chem. 50, 4048–4054.

Ferreira, V., Pet'ka, J., Aznar, M., 2002. Aroma extract dilution analysis. Precision and optimal experimental design. J. Agric. Food Chem. 50, 1508–1514.

Ferreira, V., Pet'ka, J., Aznar, M., Cacho, J., 2003. Quantitative gas chromatography-olfactometry. Analytical characteristics of a panel of judges using a simple quantitative scale as gas chromatography detector. J. Chromatogr. A 1002, 169–178.

Ferreira, V., Jarauta, I., Ortega, L., Cacho, J., 2004. Simple strategy for the optimization of solid-phase extraction procedures through the use of solid-liquid distribution coefficients. Application to the determination of aliphatic lactones in wine. J. Chromatography A 1025, 147–156.

Fletcher, P.C., Frith, C.D., Grasby, P.M., Shallice, T., Frackowiak, R.S.J., Dolan, R.J., 1995. Brain systems for encoding and retrieval of auditory-verbal memory. Brain 118, 401–416.

Fors, S., 1983. Sensory properties of volatile Maillard reaction products and related compounds: a literature review. In: Waller, G.R., Feather, M.S. (Eds.), The Maillard Reaction in Food and Nutrition. American Chemical Society, Washington, DC, pp. 185–286.

Francis, I.L., Newton, J.L., 2005. Determining wine aroma from compositional data. Aust. J. Grape Wine Res. 11, 114–126.

Frank, R.A., Bryam, J., 1988. Taste–smell interactions are tastant and odorant dependent. Chem. Senses 13, 445–455.

Frank, R.A., van der Klaauw, N.J., Schifferstein, S.J., 1993. Both perceptual and conceptual factors influence taste-odor and taste-taste interactions. Percept. Psychophys. 54, 343–354.

Frank, S., Wollmann, N., Schieberle, P., Hofmann, T., 2011. Reconstitution of the flavor signature of Dornfelder red wine on the basis of the natural concentrations of its key aroma and taste compounds. J. Agric. Food Chem. 59, 8866–8874.

Frasnelli, J., Heilmann, S., Hummel, T., 2004. Responsiveness of human nasal mucosa to trigeminal stimuli depends on the site of stimulation. Neurosci. Lett. 362, 65–69.

Frasnelli, J., van Ruth, S., Kriukova, I., Hummel, T., 2005. Intranasal concentrations of orally administered flavors. Chem. Senses 30, 575–582.

Freeman, W.J., 1991. The physiology of perception. Sci. Am. 264 (2), 78–85.

Friel, E.N., Linforth, R.S.T., Taylor, A.J., 2000. An empirical model to predict the headspace concentration of volatile compounds above solutions containing sucrose. Food Chem. 71, 309–317.

Frye, R.E., Schwartz, B.S., Doty, R.L., 1990. Dose-related effects of cigarette smoking on olfactory function. J. Am. Med. Assoc. 263, 1233–1236.

Fuchs, T., Glusman, G., Horn-Saban, S., Lancet, D., Pilpel, Y., 2001. The human olfactory subgenome: from sequence to structure and evolution. Hum. Genet. 108, 1–13.

Fujimaru, T., Lim, J., 2013. Effects of stimulus intensity on odor enhancement by taste. Chem. Percept. 6, 1–7.

Furudono, Y., Cruz, G., Lowe, G., 2013. Glomerular input patterns in the mouse olfactory bulb evoked by retronasal odor stimuli. BMC Neurosci. 14 (45), 14.

Gane, S., Georganakis, D., Maniati, K., Vamvakias, M., Ragoussis, N., Skoulakis, E.M.C., et al., 2013. Molecular vibration-sensing component in human olfaction. PLos One 8, e55789. (1–7).

Garb, J.L., Stunkard, A.J., 1974. Taste aversions in man. Am. J. Psychiat. 131, 1204–1207.

Gardiner, J.M., Java, R.I., Richardson-Klavehn, A., 1996. How level of processing really influences awareness in recognition memory. Can J. Exp. Psychol. 50, 114–122.

Garibotti, M., Navarrini, A., Pisanelli, A.M., Pelosi, P., 1997. Three odorant-binding proteins from rabbit nasal mucosa. Chem. Senses 22, 383–390.

Geithe, C., Andersen, G., Malki, A., Krautwurst, D., 2015. A butter aroma recombinate actives human class-1 odorant receptors. J. Agric. Food Chem. 63, 9410–9420.

Genovese, A., Gambuti, A., Piombino, P., Moio, L., 2007. Sensory properties and aroma compounds of sweet Fiano wine. Food Chem. 103, 1228–1236.

Genovese, A., Piombino, P., Gambuti, A., Moio, L., 2009. Simulation of retronasal aroma of white and red wine in a model mouth system. Investigating the influence of saliva on volatile compound concentrations. Food Chem. 114, 100–107.

Gilad, Y., Wiebe, V., Przeworski, M., Lancet, D., Paabo, S., 2004. Loss of olfactory receptor genes coincides with the acquisition of full trichromatic vision in primates. PLoS Biol. 2 (1), E5.

Godfrey, P.A., Malnic, B., Buck, L.B., 2004. The mouse olfactory receptor gene family. Proc. Natl. Acad. Sci. 101, 2156–2161.

Goldner, M.C., Zamora, M.C., di Leo Lira, P., Gianninoto, H., Bandoni, A., 2009. Effect of ethanol level in the perception of aroma attributes and the detection of volatile compounds in red wine. J. Sens. Stud. 24, 243–257.

Goniak, O.J., Noble, A.C., 1987. Sensory study of selected volatile sulfur compounds in white wine. Am. J. Enol. Vitic. 38, 223–227.

González, J., Barros-Loscertales, A., Pulvermuller, F., Meseguer, V., Sanjuán, A., Belloch, V., et al., 2006. Reading *cinnamon* activates olfactory brain regions. NeuroImage 32, 906–912.

Goodstein, E.S., Bohlscheid, J., Evans, M., Ross, C.F., 2014. Perception of flavor finish in model white wine: A time-intensity study. Food Qual. Pref. 36, 50–60.

Goubet, I., Le Quere, J.-L., Voilley, A.J., 1998. Retention of aroma compounds by carbohydrates: influence of their physicochemical characteristics and their physical state. A review. J. Agric. Food Chem. 46, 1981–1990.

Goyert, H., Frank, M.E., Gent, J.F., Hettinger, T.P., 2007. Characteristic component odors emerge from mixtures after selective adaptation. Brain Res. Bull. 72, 1–9.

Grbin, P.R., Costello, P.J., Herderich, M., Markides, A.J., Henschke, P.A., Lee, T.H., 1996. Developments in the sensory, chemical and microbiological basis of mousy taint in wine. In: Stockley, C.S., Sas, A.N., Johnson, R.S., Lee, T.H. (Eds.), Proc. 9th Aust. Wine Ind. Tech. Conf. Winetitles, Adelaide, Australia, pp. 57–61.

Green, B.G., Nachtigal, D., Hammond, S., Lim, J., 2012. Enhancement of retronasal odors by taste. Chem. Senses 37, 77–86.

Grosch, W., 2001. Evaluation of the key odorants of foods by dilution experiments, aroma models and omission. Chem. Senses 26, 533–545.

Gross-Isseroff, R., Lancet, D., 1988. Concentration-dependent changes of perceived odor quality. Chem. Senses 13, 191–204.

Guillou, I., Bertrand, A., De Revel, G., Barbe, J.C., 1997. Occurrence of hydroxypropanedial in certain musts and wines. J. Agric. Food Chem. 45, 3382–3386.

Guth, H., 1997a. Objectification of white wine aromas. Thesis. Technical University, Munich., (in German).

Guth, H., 1997b. Identification of character impact odorants of different white wine varieties. J. Agric. Food Chem. 45, 3022–3026.

Guth, H., (1998). Comparison of different white wine varieties by instrumental and analyses and sensory studies. In Chemistry of Wine Flavor. (L. A. Waterhouse and S. E. Ebeler, eds.), pp. 39–52. ACS Symposium Series #714., American Chemical Society, Washington, DC.

Guth, H., Sies, A., 2002. Flavour of wines: towards an understanding by reconstitution experiments and an analysis of ethanol's effect on odour activity of key compounds. In: Blair, R.J., Williams, P.J., Høj, P.B. (Eds.), 11th Aust. Wine Ind. Tech. Conf., Oct. 7–11, 2001, Adelaide, South Australia. Winetitles, Adelaide, Australia, pp. 128–139.

Hahn, I., Scherer, P.W., Mozell, M.M., 1993. Velocity profiles measured for airflow through a large-scale model of the human nasal cavity. J. Appl. Physiol. 75, 2273–2287.

Hall, L., Johansson, P., Tärning, B., Sikström, S., Deutgen, T., 2010. Magic at the marketplace: Choice blindness for the taste of jam and the smell of tea. Cognition 117, 54–61.

Hamatschek, J., 1982. Aromastoffe im Wein und deren Herkunft *Dragoco* Rep. (*Ger. Ed.*), 2759–2771.

Hashizume, K., Samuta, T., 1997. Green odorants of grape cluster stem and their ability to cause a stemmy flavor. J. Agric. Food Chem. 45, 1333–1337.

Heilmann, S., Hummel, T., 2004. A new method for comparing orthonasal and retronasal olfaction. Behav. Neurosci. 118, 412–419.

Heimann, W., Rapp, A., Völter, J., Knipser, W., 1983. Beitrag zur Entstehung des Korktons in Wein. Dtsch. Lebensm.-Rundsch 79, 103–107.

Hemingway, K.M., Alston, M.J., Chappell, C.G., Taylor, A.J., 1999. Carbohydrate-flavour conjugates in wine. Carbohydrate Polymers 38, 283–286.

Henick-Kling, T., Gerling, C., Martinson, T., Cheng, L., Lakso, A., Acree, T., 2005. Atypical aging flavor defect in white wines: sensory description, physiological causes, and flavor chemistry. Am. J. Enol. Vitic. 56, 420A.

Hérent, M.F., Collin, S., Pelosi, P., 1995. Affinities of nutty-smelling pyraxines and thiaxoles to odorant-binding proteins, in relation with their lipophilicity. Chem. Senses 20, 601–608.

Heresztyn, T., 1986. Formation of substituted tetrahydropyridines by species of *Brettanomyces* and *Lactobacillus* isolated from mousy wines. Am. J. Enol. Vitic. 37, 127–131.

Herz, R.S., Engen, T., 1996. Odor memory: review and analysis. Psychon. Bull. Rev. 3, 300–313.

Herz, R.S., McCall, C., Cahill, L., 1999. Hemispheric lateralization in the processing of odor pleasantness versus odor names. Chem. Senses 24, 691–695.

Herz, R.S., von Clef, J., 2001. The influence of verbal labeling on the perception of odors: Evidence for olfactory illusions? Perception 30, 381–391.

Hettinger, T.P., Myers, W.E., Frank, M.E., 1990. Role of olfaction in perception of non-traditional 'taste' stimuli. Chem. Senses 15, 755–760.

Hodgson, M.D., Langridge, J.P., Linforth, R.S.T., Taylor, A.J., 2005. Aroma release and delivery following the consumption of beverages. J. Agric. Food Chem. 53, 1700–1706.

Hollowood, T.A., Linforth, R.S.T., Taylor, A.J., 2002. The effect of viscosity on the perception of flavour. Chem. Senses 28, 11–23.

Hort, J., Hollowood, T.A., 2004. Controlled continuous flow delivery system for investigation taste-aroma interactions. J. Agric. Food Chem. 52, 4834–4843.

Hughson, A.L., Boakes, R.A., 2002. The knowing nose: the role of knowledge in wine expertise. Food Qual. Pref. 13, 463–472.

Iannilli, E., Bult, J.H.F., Roudnitzky, N., Gerber, J., de Wijk, R.A., Hummel, T., 2014. Oral texture influences the neural processing of ortho- and retronasal odors in humans. Brain Res. 1587, 77–87.

Iland, P.G., Marquis, N., 1993. Pinot noir – Viticultural directions for improving fruit quality. In: Williams, P.J., Davidson, D.M., Lee, T.H. (Eds.), Proc. 8th Aust. Wine Ind. Tech. Conf. Adelaide, 13-17 August, 1992. Winetitles, Adelaide, Australia, pp. 98–100.

Jackson, R.S., 2014. Wine Science: Principles and Applications, 4th ed. Academic Press, San Diego, CA.

Jaeger, S.R., McRae, J.F., Bava, C.M., Beresford, M.K., Hunter, D., Jia, Y., et al., 2013. A Mendelian trait for olfactory sensitivity affects odor experience and food selection. Curr. Biol. 23, 1601–1605.

Jinks, A., Laing, D.G., 2001. The analysis of odor mixtures by humans: evidence for a configurational process. Physiol. Behav. 72, 51–63.

Johnson, B.A., Leon, M., 2007. Chemotopic odorant coding in a mammalian olfactory system. J. Comp. Neurol. 503, 1–34.

Jones, F.N., 1968. Informational content of olfactory quality. In: Tanyolac, N. (Ed.), Theories of Odor and Odor Measurement. N. Robert College Research Center, Bedak, Istanbul, pp. 133–141.

Kadohisa, M., Wilson, D.A., 2006. Olfactory cortical adaptation facilitates detection of odors against background. J. Neurophysiol. 95, 1888–1896.

Kennedy, J. (2013) https://jameskennedymonash.wordpress.com/2013/12/13/infographic-table-of-esters-and-their-smells/.

Kennison, K.R., Gibberd, M.R., Pollnitz, A.P., Wilkinson, K.L., 2008. Smoke-derived taint in wine: The release of smoke-derived volatile phenols during fermentation of Merlot juice following grapevine exposure to smoke. J. Agric. Food Chem. 56, 7379–7383.

Kermen, F., Chakirian, A., Sezille, C., Joussain, P., Le Goff, J., Ziessel, A., et al., 2011. Molecular complexity determines the number of olfactory notes and the pleasantness of smells. Sci. Reports 1, 206. (pp. 5).

Key, B., St John, J., 2002. Axon navigation in the mammalian primary olfactory pathway: where to next? Chem. Senses 27, 245–260.

Keyhani, K., Scherer, P.W., Mozell, M.M.A., 1997. Numerical model of nasal odorant transport for the analysis of human olfaction. J. Theor. Biol. 186, 279–301.

King, E.S., Heymann, H., 2014. The effect of reduced alcohol on the sensory profiles and consumer preferences of white wine. J. Sens. Stud. 29, 33–42.

Kleene, S.J., 1986. Bacterial chemotaxis and vertebrate olfaction. Experientia. 42, 241–250.

Knaapila, A., Tuorila, H., Silventoinen, K., Wright, M.J., Kyvik, K.O., Cherkas, L.F., et al., 2008. Genetic and environmental contributions to perceived intensity and pleasantness of androstenone odor: An international twin study. Behav. Genet. 38, 484–492.

Kobal, G., Hummel, C., 1988. Cerebral chemosensory evoked potentials elicited by chemical stimulation of the human olfactory and respiratory nasal mucosa. Electroencephalo. Clin. Neurophysiol. 71, 241–250.

Kolor, M.K., 1983. Identification of an important new flavor compound in Concord grape, ethyl 3-mercaptopropionate. J. Agric. Food Chem. 31, 1125–1127.

Kotseridis, Y., Razungles, A., Bertrand, A., Baumes, R., 2000. Differentiation of the aromas of Merlot and Cabernet Sauvignon wines using sensory and instrumental analysis. J. Agric. Food Chem. 48, 5383–5388.

Koza, B.J., Cilmi, A., Dolese, M., Zellner, D.A., 2005. Color enhances orthonasal olfactory intensity and reduces retronasal olfactory intensity. Chem. Senses 30, 643–649.

Labbe, D., Rytz, A., Morgenegg, C., Ali, S., Martin, N., 2007. Subthreshold olfactory stimulation can enhance sweetness. Chem. Senses 32, 205–214.

Labbe, D., Gilbert, F., Martin, N., 2008. Impact of olfaction on taste, trigeminal, and texture perceptions. Chem. Percept. 1, 217–226.

Lacey, M.J., Allen, M.S., Harris, R.L.N., Brown, W.V., 1991. Methoxypyrazines in Sauvignon blanc grapes and wines. Am. J. Enol. Vitic. 42, 103–108.

Laing, D.G., 1983. Natural sniffing gives optimum odour perception for humans. Perception 12, 99–117.

Laing, D.G., 1986. Optimum perception of odours by humans Proc. 7 World Clean Air Congress, Vol. 4. Clear Air Society of Australia and New Zealand, 110–117.

Laing, D.G., 1994. Perceptual odour interactions and objective mixture analyses. Food Qual. Pref. 5, 75–80.

Laing, D.G., 1995. Perception of Odor Mixtures. In: Doty, R.L. (Ed.), Handbook of Olfaction and Gustation. Marcel Dekker, New York, pp. 283–297.

Laing, D.G., Eddy, A., Best, D.J., 1994. Perceptual characteristics of binary, trinary, and quaternary odor mixtures consisting of unpleasant constituents. Physiol. Behav. 56, 81–93.

Laing, D.G., Link, C., Jinks, A.L., Hutchinson, I., 2002. The limited capacity of humans to identify the components of taste mixtures and taste-odour mixtures. Perception 31, 617–635.

Lamikanra, O., Grimm, C.C., Inyang, I.D., 1996. Formation and occurrence of flavor components in Noble muscadine wine. Food Chem. 56, 373–376.

Lancet, D., 1986. Vertebrate olfactory reception. Annu. Rev. Neurosci. 9, 329–355.

Landis, B.N., Frasnelli, J., Reden, J., Lacroix, J.S., Hummel, T., 2005. Differences between orthonasal and retronasal olfactory functions in patients with loss of the sense of smell. Arch. Otolaryngol. – Head Neck Surg. 131, 977–981.

Lange, C., Martin, C., Chabanet, C., Combris, P., Issanchou, S., 2002. Impact of the information provided to consumers on their willingness to pay for Champagne: comparison with hedonic scores. Food Qual. Pref. 13, 597–608.

Laska, M., Ayabe-Kanamura, S., Hübener, F., Saito, S., 2000. Olfactory discrimination ability for aliphatic odorants as a function of oxygen moiety. Chem. Senses 25, 189–197.

Lattey, K.A., Bramley, B.R., Francis, I.L., 2010. Consumer acceptability, sensory properties and expert quality judgements of Australian Cabernet Sauvignon and Shiraz wines. Aust. J. Grape Wine Res. 16, 189–202.

Lawless, H.T., 1984a. Oral chemical irritation: psychophysical properties. Chem. Senses 9, 143–155.

Lawless, H.T., 1984b. Flavor description of white wine by "expert" and nonexpert wine consumers. J. Food Sci. 49, 120–123.

Lazard, D., Zupko, K., Poria, Y., Nef, P., Lazarovits, J., Horn, S., et al., 1991. Odorant signal termination by olfactory UDP glucuronosyl transferase. Nature 349 (6312), 790–793.

Le Berre, E., Atanasova, B., Langlois, D., Etiévant, P., Thomas-Danguin, T., 2007. Impact of ethanol on the perception of wine odorant mixtures. Food Qual. Pref. 18, 901–908.

Lee, D.-H., Kang, B.-S., Park, H.-J., 2011. Effect of oxygen on volatile and sensory characteristics of Cabernet Sauvignon during secondary shelf life. J. Agric. Food Chem. 59, 11657–11666.

Lee, J., Halpern, B.P., 2013. High-resolution time-intensity tracking of sustained human orthonasal and retronasal smelling during natural breathing. Chem. Percept. 6, 20–35.

Lee, K. (1989). Perception of Irritation from Ethanol, Capsaicin and Cinnamyl Aldehyde via Nasal, Oral and Retronasal Pathways. M.S. thesis, University of California, Davis. (reproduced in Noble, A. C. (1995). Application of time-intensity procedures for the evaluation of taste and mouthfeel. Am. J. Enol. Vitic. 46, 128–133.)

Lee, R.J., Cohen, N.A., 2015. Taste receptors in innate immunity. Cell. Mol. Life Sci. 72, 217–236.

Le Fur, Y., Lesschaeve, I., Etiévant, P., 1996. Analysis of four potent odorants in Burgundy Chardonnay wines: Partial quantitative descriptive sensory analysis and optimization of simultaneous extraction method. In: Henick-Kling, T. (Ed.), Proc. 4 th Int. Symp. Climate Vitic. Enol. NY State Agricultural Experimental Station, Geneva, NY, pp. VII-53–56.

Lehrer, A., 1975. Talking about wine. Language 51, 901–923.

Lehrner, J.P., Glück, J., Laska, M., 1999a. Odor identification, consistency of label use, olfactory threshold and their relationships to odor memory over the human lifespan. Chem. Senses 24, 337–346.

Lehrner, J.P., Walla, P., Laska, M., Deecke, L., 1999b. Different forms of human odor memory: a developmental study. Neurosci. Letts. 272, 17–20.

Li, W., Luxenberg, E., Parrish, T., Gottfried, J.A., 2006. Learning to smell the roses: Experience-dependent neural plasticity in human piriform and orbitofrontal cortices. Neuron 52, 1097–1108.

Li, W., Howard, J.D., Parrish, T.B., Gottfried, J.A., 2008. Aversive learning enhances perceptual and cortical discrimination of indiscriminable odor cues. Science 319, 1842–1845.

Licker, J.L., Acree, T.E., Henick-Kling, T., 1999. What is "Brett" (Brettanomyces) flavor?. In: Waterhouse, A.L., Ebeler, S.E. (Eds.), Chemistry of Wine Flavor ACS Symposium Series Vol. 714. American Chemical Society, Washington, DC, pp. 96–115.

Lin, D.Y., Shea, S.D., Katz, L.C., 2006. Representation of natural stimuli in the rodent main olfactory bulb. Neuron 50, 937–949.

Linforth, R., Taylor, A.J., 2000. Persistence of volatile compounds in the breath after their consumption in aqueous solutions. J. Agric. Food Chem. 48, 5419–5423.

Linforth, R., Martin, F., Carey, M., Davidson, J., Taylor, A.J., 2002. Retronasal transport of aroma compounds. J. Agric. Food Chem. 50, 1111–1117.

Livermore, A., Laing, D.G., 1998. The influence of odor type on the discrimination and identification of odorants in multicomponent odor mixtures. Psychol. Behav 65, 311–320.

Livingstone, M., 2002. Vision and Art: The Biology of Seeing. Abrams, New York, NY, 75.

Lledo, P.-M., Gheusi, G., 2003. Olfactory processing in a changing brain. Neuroreport. 14, 1655–1663.

Lorig, T.S., 2012. Beyond self-report: Brain imaging at the threshold of odor perception. Chem. Percept. 5, 46–54.

Lorrain, B., Ballester, J., Thomas©Danguin, T., Blanquet, J., Meunier, J.M., Le Fur, Y., 2006. Selection of potential impact odorants and sensory validation of their importance in typical Chardonnay wines. J. Agric. Food Chem. 54, 3973–3981.

Lubbers, S., Voilley, A., Feuillat, M., Charpontier, C., 1994. Influence of mannoproteins from yeast on the aroma intensity of a model wine. Lebensm.–Wiss. u. Technol. 27, 108–114.

Lund, C.M., Nicolau, L., Gardner, R.C., Kilmartin, P.A., 2009. Effect of polyphenols on the perception of key aroma compounds from Sauvignon blanc wine. Aust. J. Grape Wine Res. 15, 18–26.

Lundström, J.N., Boyle, J.A., Zatorre, R.J., Jones-Gotman, M., 2009. The nuronal substrates of human olfactory based kin recognition. Hum. Brain Map. 30, 2571–2580.

Lytra, G., Tempere, S., Le Flosch, A., de Revel, G., Barbe, J.-C., 2013. Study of sensory interactions among red wine fruity esters in a model solution. J. Agric. Food Chem. 61, 8504–8513.

Mackay-Sim, A., Johnston, A.N.B., Owen, C., Burne, T.H.J., 2006. Olfactory ability in the healthy population: Reassessing presbyosmia. Chem. Senses 31, 736–771.

MacRostie, S.W., 1974. Electrode Measurement of Hydrogen Sulfide in Wine. M.S. thesis. University California, Davis.

Mahaney, P., Frey, S., Henry, T., Paris, P., 1998. Influence of the barrel on the growth of *Brettanomyces* yeast in barreled red wine Proceedings from the 5th Intervitis Interfructa International Symposium. Stuttgart, Germany.260–269.

Maier, H.G., 1970. Volatile flavoring substances in foodstuffs. Angew. Chem. Internat. Edit 9, 917–926.

Mainland, J., Sobel, N., 2006. The sniff is part of the olfactory percept. Chem. Senses 31, 181–196.

Majid, A., Burenhult, N., 2014. Odors are expressible in language, as long as you speak the right language. Cognition 130, 266–270.

Malnic, B., Godfrey, P.A., Buck, L.B., 2004. The human olfactory receptor gene family. Proc. Natl. Acad. Sci. 101, 2584–2589.

Marais, J., van Rooyen, P.C., du Plessis, C.S., 1979. Objective quality rating of Pinotage wine. Vitis 18, 31–39.

Martin, B., Etiévant, P.X., Le Quéré, J.L., Schlich, P., 1992. More clues about sensory impact of Sotolon in some flor sherry wines. J. Agric. Food Chem. 40, 475–478.

Martineau, B., Acree, T.E., Henick-Kling, T., 1995. Effect of wine type on threshold for diacetyl. Food Res. Int. 28, 139–143.

Masuda, J., Okawa, E., Nishimura, K., Yunome, H., 1984. Identification of 4,5-dimethyl-3-hydroxy-2(5H)-furanone (Sotolon) and ethyl 9-hydroxynonanoate in botrytised wine and evaluation of the roles of compounds characteristic of it. Agric. Biol. Chem. 48, 2707–2710.

Matthews, M.A., 2015. Terroir and Other Mythis of Wine Growing. University of California Press, Oakland, CA, (see pp. 162–162, 177–184).

Mattivi, F., Caputi, L., Carlin, S., Lanza, T., Minozzi, M., Nanni, D., et al., 2011. Effective analysis of rotundone at below-threshold levels in red and white wines using solid-phase microextraction gas chromatography/tandem mass spectrometry. Rapid Commun. Mass Spectrom. 25, 483–488.

Mayr, C.M., Parker, M., Baldoxk, G.A., Black, C.A., Pardon, K.H., Williamson, P.O., et al., 2014. Determination of the importance of in-mouth release of volatile phenol glycoconjugates to the flavor of smoke-tainted wines. J. Agric. Food Chem. 62, 2327–2336.

Mayr, C.M., Capone, D.L., Pardon, K.H., Black, C.A., Pomeroy, D., Francis, I.L., 2015. Quantitative analysis by GC-MS/MS of 18 aroma compounds related to oxidative off-flavor in wines. J. Agric. Food Chem. 63, 3394–3401.

Mazzoleni, V., Maggi, L., 2007. Effect of wine style on the perception of 2,4,6-trichloroanisole, a compound related to cork taint in wine. Food Res. Int. 40, 694–699.

McClure, S.M., Li, J., Tomlin, D., Cypert, K.S., Montague, L.M., Montague, P.R., 2004. Neural correlates of behavioral preference for culturally familiar drinks. Neuron 44, 379–387.

McRae, J.F., Mainland, J.D., Jaeger, S.R., Adipietro, K.A., Matsunami, H., Newcomb, R.D., 2012. Genetic variation in the odorant receptor OR3J3 is associated with the ability to detect the "grassy" smelling odor, cis-3-hexen-1-ol. Chem. Senses 37, 585–593.

McRae, J.F., Jaeger, S.R., Bava, C.M., Beresford, M.K., Hunter, D., Jia, Y., et al., 2013. Identification of regions associated with variation in sensitivity to food-related odors in the human genome. Curr. Biol. 26, 1596–1600.

Meilgaard, M.C., Reid, D.S., 1979. Determination of personal and group thresholds and use of magnitude estimation in flavour chemistry. In: Land, D.G., Nursten, H.E. (Eds.), Progress in Flavour Research. Applied Science Publishers, London, pp. 67–77.

Meilgaard, M.C., Reid, D.C., Wyborski, K.A., 1982. Reference standards for beer flavor terminology system. J. Am. Soc. Brew. Chem. 40, 119–128.

Menashe, I., Man, O., Lancet, D., Gilad, Y., 2003. Different noses for different people. Nature Genetics 34, 143–144.

Mennella, J.A., Jagnow, C.P., Beauchamp, G.K., 2001. Prenatal and postnatal flavor learning by human infants. Pediatrics 107, 1–6.

Moio, L., Etiévant, P.X., 1995. Ethyl anthranilate, ethyl cinnamate, 2,3-dihydrocinnamate, and methyl anthranilate: Four important odorants identified in Pinot noir wines of Burgundy. Am. J. Enol. Vitic. 46, 392–398.

Moncrieff, R.W., 1967. Introduction to the symposium. In: Schultz, H.W. (Ed.), The Chemistry and Physiology of Flavors. AVI, Westport, CN, pp. 3–22.

Moran, D.T., Rowley, J.C., Jafek, B.W., Lovvell, M.A., 1982. The fine structure of the olfactory mucosa in man. J. Neurocytol. 11, 721–746.

Morrot, G., 2004. Cognition et vin. Rev. Oenologues 111, 11–15.

Morrot, G., Brochet, F., Dubourdieu, D., 2001. The color of odors. Brain Lang. 79, 309–320.

Moulton, D.G., 1977. Minimum odorant concentrations detectable by the dog and their implications for olfactory receptor sensitivity. In: Müller-Schwarze, D., Mozell, M.M. (Eds.), Chemical Signals in Vertebrates. Plenum, New York, NY, pp. 455–464.

Moulton, D.E., Ashton, E.H., Eayrs, J.T., 1960. Studies in olfactory acuity, 4. Relative detectability of *n*-aliphatic acids by the dog. Anim. Behav. 8, 117–128.

Muñoz-González, C., Feron, G., Guichard, E., Rodríguez-Bencomo, J., Martín-Alvarez, P.J., Moreno-Arribas, M.V., et al., 2014. Understanding the role of saliva in aroma release from wine by using static and dynamic headspace conditions. J. Agric. Food Chem. 62, 8274–8288.

Munõz-González, C., Sémon, E., Martín-Álvarez, P.J., Guichard, E., Moreno-Arribas, M.V., Feron, G., et al., 2015. Wine matrix composition affects temporal aroma release as measured by protoon transfer reaction – time-of-flight mass spectrometry. Aust. J. Grape Wine Res. 21, 367–375.

Murat, M.-L., 2005. Recent findings on rosé wine aromas. Part I: identifying aromas studying the aromatic potential of grapes and juice. Aust. NZ Grapegrower Winemaker 497a 64–65, 69, 71, 73–74, 76.

Murphy, C., 1995. Age-associated differences in memory for odors. In. In: Schab, F.R., Crowder, R.G. (Eds.), Memory of Odors. Lawrence Erlbaum, Mahwah, New Jersey, pp. 109–131.

Murphy, C., Cain, W.S., Bartoshuk, L.M., 1977. Mutual action of taste and olfaction. Sensory Processes 1, 204–211.

Murray, J.E., 2004. The ups and downs of face perception: Evidence for holistic encoding of upright and inverted faces. Perception 33, 387–398.

Murrell, J.R., Hunter, D.D., 1999. An olfactory sensory neuron line, odora, properly targets olfactory proteins and responds to odorants. J. Neurosci. 19, 8260–8270.

Nawar, W.W., 1971. Some variables affecting composition of headspace aroma. J. Agric. Food Chem. 19, 1057–1059.

Negoias, S., Visschers, R., Boelrijk, A., Hummel, T., 2008. New ways to understand aroma perception. Food Chem. 108, 1247–1254.

Nespoulous, C., Briand, L., Delage, M.M., Tran, V., Pernollet, J.C., 2004. Odorant binding and conformational changes of a rat odorant-binding protein. Chem. Senses 29, 189–198.

Noble, A.C., Williams, A.A., Langron, S.P., 1984. Descriptive analysis and quality ratings of 1976 wines from four Bordeaux communes. J. Sci. Food Agric. 35, 88–98.

Nordin, S., Almkvist, O., Berglund, B., 2012. Is loss in odor sensitivity inevitable to the gaing individual? A study of "successfully aged" elderly. Chem. Percept. 5, 188–196.

Nozawa, M., Kawahara, Y., Nei, M., 2007. Genomic drift and copy number variation of sensory receptor genes in humans. PNAS 104, 20421–20426.

Nygren, I.T., Gustafsson, I.-B., Johansson, L., 2003a. Effects of tasting technique – sequential tasting vs. mixed tasting – on perception of dry white wine and blue mould cheese. Food Service Technol. 3, 61–69.

Nygren, I.T., Gustafsson, I.-B., Johansson, L., 2003b. Perceived flavour changes in blue mould cheese after tasting white wine. Food Service Technol. 3, 143–150.

Olender, T., Fuchs, T., Linhart, C., Shamir, R., Adams, M., Kalush, F., et al., 2004. The canine olfactory subgenome. Genomics. 83, 361–372.

Olender, T., Lancet, D., 2012. Evolutionary grass roots for odor recognition. Chem. Senses 37, 581–584.

Ong, P., Acree, T.E., 1999. Similarities in the aroma chemistry of Gewürztraminer variety wines an lychee (*Litchi chinesis* Sonn.) Fruit. J. Agric. Food Chem. 47, 667–670.

Pangborn, R.M., 1981. Individuality in responses to sensory stimuli. In: Solms, J., Hall, R.L. (Eds.), Criteria of Food Acceptance. Forster, Zurich, pp. 177–219.

Parr, W.V., White, K.G., Heatherbell, D., 2003. The nose knows: Influence of colour on perception of wine aroma. J. Wine Res. 14, 99–121.

Patterson, M.Q., Stevens, J.C., Cain, W.S., Cometto-Muñiz, J.E., 1993. Detection thresholds for an olfactory mixture and its three constituent compounds. Chem. Senses 18, 723–734.

Pfeiffer, J.C., Hollowood, T.A., Hort, J., Taylor, A.J., 2005. Temporal synchrony and integration of sub-threshold taste and smell signals. Chem. Senses 30, 539–545.

Picard, M., Lytra, G., Tempere, S., Barbe, J.-C., de Revel, G., Marchand, S., 2016. Identification of piperitone as an aroma compound contributing to the positive mint nuances perceived in aged red bordeaux wines. J. Agric. Food Chem. 64, 451–460.

Pierce, J.D., Zeng, X., Aronov, E.V., Preti, G., Wysocki, C.J., 1995. Cross-adaptation of sweaty-smelling 3-methyl-2-henenoic acid by a structurally-similar, pleasant smelling odorant. Chem. Senses 20, 401–411.

Piggott, J.R., and Findlay, A.J.F. (1984). Detection thresholds of ester mixtures. In Proc. Alko Symp. Flavour Res. Alcoholic Beverages Helsinki 1984 (L. Nykänen and P. Lehtonen, eds.), Foundation Biotech. Indust. Ferm. 3, 189–197.

Pineau, B., Barbe, J.-C., van Leeuwen, C., Dubourdieu, D., 2007. Which impact for β-damascenone on red wines aroma? J. Agric. Food Chem. 55, 5214–5219.

Plassmann, H., O'Doherty, J., Shiv, B., Rangel, A., 2008. Marketing actions can modulate neural representations of experienced pleasantness. PNAS 105, 1050–1054.

Pons, A., Lavigne, V., Eric, R., Darriet, P., Dubourdieu, D., 2008. Identification of volatile compounds responsible for prune aroma in prematurely aged red wines. J. Agric. Food Chem. 56, 5285–5290.

Pons, A., Lavigne, V., Darriet, P., Dubourdieu, D., 2013. Role of 3-methyl-2,4-nonanedione in the flavor of aged red wines. J. Agric. Food Chem. 61, 7373–7380.

Rapp, A., 1998. Volatile flavour of wine: correlation between instrumental analysis and sensory perception. Nahrung 42, 351–363.

Rapp, A., Güntert, M., 1986. Changes in aroma substances during the storage of white wines in bottles. In: Charalambous, G. (Ed.), The Shelf Life of Foods and Beverages. Elsevier, Amsterdam, pp. 141–167.

Rapp, A., Versini, G., 1996. Vergleichende Untersuchungen zum Gehalt von Methylanthranilat ("Foxton") in Weinen von neueren pilzresistenten Rebsorten und *vinifera*-Sorten. Vitis 35, 215–216.

Rapp, A., Versini, G., Ullemeyer, H., 1993. 2-Aminoacetophenon: Verursachende Komponente der "Untypischen Alterungsnote" (Naphtalinton, Hybridton) bei Wein. Vitis 32, 61–62.

Read, E.A., 1908. A contribution to the knowledge of the olfactory apparatus in dog, cat and man. Am. J. Anatomy (Developmental Dynamics). 8, 17–47.

Richards, P.M., Johnson, E.C., Silver, W.L., 2010. Four irritating odorants target the trigeminal chemoreceptor TRPA1. Chem. Percept. 3, 190–199.

Rigou, P., Triay, A., Razungles, A., 2014. Influence of volatile thiols in the development of blackcurrant aroma in red wine. Foods Chem. 142, 242–248.

Robinson, A.L., Ebeler, S.E., Heymann, H., Boss, P.K., Solomon, P.S., Trengove, R.D., 2009. Interactions between wine volatile compounds and grape and wine matrix components influence aroma compound headspace partitioning. J. Agric. Food Chem. 57, 10313–10322.

Robinson, W.B., Shaulis, N., Pederson, C.S., 1949. Ripening studies of grapes grown in 1948 for juice manufacture. Fruit Prod. J. 29 36–37, 54, 62.

Rolls, E.T., 2005. Taste, olfactory, and food texture processing in the brain, and the control of food intake. Physiol. Behav. 85, 45–56.

Rolls, E.T., Critchley, H.D., Mason, R., Wakeman, E.A., 1996. Orbitofrontal cortex neurons: Role in olfactory and visual association learning. J. Neurophysiol. 75, 1970–1981.

Ross, C.F., Zwink, A.C., Castro, L., Harrison, R., 2014. Odour detection threshold and consumer rejection of 1,1,6-trimethyl-1,2-dihydronaphtha-lene in 1-year-old Riesling wines. Aust. J. Grape Wine Res. 20, 335–339.

Ross, P.E., 2006. The expert mind. Sci. Amer. 295 (2), 64–71.

Roujou de Boubée, D., Cumsille, A.M., Pons, M., Dubourdieu, D., 2002. Location of 2-methoxy-3-isobutylpyrazine in Cabernet Sauvignon grape bunches and its extractability during vinification. Am. J. Enol. Vitic. 53, 1–5.

Roussis, I.G., Lambropoulos, I., Papadopoulou, D., 2005. Inhibition of the decline of volatile esters and terpenols during oxidative storage of Muscat-white and Xinomavro-red wine by caffeic acid and N-acetyl-cysteine. Food Chem. 93, 485–492.

Russell, K., Zivanovic, S., Morris, W.C., Penfield, M., Weiss, J., 2005. The effect of glass shape on the concentration of polyphenolic compounds and perception of Merlot wine. J. Food Qual. 28, 377–385.

Ryan, D., Prenzler, P.D., Saliba, A.J., Scollary, G.R., 2008. The significance of low impact odorants in global odor perception. Trends Food Sci. Technol. 19, 383–389.

Sachs, O., 1985. The dog beneath the skin. In The Man Who Mistook His Wife for a Hat. Duckworth, London, 149–153

Sáenz-Navajas, M.-P., Campo, E., Culleré, L., Fernández-Zurbano, P., Valentin, D., Ferreira, V., 2010. Effects of the nonvolatile matrix on the aroma perception of wine. J. Agric. Food Chem. 58, 5574–5585.

Sakai, N., Imada, S., Saito, S., Kobayakawa, T., Deguchi, Y., 2005. The effect of visual images on perception of odors. Chem. Senses 30 (Suppl 1), i244–1245.

San-Juan, F., Cacho, J., Ferreira, V., Escudero, A., 2014. Differences in Chemical composition of aroma among red wines of different price category. In: Ferreira, V., Lopez, R. (Eds.), Favour Science: Proceeding from XIII Weurman Flavour Research Symposium. Sept 26, 2013. Elsevier, San Diego, CA, pp. 117–121.

San-Juan, F., Ferreira, V., Cacho, J., Escudero, A., 2011. Quality and aromatic sensory descriptors (mainly fresh and dry fruit character) of Spanish red wines can be predicted from their aroma-active chemical composition. J. Agric. Food Chem. 59, 7916–7924.

San-Juan, F., Pet'ka, J., Cacho, J., Ferreira, V., Escudero, A., 2010. Producing headspace extracts from the gas chromatography-olfactometric evaluation of wine aroma. Food Chem. 123, 188–195.

Sanz, G., Schlegel, C., Pernollet, J.-C., Briand, L., 2006. Evidence for antagonism between odorants at olfactory receptor binding in humans. In: Bredie, W.L.P., Petersen, M.A. (Eds.), Flavor Science. Vol 43. Recent Advances and Trends. Elsevier, Amsterdam, The Netherlands, pp. 9–12.

Savic, I., Gulyás, B., Berglund, H., 2002. Odorant differentiated pattern of cerebral activation: Comparison of acetone and vanillin. Human Brain Mapping 17, 17–27.

Schemper, T., Voss, S., Cain, W.S., 1981. Odor identification in young and elderly persons: sensory and cognitive limitations. J. Gerontol. 36, 446–452.

Schiffman, S., Orlandi, M., Erickson, R.P., 1979. Changes in taste and smell with age, biological aspects. In. In: Ordy, J.M., Brizzee, K. (Eds.), Sensory Systems and Communication in the Elderly. Raven, New York, pp. 247–268.

Schneider, V., 2014. Atypical aging defect: sensory discrimination, viticultural causes, and enological consequences. A review. Am. J. Enol. Vitic. 65, 277–284.

Schreier, P., Drawert, F., 1974. Gaschromatographisch-massenspektometrische Untersuchung flüchtiger Inhaltsstoffe des Weines, V. Alkohole, Hydroxy-Ester, Lactone und andere polare Komponenten des Weinaromas. Chem. Mikrobiol. Technol. Lebensm. 3, 154–160.

Schreier, P., Paroschy, J.H., 1981. Volatile constituents from Concord, Niagara (*Vitis labrusca*) and Elvira (*V. labrusca* x *V. riparia*) grapes. Can. Inst. Food Sci. Technol. J. 14, 112–118.

Sefton, M.A., Francis, I.L., Williams, P.J., 1993. The volatile composition of Chardonnay juices: A study by flavor precursor analysis. Am. J. Enol. Vitic. 44, 359–370.

Segurel, M.A., Razungles, A.J., Riou, C., Salles, M., Baumes, R.L., 2004. Contribution of dimethyl sulfide to the aroma of Syrah and Grenache noir wines and estimation of its potential in grapes of these varieties. J. Agric. Food Chem. 52, 7084–7093.

Senf, W., Menco, B.P.M., Punter, P.H., Duyvesteyn, P., 1980. Determination of odour affinities based on the dose-response relationships of the frog's electro-olfactogram. Experientia 36, 213–215.

Seow, Y.-X., Ong, P.K.C., Huang, D., 2016. Odor-specific loss of smell sensitivity with age as revealed by the specific sensitivity test. Chem. Senses 41, 487–495.

Sezille, C., Ferdenzi, C., Chakirian, A., Fournel, A., Thevenet, M., Gerber, J., et al., 2015. Neuroscience 287, 23–31.

Shepard, G.M., 2006. Smell images and the flavour system in the human brain. Nature 444, 316–321.

Simpson, R.F., 1978. 1,1,6-Trimethyl-1,2-dihydronaphthalene, an important contribution to the bottle aged bouquet of wine. Chem. Ind. (London) 1, 37.

Simpson, R.F., 1990. Cork taint in wine: a review of the causes. Aust. N.Z. Wine Ind. J. Nov. 286–287, 289, 293–296.

Simpson, R.F., Miller, G.C., 1984. Aroma composition of Chardonnay wine. Vitis 23, 143–158.

Simpson, R.F., Capone, D.L., Sefton, M.A., 2004. Isolation and identification of 2-methoxy-3,5-dimethylpyrazine, a potent musty compound from wine corks. J. Agric. Food Chem. 52, 5425–5430.

Sinding, C., Puschmann, L., Hummel, T., 2014. Is the age-related loss in olfactory sensibility similar for light and heavy molecules. Chem. Senses 39, 383–390.

Singleton, V.L., 1987. Oxygen with phenols and related reactions in musts, wines, and model systems, observation and practical implications. Am. J. Enol. Vitic. 38, 69–77.

Small, D.M., Gerber, J.C., Mak, Y.E., Hummel, T., 2005. Differential neural responses evoked by orthonasal versus retronasal odorant perception in humans. Neuron 47, 593–605.

Smith, D.V., Duncan, H.J., 1992. Primary olfactory disorders: Anosmia, hyperosmia, and dysosmia. In: Serby, M.J., Chobor, K.L. (Eds.), Science of Olfaction. Springer-Verlag, New York, pp. 439–466.

Sobel, N., Kahn, R.M., Hartley, C.A., Sullivan, E.V., Gabrieli, J.D., 2000. Sniffing longer rather than stronger to maintain olfactory detection threshold. Chem. Senses 25, 1–8.

Solomon, G.E.A., 1988. Great Expectations: The Psychology of Novice and Expert Wine Talk. Doctoral Thesis. Harvard university., (cited in Hughson and Boakes, 2002).

Spehr, M., Kelliher, K.R., Li, X.-H., Boehm, T., Leinders-Zufall, T., Zufall, F., 2006. Essential role of the main olfactory system in social recognition of major histocompatibility complex peptide ligands. J. Neurosci. 26, 1961–1970.

Sponholz, W.R., Hühn, T., 1996. Aging of wine: 1,1,6-Trimethyl-1,2-dihydronaphthalene (TDN) and 2-aminoacetophenone. In: Henick-Kling, T. (Ed.), Proc. 4th Int. Symp. Cool Climate Vitic. Enol. New York State Agricultural Experiment Station, Geneva, New York, pp. VI-37–57.

Stamatopoulos, P., Fréot, E., Tempère, S., Pons, A., Darriet, P., 2014. Identification of a new lactone contributing to overripe orange aroma in Bordeaux dessert wines via perceptual interaction phenomena. J. Agric. Food Chem. 62, 2469–2478.

Starkenmann, C., Chappuis, C.J.-F., Niclass, Y., Deneulin, P., 2016. Identification of hydrogen disulfanes and hydrogen trisulfanes in H2S bottle, in flint, and in dry mineral white wine. J. Agric. Food Chem. 64, 9033–9040.

Stevens, D.A., O'Connell, R.J., 1995. Enhanced sensitivity to androstenone following regular exposure to pemenone. Chem. Senses 20, 413–420.

Stevenson, R.J., Tomiczek, C., 2007. Olfactory-induced synesthesias: A review and model. Psychol. Bull. 133, 294–309.

Stevenson, R.J., Prescott, J., Boakes, R.A., 1999. Confusing tastes and smells: how odours can influence the perception of sweet and sour tastes. Chem. Senses 24, 627–635.

Stevenson, R.J., Case, T.I., Boakes, R.A., 2003. Smelling what was there: Acquired olfactory percepts are resistant to further modification. Learn. Motivat. 34, 185–202.

Stoddart, D.M., 1986. The role of olfaction in the evolution of human sexual biology: An hypothesis. Man 21, 514–520.

Strauss, C.R., Wilson, B., Gooley, P.R., and Williams, P.J. (1986). The role of monoterpenes in grape and wine flavor – A review. In Biogeneration of Aroma Compounds (T. H. Parliament and R. B. Croteau, eds.), pp. 222–242. ACS Symposium Series No. 317. American Chemical Society, Washington, DC.

Strauss, C.R., Wilson, B., Anderson, R., Williams, P.J., 1987a. Development of precursors of C_{13} nor-isoprenoid flavorants in Riesling grapes. Am. J. Enol. Vitic. 38, 23–27.

Strauss, C.R., Wilson, B., Williams, P.J., 1987b. Flavour of non-muscat varieties. In: Lee, T. (Ed.), Proc. 6th Aust. Wine Ind. Tech. Conf. Australian Industrial Publishers, Adelaide, Australia, pp. 117–120.

Sudraud, P., 1978. Évolution des taux d'acidité volatile depuis le début du siècle. Ann. Technol. Agric. 27, 349–350.

Suprenant, A., Butzke, C.E., 1996. Implications of odor threshold variations on sensory quality control of cork stoppers. In: Henick-Kling, T. (Ed.), Proc. 4th Int. Symp. Cool Climate Vitic. Enol.. New York State Agricultural Experimental Station, Geneva, NY, pp. VII-70–74.

Takeuchi, H., Ishida, H., Hikichi, S., Kurahashi, T., 2009. Mechanisms of olfactory masking in the sensory cilia. J. Gen. Physiol. 133, 583–601.

Takeuchi, H., Kato, H., Kurahashi, T., 2013. 2,4,6-Trichloroanisole is a potent suppressor of olfactory signal transduction. PNAS 110, 16235–16240.

Taylor, A.J., Linforth, R.S.T., Harvey, B.A., Blake, A., 2000. Atmospheric pressure chemical ionisation mass spectrometry for in vivo analysis of volatile flavor release. Food Chem. 71, 327–338.

Taylor, A.J., Cook, D.J., Scott, D.J., 2008. Role of odorants binding proteins: comparing hypothetical mechanisms with experimental data. Chem. Percept. 1, 153–162.

Tempere, S., Cuzange, E., Malik, J., Cougeant, J.C., de Revel, G., Sicard, G., 2011. The training level of experts influences their detection thresholds for key wine compounds. Chem. Percept. 4, 99–115.

Tempère, S., Cuzange, E., Schaaper, M.H., de Lescar, R., de Revel, G., Sicard, G., 2014. "Brett character" in wine: Is that a consensus among professional assessors? A perceptual and conceptual approach. Food Qual. Pref. 34, 29–34.

Tempere, S., Schaaper, M.H., Cuzange, E., de Lescar, R., de Revel, G., Sicard, G., 2016. The olfactory masking effect of ethylphenols: Characterization and elucidation of its origin. Food Qual. Pref. 50, 135–144.

Thomas-Danguin, T., Sinding, C., Romagny, S., El Mountassir, F., Atanasova, B., Le Berre, E., et al., 2014. The perception of odor objects in everyday life: a review on the processing of odor mixtures. Front. Psychol. 5 (504), 18. http://dx.doi.org/10.3389/fpsyg.2014.00504.

Tominaga, T., Darriet, P., Dubourdieu, D., 1996. Identification de l'acétate de 3-mercaptohexanol, composé à forte odeur de buis, intervenant dans l'arôme des vins de Sauvignon. Vitis 35, 207–210.

Tominaga, T., Furrer, A., Henry, R., Dubourdieu, D., 1998. Identification of new volatile thiols in the aroma of Vitis vinifera L. var. Sauvignon blanc wines. Flavour Fragr. J. 13, 159–162.

Tominaga, T., Baltenweck-Guyot, R., Peyrot des Gachons, C., Dubourdieu, D., 2000. Contribution of volatile thiols to the aromas of white wines made from several Vitis vinifera grape varieties. Am. J. Enol. Vitic. 51, 178–181.

Tominaga, T., Guimbertau, G., Dubourdieu, D., 2003. Contribution of benzenemethanethiol to smoky aroma of certain Vitis vinifera L. wines. J. Agric. Food Chem. 51, 1373–1376.

Tominaga, T., Masneuf, I., and Dubourdieu, D. (2004). Powerful aromatic volatile thiols in wines made from several Vitis vinifera L. grape varieties and their releasing mechanism. In Nutraceutical Beverages: Chemistry, Nutrition, and Health Effects. (F. Shahidi and D. K. Weerasinghe, eds.), pp. 314–337. ACS Symp. Ser. # 871, American Chemical Society, Washington, DC.

Tong, J., Mannea, E., Aimé, P., Pfluger, P.T., Yi, C.X., Castaneda, T.R., et al., 2011. Ghrelin enhances olfactory sensitivity and exploratory sniffing in rodents and humans. J. Neurosci. 31, 5841–5846.

Tozaki, H., Tanaka, S., Hirata, T., 2004. Theoretical consideration of olfactory axon projection with an activity-dependent neural network model. Mol. Cell Neurosci. 26, 503–517.

Trant, A.S., Pangborn, R.M., 1983. Discrimination, intensity, and hedonic responses to color, aroma, viscosity, and sweetness of beverages. Lebensm. Wiss. Technol. 16, 147–152.

Tsachaki, M., Linforth, R.S.T., Taylor, A.J., 2005. Dynamic headspace analysis of the release of volatile organic compounds from ethanolic systems by direct APCI-MS. J. Agric. Food Chem. 53, 8328–8333.

Tsachaki, M., Linforth, R.S.T., Taylor, A.J., 2009. Aroma release from wines under dynamic conditions. J. Agric. Food Chem. 57, 6976–6981.

Turin, L., Gane, S., Georganakis, D., Maniati, K., Skoulakis, E.M.C., 2015. Plausibility of the vibrational theory of olfaction. Proc. Natl. Acad. Sci. 112, E3154.

van Wyk, C.J., Augustyn, O.P.H., de Wet, P., Joubert, W.A., 1979. Isoamyl acetate—A key fermentation volatile of wines of Vitis vinifera cv Pinotage. Am. J. Enol. Vitic. 30, 167–173.

Vernin, G., Metzger, J., Rey, C., Mezieres, G., Fraisse, D., Lamotte, A., 1986. Arômes des cépages et vins du sud-est de la France. Prog. Agric. Vitic. 103, 57–98.

Versini, G., 1985. Sull'aroma del vino "Traminer aromatico" o "Gewürztraminer.". Vignevini 12, 57–65.

Voilley, A., Beghin, V., Charpentier, C., Peyron, D., 1991. Interactions between aroma substances and macromolecules in a model wine. Lebensm. Wiss. Technol. 24, 469–472.

Wedral, D. (2007) Presence of Brettanomyces bruxellensis in North Georgia Wines and Chemical Interaction Resulting Flavor Metabolites and Acetic Acid. Thesis, Cornell University, Ithaca, NY.

Wells, D.L., Hepper, P.G., 2000. The discrimination of dog odors by humans. Perception 29, 111–115.

Wenzel, K., Dittrich, H.H., Seyffard, H.P., Bohnert, J., 1980. Schwefelrückstände auf Trauben und im Most und ihr Einfluß auf die H_2S-Bildung. Wein Wissenschaft 35, 414–420.

Williams, A.A., Rosser, P.R., 1981. Aroma enhancing effects of ethanol. Chem. Senses 6, 149–153.

Williams, A.A., Bains, C.R., Arnold, G.M., 1982. Towards the objective assessment of sensory quality in less expensive red wines. In: Webb, A.D. (Ed.), Grape Wine Centennial Symp. Proc. University of California, Davis, pp. 322–329.

Williams, A.A., Rogers, C., and Noble, A.C. (1984). Characterization of flavour in alcoholic beverages. In Proc. Alko Symp. Flavour Res. Alcoholic Beverages. Helsinki, 1984. (L. Nykänen and P. Lehonen, eds.). Found. Biotech Indust. Ferment. Res. 3, 235–253.

Wilson, D.A., Stevenson, R.J., 2003. The fundamental role of memory in olfactory perception. Trends Neurosci. 26, 243–247.

Wilson, D.A., Fletcher, M.L., Sullivan, R.M., 2004. Acetylcholine and olfactory perceptual learning. Learn. Mem. 11, 28–34.

Winton, W., Ough, C.S., Singleton, V.L., 1975. Relative distinctiveness of varietal wines estimated by the ability of trained panelists to name the grape variety correctly. Am. J. Enol. Vitic. 26, 5–11.

Wollan, D., 2005. Controlling excess alcohol in wine. Aust. NZ Wine Ind. J. 20, 48–50.

Wood, C., Siebert, T.E., Parker, M., Capone, D.L., Elsey, G.M., Pollnitz, A.P., et al., 2008. From wine to pepper: Rotundone, an obscure sesquiterpene, is a potent spicy aroma compound. J. Agric. Food Chem. 56, 3738–3744.

Wright, K.M., Hills, B.P., Hollowood, T.A., Linforth, R.S.T., Taylor, A.J., 2003. Persistence effects in flavour release from liquids in the mouth. Int. J. Food Sci. Technol. 38, 343–350.

Wysocki, C.J., Dorries, K.M., Beauchamp, G.K., 1989. Ability to perceive androsterone can be acquired by ostensibly anosmic people. Proc. Natl. Acad. Sci. USA 86, 7976–7978.

Yabuki, M., Scott, D.J., Briand, L., Taylor, A.J., 2011. Dynamics of odorant binding to thin aqueous films of rat-OBP3. Chem. Senses 36, 659–671.

Yoder, W.M., Stratis, K., Pattanail, S., Molina, S., Nguyen, J., Weisberg, S., et al., 2013. Time course of perceptual adaptation differs among odorants. J. Sens. Stud. 28, 495–503.

Young, J.M., Endicott, R.M., Parghi, S.S., Walker, M., Kidd, J.M., Trask, B.J., 2008. Extensive copy-number variation of the human olfactory receptor gene family. Am. J. Hum. Genet. 83, 228–242.

Yousem, D.M., Maldjian, J.A., Siddiqi, F., Hummel, T., Alsop, D.C., Geckle, R.J., et al., 1999. Gender effects on odor-stimulated functional magnetic resonance imaging. Brain Res. 818, 480–487.

Zampini, M., Wantling, E., Phillips, N., Spence, C., 2008. Multisensory flavor perception: Assessing the influence of fruit acids and color cues on the perception of fruit-flavored beverages. Food Qual. Pref. 19, 335–343.

Zellner, D.A., Whitten, L.A., 1999. The effect of color intensity and appropriateness on color-induced odor enhancement. Am. J. Psychol. 112, 585–604.

Zhao, K., Scherer, P.W., Hajiloo, S.A., Dalton, P., 2004. Effect of anatomy on human nasal air flow and odorant transport patters: Implications for olfaction. Chem. Senses 29, 365–379.

Zou, Z., Buck, L.B., 2006. Combinatorial effects of odorant mixes in olfactory cortex. Science 311, 1477–1481.

Zufall, F., Leinders-Zufall, T., 2000. The cellular and molecular basis of odor adaptation. Chem. Senses 25, 369–380.

4

Oral Sensations (Taste and Mouth-Feel)

In comparison with the incredible diversity of odor qualities, there are only five (or possibly six to seven) taste modalities and several mouth-feel (chemesthesis) sensations. These attributes tend to retain their modal quality, independent of concentration or combination. In addition, emotional responses to these sensations are hardwired, potentially changing only with experience. In contrast, there are no known hardwired emotional responses for odorants, except those classed as pungent, putrid, or in some other way irritant. Finally, wine-induced taste and mouth-feel sensations are responses to a wine's major chemical constituents, i.e., sugars, acids, alcohol, and phenolics. These usually occur in the order of parts per thousand or more, whereas odorants tend to be minor constituents, being occasionally detectable at parts per trillion.

Taste and mouth-feel sensations arise from two distinct types of chemoreceptors. One set involves specialized receptor cells that generate gustatory sensations, notably sweet, sour, salt, bitter, and umami (savory). They are found primarily in cavities within taste buds. Some also occur scattered in the endothelial layer of the small intestine (San Gabriel, 2015), as well as the nose (Lee et al., 2014). The other set of receptors involves free nerve endings of branches of the trigeminal nerve. They generate the mouth-feel sensations of chemesthesis in the oral cavity. They include the perceptions of astringency, dryness, viscosity, burning, heat, coolness, touch, prickling, and pain. These nerve endings are scattered throughout the oral cavity, as well as to a lesser degree in the nasal passages. The combination of both sets of oral modalities, with perceptions derived from the nasal passages (olfactory and chemesthesis), produces the cerebral construct termed flavor. The olfactory component of flavor comes from volatile compounds that enter the nasal passages primarily via the back of the throat (retronasal odor) (Chapter 3, Olfactory Sensations).

By themselves, taste and mouth-feel modalities are comparatively monolithic. This feature is particularly evident when a head cold clogs the nasal passages. Because the passage of odorants from food (or wine) to the olfactory patches is impeded, both lose most of their sensory appeal.

Achieving balance between the various olfactory and gustatory sensations of a wine is one of a winemaker's most demanding tasks. Harmony among the diverse sensations in a wine is a distinguishing feature of a superior wine. Imbalances created by excessive acidity, astringency, bitterness, etc. are often the first failing noted by a taster. Thus, although relatively sensorially simple, gustatory and mouth-feel sensations are critical to the perception of wine quality.

TASTE

As noted above, gustatory sensations are detected by specialized epithelial receptor cells, located in flask-shaped depressions called taste buds (Fig. 4.1). They may contain a few to over 100 slender, columnar, receptor cells, specialized supporting cells, and basal cells. Basal cells differentiate into receptor cells on a regular basis, the life of which averages about 8–12 days. About two-thirds of all taste buds are found on the tongue, where they are found on the sides of raised growths called papillae. The remainder occur primarily on the soft palate and epiglottis (Schiffman, 1983). A few taste buds are also found on the pharynx, larynx, and upper portions of the esophagus.

Taste buds are associated with three of the four classes of papillae (Fig. 4.2). Fungiform papillae occur primarily on the front two-thirds of the tongue, especially at the tip and along the sides. Each contains up to about 20 taste buds. Fungiform papillae are sufficiently critical to gustation that their density has been correlated with taste acuity (Miller and Reedy, 1990; Zuniga et al., 1993). Nonetheless, fungiform density varies markedly from person to person,

Wine Tasting.
DOI: http://dx.doi.org/10.1016/B978-0-12-801813-2.00004-5

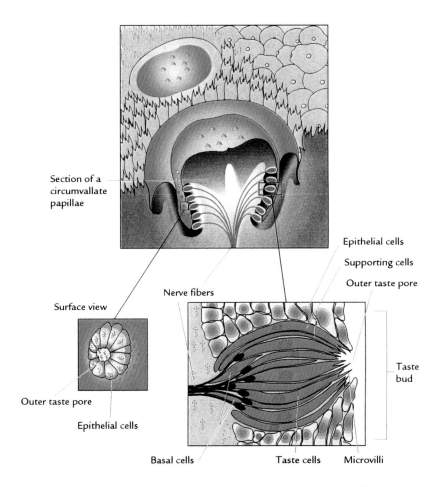

FIGURE 4.1 Schematic drawing of a taste bud within a circumvallate papilla. The top portion of the figure shows the location of the taste buds within the papillae, and the bottom figures show the taste bud itself. *From Levine, M.W., Shefner, J.M. 1991. Fundamentals of Sensation and Perception, second ed. Brooks/Cole Publishing, Pacific Grove, CA, reproduced by permission.*

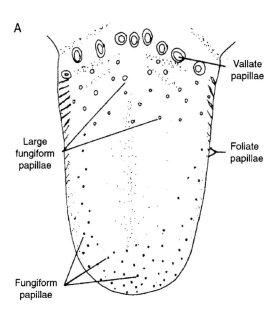

FIGURE 4.2 Schematic drawing of the tongue indicating the location of the major types of papillae. *From Jackson, 2014, reproduced by permission.*

depending on their genetics, sex, age, and modifiable factors such as smoking (Fischer et al., 2013). A few large circumvallate papillae develop along a V-shaped zone across the back of the tongue. They may each contain hundreds of taste buds. Foliate papillae are restricted to ridges between folds, along the edges of the posterior portion of the tongue, each containing 10 or more taste buds. Filiform papillae are the most common class, but contain no taste buds. Their tapering, fibrous extensions give the tongue its characteristic rough texture. The absence of significant numbers of taste buds in the central regions of the tongue means that this zone is essentially insensitive to gustatory sensations.

Gustatory receptor cells located in taste buds are specialized epithelial cells that possess characteristics typical of nerve cells. Each receptor cell culminates in a receptive dendrite (microvillus), or many fine extensions (microvilli). These project into the saliva, filling the lumen of the taste bud. Microvilli possess multiple copies of a receptor protein. When sufficient tastant reacts with these proteins, a cascade of events changes the cell's membrane potential, leading to the release of neurotransmitters from the base of the cell body (Nagai et al., 1996). After diffusion across a synapse, neurotransmitters incite depolarization of associated afferent nerve cells, generating an action potential that propagates an impulse along the nerve toward the brain. Each afferent nerve cell may synapse with many receptor cells, in several adjacent taste buds. Nerve stimulation not only generates the impulses sent to the brain, but also maintains taste-bud integrity. Although taste buds tend to possess receptors to all tastant modalities, there is partial localization of sensitivity across the tongue (Zhao et al., 2003; Fig. 4.3). In addition, the brain appears to respond to signals from certain regions in programmed patterns, regardless of the actual stimulus (Chen et al., 2011).

The neurons that enervate taste buds originate from one of three cranial nerves, each associated with separate areas of the tongue, palate, and throat. Impulses from the nerves pass initially to the solitary nucleus in the brain stem. Subsequently, most impulses pass to the thalamus and taste centers in the cortex. Additional neurons transmit signals to the hypothalamus and amygdala, evoking the emotional response to the stimulus.

Of the five taste modalities currently recognized—sweet, bitter, sour, salty, and umami (savory)—only the first three have direct relevance to wine quality and its combination with food. A metallic "taste" sensation, occasionally detected in wine, is in reality a retronasal odor perception, misinterpreted as a taste sensation (Epke et al., 2009).

Sensitivity to the various tastants is associated with specific receptor proteins, or their combination (Gilbertson and Boughter, 2003). Individual receptor cells produce only one or a select pair of receptor proteins. Thus, each receptor may generate impulses to one or a few modalities. Intriguingly, the release of neurotransmitters from one receptor cell appears to activate afferent neurons associated with adjacent receptors, which may be responsive to other modalities.

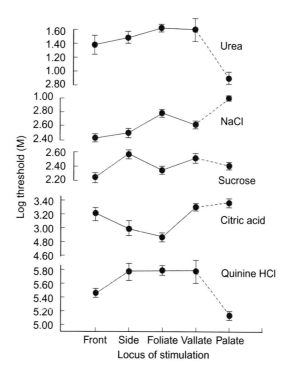

FIGURE 4.3   Threshold of five compounds as a function of locus on the tongue and soft palate. *From Collings, V.B., 1974. Human taste response as a function of focus on the tongue and soft palate. Front and Side refer to fungiform papillae. The higher the value along the y-axis denotes lower the taste threshold. Percept. Psychophys. 16, 169–174, reproduced by permission.*

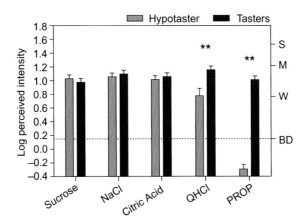

FIGURE 4.4   Log means of perceived intensity ± standard error for the sweetness of sucrose, saltiness of NaCl, sourness of citric acid, bitterness of quinine chloride (QHCl), and bitterness of PROP, grouped by PROP taster status. Asterisk indicates significant differences on perceived intensities by the t-test ($^{%%}p < .005$). *Reprinted from Lim, J., Wood, A., and Green, B. G. (2009) Derivation and evaluation of a labeled hedonic scale. Chem. Senses* **34**, *739–751., by permission of Oxford University Press.*

Response to sweet, umami, and bitter tastants is associated with a group of about 30 related *TAS* (taste) genes. Sour and salty gustatory responses are associated with an unrelated group of genes. They encode ion-channel transporters, responding respectively to cations, either H^+ or metallic ions, notably Na^+.

Taste acuity has been correlated with the number of taste pores on the tongue. Most people possess about 70 fungiform papillae per cm^2 (at the tip of the tongue). In contrast, individuals with more densely packed fungiform papillae ($>100/cm^2$) show enhanced taste sensitivity (hypertasters). Conversely, those with significantly reduced tastant sensitivity (hypotasters) have about 50 fungiform papillae per cm^2 (Bartoshuk et al., 1994). In addition, the fungiform papillae of hypertasters are smaller, but have more taste buds per papilla. For example, hypertasters possess about 670 taste buds per cm^2, whereas average and hypotaster categories possess about 350, and 120 taste buds per cm^2 tongue surface). The taste buds of hypertasters may also have significantly greater enervation by trigeminal nerves, and are correspondingly more responsive to chemesthesis sensations.

The number of fungiform papillae at the tip of the tongue also correlates with sensitivity to the bitter tastants (6-*n*-propylthiouracil [PROP] and phenylthiocarbamide [PTC]). However, sensitivity to these compounds does not necessarily correlate with sensitivity to other bitter compounds (Roura et al., 2015), or other taste modalities, especially for hypotasters (Fig. 4.4). In addition, sensitivity to bitter tastants is poor at the tip of the tongue. Most cells sensitive to bitter compounds are located at the back of the tongue, where few fungiform papillae occur. The genetic basis underlying variation in the number of taste buds/papilla, or the number of receptors/taste bud appears not to have been investigated. Thus, the origin of the differences in acuity may be more complex than suspected, being also influenced by factors such as sex (Fig. 4.5), saliva production (Matsuo, 2000), and age.

Because hypertasters are more sensitive to oral sensations than average, it depends on the function of a panel of tasters whether such atypically sensitive individuals are preferable or not. Being especially sensitive to sapid substances, they might reject or rank poorly those wines marginally sour, bitter, or sweet. The same issue applies to those particularly sensitive to odors. If the intention is analytic or discriminatory, then panelists possessing especially acute sensitivity to sapid and olfactory sensations would likely be preferable. However, if the data are to relate to average consumer acceptance, then tasters reflecting the norm of human sensitivity would be more appropriate. At present, there appears to be no evidence that gustatory sensitivity is correlated with olfactory acuity.

Sweet, Umami, and Bitter Tastes

Although the modalities of sweet, umami, and bitter appear perceptively unrelated, sensory response is based on proteins encoded by a pair of related genes: *TAS1R* and *TAS2R*.

The *TAS1R* group consists of three genes. They encode proteins responsive to sweet and/or savory (umami) tastants. The binary product of TAS1R2 and TAS1R3 proteins provides high sensitivity to sugars, artificial sweeteners, and a few amino acids (Li et al., 2002), whereas the binary product of TAS1R1 and TAS1R3 produces a receptor responsive to umami activators, such as L-amino acids, notably monosodium glutamate (MSG) (Nelson et al., 2002; Matsunami and Amrein, 2004). Allelic variants of *TAS1R* genes are thought to be the source of the various qualitative differences experienced among different sugars and sugar substitutes.

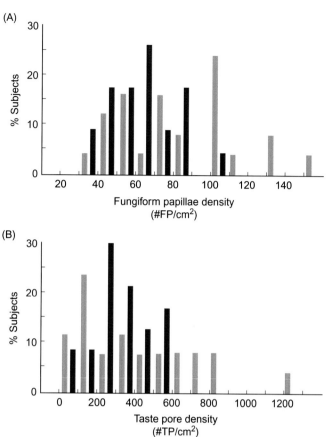

FIGURE 4.5 Distribution of (A) fungiform papillae (FP) density, and (B) taste pore (TP) density on the anterior tongue. Frequencies are expressed as percent of each sex: *pale bar*, female; *dark bar*, male. *From Bartoshuk, L.M., Duffy, V.B., Miller, I.J., 1994. PTC/PROP tasting: Anatomy, psychophysics and sex effects. Physiol. Behavior 56, 1165–1171, reproduced by permission.*

The *TAS2R* gene group consists of about 25 functional genes (not counting nonfunctional pseudogenes). Each encodes a receptor protein responsive to one or more bitter tastants (Meyerhof et al., 2010). These can include certain amino acids and acetylated sugars, alkaloids, amines, carbamates, ionic salts, isohumulones, phenolics, and ureas/thioureas. Receptor cells responsive to bitter tastants occur primarily in taste buds at the back of the tongue and palate, but sporadically elsewhere in the oral cavity. Individual receptors cells may coexpress several *TAS2R* gene transcripts, and thereby may respond to several bitter tastants. Nevertheless, there appears to be at least some specialization in receptor sensitivity to particular chemical subgroups. For example, alkaloid-induced bitterness is detected primarily at the tip of the tongue, but not the bitterness associated with iso-α-amino acids. All cells expressing *TAS2R* genes also express a gustatory G-protein, β-gustaducin, important in sensing bitterness. There may also exist a second bitter sensing mechanism. It appears to involve compounds that can penetrate taste receptor cells, directly activating the cell, and generating a lingering bitter aftertaste (Sawano et al., 2005). Age-related loss in bitterness sensitivity appears to be chemically specific (Cowart et al., 1994).

There is considerably more allelic variation within the *TAS2R* loci than would be expected, based on variation in other human genes (Kim et al., 2005). This situation seems to mimic variation in olfactory receptor (OR) genes.

TAS1R and TAS2R receptor proteins are associated with guanosine triphosphate (GTP) and, correspondingly, termed G-proteins. Through a sequence of reactions, tastants indirectly induce depolarization of the cell membrane. The initial stage involves activation of adenylate cyclase. This modifies K^+ conductance and intracellular free-calcium concentration (Gilbertson and Boughter, 2003). This, in turn, results in the release of neurotransmitters that initiates activation of associated afferent neurons, and impulse transmission to the brain.

Slight structural changes in many sweet- and bitter-tasting compounds can change their taste quality from sweet to bitter, or vice versa. The change in taste quality appears partially due to their ability to jointly activate members of both receptor groups. For example, the sweet tastant, saccharine, is often perceived to have a bitterish aspect. This presumably results from its activation of the bitter receptors TAS2R43 and TAS2R44 (Kuhn et al., 2004). Bitter- and sweet-tasting compounds are also well known to mask the perception of each other's intensity, without modifying their individual sensory modality. The molecular origins of this phenomenon are presently unknown.

The primary sources of sweetness in wine are glucose and fructose. Their perception is enhanced in the presence of glycerol and ethanol. There are no significant umami tastants found in wine.

Flavonoids are the constituents most associated with wine bitterness (Kielhorn and Thorngate, 1999; Vidal et al., 2003; Hufnagel and Hofmann, 2008). Separate compounds appear to activate different TAS2R receptors. For example, (–)-epicatechin activates three (TAS2R4, TAS2R5, and TAS2R39), pentagalloylglucose (a hydrolyzable tannin) activates two (TAS2R4 and TAS2R5), whereas malvidin-3-glucoside and a procyanidin trimer activate only TAS2R7 and TAS2R5, respectively (Soares et al., 2013). Because the same compounds may generate, to varying degrees, both bitterness and astringency, these attributes can be confused (Lee and Lawless, 1991), or potentially mask one another (Arnold and Noble, 1978), notably if one of these modalities is more intense than the other.

During aging, red wines often become smoother in taste. This partially relates to flavonoids polymerizing into tannins. Large tannin polymers lose their ability to effectively fit the receptor sites in taste buds. However, hydrolyzable tannins from oak, which break down into their monomers in wine, may increase bitterness during aging. In addition to wine phenolics, several glycosides, amines, triterpenes, and alkaloids may elicit bitter sensations. Examples include some terpene glycosides found in Muscat wines (Noble et al., 1988), and the flavanone glycoside, naringin, found in Riesling wines. Tyrosol, produced by yeasts during fermentation, is bitter but unlikely to occur in sufficient concentration to be detectable. Whether some of these compounds act in conjunction to induce bitterness is unknown.

Although white wines are not noted for bitterness, if aged in oak cooperage they may extract sufficient nonflavonoids, such as gallic acid, to generate detectable bitterness as well as astringency. In addition, wine can extract a variety of nonflavonoid lignans and quercotriterpenoside flavorants from oak to affect the wine's oral attributes. Both can reach concentrations sufficient to generate bitter and sweet sensations, respectively (Marchal et al., 2011, 2014).

Wine bitterness may also increase as a consequence of spoilage bacteria that can produce acrolein (Rentschler and Tanner, 1951). Additional sources of bitter tastants can be derived from the addition of pine resins (as in Retsina), or the herbs and barks used to flavor vermouth.

Many amino acids possess a variety of taste sensations, but occur in wine at concentrations unlikely to have a perceptible influence. Fatty acids also activate particular receptors cells in the mouth, but as with amino acids, they may occur in wine in amounts below detectable levels.

Sour and Salty Tastes (ASIC and ENaC Channels)

Sourness and saltiness are commonly called the electrolytic tastes. In both instances, small soluble inorganic cations (positively charged ions) are the stimulants. For acid receptors, H^+ ions are selectively transported across the cell membrane (ion channels), whereas for salt receptors, Na^+ is the primarily active ion. In both cases, ion influx induces membrane depolarization and a cascade of reactions leading to the release of neurotransmitters from axonal endings. Since related ion channels are involved in both instances, acids frequently express some saltiness, and some salts show a mild sour aspect (Ossebaard et al., 1997).

Acid and salt detection appear to be associated with distinct, but related, genes. Those that respond primarily to acids possess a receptor protein encoded by the acid-sensing ion channel gene (ASIC2) (Gilbertson and Boughter, 2003; Gonzales et al., 2009), and/or the PKD2L1 gene (Chandrashekar et al., 2009), whereas salt detection seems to be most associated with an ENaC gene (Chandrashekar et al., 2010).

The tendency of acids to dissociate into their component ionic constituents is influenced primarily by the anionic (negatively charged) component and pH. Thus, both factors significantly affect H^+ ion liberation, and the acid's perceived sourness. Although undissociated acid molecules do not effectively stimulate receptor neurons, their presence can affect the perceived acidity of other organic acids (Ganzevles and Kroeze, 1987). The major sourness-inducing acids in wine are tartaric, malic, and lactic acids. These acids can also induce astringency, possibly by denaturing saliva (Sowalsky and Noble, 1998) and/or membrane proteins of oral epithelial cells, as well as by enhancing the astringency of phenolics (Siebert and Euzen, 2008). Conversely, phenolics, the wine's principal astringent agents, significantly enhance the sourness of acids (Peleg et al., 1998). Additional acids occur in wine, but, except for acetic acid, rarely occur in sufficient amounts to affect perceived acidity.

In addition to activating H^+ ion receptors in taste buds, acids can also stimulate nocioreceptors (pain receptors) connected to trigeminal nerve endings. This produces the sharp sensation found in excessively acidic wines.

Acid-receptor cells also react to carbon dioxide. However, this requires oral epithelial cells to possess carbonic anhydrase (CA4) in their membranes. The enzyme rapidly converts CO_2 into carbonic acid. H^+ ions liberated on ionization of the acid activate acid-responsive receptor cells in taste buds (Chandrashekar et al., 2009).

Salts that dissociate sufficiently under wine conditions can induce saltiness. The most frequently associated salt cations in wine are K^+, Ca^{2+}, Na^+. Their corresponding anions tend to be either Cl^- or bitartrate. As with sourness, salt perception is not solely influenced by the activating salt cation. The ability of a salt to ionize under wine conditions strongly affects perceived saltiness. For example, large organic anions (e.g., bitartrate) limit the sensation of saltiness by restricting dissociation, as well as delaying reaction time (Delwiche et al., 1999). Because the major wine salts (tartrates and bitartrates) possess large organic anions, weak dissociation at wine pH values limits their perceived saltiness. In addition, because the typical cation in wine is K^+ (not the primary salt activating ion, Na^+), saltiness is rarely detected in wine. If detected, it is probably due to H^+ ions activating salt receptors. Nonetheless, in situations of poor soil-water quality and insufficient rainfall, soil salt contents can reach levels resulting in sufficient sodium uptake to donate detectable saltiness, and negatively affect wine quality (de Loryn et al., 2014).

FACTORS INFLUENCING TASTE PERCEPTION

Multiple factors affect the ability to detect and identify taste sensations. These can be conceptually divided into four categories: physicochemical, chemical, biologic, and psychologic.

Physicochemical

After a century of investigation, the role of temperature on taste perception is still uncertain. The general view is that perception is optimal at normal mouth temperature. For example, cooling reduced the sweetness of sugars (Fig. 4.6) and the bitterness of caffeine (Green and Frankmann, 1987). However, temperature had limited effects on

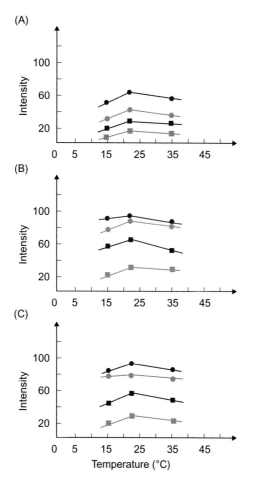

FIGURE 4.6 Variation in the sweetness intensity of (A) D-glucose, (B) D-fructose, and (C) sucrose as a function of temperature and concentration (● 9.2 g/mL; ○ 6.9 g/mL; ■ 4.6 g/mL; □ 2.3 g/mL). *From Portmann, M.-O., Serghat, S., Mathlouthi, M., 1992. Study of some factors affecting intensity/time characteristics of sweetness. Food Chem. 44, 83–92, reproduced by permission.*

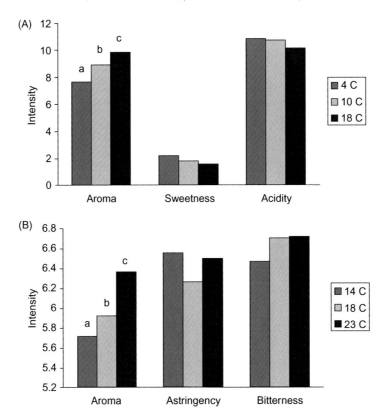

FIGURE 4.7   The effect of serving temperature of several attributes of wine: (A) white wine, (B) red wine. Within each attribute different letters indicate significant differences ($p \leq .05$; n = 72). *From Ross, C.F., and Weller, K., 2008. The effect of serving temperature on the sensory attributes of red and white wines. J. Sens. Stud. 23, 398–416, reproduced by permission.*

the taste attributes of white and red wines, over the range of temperatures at which these wines might usually be served (Fig. 4.7). One of the difficulties in these studies is differentiation between the effects of the temperature of the wine and that of the receptors in the mouth over the course of the experiment.

Another factor affecting taste perception is pH. It not only directly influences salt and acid ionization, but also can indirectly affect the shape and biological activity of receptor and other cellular membrane proteins. However, any resultant distortion of taste responsiveness due to pH may be limited, at least initially, by the buffering activities of saliva.

Chemical

Sapid substances not only directly activate specific taste receptors, presumably in a key-in-lock manner, but also influence the perception of other tastants. For example, mixtures of different sugars can suppress the perceived intensity of sweetness, especially at high concentrations (McBride and Finlay, 1990). In addition, members of one modality can affect the perception of another. At low and moderate concentrations, the effects often tend to be additive, whereas at high concentration, suppression is the norm. Nevertheless, generalizations are tenuous, as the interactions are often complex and vary with the compounds involved (Keast and Breslin, 2003; Sáenz-Navajas et al., 2012).

A common example of intermodal interaction is the suppression of bitterness, astringency, and sourness by sugar (Lyman and Green, 1990; Smith et al., 1996). The saltiness of some cheeses can equally suppress bitterness (Breslin and Beauchamp, 1995), thereby rationalizing the common presence of hard cheeses at most wine tastings. Such interactions may not be reciprocal. For example, tannins had only a minor effect of the perceived sweetness of sucrose (Fig. 4.8), whereas sucrose progressively suppressed tannin-induced astringency (Fig. 4.9). Although suppression of another modality is common, the reverse is possible, e.g., the enhancement of sweetness, bitterness (Noble, 1994, Vidal et al., 2004c), and astringency by ethanol (Gawel et al., 2013). Another example is the increase in tannin-induced bitterness and astringency by acidity.

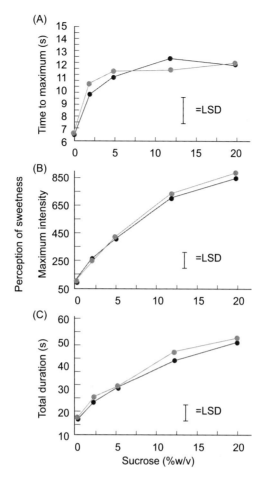

FIGURE 4.8 Effect of the astringency of tannins (○ low, 1300 GAE; ● high, 1800 GAE) on sweetness perception: (A) maximum intensity, (B) time to maximum intensity, and (C) total duration. (*GAE*, gallic acid equivalents). *From Ishikawa, T., Noble, A.C., 1995. Temporal perception of astringency and sweetness in red wine. Food Qual. Pref. 6, 27–34, reproduced by permission.*

Similar to the effect of aromatic compounds on flavor, sapid substances can significantly influence olfactory perception (Voilley et al., 1991; Dufour and Bayonove, 1999a,b; Sáenz-Navajas et al., 2010a, Frank et al., 2011). Cross-modal interactions seem to be the rule (Delwiche and Heffelfinger, 2005), especially at subthreshold concentrations.

Many sapid substances have more than one sensory modality. For example, small phenol polymers may be both bitter and astringent, as well as possess fragrance; glucose can be sweet and mildly tart; acids can show both sour and astringent properties; potassium salts are both salty and bitter; and alcohol can generate perceptions of heat, weight, and sweetness. In mixtures, these secondary (side-tastes) can significantly affect overall taste quality (Kroeze, 1982).

Nonetheless, the perceived intensity of a mixture generally reflects the intensity of the dominant component, not a simple summation of their individual effects (McBride and Finlay, 1990). The origin of these interactions may be various and complex (Avenet and Lindemann, 1989).

The interaction of sapid compounds in wine is further complicated by its changing chemical composition in the mouth. Wine stimulates saliva flow (Fig. 4.10), which both dilutes and modifies wine chemistry. Saliva contains proteins, of which about 70 percent are rich in proline (25–42 percent), as well as glutamine and glycine (Bennick, 1982). This facilitates binding with phenolics, notably catechins, gallic acid, and proanthocyanidins (Obreque-Slíer et al., 2011). Saliva histatins also react with polyphenolics (Wróblewski et al., 2001). Such bonding not only reduces bitterness (by limiting their ability to react with bitter receptors), but also astringency (Glendinning, 1992) (by limiting the binding of phenolics to receptors of trigeminal nerves as well as membrane proteins of epithelial cells). Because the chemical composition of saliva changes throughout the day, and often varies among individuals, it is difficult to predict the specific effects of saliva on oral perception. People also differ in their saliva flow rates, affecting how individuals perceive tastants (Fischer et al., 1994). Increased gustatory sensitivity appears to be partially correlated with reduced saliva flow rate (Condelli et al., 2006).

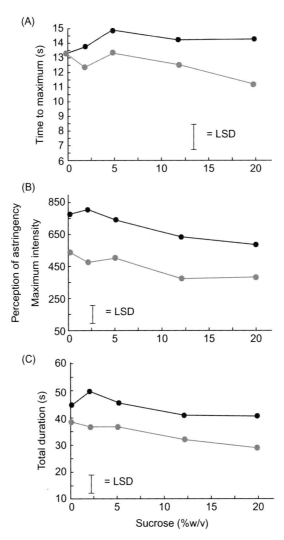

FIGURE 4.9   Effect of sweetness on the perception of astringency (A) time to maximum intensity, (B) maximum intensity, and (C) total duration (○ low, 1300 GAE; ● high, 1800 GAE). *From Ishikawa, T., Noble, A.C., 1995. Temporal perception of astringency and sweetness in red wine. Food Qual. Pref. 6, 27–34, reproduced by permission.*

Biologic

Several studies have noted a loss in sensory acuity with age (Bartoshuk et al., 1986; Stevens and Cain, 1993). The number of papillae on the tongue reaches a maximum in midchildhood, declining slowly thereafter. The number of taste buds and sensory receptors per taste bud also declines, notably after middle age. Losses in taste acuity are not solely directly age-related, but partially are a consequence of a loss of independence (Sulmont-Rossé et al., 2015). Nevertheless, age-related sensory loss is not known to seriously limit wine-tasting ability; experience and concentration seemingly offset diminished acuity.

Certain medications can significantly impair acuity (Schiffman, 1983) by generating their own tastes, and/or producing distorted taste responses (Doty and Bromley, 2004). For example, acetazolamide blocks the response of nerve fibers to carbonation, eliminating the prickling sensation of sparkling wine (Komai and Bryant, 1993). In addition, household products can disrupt taste perception. A familiar example is the effect of sodium lauryl sulfate (sodium dodecyl sulfate), a common ingredient in toothpaste (DeSimone et al., 1980).

Chronic oral and dental ailments may create lingering oral sensations, complicating discrimination at low concentrations (Bartoshuk et al., 1986). This could explain why detection thresholds are usually higher in elderly people with natural dentation than those with dentures. Acuity loss further complicates identification of the individual components of tastant mixtures (Stevens and Cain, 1993).

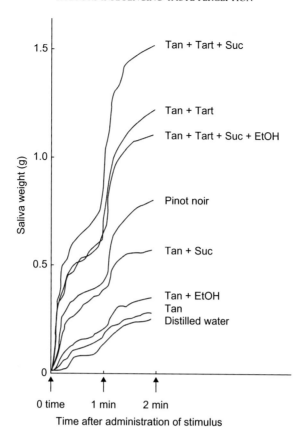

FIGURE 4.10 Amount of parotid saliva secreted in response to tasting Pinot noir wine and selected constituents of the wine, singly and in combination. *EtOH*, ethanol; *Suc*, sucrose; *Tan*, tannin; *Tart*, tartaric acid. *From Hyde, R.J., Pangborn, R.M., 1978. Parotid salivation in response to tasting wine. Am. J. Enol. Vitic. 29, 87–91, reproduced by permission.*

Allelic variants in some of the 25 *TAS2R* genes are associated with taste deficiencies, e.g., the association of certain alleles of *TAS2R38* and reduced sensitivity to PTC and PROP (Behrens et al., 2013). Two functional alleles donate the property of hypertaster, one functional allele is associated with moderate sensitivity, and two nonfunctional alleles result in hypotaster status. The degree of sensitivity to PTC and PROP has been correlated with the degree of sensitivity to the acidity and astringency of red wines (Pickering et al., 2004). These influences may be indirect, i.e., by affecting the cerebral integration of perception, rather than directly affecting receptor activity (Keast et al., 2004). Individual taste modalities are partially assessed locally in the brain, but there is overlap, and considerable interindividual variation (Schoenfeld et al., 2004). Consequently, there can be marked variation in taste acuity (Fig. 4.11; Delwiche et al., 2002; Drewnowski et al., 2001), presumably accounting for some of the dissension usually expressed at tastings.

Acuity can also vary temporally. For example, sensitivity to PTC can vary by a factor of 100 over several days (Blakeslee and Salmon, 1935). Adaptation can also result in temporary, short-term decreases in acuity. At moderate concentrations, adaptation can be complete. It is for this, if no other reason, that tasters should cleanse their palates between samples.

Psychologic

The context of a tasting has long been known to influence perception, with its role in relation to food perception and acceptance having been the subject of extensive investigation (Spence and Piqueras-Fiszman, 2014).

The development of cross-modal associations often have a cultural origin, expressed in ethnic differences in odor/taste judgments (Barker, 1982; Clurea et al., 2004). The oft-mentioned affinity between regional cuisines and local wines is probably an example, being the embodiment of cultural habituation, or the influence of vacationing in a charming locale.

Contextual factors can also play a role on a smaller scale, with many cross-modal associations apparently imposed by the orbitofrontal cortex, based on hardwired percepts modified by experience (Fig. 3.11). Even verbal suggestions can affect perception (Djordjevic et al., 2004; Aqueveque, 2015). Nonetheless, assessing the fragrant and gustatory aspects of a mixture separately (by nose closure or blindfolding) can minimize or eliminate most cross-modal influences.

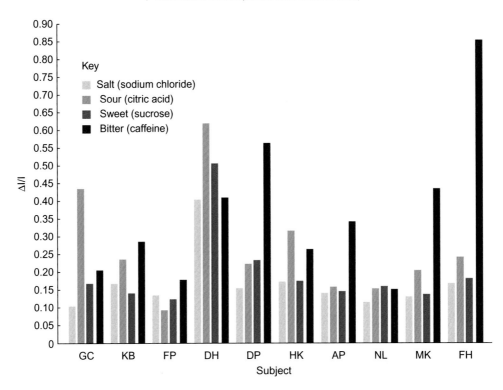

FIGURE 4.11   Illustration of the variability in the differential sensitivity ($\Delta I/I$) for the principal taste qualities in each of 10 subjects. *From Schutz, H.G., Pilgrim, F.J., 1957. Differential sensitivity in gustation. J. Exp. Psychol. 54, 41–48, (Schutz and Pilgrim, 1957), reproduced with permission.*

MOUTH-FEEL

Mouth-feel refers to sapid sensations activated by free nerve endings of the trigeminal nerve. Trigeminal nerve endings surround the taste buds (Whitehead et al., 1985), but are also located throughout the oral cavity. The distribution of different classes of trigeminal receptors appears to be somewhat localized throughout the oral cavity (Rentmeister-Bryant and Green, 1997), e.g., receptors particularly sensitive to burning irritants are clustered primarily at the tip and along the sides of the tongue. Nonetheless, somatosensory receptors typically generate diffuse, poorly defined sensations throughout the mouth. Relative to wine, significant mouth-feel perceptions include astringency, burning, prickling, viscosity, body, and heat/cold.

Most mouth-feel perceptions arise from stimulation of one or more of the four categories of trigeminal receptors. These are mechanoreceptors (touch, pressure, vibration), thermoreceptors (heat and cold), nociceptors (pain), and proprioreceptors (texture, movement, and position). Although there is a degree of specialization, it is incomplete. For example, nociceptors may respond to high concentrations of ethanol (eliciting a burning sensation) and carbon dioxide (inducing pain) (see Green, 2004). The relative activation of several receptor types is likely to generate the various modalities of astringency.

Unlike most gustatory and olfactory sensations, the sensation of astringency develops comparatively slowly (Fig. 4.12). This probably arises because trigeminal receptors are often buried within or beneath the mucosal epithelium. Adaptation is also slow, or may not develop. The latter is particularly evident in the increased intensity of astringent sensations on repeat exposure (Fig. 5.56); this is one of the reasons why palate cleansers that are effective should be used between each sampling.

Astringency

Astringency refers to a range of sensations, variously described as dry, grainy, rough, puckery, and occasionally velvety. These attributes typify most red wines (Lawless et al., 1994; Francis et al., 2002), but not white wines. Astringency is primarily induced by flavonoid constituents, derived coming principally from grape seeds and skins (for a summary of terminology relative to grape phenolics see Table 4.1). Anthocyanins can enhance the perceived astringency of tannins, but do not in themselves contribute significantly to wine astringency or bitterness (Brossaud

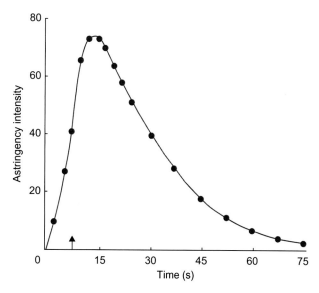

FIGURE 4.12   Average time–intensity curve for astringency of 5 g/L tannic acid in white wine. The sample was held in the mouth for 5 s (↑) before being expectorated. *From Guinard, J.-X., Pangborn, R.M., Lewis, M.J., 1986a. The time–course of astringency in wine upon repeated ingestion. Am. J. Enol. Vitic. 37, 184–189, reproduced by permission.*

TABLE 4.1   Phenolic and related substances in grapes and wine[a]

General type	General structure	Examples	Major source[b]
NONFLAVONOIDS			
Benzoic acid	COOH	Benzoic acid	G, O
		Vanillic acid	O
		Gallic acid	G, O
		Protocatechuic acid	G, O
		Hydrolyzable tannins	G
Benzaldehyde	CHO	Benzaldehyde	G, O, Y
		Vanillin	O
		Syringaldehyde	O
Cinnamic acid	CH=CHCOOH	p-Coumaric acid	G, O
		Ferulic acid	G, O
		Chlorogenic acid	G
		Caffeic acid	G

(Continued)

TABLE 4.1 (Continued)

General type	General structure	Examples	Major source[b]
Cinnamaldehyde	CH=CHCHO	Coniferaldehyde	O
		Sinapaldehyde	O
Tyrosol	CH$_2$CH$_2$OH OH	Tyrosol	Y

FLAVONOIDS

General type	General structure	Examples	Major source[b]
Flavonols		Quercetin	G
		Kaempferol	G
		Myricetin	G
Anthocyanins		Cyanin	G
		Delphinin	G
		Petunin	G
		Peonin	G
		Malvin	G
Flavan-3-ols		Catechin	G
		Epicatechin	G
		Gallocatechin	G
		Procyanidins	G
		Condensed tannins	G

[a]Data from Amerine and Ough (1980) and Ribéreau-Gayon (1964).
[b]G, grape; O, oak; Y, yeast.

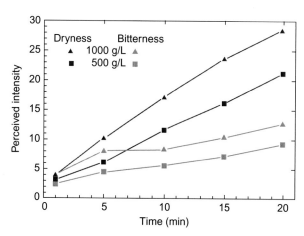

FIGURE 4.13 Growth in perceived intensity of sensations of dryness and bitterness caused by repeated, whole-mouth rinses with two concentrations of tannic acid. *Reprinted from Green, B.G., 1993. Oral astringency: A tactile component of flavor. Acta Psychol. 84, 119–125., adapted from Lyman, B.J., Green, B.G., 1990. Oral astringency: Effects of repeated exposure and interactions with sweeteners. Chem. Senses 15, 151–164, with permission of Elsevier.*

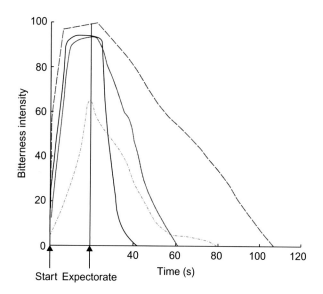

FIGURE 4.14 Individual time–intensity curves for four judges in response to bitterness of 15 ppm quinine in distilled water. *From Leach, E.J., Noble, A.C., 1986. Comparison of bitterness of caffeine and quinine by a time–intensity procedure. Chem. Senses 11, 339–345. Reproduced by permission of Oxford University Press.*

et al., 2001). During maturation and aging, anthocyanins bond with catechins and their polymers, participating in condensed tannin generation. Although the proportion of these pigmented polymers generally increases with time (Brossaud et al., 2001), it may also decrease (Vidal et al., 2004b; Weber et al., 2013). Nonflavonoid phenolics from grapes (notably flavonol glycosides and dihydroflavonol rhamnosides) may also participate in the generation of astringency. Additional astringent nonflavonoids (notably hydrolyzable tannins) can be absorbed from oak cooperage. Additional, but less significant, wine constituents potentially associated with astringency are organic acids, notably tartaric acid.

Astringency can be confused with bitterness (Lee and Lawless, 1991)—both perceptions being typically induced by related, if not identical, compounds (Fig. 4.13). Adding to the potential confusion is the similar nature of their time–intensity profiles, both perceptions developing comparatively slowly and possessing lingering aftertastes (Figs. 4.12 and 4.14). The strong astringency of many phenolic compounds may partially mask their bitterness (Arnold and Noble, 1978). When requested, trained tasters appear to differentiate between these sensations. Without an objective and definitive measure for both sensations, how well tasters succeed in this ability is a moot point. Data from Ross and Weller (2008) suggest that astringency is more often confused with bitterness than the reverse. Fig. 4.15 illustrates the relative intensities of the astringent versus bitter attributes of several wine phenolics. Differentiation

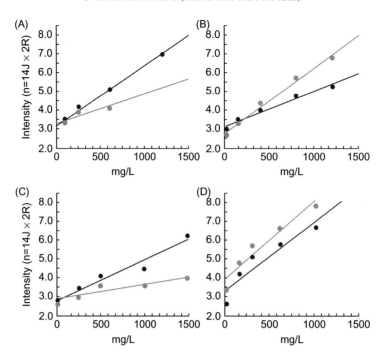

FIGURE 4.15   Mean intensity rating for bitterness (●) and astringency (○) as a function of concentration for (A) catechin, (B) grape seed tannin, (C) gallic acid, and (D) tannic acid. *From Robichaud, J. L., and Noble, A. C. 1990. Astringency and bitterness of selected phenolics in wine. J. Sci. Food Agric. 53, 343–353. Copyright Society of Chemical Industry. Reproduced with permission. Permission is granted by John Wiley & Sons Ltd on behalf of the SCI.*

between flavonoid-induced astringency and bitterness can be achieved for some phenolics by adding trigeminal suppressive compounds such as *trans*-pellitorine. It lowers the astringency of epigallocatechin gallate without influencing its bitterness (Obst et al., 2013).

The bitter-astringency of red wines is probably the principal cause for many people disliking red wines when first experienced, similar to unsweetened coffee or tea. Nonetheless, for many consumers, the dislike of the bitter-astringent aspect of most red wines is overcome, and maybe even come to be appreciated with experience. For those for whom this does not occur, or is resisted, an increasing number of wineries are releasing wines blended to be very flavorful, with intense color, but smooth in taste. Whether this will counter the second most common reason for avoiding red wines—headache induction—is unknown.

Astringency is usually attributed to the binding and precipitation of salivary proline-rich proteins (PRPs) with phenolic compounds (Haslam and Lilley, 1988; Kielhorn and Thorngate, 1999; Fig. 4.16). Because tannins precipitate PRPs, but not mucins, whereas acid precipitates mucins (glycoproteins), but not PRPs, the mechanism of astringency involving salivary proteins appears to depend on the astringent group (Lee et al., 2012). Even subclasses of PRPs react differently to tannins during the course of wine assessment (Brandão et al., 2014). In addition, phenolics differ markedly in their astringency, with epigallocatechin-3-gallate being particularly active in precipitating saliva proteins, whereas epicatechin has little effect in this regard (Rossetti et al., 2009).

The binding of phenolics to PRPs seems partially associated with their amino acid pyrrolidine ring structures (Charlton et al., 2002; Jöbstl et al., 2004). They provide multiple "hydrophobic sticky sites" for stacking with the benzene rings of tannins. The main reactions appear to involve the $-NH_2$ and $-SH$ groups of amino acids and the *o*-quinone sites on tannins (Haslam et al., 1992). As they associate, their mass, shape, and electrical properties change, frequently resulting in precipitation. Other tannin–protein reactions are known, but apparently are of little significance in wine (Guinard et al., 1986b).

An important factor influencing astringency is pH (Fontoin et al., 2008; Fig. 4.17). Hydrogen ion concentration affects protein hydration, and both phenol and protein ionization. In very acidic wines, low pH can alone induce salivary glycoprotein precipitation (Dawes, 1964), eliciting astringency (Corrigan Thomas and Lawless, 1995). Typical wine ethanol contents appear to reduce astringency, but enhance the bitterness of oligomeric tannins (Fig. 4.18). These may accrue from alcohol limiting the binding of tannins to saliva proteins (Serafini et al., 1997). Nonetheless, this view appears to be contradicted by Obreque-Slíer et al. (2010). Wine aroma can also influence the perception of astringency, with studies either showing an increase (Ferrer-Gallego et al., 2014) or decrease (Sáenz-Navajas et al., 2010b).

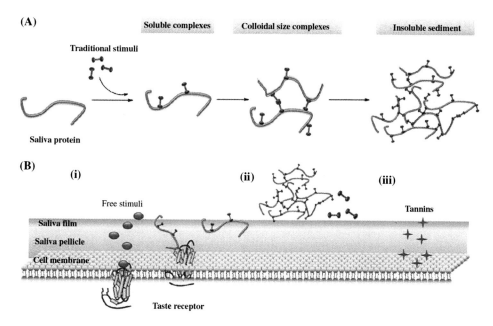

FIGURE 4.16 Schematic representation of possible astringency mechanisms: (A) A 3-stage model of the interaction between stimuli and proteins; (B) Astringency stimulation: (i) "Free" stimuli and soluble stimuli–protein complexes deplete the protective salivary film and eventually bind to the pellicle or even to the receptors exposed; (ii) Insoluble stimuli–protein complex and traditional stimuli are rejected against salivary film. Insoluble stimuli–protein complexes trigger astringency sensation via increasing friction. (iii) Tannins interact with oral cavity membrane. *Reprinted from Ma, W., Guo, A., Zhang, Y., Wang, H., Liu, Y., Li, H., 2014. A review on astringency and bitterness perception of tannins in wine. Food Sci. Technol. 40, 6–19, with permission from Elsevier.*

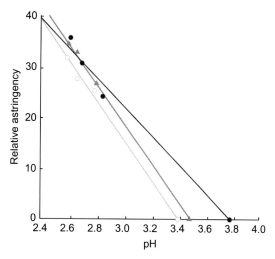

FIGURE 4.17 Effect of pH on perceived intensity of astringency in model wine solutions at three tannic acid concentrations: ●, 0.5 g/L; ▲, 1 g/L; ○, 2 g/L. *From Guinard, J., Pangborn, R. M., Lewis, M. J., 1986b. Preliminary studies on acidity-astringency interactions in model solutions and wines. J. Sci. Food Agric. 37, 811–817. Copyright Society of Chemical Industry. Reproduced with permission. Permission is granted by John Wiley & Sons Ltd on behalf of the SCI.*

One of the principal consequences of the reaction between condensed tannins and saliva proteins is reduced saliva viscosity and increased oral friction. Both factors reduce the lubricating properties of saliva, inducing a rough sensation (Prinz and Lucas, 2000). In addition, precipitation of the complexes likely forces water away from the cell surface, simulating dryness. Precipitated salivary proteins also coat the teeth, producing the characteristic rough texture often associated with astringency. Inadequately investigated are reactions between tannins (and/or low pH) on denaturing the glycoproteins and phospholipids of the mucous membrane. Malfunctioning of the plasma membrane, such as disruption of catechol amine methylation, could play a role in the perception of astringency.

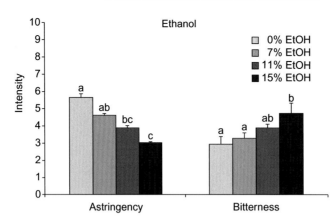

FIGURE 4.18   Mean astringency and bitterness intensities as a function of ethanol level. Bars represent standard errors over the two repetitions. A significant difference between mean ratings ($p < .05$) is indicated within an attribute (a, b, or c). *Reproduced from Fontoin, H., Saucier, C., Teissedre, P.-L., Glories, Y., 2008. Effect of pH, ethanol and acidity on astringency and bitterness of grape seed tannin oligomers in model wine solution. Food Qual. Pref. 19, 286–291, with permission from Elsevier.*

This might explain why catechin content correlates more with astringency than saliva protein precipitation (Kallithraka et al., 2000). However, data from Guest et al. (2008) do not support this interpretation. In addition, preliminary investigations suggest that condensed tannins seem likely to disturb the structural integrity of the membrane's bilipid layer (Yu et al., 2011). Although hydrolyzable tannins could act similarly, their breakdown in wine presumably limits their influence to young wines. The relatedness of certain tannin constituents to adrenaline and noradrenaline could be another mechanism by which tannins might stimulate localized blood vessel constriction, enhancing a dry, puckery, sensation.

Astringency appears to have both direct and indirect tactile effects on trigeminal receptors. For example, tannins with one to several gallic acid moieties can activate the G-protein receptors on some trigeminal neurons, in a manner reminiscent of gustatory activation (Schöbel et al., 2014). Oxidized epigallocatechin gallate is also more effective than other oxidized catechins in activating TRPV1 and TRPA1 membrane-bound receptors (Kurogi et al., 2015). Both membrane receptors react to a range of chemical and physical irritants (Ramsey et al., 2006). This suggests that astringency may possess both chemical as well as mechanosensory modes of activation. Precipitation of saliva or other proteins can also indirectly activate touch sensors in the mouth. Because of the widespread distribution of trigeminal receptors throughout the oral cavity, this probably explains why actively moving wine in the mouth enhances the perception of astringency. As well, multiple G-protein variants, involving TRP channel–induced membrane depolarization, have been found associated with trigeminal neurons, further helping to explain some of the qualitative differences between astringent compounds (Fig. 4.19). In addition, unlike in Yu et al. (2011), monomeric flavanols were detected to react with the cellular lipids directly, and thus likely to perturb membrane structure and function (Furlan et al., 2015).

Astringency is one of the slowest in-mouth sensations to develop. Depending on the concentration, chemical nature, and repeat exposure to the compound, astringency can take up to 15 s before reaching maximal intensity (Fig. 4.14). The decline in perceived intensity occurs even more slowly. The response of individuals varies slightly as does their qualitative perception of different astringent agents. These features tend to remain stable over time (Valentová et al., 2002).

The intensity and duration of astringency tend to increase with repeat sampling (Guinard et al., 1986a, Fig. 5.56). This is less likely to occur when the wine is consumed with food, due to reactions between tannins and food proteins and lipids, as well as by dilution. However, when assessing wine, the increase in apparent astringency could produce sequence errors, especially if wines are sampled in quick succession, without adequate palate cleansing. Sequence errors are differences in perception owing to the order in which objects are sampled. Although tannins stimulate the secretion of saliva (Fig. 4.10), production is usually insufficiently rapid or marked to limit an increase in perceived astringency.

One of the more important factors affecting phenolic-induced astringency is molecular size. Thus, catechins (flavan-3-ol monomers) bond weakly to proline-rich salivary proteins, and not to α-amylase (de Freitas and Mateus, 2003). Bonding with proteins appears to be roughly correlated with molecular size (the number of potential binding sites). The puckery astringency so induced is amplified in the presence of organic acids (Frank et al., 2011). However, at above 3400 Da, tannin polymers begin to lose conformational flexibility. Steric hindrance progressively limits their

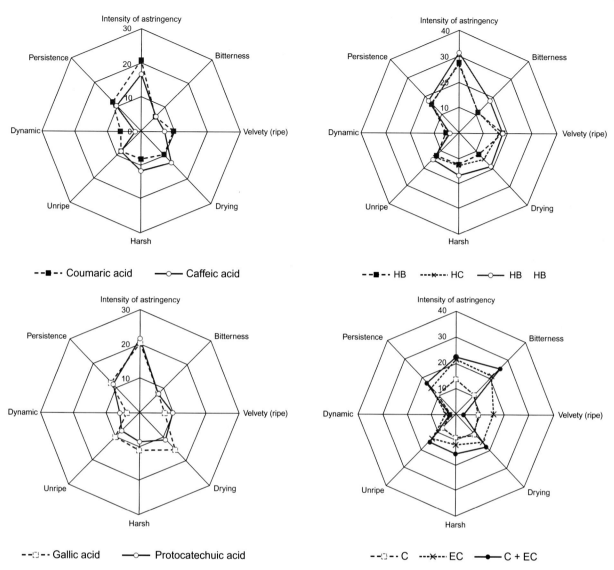

FIGURE 4.19 Visual representation of the qualitative differences in the sensory attributes of several monomeric phenolics: hydroxycinnamic acids (coumaric and caffeic acids), hydroxybenzoic acids (gallic and protocatechuic acids), and catechins and epicatechin. *C*, catechin; *EC*, epicatechin; *HB*, hydroxybenzoic acids; *HC*, hydroxycinnamic acids. *From Ferrer-Gallego, R., Hernández-Hierro, J.M., Rivas-Gonzalo, J.C., Escribano-Bailón, M.T., 2014. Sensory evaluation of bitterness and astringency sub-qualities of wine phenolic compounds: Synergistic effect and modulation by aromas. Food Res. Int. 62, 1100–1107, reproduced with permission.*

ability to bind with proteins. Thus, phenolic polymerization is correlated with increased protein-associated astringency only up to the point where the polymers precipitate, or are no longer able to bond with proteins. Although flavonoid polymerization and precipitation during aging have received most of the attention, acid-induced tannin cleavage also occurs (Vidal et al., 2002).

The precipitation of saliva proteins has been the most studied aspect of astringency. However, by itself, it does not explain all aspects of astringency. For example, proanthocyanidins and their catechin monomers poorly precipitate saliva proteins, but generate astringency. However, Kielhorn and Thorngate (1999) question the use of the term astringency for the sensation induced by low-molecular-weight flavonols. Furthermore, several authors have described distinct astringent modalities (Gawel et al., 2000, 2001). Increasing polymerization of proanthocyanidins augments drying, chalky, grainy, puckery attributes, whereas increased galloylation (flavanols esterified with gallic acid) accentuates rough/coarse attributes and dryness (Vidal et al., 2003). Galloylated tannins are particularly common in seed tannins. Although astringency is generally correlated with phenolic content, astringency is also influenced by other wine constituents, e.g., wine polysaccharides, mannoproteins, acidity, and ethanol content (Vidal et al., 2004a).

In spite of what is known about the various roles of phenolic subgroups relative to astringency, much still needs to be learned before a clear picture of its complexities develops. This reflects the extreme structural complexity of tannins, their age-related modifications, and the variety of sensory modalities generated. For example, flavonoid and nonflavonoid monomers can polymerize to form dimers, trimers, and a diversity of higher oligomers and polymers. They also bond with anthocyanins, acetaldehyde, pyruvic acid derivatives, other yeast byproducts, and various sugars. How these products relate to the so-called "soft" and "hard" tannins of winemakers is only now becoming somewhat clearer. For example, most of the attention relating to astringency has involved studies on flavanoid monomers (catechins, flavan-3-ols) and their polymers, the more desirable, velvety, astringent attribute may be due to nonflavonoid and flavonol glycosides and dihydroflavonol rhamnosides (Hufnagel and Hofmann, 2008a; Frank et al., 2011). If confirmed, this could prove profitable to producers producing red wines more attuned with the preferences of most wine consumers. What is needed are investigations on how to achieve this goal. Understanding the origins of the rougher aspects of astringency may also facilitate the development of means to predict this characteristic at crush stage, thereby limiting their presence, at least in wines designed for early drinking.

Although phenolics and organic acids are the main inducers of wine astringency, other compounds can be directly involved, e.g., high ethanol concentrations. Ethanol-induced astringency is suspected to result from its disruption of cell membranes. Conversely, ethanol can also limit tannin-induced protein precipitation (Lea and Arnold, 1978; Yokotsuka and Singleton, 1987). This feature, combined with the ameliorating influence of the sugar content in port, may account for its lower-than-expected astringency.

Burning

High ethanol contents produce a burning sensation in the mouth, but this is more associated with brandies and other distilled beverages than wines. Some wine phenolics and sesquiterpenes also produce a peppery, burning sensation. These perceptions probably result from the activation of polymodal, trigeminal, vanilloid receptors (TRPV1). These respond to a variety of stimuli, including heat, cuts, pinching, acids, and chemicals such as capsaicin. Although the burning sensation arises from the activation of several types of primarily heat-sensitive pain receptors, there are differences in receptor sensitivity throughout the buccal cavity. These probably produce the qualitative differences between the sensations of various irritants, such as mustard, horseradish, chilies, and black pepper. Wines particularly high in sugar content (e.g., icewines and Tokaji Eszencia) may generate a burning sensation.

Temperature

Another trigeminal receptor important to wine assessment, TRPM8, responds to cool temperatures and chemicals such as menthol. It may also be associated with the burning and stinging sensation associated with cold beverages. This should not be too surprising, as the receptor is closely related to heat-activated receptors.

The cool mouth-feel produced by chilled white and sparkling wine adds to the interest and pleasure of wines of subtle flavor. Temperature also affects the perception of a wine's taste and mouth-feel attributes, e.g., cooling a sugar solution down to 5–12°C reduces its sweetness (Green and Nachtigal, 2015). Nonetheless, the effect of temperature on taste and mouth-feel seems relatively minor, within the normal range at which wines are served. This is in contrast to the stronger influence temperature has on volatility (Fig. 4.7). Another influence of cold temperatures is to enhance the prickling sensation of carbonated beverages and sparkling wine (Fig. 4.20).

Some of the effects previously reported for temperature on gustation may arise from direct effects on the selective stimulation of taste receptors, notably sweetness (Cruz and Green, 2000). Warming or cooling the tongue can also induce weak taste sensations (termed thermal taste), depending on the part of the tongue treated. In addition, those who experience thermal taste tend to be more sensitive to other tastants (Green and George, 2004).

None of these results convincingly confirms the typical temperatures recommended for serving different wines. Thus, standard recommendations may largely reflect habituation (Zellner et al., 1988), possibly explaining how the 19th century preference for serving red wines at cool temperatures (Saintsbury, 1920) could have arisen.

Prickling and Related Carbon Dioxide Effects

Carbon dioxide bubbles bursting on the tongue activates nociceptors, producing a prickling, irritating, and at high concentrations a painful sensation. These sensations seen not to be directly generated by CO_2, but by its conversion to bicarbonate and H^+ ions (Dessirier et al., 2000). This conversion is catalyzed by the action of carbonic anhydrase, located on acid-sensitive receptor cells (Chandrashekar et al., 2009). A subset of trigeminal receptors possessing a

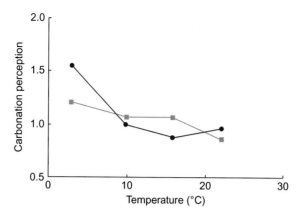

FIGURE 4.20   Temperature versus carbonation perception by trained (■) and naive (●) panels. The *y*-axis is rescaled magnitude estimates of response. *From Yau, N.J.N., McDaniel, M.R., 1991. The effect of temperature on carbonation perception. Chem. Senses 16, 337-348. Reproduced by permission of Oxford University Press.*

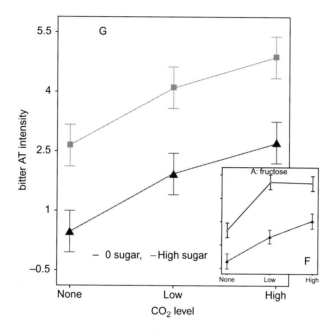

FIGURE 4.21   The influence of carbon dioxide and sugars (*F*, fructose; *G*, glucose;) on the bitter aftertaste (AT) of carbon dioxide in a solution of citric acid (0.5 g/L). Low carbon dioxide (~ 1.5 vol.); high carbon dioxide (~ 3.6 vol); high glucose (150 g/L); high fructose (64 g/L). *Reprinted from Hewson, L., Hollowood, T., Chandra, S., and Hort, J. (2009) Gustatory, olfactory and trigeminal interactions in a model carbonated beverage. Chem. Percept. 2, 94–107, with kind permission of Springer Science + Business.*

functional *TRPA1* gene (Wang et al., 2010) appear to respond to H⁺ ions released as carbonic acid ionizes. Ionization can also generate a slightly sour taste (Hewson et al., 2009). Additional sensations associated with CO_2 include bitter and salty side-tastes (Cowart, 1998). These sensations arise primarily in wines containing more than 3–5‰ carbon dioxide, and are affected by bubble size and temperature. In addition, bubbles donate a textural sensation as they burst, activating proprioreceptors on the tongue.

$$CO_2 + H_2O \rightarrow H_2CO_3 \rightarrow HCO_3^- + H^+$$

Carbon dioxide can modify the perception of other taste modalities (Cometto-Muñiz et al., 1987; Cowart, 1998), notably suppressing sweetness (Di Salle et al., 2013), whereas it enhances sourness, saltiness, and bitterness (Hewson et al., 2009). These effects are often complex, depending on the concentration of the carbon dioxide and other matrix influences (Fig. 4.21). In addition, carbonation significantly augments the perception of coolness in the mouth, while cold enhances the sensory effects of carbon dioxide (Green, 1992).

Although carbon dioxide promotes the evaporation of some volatile compounds, olfaction can be suppressed by the prickling sensation of CO_2 in the nose (Cain and Murphy, 1980). This may explain the reduced foxiness of sparkling wines produced from Concord grapes.

Bubble bursting also produces auditory stimuli. These can enhance the perception of carbonation of a soft drink (Zampini and Spence, 2005). However, other than the "pop" or "sigh" produced during opening, it is unknown if the sound of bubbles bursting in the glass has any effect on wine perception.

Body (Weight)

Despite the seeming importance of body to overall wine quality, precisely what is meant by the term remains vague. Gawel et al. (2007) reported a correlation between body and higher ratings for flavor and/or perceived viscosity. In sweet wines, body is often viewed as being roughly correlated with sugar content. In dry wines, it has often been associated with alcohol content, but the effect is largely restricted to ranges beyond those typically found in wines (Pickering et al., 1998; Runnebaum et al., 2011). Glycerol content has often been viewed as being involved with the concept of body, but no clear correlation has been found within its natural range in wine (1.23 and 1.32 mPs.s) (Runnebaum et al., 2011), nor any correlation with wine quality (Nieuwoudt et al., 2002). In addition, wine glycerol contents are generally too low to affect perceived viscosity directly (Gawel and Waters, 2008). Ethanol was found to affect viscosity more than glycerol, within the normal ranges found in wine (Yanniotis et al., 2007). Nevertheless, the glycerol, 1,2-propanediol, and *myo*-inositol contents of an Amarone wine were correlated the perception of body (Hufnagel and Hofmann, 2008b). Body, assessed as perceptible viscosity, was correlated with the osmotic potential, magnesium content, and wine extract for several white wines (Runnebaum et al., 2011). Another possibility is that body is a term equivalent to the newly identified taste enhancer, *kokumi*, roughly translated as mouthfullness. It is based on the response of a calcium-sensitive transporter found on some gustatory receptors (Maruyama et al., 2012; Kuroda and Miyamura, 2015). Several γ-glutamyl peptides found in aged cheeses, such as Roquefort, Gouda, and Parmesan, are key activators of the *kokumi* response. Whether such peptides occur in significant quantities in wine sufficient to induce *kokumi* seems as yet uninvestigated. Another possibility is the activity of the fatty acids in wine, activating what appears to be special receptors for several fatty acids (Gilbertson and Khan, 2014; Running et al., 2015).

Experiments with model wines, composed of the major chemical constituents in wines, i.e., ethanol, acids, and tannins, do not generate sensations similar to real wines. Thus, other typical constituents, such as yeast proteins and polysaccharides (Vidal et al., 2004c), may be involved in generating the perception of body. However, still, without a clear definition of what body means, little progress is likely to be made, and it will remain as nebulous a vinous concept as quality itself.

Metallic

A metallic perception has occasionally been ascribed to dry wines, particularly the aftertaste of some sparkling wines. The precise origin of this perception is unclear. Iron and copper ions can generate a metallic taste, but at concentrations at or above those typically found in wine (>20 and 2 mg/L, respectively). Nonetheless, detection of a metallic sensation, associated with Cu^{2+} content, is apparently augmented in the presence of tannins (Moncrieff, 1964). More significantly, though, may be the generation of a metallic sensation in wines at low concentrations by ferrous ions. This perception is significantly reduced when the nose is occluded, or the solution does touch the tongue (Fig. 4.22). Thus, the metallic "taste" appears to be a retronasal smell, but interpreted as orally derived. Because metals can catalyze lipid oxidation, forming byproducts such as oct-1-en-3-one (Forss, 1969), octa-1, *cis*-5-dien-3-one (Swoboda and Peers, 1978), and hex-1-en-e-one (Lorber et al., 2014), these could be the volatile compounds generating the metallic perception (Lawless et al. (2004). Acetamides have also been reported to have a metallic aspect (Rapp, 1987). None of these perceptions seem correlated with the current fad of ascribing minerality to many wines.

CHEMICAL COMPOUNDS INVOLVED

Sugars

All wines possess residual amounts of unfermented sugars. The most common are fructose and glucose, but trace amounts of arabinose, galactose, rhamnose, ribose, and xylose also occur. Typically, none of the latter occur at

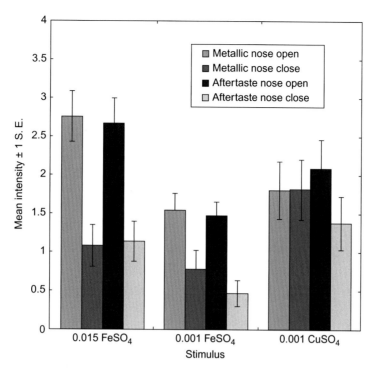

FIGURE 4.22 Mean metallic taste and aftertaste intensity for ferrous and cupric sulfate solutions under two nasal conditions. *Reprinted from Epke, E.M., McClure, S.T., Lawless, H.T., 2009. Effects of nasal occlusion and oral contact on perception of metallic taste from metal salts. Food Qual. Pref. 20, 133–137, with permission from Elsevier.*

concentrations sufficient to induce detectable sweetness in table wines. Only in dessert wines is the sugar content typically desired to donate a sweet taste.

The sugar concentration must exceed about 0.2% to exhibit perceptible sweetness in hypertasters. Because the residual sugar content of most table wines is less than 0.2%, they generally appear dry. When sweetness is detected in dry wines, it is usually due to the presence of noticeable fruitiness in the fragrance. Fruity odors are often perceived to possess sweetness (Prescott, 2004), even in absence of sensorially detectable amounts of sugar.

Sugars usually begin to have a detectable influence on perceptible sweetness at concentrations at or above 0.5%. At this level, sugar also begins to affect the perception of body. At high concentrations, sugars can generate a cloying effect, as well as a burning mouth-feel. This is particularly noticeable without sufficient acidity. Thus, it is important, especially for sweet wines (such as icewines), to possess sufficient acidity to preserve balance. In sweet wines, such as port, the high residual sugar content helps to diminish the harshness of its tannin content.

Alcohols

Several alcohols occur in wine, but only ethanol occurs in sufficient amounts to produce gustatory sensations. Although ethanol possesses a sweet aspect, the acidity of wine diminishes its sensory significance. Ethanol does, however, slightly enhance the sweetness of sugars. Ethanol also reduces the perception of acidity, making acidic wines appear mildly less sour and more balanced. At high concentrations (above 14%), alcohol increasingly generates a burning sensation, and may contribute to the feeling of weight or body, especially in dry wines. Ethanol also can augment the perceived intensity of bitter phenolic compounds, while decreasing the sensation of tannin-induced astringency (Fig. 4.18).

Of wine polyols, glycerol is the most common. In dry wine, it can also be the most abundant compound, after water and ethanol. Because of its viscosity, glycerol has often been assumed to generate a smooth mouth-feel and contribute to the perception of body. Although likely contributing to these perceptions, glycerol rarely reaches a concentration that perceptibly affects viscosity (≥26g/L) (Noble and Bursick, 1984; Nurgel and Pickering, 2005). Glycerol content may, however, be sufficient to play a minor role in suppressing the perception of acidity, bitterness, and astringency. The slightly sweet taste of glycerol may also play a minor role in dry wines, where its concentration often surpasses its sensory threshold for sweetness (>5g/L) (Ough et al., 1972). Although the glycerol content of dessert wines is often relatively high (≥12 g/L), it is unlikely that glycerol contributes detectably to the sweetness of wines possessing sugar contents often well above 75 g/L.

Several sugar alcohols, such as alditol, arabitol, erythritol, mannitol, *myo*-inositol, as well as sorbitol and 1,2-propanediol, also occur in wine. Combined, their individual effects may influence the sensation of body.

Acids

As a group, carboxylic acids are as essential to the sensory attributes of wines as alcohol. They generate its refreshingly tart taste (or sourness, if in excess); can induce both sharp and astringent sensations; influence wine coloration (and stability); and modify the perception of other sapid compounds. The effect of acidity in diminishing perceived sweetness appears less marked than its converse, that of sugar suppressing the perception of acidity (Ross and Weller, 2008). In addition, the release of acids from cell vacuoles during crushing can initiate an acid-induced hydrolytic release of aromatics from the fruit (Winterhalter et al., 1990). Several important aroma compounds, such as monoterpenes, phenolics, C_{13} norisoprenoids, benzyl alcohol, and 2-phenylethanol, often occur in grapes as acid-labile, nonvolatile glycosides (Strauss et al., 1987).

Acids are crucial to the color stability of red wines by maintaining a low pH. As the pH rises, anthocyanins begin to decolorize and may eventually turn blue. In addition, wine acids limit the occurrence of readily oxidizable phenolate phenolics. Accordingly, wines of high pH (≥3.9) are especially susceptible to oxidization and loss of their fresh aroma and young color (Singleton, 1987).

The influences of wine acids depend primarily on their potential to ionize, a complex function of the acid's structure, the pH, and the concurrent presence of metal and metalloid anions (notably K^+ and Ca^{2+}). Because of the partial nonspecificity of gustatory receptors, and variability in the reaction of trigeminal nerve endings to acids, it is not surprising that people respond idiosyncratically to wine acidity, and that perceived sourness cannot easily be predicted from a wine's pH or acid content.

Of common vinous carboxylic acids, malic acid has the most intense taste, lactic acid the least, with tartaric acid somewhere in between (Amerine et al., 1965). Thus, one of the prime advantages of malolactic fermentation is the decarboxylation of malic to lactic acid, and the corresponding reduction in sourness. Because malic acid does not markedly affect the pH, its decarboxylation does not undesirably increase the pH.

Most wine acids are derived from grapes and are nonvolatile, and generate a wine's fresh, acidic taste. In contrast, a potentially significant carboxylic acid, acetic acid, is of microbial origin. Although possessing a sharp, vinegary odor, acetic acid can be beneficial in increasing the perceived intensity of aromatic compounds at subthreshold concentrations (Miyazawa et al., 2008). At much above threshold levels, acetic acid is considered a fault, being also associated with the odor of ethyl acetate (the ester formed between acetic acid and ethanol).

Acetic acid can be a byproduct of acetic acid bacteria growing on infected grapes, can be produced during yeast fermentation, and can arise from the breakdown of oak constituents during in-barrel maturation. Nonetheless, at clearly detectable levels, it is usually the consequence of acetic acid bacterial metabolism in wine. Lactic acid bacteria may be involved, but rarely to any significant extent.

Phenolics

The predominant phenolics in wine are either flavonoids (two phenolics joined via a pyran ring) or nonflavonoids (phenolics possessing at least one hydroxyl group). Their basic differences are illustrated in Table 4.1. Both groups can exert a marked influence on taste and mouth-feel. Many of these compounds generate an enormous and structurally complex group of polymers, grouped together under the term tannins.

Most wine (condensed) tannins originate from grapes. They are composed primarily of subunits, generically termed catechins (flavan-3-ols). These bond directly with each other in various linkages, other phenolics, and/or via other compounds such as acetaldehyde. Because the connections are covalent, the polymers formed are relatively resistant to breakdown under wine conditions. Smaller polymeric condensed tannins are often referred to as oligomeric procyanidins or proanthocyanidins (procyanidins being a particular subgroup of proanthocyanidins). Condensed tannins, possessing from about 2 to 10 catechin subunits, are inherently soluble, except when bonded to cellular constituents in grapes. Larger polymers become progressively insoluble.

The other main group of tannins are termed hydrolyzable tannins. They come primarily from wood cooperage, in which the wine may have been fermented and/or matured. As the term suggests, hydrolyzable tannins fairly readily break down (hydrolyze) under wine conditions. They consist primarily of nonflavonoid, gallic acid subunits.

In wine, various phenolics can, to varying degrees, elicit bitter and astringent sensations, as well as contribute to wine color, body, and flavor. Their individual influences depend on the constituent phenolic, their oxidized or ionized state, and degree of polymerization with other phenolics, polysaccharides, proteins, acetaldehyde, sulfur dioxide, or other wine constituents. Flavonoid (condensed) tannins constitute the major phenolic compounds in red wines, whereas nonflavonoids constitute the major phenolic group in white wines. Flavonoid tannins come primarily from skins and seeds of the grapes, whereas monomeric flavonoids and nonflavonoids principally occur in grape-cell vacuoles, from which they escape during crushing and pressing. Additional nonflavonoids come from wood cooperage.

Seed tannins are primarily polymers of three flavanols (catechin, epicatechin, and gallate epicatechin), whereas skin tannins are characterized by higher degrees of polymerization and greater epigallocatechin contents. In contrast, anthocyanins and flavonols, such as quercetin, commonly collect in the cell vacuoles of the skin. Flavonols may also be extracted from stem tissue.

Catechins and their polymers (condensed tannins) have generally been viewed as generating the majority of bitter and astringent sensations in red wines. However, the isolation of individual compounds, calculation of their dose-over-threshold (DoT) ratios, and omission studies (Hufnagel and Hofmann, 2008b) have shed new light on the relative importance of different phenolics in the generation of bitterness and the modalities of astringency (Table 4.2).

TABLE 4.2 Taste qualities, taste thresholds, concentrations, and dose-over-threshold (DoT) factors of the major nonvolatiles involved in the astringent and bitter attributes of an Amarone red wine

Taste compound	Threshold[a] (μmol/L)	Conc. (μmol/L)	DoT factor[b]
GROUP 1: NONBITTER ASTRINGENT COMPOUNDS			
Flavonol-3-ol glycosides (*velvety, silky astringency*)	0.20	5.38	27.0
Syringetin-3-*O*-β-D-glucopyranoside	2.48	8.42	3.4
Isorhamnetin-3-*O*-β-D-glucopyranoside	3.70	5.71	1.5
Dihydroquercetin-3-*O*-β-D-rhamnopyranoside	0.43	0.31	0.7
Quercetin-3-*O*-β-D-galactopyranoside	4.81	1.50	0.3
PHENOLIC ACIDS AND FURAN ACIDS (PUCKERING ASTRINGENT)			
(*E*)-caftaric acid	1.6	130	8.1
Gallic acid	292	765	2.7
Furan-2-carboxylic acid	160	220	1.4
Caffeic acid	72	54.8	0.8
p-Coumaric acid	139	80.4	0.6
POLYMERS (PUCKERING ASTRINGENT)			
High-molecular-weight tannins (>5kDa)	22 (mg/L)	5.45 (g/L)	247.7
GROUP 2: BITTER COMPOUNDS			
Flavan-3-ols (bitter and puckering astringent)			
(+) Catechin	410[c]/1000[d]	57.6	0.2[c]/<0.1[d]
(−) Epicatechin	930[c]/930[d]	27.6	0.1[c]/<0.1[d]
PHENOLIC ACID ETHYL ESTERS (BITTER AND PUCKERING ASTRINGENT)			
Gallic acid ethyl ester	185[c]/2200[d]	153	0.2[c]/<0.1[d]
p-Coumaric acid ethyl ester	143[c]/715[d]	24.6	0.4[c]/<0.1[d]

Data from Hufnagel and Hofmann, 2008a.
[a]*Taste threshold concentrations (TC) were determined in bottled water by means of a triangle test for bitter, sweet, sour, salty, and umami compounds and by means of the half-tongue test for astringent compounds.*
[b]*The DoT factor is calculated as the ratio of the concentration and taste threshold.*
[c]*Taste threshold for astringency is 0.5 mol/L. The DoT factors for astringency are given in parentheses.*
[d]*Taste threshold or DoT factor for bitterness.*

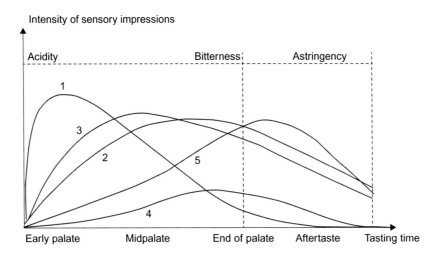

FIGURE 4.23 Influence of flavonoid polymerization on their sensory attributes: 1, catechins and simple procyanidins; 2, oligomeric procyanidins; 3, procyanidin polymers; 4, anthocyanins; 5, stalk tannins. *From Glories, 1981, unpublished, reproduced in Ribéreau-Gayon, P., Glories, Y., Maujean, A., Dubourdieu, D., 2006. Handbook of Enology, Vol.2, The Chemistry of Wine Stabilization and Treatments, second ed. John Wiley & Sons, Ltd., Chichester, UK, reproduced by permission.*

When the DoT is >1, it is likely that the compound may significantly influence one or more gustatory attributes, in a manner similar to the predictive sensory impact of OAVs >1 for aromatic compounds. Such studies have indicated the importance of flavonol and dihydroflavonol glycosides, and supported the central role of condensed tannins in the generation of the puckery aspect of astringency and the relative insignificance of some nonflavonoids (phenolic acids) to astringency. In comparison to flavan-3-ols and their polymers, flavonol-3-ol glycosides (Table 4.2) appear to be particularly significant to the expression of a velvety/silky astringency (Hufnagel and Hofmann, 2008a), despite their constituting only about 1% of the total phenolic content. Various catechins and their polyphenolic fractions generate astringency with distinct attributes, described as graininess, chalky, dry, adhesive, or puckery (e.g., Vidal et al., 2003). Epicatechin gallate, derived from seed tannins, seems to generate a coarseness and dryness not found in skin tannins (Francis et al., 2002).

Data from Sáenz-Navajas et al. (2012) support most of these findings, but suggest a greater role for aconitic acid (a nonflavonoid) to the astringency of Spanish red wines. In another study, Ferrer-Gallego et al. (2015) found catechins and procyanidins containing dihydroxylated monomers (e.g., catechin and epicatechin) possessed a more bitter, dry, rough, and persistent astringency than those containing more trihydroxylated monomers (e.g., epigallocatechin). The latter expressed astringency with smoother, more velvety, viscous attributes. The differences were consistent with the faster, more persistent binding of catechin to PRPs than gallocatechin. Fig. 4.23 illustrates one view of the relative effects of different phenolics on taste and mouth-feel.

Nonetheless, it is important to realize that the flavonoid content of a wine is dynamic, changing throughout maturation and during aging, largely from catechin monomers extracted during fermentation. Comparatively few phenolics are derived from the tannin polymers in seeds and skins during short macerations, most of which remain tightly bound to grape cell-wall constituents, or precipitate during fermentation. The seed and skin tannins present in wines are extracted primarily during prolonged joint maceration and fermentation. The largest grape tannins are too massive to be dissolved. Because flavonoid polymerization occurs comparatively slowly during maturation and aging, the progressive formation of massive tannin polymers (that react poorly with saliva, membrane proteins, or lipids) probably explains much of the eventual loss in the rough taste of most red wines. If they precipitate after bottling, due to binding with residual soluble proteins or their enlarged size, the sediment generated affects mouth-feel only if resuspended by agitation. The more common and effective use of fining today has largely eliminated the formation of sediment in most red wines.

Because the flavonoid content of most modern white wines tends to be low, it seldom has a detectable influence on their taste or flavor. It is only when the juice is left in contact with the seeds and skins (maceration) for several hours, before fermentation, that the flavonoid content begins to rise. As such, they may contribute to the wine's sensation of body. Nevertheless, because flavonoids are involved in browning and the generation of bitterness (>40 mg/L), there is a limit to their desirability. Other flavonoids potentially affecting the perception of white wines include flavanone glycosides (generally uncommon in grape varieties), such as naringin. They can, however, contribute a slight bitterness to Riesling and Silvaner wines (Drawert, 1970). Also, in late-harvest white wines, the flavan-3,4-ol content increases in relation to the degree of botrytization and/or partial dehydration (Dittrich et al., 1975). Nevertheless, the predominant phenolic tastants in white wines are nonflavonoids. Examples are caffeoyl tartrate (caftaric acid), which can induce a bitter sensation (Ong and Nagel, 1978), and benzoic acid derivatives, which can confer a range of perceptions, including astringency, bitterness, sourness, and even sweetness (Peleg and Noble, 1995). Fermentation and/or maturation in oak cooperage is an additional source of nonflavonoids in white wines.

The primary nonflavonoids in wines not matured in oak are derivatives of hydroxycinnamic and hydroxybenzoic acids. They are readily released from cell vacuoles during crushing. The most numerous, and variable in composition, are the hydroxycinnamic acid derivatives, notably caftaric, coutaric, and fertaric acids. Individually, most occur at concentrations below their detection thresholds (Singleton and Noble, 1976). In combination, though, they may influence the perception of a wine's bitterness (Gawel et al., 2016). In addition, hydrolysis of their tartaric acid esters during maturation may liberate sufficient free acids to augment the bitterness of white wines (Ong and Nagel, 1978). This potential increases relative to the alcohol content of the wine.

Elevated levels of hydroxybenzoic acid derivatives, notably ellagic acid, occur frequently in wines aged in oak cooperage. Ellagic acid comes from the breakdown of ellagitannins, the principal group of oak-hydrolyzable tannins. Castalagin and vescalagin (polymers of three ellagic acids) are the forms most frequently found in oak. Degradation of oak lignins liberates various cinnamaldehyde and benzaldehyde derivatives.

Hydrolyzable tannins apparently seldom play a direct role in the sensory quality of wine (Pocock et al., 1994). Although more astringent than condensed tannins, their low concentration and early degradation in wine generally limits their sensory impact. Nonetheless, the ready solubility of their gallic acid moieties may contribute to the persistent bitterness of some red wines.

Tyrosol production by yeast metabolism can also contribute to wine bitterness. This effect is particularly noticeable in sparkling wines. Tyrosol level increases considerably during the second in-bottle fermentation that characterizes the production of most sparkling wines. Even in still white wines, tyrosol may contribute to bitterness at concentrations above ~25 mg/L.

In addition to direct influences on bitterness and astringency, phenolics influence the perception of sweetness, acidity, and the release of aromatics. Phenolics are also considered to have a direct impact on the perception of body and balance.

Polysaccharides

The majority of wine polysaccharides are grape- and yeast-derived. Although present in small amounts, they play a significant role in delaying (extending) the release of aromatics during tasting (Chapter 3, Olfactory Sensations). Nonetheless, they may also play a role in reducing the astringency of wine phenolics. Exactly how they have this action is unclear, but it may involve solubilization of protein–tannin aggregates (de Freitas et al., 2003), and/or disruption of the binding of polyphenolics with proteins (Ozawa et al., 1987). Yeast-derived mannoproteins are also considered important in the slow release of carbon dioxide from sparkling wine (Senée et al., 1999), helping to give it the effervescent attributes for which the style of wine is famous.

Nucleic Acids

During extended contact with autolyzing yeast cells, such as during *sur lies* or sparkling wine maturation, nucleotides are released, especially early on during autolysis (Charpentier et al., 2005). Several ribonucleotides, notably in the presence of monosodium glutamate, enhance flavor perception. Glutamate acts in conjunction with olfactory compounds to elicit flavor enhancement (McCabe and Rolls, 2007). Thus, some interest has been shown in the potential sensory significance of nucleotides in wine flavor (Courtis et al., 1998; Charpentier et al., 2005), despite their low concentration. If significant, enhancement of wine flavor may be more relevant in association with food, where higher concentrations of free nucleotides are likely to occur.

TASTE AND MOUTH-FEEL SENSATIONS IN WINE TASTING

Where considered relevant, tasters can distinguish between the various taste and mouth-feel sensations, by concentrating sequentially on their separate perceptions. Their distinct time–intensity curves can be useful in confirming identification (Kuznicki and Turner, 1986). Spatial localization on the tongue or throughout the mouth can occasionally help affirm taste identification. Wine fragrance can not only influence the apparent presence of taste sensations, but also reduce the time taken to identify specific sensations (White and Prescott, 2001).

The perception of sweetness is probably the most rapidly detected taste sensation. If detected in a dry wine, the perception is likely to be of very short duration (Portmann et al., 1992). If mild but prolonged, it is more likely due to oral referral due to the wine's fruity fragrance. Persistent sweetness is typical only in dessert or some fortified wines. Sensitivity to sweetness is optimal at the front/sides of the tongue, but occurs elsewhere in the mouth (Fig. 4.3). Perceived sweetness is reduced in the presence of wine acids and phenolics. The latter are important in avoiding a cloying sensation in particularly sweet wines.

Sourness is also detected early. The rate of adaptation is usually slower than for sugars, and may represent the primary aftertaste in dry white wines. Acid detection is commonly strongest along the sides of the tongue, but varies considerably among individuals. For example, some people detect sourness more distinctly on the insides of the cheeks or lips, due to the astringent side-taste of acidity. The perception of sourness is primarily associated with stimulation of acid-responsive taste receptors in taste buds, but may also involve the activation of a few types of trigeminal receptors. Strongly acidic wines can also induce astringency, giving the teeth a rough feel, and accentuate the astringency of tannins. Although detectable viscosity can suppress the perceptions of a wine's sour and astringent attributes (Smith and Noble, 1998), wines rarely reach levels of viscosity sufficient to have this effect.

Bitterness is usually the next oral sensation to be detected, if present, often taking several seconds to commence. Bitterness may take 10–15 s to reach peak intensity (Fig. 4.14). On expectoration, the sensation declines gradually and may linger for several minutes. Most bitter-tasting wine phenolics are detected at the back-central portion of the tongue. In contrast, the bitterness of some compounds, such as alkaloids, is more (or initially) perceived on the soft palate and front of the tongue (Boudreau et al., 1979). This feature is rarely noted in wine because the bitterness of phenolics develops more rapidly (McBurney, 1978). In addition, few if any bitter alkaloids occur in wines, except where extracted from herbs used to flavor vermouth.

Wine bitterness is more difficult to assess with confidence when the wine is also markedly astringent. High levels of astringency seem to diminish or partially mask the perception of bitterness. Increasing alcohol content enhances the perception of bitterness (Noble, 1994), whereas detectable sweetness suppresses it (Schiffman et al., 1994). Because most table wines are dry, the suppressive influence of sugar on bitterness is only of relevance to dessert or sweet fortified wines.

The perception of astringency is also slow to develop (Fig. 4.12), often being the last principal in-mouth sensation to be detected. On expectoration, perception slowly declines over a period of several minutes. Astringency is poorly localized, probably due to the random precipitation of saliva protein in the mouth, as well as the dispersion of trigeminal receptors throughout the oral cavity. Because perceived intensity and duration increase with repeat sampling, some authorities recommend that the assessment of astringency be based on the first taste. This would give a perception more closely approximating the sensation detected on consuming wine with a meal. Others consider that assessing astringency should occur only after several samplings, when the modifying affects of saliva have been reduced.

The increase in perceived astringency when tasting (Guinard et al., 1986a) can significantly affect assessment, notably in red wines; the first wine often appears smoother and more harmonious. A similar situation can arise when a series of dry acidic wines, or very sweet wines, is tasted. The effect of such sequence errors can be partially offset by presenting the wines in different, but randomized, order to each assessor. Lingering taste effects can be reduced by palate cleansing between samples.

Although the in-mouth sensations of wines are relatively few, they appear to be particularly important in wine acceptance by the majority of consumers. Unlike professionals, many consumers seemingly disregard the wine's fragrance. Thus, taste and haptic sensations are far more important to initial appreciation. Even for the connoisseur, one of the ultimate expressions of finesse is the holistic impression of elegance and balance. Producing a wine with a fine, complex, and interesting fragrance is often a significant challenge for the winemaker. Even more demanding is the production of a wine that also possesses a rich, full, and balanced in-mouth sensation.

Postscript

Wine may show its greatest diversity, elegance, and intrigue through fragrance, but it is flavor that most consumers notice. Thus, although the basic in-mouth sensations are few, it is their association with retronasal smell, combined with color, that are what most often influence consumer opinion. Except for sweet wines, the taste aspects of greatest significance are perceived sourness, bitterness, and astringency. Of these, it is acidity that is the simplest for the winemaker to adjust. Bitterness and astringency are far more complex in chemical origin, and they change during maturation and aging. That astringency expresses itself in diverse ways suggests that the term incorporates separate phenomena with distinct derivations. Although much has been learned, the nature of a wine's phenolic composition is still imprecisely known, our understanding changes with new discoveries, and thus, is difficult to regulate or predict with precision. In the absence of precise knowledge, blending to accentuate the positive and eliminate the negative influences of gustatory sensations remains more an art than a science. Thus, it is crucial for any winemaker to develop an extensive memory and astute sensory skills to adjust wine attributes to match consumer preferences. In the end, wine should be produced for the enjoyment of the average consumer and connoisseur alike, not to enhance the ego of the producer or the social and economic standing of some financial titan.

Suggested Readings

Auvray, M., Spence, C., 2008. The multisensory perception of flavor. Conscious. Cognit. 17, 1016–1031.

Carstens, E., Carstens, M.I., Dessirier, J.-M., O'Mahony, M., Simons, C.T., Sudo, M., Sudo, S., 2002. It hurts so good: Oral irritation by spices and carbonated drinks and the underlying neural mechanisms. Food Qual. Pref. 13, 431–443.

Francis, I.L., Gawel, R., Iland, P.G., Vidal, S., Cheynier, V., Guyot, S., Kwiatkowski, M.J., Waters, E.J., 2002. Characterising mouth-feel properties of red wines. In: Blair, R., Williams, P., Høj, P. (Eds.) Proc. 11th Aust. Wine Indust. Tech. Conf., Aust. Wine Indust. Tech. Conf. Inc., Urrbrae, Australia, pp. 123–127.

Haslam, E., 2007. Vegetable tannins — Lessons of a phytochemical lifetime. Phytochemistry 68, 2713–2721.

Ma, W., Guo, A., Zhang, Y., Wang, H., Liu, Y., Li, H., 2014. A review on astringency and bitterness perception of tannins in wine. Food Sci. Technol. 40, 6–19.

Pelchat, M.L., Bryant, B., Cuomo, R., Di Salle, F., Fass, R., Wise, P., 2014. Carbonation. A review of sensory mechanisms and health effects. Nutr. Today 49 (6), 308–312.

Reed, D.R., Tanaka, T., McDaniel, A.H., 2006. Diverse tastes: Genetics of sweet and bitter perception. Physiol. Behav. 88, 215–226.

Schoenfeld, M.A., Cleland, T.A., 2005. The anatomical logic of smell. Trends Neurosci. 28, 620–627.

Scollary, G.R., Pásti, G., Kállay, M., Blackman, J., Clark, A.C., 2012. Astringency response of red wines: Potential role of molecular assembly. Trend. Food Sci. Technol 27, 25–36.

Silver, W.L., Maruniak, J.A., 2004. Trigeminal chemoreception in the nasal and oral cavities. Chem. Senses 6, 295–305.

Spence, C., 2016. Oral referral: On the mislocalization of odours to the mouth. Food Qual. Pref. 50, 117–128.

Taylor, A.J., Roberts, D.D. (Eds.), 2004. Flavor Perception, Blackwell Publ. Ltd., Oxford, UK.

References

Amerine, M.A., Ough, C.S., 1980.). Methods for Analysis of Musts and Wines. John Wiley, New York.

Amerine, M.A., Roessler, E.B., Ough, C.S., 1965. Acids and the acid taste. I. The effect of pH and titratable acidity. Am. J. Enol. Vitic. 16, 29–37.

Aqueveque, C., 2015. The influence of experts' positive word-of-mouth on a wine's perceived quality and value: The moderator role of consumers' expertise. J. Wine Res. 26, 181–191.

Arnold, R.A., Noble, A.C., 1978. Bitterness and astringency of grape seed phenolics in a model wine solution. Am. J. Enol. Vitic. 29, 150–152.

Avenet, P., Lindemann, B., 1989. Perspective of taste reception. J. Membrane Biol. 112, 1–8.

Barker, L.M. (Ed.), 1982. The Psychobiology of Human Food Selection. AVI, Westport, CT.

Bartoshuk, L.M., Rifkin, B., Marks, L.E., Bars, P., 1986. Taste and aging. J. Gerontol. 41, 51–57.

Bartoshuk, L.M., Duffy, V.B., Miller, I.J., 1994. PTC/PROP tasting: Anatomy, psychophysics and sex effects. Physiol. Behavior 56, 1165–1171.

Behrens, M., Gunn, H.C., Ramos, P.D.M., Meyerhof, W., Wooding, S.P., 2013. Genetic, functional, and phenotypic diversity in TAS2R38-mediated bitter taste perception. Chem. Senses 38, 475–484.

Bennick, A., 1982. Salivary proline-rich proteins. Mol. Cell. Biochem. 45, 83–99.

Blakeslee, A.F., Salmon, T.N., 1935. Genetics of sensory thresholds, individual taste reactions for different substances. Proc. Natl. Acad. Sci. U.S.A. 21, 84–90.

Boudreau, J.C., Oravec, J., Hoang, N.K., and White, T.D. (1979). Taste and the taste of foods. In Food Taste Chemistry (J. C. Boudreau, ed.), pp. 1–30. ACS Symposium Series No. 115. American Chemical Society, Washington, DC.

Brandão, E., Soares, S., Mateus, N., de Freitas, V., 2014. In vivo interactions between procyanidins and human saliva proteins: Effect of repeated exposures to procyanidins solution. J. Agric. Food Chem. 62, 9562–9568.

Breslin, P.A.S., Beauchamp, G.K., 1995. Suppression of bitterness by sodium: Variation among bitter taste stimuli. Chem. Senses 20, 609–623.

Brodal, A., 1981.). Neurological Anatomy in Relation to Clinical Medicine, third ed Oxford University Press, New York.

Brossaud, F., Cheynier, V., Noble, A.C., 2001. Bitterness and astringency of grape and wine polyphenols. J. Grape Wine Res. 7, 33–39.

Cain, W.S., Murphy, C.L., 1980. Interaction between chemoreceptive modalities of odour and irritation. Nature 284, 255–257.

Chandrashekar, J., Yarmolinsky, D., von Buchholtz, L., Oka, Y., Sly, W., Ryba, N.J.P., et al., 2009. The taste of carbonation. Science 326, 443–445.

Chandrashekar, J., Kuhn, C., Oka, Y., Yarmolinsky, D.A., Hummler, E., Ryba, N.J.P., et al., 2010. The cells and peripheral representation of sodium taste in mice. Nature 464, 297–301.

Charlton, A.J., Baxter, N.J., Khan, M.L., Moir, A.J.G., Haslam, E., Davies, A.P., et al., 2002. Polyphenol/peptide binding and precipitation. J. Agric. Food Chem. 50, 1593–1601.

Charpentier, C., Aussenac, J., Charpentier, M., Prome, J.-C., Duteurtre, B., Feuillat, M., 2005. Release of nucleotides and nucleosides during yeast autolysis: Kinetics and potential impact on flavor. J. Agric. Food Chem. 53, 3000–3007.

Chen, X., Gabitto, M., Peng, Y., Ryba, N.J.P., Zuker, C.S., 2011. A gustotopic map of taste qualities in the mammalian brain. Science 333, 1262–1266.

Chrea, C., Valentin, D., Sulmont-Rossé, C., Mai, H.L., Nguyen, D.H., Adbi, H., 2004. Culture and odor categorization: Agreement between cultures depends upon the odors. Food Qual. Pref. 15, 669–679.

Collings, V.B., 1974. Human taste response as a function of focus on the tongue and soft palate. Front and Side refer to fungiform papillae. The higher the value along the y-axis denotes lower the taste threshold. Percept. Psychophys. 16, 169–174.

Cometto-Muñiz, J.E., Garcia-Media, M.R., Calvino, A.M., Noriega, G., 1987. Interactions between CO_2 oral pungency and taste. Perception 16, 629–640.

Condelli, N., Dinnella, C., Cerone, A., Monteleone, E., Bertuccioli, M., 2006. Prediction of perceived astringency induced by phenolic compounds II: Criteria for panel selection and preliminary application on wine samples. Food Qual. Pref. 17, 96–107.

Corrigan Thomas, C.J., Lawless, H.T., 1995. Astringent subqualities in acids. Chem. Senses. 20, 593–600.

Courtis, K., Todd, B., Zhao, J., 1998. The potential role of nucleotides in wine flavour. Aust. Grapegrower Winemaker 409, 31–33.

Cowart, B.J., 1998. The addition of CO_2 to traditional taste solutions alters taste quality. Chem. Senses 23, 397–402.

Cowart, B.J., Yokomakai, Y., Beauchamp, G.K., 1994. Bitter taste in again: Compound-specific decline in sensitivity. Physiol Behav. 56, 1237–1241.

Cruz, A., Green, B., 2000. Thermal stimulation of taste. Nature 403, 889–892.

Dalton, P., Doolittle, N., Nagata, H., Breslin, P.A.S., 2000. The merging of the senses: Integration of subthreshold taste and smell. Nature Neurosci. 3, 431–432.

Dawes, C., 1964. Is acid-precipitation of salivary proteins a factor in plaque formation? Arch. Oral Biol. 9, 375–376.

de Freitas, V., Mateus, N., 2003. Nephelometric study of salivary protein-tannin aggregates. J. Sci. Food Agric. 82, 113–119.

de Freitas, V., Carvalho, E., Mateus, N., 2003. Study of carbohydrate influence on protein-tannin aggregation by nephelometry. Food Chem. 81, 503–509.

de Loryn, L.C., Petri, P.R., Hasted, A.M., Johnson, T.E., Collins, C., Bastian, S.E.P., 2014. Evaluation of sensory threshold and perception of sodium chloride in grape juice and wine. Am J. Enol. Vitic. 65, 124–133.

Delwiche, J.F., Heffelfinger, A.L., 2005. Cross-modal additivity of taste and smell. J. Sens. Stud. 20, 512–525.

Delwiche, J.F., Halpern, B.P., Desimone, J.A., 1999. Anion size of sodium salts and simple taste reaction times. Physiol. Behav. 66, 27–32.

Delwiche, J.F., Buletic, Z., Breslin, P.A.S., 2002. Clustering bitter compounds via individual sensitivity differences: Evidence supporting multiple receptor–transduction mechanisms. In: Given, P., Paredes, D. (Eds.), Chemistry of Taste: Mechanisms, Behaviors, and Mimics. American Chemical Society, Washington, DC, pp. 66–77.

DeSimone, J.A., Heck, G.L., Bartoshuk, L.M., 1980. Surface active taste modifiers, a comparison of the physical and psychophysical properties of gymnemic acid and sodium lauryl sulfate. Chem. Senses 5, 317–330.

Dessirier, J.M., Simons, C.T., Carstens, M.I., O'Mahony, M., Carstens, E., 2000. Psychophysical and neurobiological evidence that the oral sensation elicited by carbonated water is of chemogenic origin. Chem. Senses 25, 277–284.

Di Salle, F., Cantone, E., Savarese, M.F., Aragri, A., Prinster, A., Nicolai, E., et al., 2013. Effect of carbonation on brain processing of sweet stimuli in humans. Gastroenterology 145, 537–539.

Dittrich, H.H., Sponholz, W.R., Kast, W., 1975. Vergleichende Untersuchungen von Mosten und Weinen aus gesunden und aus Botrytis-infizierten Traubenbeeren. Vitis 13, 336–347.

Djordjevic, J., Zatorre, R.J., Jones-Gotman, M., 2004. Effects of perceived and imagined odors on taste detection. Chem. Senses 29, 199–208.

Doty, R.L., Bromley, S.M., 2004. Effects of drugs on olfaction and taste. Otolaryngol. Clin. N. Am. 37, 1229–1254.

Drawert, F., 1970. Causes déterinant l'amertume de certains vins blancs. Bull. O.I.V 43, 19–27.

Drewnowski, A., Henderson, A.S., Barratt-Fornell, A., 2001. Genetic taste markers and food preferences. Drug Metab. Disposit 29, 535–538.

Dufour, C., Bayonove, C.L., 1999a. Influence of wine structurally different polysaccharides on the volatility of aroma substances in a model system. J. Agric. Food Chem. 47, 671–677.

Dufour, C., Bayonove, C.L., 1999b. Interactions between wine polyphenols and aroma substances. An insight at the molecular level. J. Agric. Food Chem. 47, 678–684.

Epke, E.M., McClure, S.T., Lawless, H.T., 2009. Effects of nasal occlusion and oral contact on perception of metallic taste from metal salts. Food Qual. Pref. 20, 133–137.

Ferrer-Gallego, R., Hernández-Hierro, J.M., Rivas-Gonzalo, J.C., Escribano-Bailón, M.T., 2014. Sensory evaluation of bitterness and astringency sub-qualities of wine phenolic compounds: Synergistic effect and modulation by aromas. Food Res. Int. 62, 1100–1107.

Ferrer-Gallego, R., Quijada-Morín, N., Brás, N.F., Gomes, P., de Freitas, V., Rivas-Gonzalo, J.C., et al., 2015. Characterization of sensory properties of flavanols–A molecular dynamic approach. Chem.Sens. 40, 381–390.

Fischer, M.E., Cruickshanks, K.J., Schubert, C.R., Pinto, A., Klein, R., Pankratz, N., et al., 2013. Factors related to fungiform papillae density: The Beaver Dam offstring study. Chem. Senses 38, 669–677.

Fischer, U., Boulton, R.B., Noble, A.C., 1994. Physiological factors contributing to the variability of sensory assessments: Relationship between salivary flow rate and temporal perception of gustatory stimuli. Food Qual. Pref. 5, 55–64.

Fontoin, H., Saucier, C., Teissedre, P.-L., Glories, Y., 2008. Effect of pH, ethanol and acidity on astringency and bitterness of grape seed tannin oligomers in model wine solution. Food Qual. Pref. 19, 286–291.

Forss, D.A., 1969. Role of lipids in flavors. J. Agr. Food Chem. 17, 681–685.

Francis, I.L., Gawel, R., Iland, P.G., Vidal, S., Cheynier, V., Guyot, S., et al., 2002. Characterising mouth-feel properties of red wines. In: Blair, R., Williams, P., Høj, P. (Eds.), Proc. 11th Aust. Wine Ind. Tech. Conf. Australian Wine Industrial Technical Conference Inc., Glen Osmond, SA, Australia, pp. 123–127.

Frank, S., Wollmann, N., Schieberle, P., Hofmann, T., 2011. Reconstruction of the flavor signature of Dornfelder red wine on the basis of the natural concentrations of its key aroma and taste compounds. J. Agric. Food Chem. 59, 8866–8874.

Furlan, A.L., Jobin, M.-L., Pianet, I., Dufourc, E.J., Géan, J., 2015. Flavanol/lipid interaction: A novel molecular perspective in the description of wine astringency & bitterness and antioxidant action. Tetrahedron 71, 3143–3147.

Ganzevles, P.G.J., Kroeze, J.H.A., 1987. The sour taste of acids, the hydrogen ion and the undissociated acid as sour agents. Chem. Senses 12, 563–576.

Gawel, R., Waters, E., 2008. The effect of glycerol on the perceived viscosity of dry white table wine. J. Wine Res. 19, 109–114.

Gawel, R., Oberholster, A., Francis, I.L., 2000. A "Mouth-feel Wheel": Terminology for communicating the mouth-feel characteristics of red wine. Aust. J. Grape Wine Res. 6, 203–207.

Gawel, R., Iland, P.G., Francis, I.L., 2001. Characterizing the astringency of red wine: A case study. Food Qual. Pref. 12, 83–94.

Gawel, R., van Sluyter, S., Waters, E.J., 2007. The effects of ethanol and glycerol on the body and other sensory characteristics of Riesling wines. Aust. J. Grape Wine Res. 13, 38–45.

Gawel, R., Van Sluyter, S.C., Smith, P.A., Waters, E.J., 2013. Effect of pH and alcohol on perception of phenolic character in white wine. Am. J. Enol. Vitic. 64, 425–429.

Gawel, R., Schulkin, A., Day, M., Barker, A., Smith, P.A., 2016. Interactions between phenolics, alcohol and acidity in determining the mouthfeel and bitterness of white wine. Wine Vitic. J. 31 (1), 30–34.

Gilbertson, T.A., Boughter Jr., J.D., 2003. Taste transduction: Appetizing times in gustation. NeuroReport 14, 905–911.

Gilbertson, T.A., Khan, N.A., 2014. Cell signaling mechanisms of oro-gustatory detection of dietary fat: Advances and challenges. Prog. Lipid Res. 53, 82–92.

Glendinning, J., 1992. Effect of salivary proline-rich proteins on ingestive responses to tannic acid in mice. Chem. Senses 17, 1–12.

Gonzales, E.B., Kawate, T., Gouaux, E., 2009. Pore architecture and ion sites in acid-sensing ion channels and P2X receptors. Nature 460, 599–605.

Green, B.G., 1992. The effects of temperature and concentration on the perceived intensity and quality of carbonation. Chem. Senses 17, 435–450.

Green, B.G., 1993. Oral astringency: A tactile component of flavor. Acta Psychol. 84, 119–125.

Green, B.G., 2004. Oral chemesthesis: And integral component of flavour. In: Taylor, A.J., Roberts, D.D. (Eds.), Flavor Perception. Blackwell Publ. Ltd., Oxford, UK, pp. 151–171.

Green, B.G., Frankmann, S.P., 1987. The effect of cooling the tongue on the perceived intensity of taste. Chem. Senses 12, 609–619.

Green, B.G., Frankmann, S.P., 1988. The effect of cooling on the perception of carbohydrate and intensive sweeteners. Physiol. Behav. 43, 515–519.

Green, B.G., George, P., 2004. Thermal taste" predicts higher responsiveness to chemical taste and flavor. Chem. Senses 29, 617–628.

Green, B.G., Nachtigal, D., 2015. Temperature effects human sweet taste via at least two mechanisms. Chem. Sens. 40, 391–399.

Guest, S., Essick, G., Young, M., Phillips, N., McGlone, F., 2008. The effect of oral drying and astringent liquids on the perception of mouth wetness. Physiol. Behav. 93, 888–896.

Guinard, J., Pangborn, R.M., Lewis, M.J., 1986b. Preliminary studies on acidity-astringency interactions in model solutions and wines. J. Sci. Food Agric. 37, 811–817.

Guinard, J.-X., Pangborn, R.M., Lewis, M.J., 1986a. The time–course of astringency in wine upon repeated ingestion. Am. J. Enol. Vitic. 37, 184–189.

Haslam, E., Lilley, T.H., 1988. Natural astringency in foods. A molecular interpretation. Crit. Rev. Food Sci. Nutr. 27, 1–40.

Haslam, E., Lilley, T.H., Warminski, E., Liao, H., Cai, Y., Martin, R., et al. (1992). Polyphenol complexation, a study in molecular recognition. In Phenolic Compounds in Food and Their Effects on Health, 1, Analysis, Occurrence, and Chemistry. (C.-T. Ho et al., eds.), pp. 8–50. ACS Symposium Series No. 506. American Chemical Society, Washington, DC.

Hewson, L., Hollowood, T., Chandra, S., Hort, J., 2009. Gustatory, olfactory and trigeminal interactions in a model carbonated beverage. Chem. Percept. 2, 94–107.

Hufnagel, J.C., Hofmann, T., 2008a. Orosensory-directed identification of astringent mouthfeel and bitter-tasting compounds in red wine. J. Agric. Food Chem. 56, 1376–1386.

Hufnagel, J.C., Hofmann, T., 2008b. Quantitative reconstruction of the nonvolatile sensometabolome of a red wine. J. Agric. Food Chem. 56, 9190–9199.

Hyde, R.J., Pangborn, R.M., 1978. Parotid salivation in response to tasting wine. Am. J. Enol. Vitic. 29, 87–91.

Ishikawa, T., Noble, A.C., 1995. Temporal perception of astringency and sweetness in red wine. Food Qual. Pref. 6, 27–34.

Jackson, R.S., 2014. Wine Science: Principles and Application, fourth ed. Academic Press, San Diego, CA.

Jöbstl, E., O'Connell, J., Fairclough, P.A., Williamson, M.P., 2004. Molecular model for astringency produced by polyphenol/protein interactions. Biomacromolecules 5, 942–949.

Kallithraka, S., Bakker, J., Clifford, M.N., 2000. Interaction of (+)-catechin, (-)-epicatechin, procyanidin B2 and procyanidin C1 with pooled human saliva in vitro. J. Sci. Food Agric. 81, 261–268.

Keast, R.S.J., Breslin, P.A.S., 2003. An overview of binary taste–taste interactions. Food Qual. Pref. 14, 111–124.

Keast, R.S.J., Canty, T.M., Breslin, T.A.S., 2004. The influence of sodium salts on binary mixtures of bitter-tasting compounds. Chem. Senses 29, 431–439.

Kielhorn, S., Thorngate, J.H., 1999. Oral sensations associated with the flavan-3-ols (+)-catechin and (-)-epicatechin. Food Qual. Pref. 10, 109–116.

Kim, U., Wooding, S., Ricci, D., Jorde, L.B., Drayna, D., 2005. Worldwide haplotype diversity and coding sequence variation at human bitter taste receptor loci. Hum. Mutat. 26, 199–204.

Kinnamon, S.C., 1996. Taste transduction: Linkage between molecular mechanism and psychophysics. Food Qual. Pref. 7, 153–160.

Komai, M., Bryant, B.P., 1993. Acetazolamide specifically inhibits lingual trigeminal nerve responses to carbon dioxide. Brain Res. 612, 122–129.

Kroeze, J.H.A., 1982. The relationship between the side taste of masking stimuli and masking in binary mixtures. Chem. Senses 7, 23–37.

Kuhn, C., Bufe, B., Winnig, M., Hofmann, T., Frank, O., Behrens, M., et al., 2004. Bitter taste receptors for saccharin and acesulfame K. J. Neurosci. 24, 10260–10265.

Kuroda, M., Miyamura, N., 2015. Mechanism of the perception of "kokumi" substances and the sensory characteristics of the "kokumi" peptide, γ-Glu-Val-Gly. Flavour 4, 11. (3 pp).

Kurogi, M., Kawau, Y.H., Nagatomo, K., Tateyama, M., Kubo, Y., Saitoh, O., 2015. Auto-oxidation products of epigallocatechin gallate activate TRPA1 and TRPV1 in sensory neurons. Chem. Senses 40, 27–46.

Kuznicki, J.T., Turner, L.S., 1986. Reaction time in the perceptual processing of taste quality. Chem. Senses 11, 183–201.

Lawless, H.T., Corrigan, C.J., Lee, C.B., 1994. Interactions of astringent substances. Chem. Senses 19, 141–154.

Lawless, H.T., Schlake, S., Smythe, J., Lim, J., Yang, H., Chapman, K., et al., 2004. Metallic taste and retronasal smell. Chem. Senses 29, 25–33.

Lea, A.G.H., Arnold, G.M., 1978. The phenolics of ciders: Bitterness and astringency. J. Sci. Food Agric. 29, 478–483.

Leach, E.J., Noble, A.C., 1986. Comparison of bitterness of caffeine and quinine by a time-intensity procedure. Chem. Senses 11, 339–345.

Lee, C.A., Ismail, B., Vickers, Z.M., 2012. The role of salivary proteins in the mechanism of astringency. J. Food Sci. 77, C381–C387.

Lee, C.B., Lawless, H.T., 1991. Time-course of astringent sensations. Chem. Senses 16, 225–238.

Lee, R.J., Kofonow, J.M., Rosen, P.L., Siebert, A.P., Chen, B., Doghramji, L., et al., 2014. Bitter and sweet taste receptors regulate human upper respiratory innate immunity. J. Clin. Invest. 124, 1393–1405.

Levine, M.W., Shefner, J.M., 1991. Fundamentals of Sensation and Perception, second ed. Brooks/Cole Publishing, Pacific Grove, CA.

Li, X., Staszewski, L., Xu, H., Durick, K., Zoller, M., Adler, E., 2002. Human receptors for sweet and umami taste. PNAS 99, 4692–4696.

Lim, J., Urban, L., Green, B.G., 2008. Measures of individual differences in taste and creaminess perception. Chem. Senses. 33, 493–501.

Lim, J., Wood, A., Green, B.G., 2009. Derivation and evaluation of a labeled hedonic scale. Chem. Senses 34, 739–751.

Lorber, K., Schieberle, P., Buettner, A., 2014. Influence of the chemical structure on odor qualities and odor thresholds in homologous series of alka-1,5-dien-3-ones, alk-1-en-3-ones, alka-1,5-dien-3-ols, and alk-1-en-3-ols. J. Agric. Food Chem. 62, 1025–1031.

Lyman, B.J., Green, B.G., 1990. Oral astringency: Effects of repeated exposure and interactions with sweeteners. Chem. Senses 15, 151–164.

Ma, W., Guo, A., Zhang, Y., Wang, H., Liu, Y., Li, H., 2014. A review on astringency and bitterness perception of tannins in wine. Trends Food Sci. Technol. 40, 6–19.

Marchal, A., Waffo-Téguo, P., Génin, E., Mérillon, J.-M., Dubourdieu, D., 2011. Identification of new natural sweet compounds in wine using centrifugal partition chromatography–gustatometry and Fourier transform Mass Spectrometry. Anal. Chem. 83, 9629–9637.

Marchal, A., Cretin, B.N., Sindt, L., Waffo-Teguo, P., Dubourdieu, D., 2015. Contribution of oak lignans to wine taste: Chemical identification, sensory characterization and quantification. Tetrahedron 71, 3148–3156.

Maruyama, Y., Yasuda, R., Kuroda, M., Eto, Y., 2012. Kokumi substances, enhancers of basic tastes, induce responses in calcium-sensing receptor expressing taste cells. PLoS ONE 7, e34489.

Matsunami, H., Amrein, H., 2004. Taste perception: How to make a gourmet mouse. Curr. Biol. 14, R118–R120.

Matsuo, R., 2000. Role of saliva in the maintenance of taste sensitivity. Crit. Rev. Oral Biol. Med. 11, 216–229.

McBride, R.L., Finlay, D.C., 1990. Perceptual integration of tertiary taste mixtures. Percept. Psychophys. 48, 326–336.

McBurney, D.H., 1978. Psychological dimensions and perceptual analyses of taste In: Carterette, E.C. Friedman, M.P. (Eds.), Handbook of Perception, Vol. 6A. Academic Press, New York, pp. 125–155.

McCabe, C., Rolls, E.T., 2007. Umami: A delicious flavour formed by convergence of taste and olfactory pathways in the human brain. Eur. J. Neurosci. 25, 1855–1864.

Meyerhof, W., Batram, C., Kuhn, C., Brockhoff, A., Chudoba, E., Bufe, B., et al., 2010. The molecular receptive ranges of human TAS2R bitter taste receptors. Chem. Senses 35, 157–170.

Miller, I.J., Reedy Jr., E.E., 1990. Variations in human taste bud density and taste intensity perception. Physiol. Behav. 46, 1213–1219.

Miyazawa, T., Gallagher, M., Preti, G., Wise, P.M., 2008. The impact of subthreshold carboxylic acids on the odor intensity of suprathreshold flavor compounds. Chem. Percept. 1, 163–167.

Moncrieff, R.W., 1964. The metallic taste. Perfumery Essent. Oil Rec. 55, 205–207.

Nagai, T., Kim, D.J., Delay, R.J., Roper, S.D., Kinnamon, S.C., 1996. Neuromodulation of transduction and signal processing in the end organs of taste. Chem. Senses 21, 353–365.

Nelson, G., Chandrashekar, J., Hoon, M.A., Feng, L., Zhao, G., Ryba, N.J.P., et al., 2002. An amino-acid taste receptor. Nature 416 (6877), 199–202.

Nieuwoudt, H.H., Prior, B.A., Pretorius, I.M., Bauer, F.F., 2002. Glycerol in South African wines: An assessment of its relationship to wine quality. S. Afr. J. Enol. Vitic. 23, 22–30.

Noble, A.C., 1994. Bitterness in wine. Physiol. Behav. 56, 1251–1255.

Noble, A.C., Bursick, G.F., 1984. The contribution of glycerol to perceived viscosity and sweetness in white wine. Am. J. Enol. Vitic. 35, 110–112.

Noble, A.C., Strauss, C.R., Williams, P.J., Wilson, B., 1988. Contribution of terpene glycosides to bitterness in Muscat wines. Am. J. Enol. Vitic. 39, 129–131.

Nurgel, C., Pickering, G., 2005. Contribution of glycerol, ethanol and sugar to the perception of viscosity and density elicited by model white wines. J. Texture Studies 36, 303–323.

Obreque-Slíer, E., Peña-Neira, A., López-Solís, R., 2010. Enhancement of both salivary protein-enological tannin interactions and astringency perception by ethanol. J. Agric. Food Chem. 58, 3729–3735.

Obreque-Slíer, E., Peña-Neira, A., López-Solís, R., 2011. Precipitation of low molecular weight phenolic compounds of grape seeds cv. Carménère (Vitis vinifera L.) by whole saliva. Eur. Food Res. Technol. 232, 113–121.

Obst, K., Paetz, S., Backes, M., Reichelt, K.V., Ley, J.P., Engel, K.-H., 2013. Evaluation of unsaturated alkanoic acid amides as markers of epigallocatechin gallate astringency. J. Agric. Food Chem. 61, 4242–4249.

Ong, B.Y., Nagel, C.W., 1978. High-pressure liquid chromatographic analysis of hydroxycinnamic acid tartaric acid esters and their glucose esters in Vitis vinifera. J. Chromatogr. 157, 345–355.

Ossebaard, C.A., Polet, I.A., Smith, D.V., 1997. Amiloride effects on taste quality: Comparison of single and multiple response category procedures. Chem. Senses 22, 267–275.

Ough, C.S., Fong, D., Amerine, M.A., 1972. Glycerol in wine, determination and some factors affecting. Am. J. Enol. Vitic. 27, 1–5.

Ozawa, T., Lilly, T.H., Haslam, E., 1987. Polyphenol interaction: Astringency and the loss of astringency in ripening fruit. Phytochemistry 26, 2937–2942.

Peleg, H., Noble, A.C., 1995. Perceptual properties of benzoic acid derivatives. Chem. Senses 20, 393–400.

Peleg, H., Bodine, K., Noble, A.C., 1998. The influence of acid on astringency of alum and phenolic compounds. Chem. Senses 23, 371–378.

Pickering, G.J., Heatherbell, D.A., Vanhaenena, L.P., Barnes, M.F., 1998. The effect of ethanol concentration on the temporal perception of viscosity and density in white wine. Am. J. Enol. Vitic. 49, 306–318.

Pickering, G.J., Simunkova, D., DiBattista, D., 2004. Intensity of taste and astringency sensations elicited by red wine is associated with sensitivity to PROP (6-n-propylthiouracil). Food Qual. Pref. 15, 147–154.

Pocock, K.F., Sefton, M.A., Williams, P.J., 1994. Taste thresholds of phenolic extracts of French and American oakwood: The influence of oak phenols on wine flavor. Am. J. Enol. Vitic. 45, 429–434.

Portmann, M.-O., Serghat, S., Mathlouthi, M., 1992. Study of some factors affecting intensity/time characteristics of sweetness. Food Chem. 44, 83–92.

Prescott, J., 2004. Psychological processes in flavour perception. In: Taylor, A.J., Roberts, D. (Eds.), Flavour Perception. Blackwell Publishing, London, pp. 256–278.

Prinz, J.F., Lucas, P.W., 2000. Saliva tannin interactions. J. Oral Rehabil. 27, 991–994.

Ramsey, I.S., Delling, M., Clapham, D.E., 2006. An introduction to TRP channels. Annu. Rev. Physiol. 68, 619–647.

Rapp, A. (1987). Veränderung der Aromastoffe während der Flaschenlagerung von Weißweinen. In "Primo Simposio Internazionale: Le Sostanze Aromatiche dell'Uva e del Vino" pp. 286–296.

Rentmeister-Bryant, H., Green, B.G., 1997. Perceived irritation during ingestion of capsaicin or piperine: Comparison of trigeminal and non-trigeminal areas. Chem. Senses 22, 257–266.

Rentschler, H., Tanner, H., 1951. Red wines turning bitter; contribution to the knowledge of presence of acrolein in beverages and its correlation to the turning bitter of red wines. Mitt. Lebensmitt. Hygiene 42, 463–475.

Ribéreau-Gayon, P., 1964. Les composés phénoloques du raisin et du vin. I, II, III. Ann. Physiol. Veg. 6, 119–147. 211–242, 259–282.

Ribéreau-Gayon, P., Glories, Y., Maujean, A., Dubourdieu, D., 2006. *Handbook of Enology, Vol.2, The Chemistry of Wine Stabilization and Treatments,* second ed. John Wiley & Sons, Ltd, Chichester, UK.

Ross, C.F., Weller, K., 2008. The effect of serving temperature on the sensory attributes of red and white wines. J. Sens. Stud. 23, 398–416.

Rossetti, D., Bongaerts, J.H.H., Wantling, E., Stokes, J.R., Williamson, Q.-M., 2009. Astringency of tea catechins: More than an oral lubrication tactile percept. Food Hydrocol. 23, 1984–1992.

Roura, E., Aldayyani, A., Thavaraj, P., Prakash, S., Greenway, D., Thomas, W.G., et al., 2015. Variability in human bitter taste sensitivity to chemically diverse compounds can be accounted for by differential TAS2R activation. Chem. Sens. 40, 427–435.

Runnebaum, R.C., Boulton, R.B., Powell, R.L., Heymann, H., 2011. Key constituents affecting wine body – an exploratory study. J. Sens. Stud. 26, 62–70.

Running, C.A., Craig, B.A., Mattes, R.D., 2015. Oleogustus: The unique taste of fat. Chem. Senses 40, 507–516.

Sáenz-Navajas, M.-P., Campo, E., Culleré, L., Fernández-Zurbano, P., Valentin, D., Ferreira, V., 2010a. Effects of the nonvolatile matrix on the aroma perception of wine. J. Agric. Food Chem. 58, 5574–5585.

Sáenz-Navajas, M.-P., Campo, E., Fernández-Zurbano, P., Valentin, D., Ferreira, V., 2010b. An assessment of the effects of wine volatiles on the perception of taste and astringency of wine. Food Chem. 121, 1139–1149.

Saintsbury, G., 1920. Notes on a Cellar-Book. Macmillan, London.

San Gabriel, A.M., 2015. Taste receptors in the gastrointestinal system. Flavor 4, 14. (4 pp).

Sawano, S., Sato, E., Mori, T., Hayashi, Y., 2005. G-protein-dependent and -independent pathways in denatonium signal transduction. Biosci. Biotechnol. Biochem. 69, 1643–1651.

Schiffman, S.S., 1983. Taste and smell in disease. N. Eng. J. Med. 308, 1275–1279.

Schiffman, S.S., Gatlin, L.A., Sattely-Miller, E.A., Graham, B.G., Heiman, S.A., Stagner, W.C., et al., 1994. The effect of sweeteners on bitter taste in young and elderly subjects. Brain Res. Bull. 35, 189–204.

Schoenfeld, M.A., Neuer, G., Tempelmann, C., Schüßler, K., Noesselt, T., Hopf, J.M., et al., 2004. Functional magnetic resonance tomography correlates of taster perception in the human primary taste cortex. Neuroscience 127, 347–353.

Schutz, H.G., Pilgrim, F.J., 1957. Differential sensitivity in gustation. J. Exp. Psychol. 54, 41–48.

Senée, J., Robillard, B., Vignes-Adler, M., 1999. Films and foams of Champagne wines. Food Hydrocolloid. 13, 15–26.

Serafini, M., Maiani, G., Ferroluzzi, A., 1997. Effect of ethanol on red wine tannin-protein (BSA) interactions. J. Agric. Food Chem. 45, 3148–3151.

Shahbake, M., Hutchinson, I., Laing, D.G., Jinks, A.L., 2005. Rapid quantitative assessment of fungiform papillae density in the human tongue. Brain Res. 1052, 196–201.

Siebert, K.J., Euzen, C., 2008. The relationship between expectorant pH and astringency perception. J. Sens. Stud. 23, 222–233.

Singleton, V.L., 1987. Oxygen with phenols and related reactions in must, wines and model systems, observations and practical implications. Am. J. Enol. Vitic. 38, 69–77.

Singleton, V.L., and Noble, A.C. (1976). Wine flavour and phenolic substances. In Phenolic, Sulfur and Nitrogen Compounds in Food Flavors. (G. Charalambous and A. Katz, eds.), pp. 47–70. ACS Symposium Series No. 26, American Chemical Society, Washington, DC.

Smith, A.K., Noble, A.C., 1998. Effects of increased viscosity on the sourness and astringency of aluminum sulfate and citric acid. Food Qual. Pref. 9, 139–144.

Smith, A.K., June, H., Noble, A.C., 1996. Effects of viscosity on the bitterness and astringency of grape seed tannin. Food Qual. Pref. 7, 161–166.

Soares, S., Kohl, S., Thalmann, S., Mateus, N., Meyerhof, W., De Freitas, V., 2013. Different phenolic compounds activate distinct human bitter taste receptors. J. Agric. Food Chem. 61, 1525–1533.

Sowalsky, R.A., Noble, A.C., 1998. Comparison of the effects of concentration, pH and anion species on astringency and sourness of organic acids. Chem. Senses 23, 343–349.

Spence, C., Piqueras-Fiszman, B., 2014. The Perfect Meal: The Multisensory Science of Food and Dining. John Wiley & Son, Oxford, UK.

Stevens, J.C., Cain, W.C., 1993. Changes in taste and flavor in aging. Crit. Rev. Food Sci. Nutr. 33, 27–37.

Strauss, C.R., Gooley, P.R., Wilson, B., Williams, P.J., 1987. Application of droplet countercurrent chromatography to the analysis of conjugated forms of terpenoids, phenols, and other constituents of grape juice. J. Agric. Food Chem. 35, 519–524.

Sulmont-Rossé, C., Maître, I., Amand, M., Symoneaux, R., van Wymelbeke, V., Caumon, E., et al., 2015. Evidence for different patterns of chemosensory alterations in the elderly population: Impact of age versus dependency. Chem. Senses 40, 153–164.

Symoneaux, R., Guichard, H., Le Quéré, J.-M., Baron, A., Chollet, S., 2015. Could cider aroma modify cider mouthfeel properties? Food Qual. Pref. 45, 11–17.

Swoboda, P.A.T., Peers, K.E., 1978. The formation of metallic taint by selective lipid oxidation, the significance of octa-1,cis-5-dien-3-one. In: Land, D.G., Nursten, H.E. (Eds.), Progress in Flavor Research. Applied Science Publ., London.

Valentová, H., Skrovánková, S., Panovská, Z., Pokorný, J., 2002. Time–intensity studies of astringent taste. Food Chem 78, 29 37.

Vidal, S., Cartalade, D., Souquet, J.M., Fulcrand, H., Cheynier, V., 2002. Changes in proanthocyanidin chain-length in wine-like model solutions. J. Agric. Food Chem. 50, 2261 2266.

Vidal, S., Francis, L., Guyot, S., Marnet, N., Kwiatkowski, M., Gawel, R., et al., 2003. The mouth-feel properties of grape and apple proanthocyanidins in a wine-like medium. J. Sci. Food Agric. 83, 564–573.

Vidal, S., Courcoux, P., Francis, L., Kwiatkowski, M., Gawel, R., Williams, P., et al., 2004a. Use of an experimental design approach for evaluation of key wine components on mouth-feel perception. Food Qual. Pref. 15, 209–217.

Vidal, S., Francis, L., Noble, A., Kwiatkowski, M., Cheynier, V., Waters, E., 2004b. Taste and mouth-feel properties of different types of tannin-like polyphenolic compounds and anthocyanins in wine. Anal. Chim. Acta 513, 57–65.

Vidal, S., Francis, L., Williams, P., Kwiatkowski, M., Gawel, R., Cheynier, V., et al., 2004c. The mouth-feel properties of polysaccharides and anthocyanins in a wine like medium. Food Chem. 85, 519–525.

Voilley, A., Beghin, V., Charpentier, C., Peyron, D., 1991. Interactions between aroma substances and macromolecules in a model wine. Lebensm. Wiss. Technol. 24, 469–472.

Wang, Y., Chang, R.B., Liman, E.R., 2010. TRPA1 is a component of the nociceptive response to CO_2. J. Neurosci. 30, 12958–12963.

Weber, F., Greve, K., Durner, D., Fischer, U., Winterhalter, P., 2013. Sensory and chemical characterization of phenolic polymers from red wine obtained by gel permeation chromatography. Am. J. Enol. Vitic. 64, 15–25.

White, T., Prescott, J., 2001. Odors influence the speed of taste naming. Chem. Senses 26, 1119.

Whitehead, M.C., Beeman, C.S., Kinsella, B.A., 1985. Distribution of taste and general sensory nerve endings in fungiform papillae of the hamster. Am. J. Anat. 173, 185–201.

Winterhalter, P., Sefton, M.A., Williams, P.J., 1990. Volatile C_{13}–norisoprenoid compounds in Riesling wine are generated from multiple precursors. Am. J. Enol. Vitic. 41, 277–283.

Wróblewski, K., Muhandiram, R., Chakrabartty, A., Bennick, A., 2001. The molecular interaction of human salivary histatins with polyphenolic compounds. Eur. J. Biochem. 268, 4384–4397.

Yanniotis, S., Kotseridis, G., Orfanidou, A., Petraki, A., 2007. Effect of ethanol, dry extract and glycerol on the viscosity of wine. J. Food Engin. 81, 399–403.

Yau, N.J.N., McDaniel, M.R., 1991. The effect of temperature on carbonation perception. Chem. Senses 16, 337–348.

Yokotsuka, K., Singleton, V.L., 1987. Interactive precipitation between graded peptides from gelatin and specific grape tannin fractions in wine-like model solutions. Am. J. Enol. Vitic. 38, 199–206.

Yu, X., Chu, S., Hagerman, A.E., Lorigan, G.A., 2011. Probing the interaction of polyphenols with lipid bilayers by solid-state NMR spectroscopy. J. Agric. Food Chem. 59, 6783–6789.

Zampini, Z., Spence, C., 2005. Modifying the multisensory perception of a carbonated beverage using auditory. Food Qual. Pref. 16, 632–641.

Zellner, D.A., Stewart, W.F., Rozin, P., Brown, J.M., 1988. Effect of temperature and expectations on liking for beverages. Physiol. Behav. 44, 61–68.

Zhao, G.Q., Zhang, Y., Hoon, M.A., Chandraskekar, J., Erlenbach, I., Ryba, N.J.P., et al., 2003. The receptors for mammalian sweet and umami taste. Cell 115, 255–266.

Zuniga, J.R., Davies, S.H., Englehardt, R.A., Miller Jr., I.J., Schiffman, S.S., Phillips, C., 1993. Taste performance on the anterior human tongue varies with fungiform taste bud density. Chem. Senses 18, 449–460.

APPENDIX 4.1

A simple measure of taste sensitivity can be obtained by counting the number of fungiform papillae on the tip of the tongue. The technique given here was developed by Linda Bartoshuk, Yale University.

Subjects swab the tip of their tongue with a cotton swab (e.g., a Q-tip) dipped in a dilute solution of methylene blue (or blue food coloring). Rinsing away the excess dye reveals the unstained fungiform papillae as pink circles against a blue background. Placing the tongue in a tongue holder flattens the tongue, making the fungiform papillae easier to view. The tongue holder consists of two plastic microscope slides held together by three small nutted screws (see the following diagram). For convenience and standardization, only fungiform papillae in the central tip region are counted. This can be achieved with the use of a small piece of waxed paper containing a hole in the center (using a hole punch). Counting is facilitated with the use of a 10× hand lens. The average number of fungiform papillae per cm^2 is obtained by dividing the number of papillae counted by the surface area (πr^2) of the observation hole.

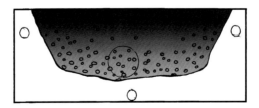

An alternative and more rapid approach is described by Shahbake et al. (2005). It uses a digital camera (4 mega-pixel or greater) to photograph the tongue. The tongue is pat-dried, stained by placing a 6-cm circular section of Whatman filter paper (No. 1) impregnated with methylene blue, and the tongue redried. After the digital photograph is transferred to a computer, the number of fungiform papillae can be counted by zooming in on the stained region of the images.

5

Quantitative (Technical) Wine Assessment

This chapter deals primarily with the technical side of wine assessment, i.e., its sensory evaluation. Typically, wine evaluation aims at understanding how wines differ in their sensory attributes, and how these may contribute to preferences. Based on such information, producers are in a better position to design wines to appeal to their intended purchasers. This is especially the situation for large wine firms that possess a dedicated sensory lab. It may also influence the practices of smaller producers, but more often winemakers probably see themselves akin to midwives, aiding and abetting in the birth of an artisanal product. Even where a firm may possess a sensory lab, most procedures are not particularly well adapted to detecting certain aspects. Wine attributes typically vary from year to year, and change in character during aging. Evaluation is best suited to detecting the effect of new or modified procedures, brand, regional or stylistic consistency, or faults.

Sensory evaluation can yield data on the attributes that distinguish wines, but rarely clarify issues concerning their relative importance to a wine's perceived quality. Equally, ranking wines can provide data on perceived qualities, but not clarify those attributes that were central to making those decisions. Novel approaches to integrate quantitative and qualitative techniques have been proposed recently by Bécue-Bertaut et al. (2008), Perrin et al. (2008), and Pineau et al. (2009). Regrettably, most procedures are labor intensive and may involve complex statistical manipulation and assumptions that are not intuitive or readily verified. A simpler approach might involve comparing graphic representations of the sensory attributes that characterize a series of wines (e.g., Figs. 8.14, 8.15, and 8.21) with the rankings generated by tasting panels. However, it must be acknowledged that there may be no direct correlation between sensory attributes and quality indicators. Clearly tasting panels, wine judges, and wine critics detect the same sensory attributes, but this does not mean that the sensory input is processed similarly into equivalent conscious perceptions. Earlier chapters have noted how experience influences sensory associations and the development of models of reality, and how these in turn can prejudice perception. Correspondingly, the views of various "experts" are often at odds with those of consumers (Lattey et al., 2010), with the results of sensory analysis potentially having little relevance to consumer preference, without serious attempts to find correlations.

Consumers rarely possess a clear or consistent concept of wine quality. Nor may they have a well-developed odor memory of wine styles or varietal attributes. Consumers too often feel that such subtleties are beyond their abilities. Alternatively, they may explain their inability to detect traditional attributes due to a poor vintage or bottle-to-bottle variation. The opinion of wine "authorities" often takes precedence (Aqueveque, 2015). Furthermore, most wine purchasers assess wines holistically, rather than dissecting them critically (attribute by attribute). Other than a cursory look at the label, a symbolic sip, and possibly a sniff, consumers rarely give a wine's sensory attributes the attention professionals consider appropriate, their attention quickly shifting to food or conversation. Although regrettable, relative to the effort extended by winemakers and grape growers, this can be to the benefit of wine retailers, since astute consumers are unlikely to accept mediocre wines. Furthermore, if one wine does not satisfy, there are thousands of others to try, and memory is often short. For the majority of purchasers, features such as convenience, price, brand recognition, and cultural identity are often of greater significance to purchase choice than sensory quality. Also, wines with too much character may compromise appreciation of the food, thus the neutral character of most restaurant house wines.

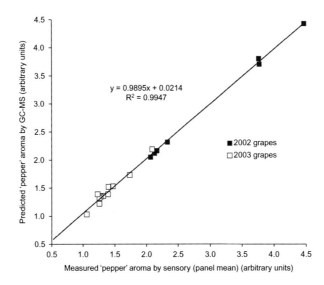

FIGURE 5.1 Relationship between the predicted pepper aroma in grapes, derived chemically by gas chromatography–mass spectrometry (GC-MS) and by a sensory panel. *From Parker, M., Pollnitz, A.P., Cozzolino, D., Francis, I.L., Herderich, M.J., 2007. Identification and quantification of a marker compound for 'pepper' aroma and flavor in Shiraz grape berries by combination of chemometrics and gas chromatography mass spectrometry. J. Agric. Food Chem. 55, 5948–5955, reproduced by permission.*

   The context of the tasting or purchasing environment can also influence perception. Even seemingly irrelevant conditions, such as the type of background music, can shape selection. For example, playing French music increased the sales of French wines, whereas playing German Bierkeller music improved German wine sales (North et al., 1999). In addition, playing classical versus top 40 music apparently can influence the value of the wine purchased, classical music enhancing the average purchase price, but not the quantity sold (Areni and Kim, 1993). Other studies have looked at associations between particular flavorants and musical instruments and pitch (Spence, 2011), as well as particular wines with specific pieces of classical music (Spence et al., 2013). While intriguing, this has no or negative relevance to wine analysis, other than to further confirm the importance of a neutral tasting environment.

   Despite a clientele that is less discriminating than those in the wine industry would appreciate, the minimum acceptable standard is the absence of faults. Because producers can become habituated to faults, or even viewing them as a desirable attribute (for example, a *Brett* taint), trained panels can act as a safety net, minimizing the likelihood of aberrant wines reaching the public. More detailed assessments have their principal value in defining the sensory features that distinguish particular wines. For example, sensory analysis may be used to investigate to what extent, and in what respects, wines represent a regional style (its typicity). The data derived can help to maintain or enhance a particular regional style, often important in international marketing. Alternatively, analysis is an opportunity to discover those procedures that distinguish the wines of one producer from neighboring wines, often considered important in direct-from-winery sales.

   Depending on the purpose for the assessment, the selection and training of panel members can change. For example, in descriptive sensory analysis, the panel is essentially functioning as if it were an analytical instrument. Fig. 5.1 illustrates how effective humans can occasionally be in this regard. If people were always consistent and precise, only one sufficiently skilled person would be required. However, individual and daily variability requires that a panel be available to neutralize this inherent variation (Fig. 5.2). Differences in detection threshold among individuals can occasionally be greater than three orders of magnitude (Tempere et al., 2011). Panel members are often trained to develop a common lexicon as well as being assessed for consistency. In contrast, in consumer tastings, member selection is aimed at representing the inherent variation of the wines' intended purchasers.

   Commercial wine competitions (discussed in Chapter 6, Qualitative Wine Assessment) are primarily designed to promote wine sales and showcase wines to the news media, supplying a profusion of gold, silver, and bronze medals. Although it would be desirable to have strict controls on how such tastings are conducted (as with descriptive sensory analysis), this is unlikely. The view that these tastings are impartial and use skilled (professionally trained) tasters are usually unjustified. Although seemingly a harsh indictment, it is not intended as a condemnation, just an acknowledgment of reality. Why go to great lengths when publicity is the principal goal? In addition, the results are unlikely to have any influence on production procedures or commercial decisions.

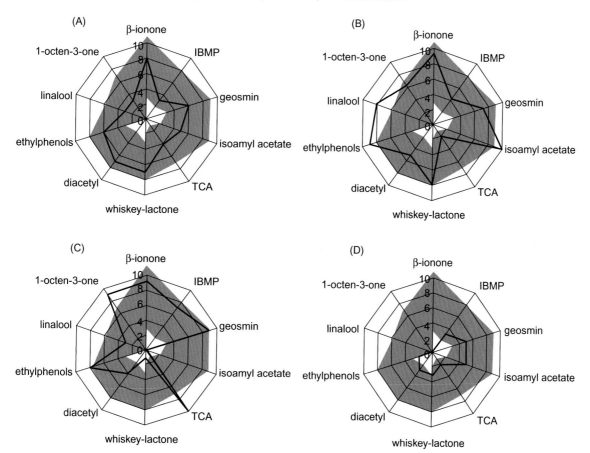

FIGURE 5.2 Variation in olfactory sensitivity of four wine experts to several aromatic compounds. (A) Midrange sensitivity profile, (B) broad-spectrum sensitivity profile, (C) an unequal profile, (D) a relatively insensitive profile. Thresholds labeled 10 correspond to the most sensitive tasters. *Solid line*, subject profile; *shaded area*, threshold area corresponding to 80% of the studied individuals. *Reprinted from Tempere, S., Cuzange, E., Malik, J., Cougeant, J. C., de Revel, G., and Sicard, G., 2011. The training level of experts influences their detection thresholds for key wine compounds. Chem. Percept. 4, 99–115, with kind permission of Springer Science + Business.*

In contrast, assessments conducted by winery staff do have utilitarian purposes. Regrettably, in small (and occasionally large) wineries, sensory assessment is the prerogative of one or a few individuals (the chief winemaker and assistants). As suggested in Fig. 5.2, this is unlikely to be adequate. Human perceptive ability is too variable and limited for one or a few individuals to assess wine attributes both consistently and accurately. Individual sampling may be acceptable for the artisan wine producer, where wines are made to their personal preferences. Even champagne and port blends are often based on the opinions of a few individuals. However, this is rapidly becoming archaic in most large, modern wineries. Leaving sensory decisions up to one or a few individuals is far too risky. Stockholders or bankers are not interested in taking that much of a gamble, as many millions of dollars are at stake. Sensory analysis also plays a major role in university and government research. The research is often designed to investigate the psychophysiology of human perception, or to study how modifying viticultural or enologic factors influences wine attributes. In addition, the results may be used to supply support for, or justify, marketing programs, notably those dealing with the typicity (uniqueness) of a region's wines.

One of the principal benefits of sensory analysis is helping to uncover the physical, chemical, biological, and psychological underpinnings of perceived quality. This provides information potentially useful to producers with the aim of improving wine quality. That expensive wine provides some people with a feeling of prestige and exclusivity is fine. They can afford to purchase such wines, usually of fine quality, but at prices that reflect more reputation than quality (Landon and Smith, 1998). However, quality wine need not, should not, be the prerogative of the wealthy. Sensory evaluation is helping to clarify issues relative to the origin of quality (e.g., Sáenz-Navajas et al., 2010), aiding the egalitarian goal of bringing fine wines within the reach of everyone. Increased exposure to superior wines is the most direct means of enhancing consumer awareness of what they have been missing and generating a more discriminative audience.

Quantitative wine assessment often consists of two independent but integral components: evaluation and analysis. Evaluation involves the differentiation and/or ranking of wines. It can vary from consumer tastings, intended to assess purchaser preference, to trained panels designed to assess differences generated by experimental treatments, or to choose the formula for a blend. Occasionally, but rarely, expert panels may go further, assigning a monetary value to the rankings. In contrast, wine analysis involves detailed investigation of a wine's sensory attributes. The intent may be to examine the magnitude of the characteristics that describe or distinguish a series of wines, the features that lead to preferences among wines, or the psychophysics (nature and dynamics) of sensory perception as it relates to wine assessment. Alternatively, analysis may function as a quality control procedure. Sensory evaluation may or may not use trained panels, whereas sensory analysis essentially always employs panels specifically trained for the task. Because quantitative wine analysis is primarily a research/developmental tool, it is essential that the assessment be conducted in a quiet, neutral environment, devoid of taster interaction, or minimal knowledge of the sample's origin.

The purpose of analysis dictates the type or types of sensory tests appropriate, the training required (if any), and the number of members desirable. If the only question is whether two or more samples are detectably different, then one of a number of discrimination tests may be performed. For consumer panels, orientation and habitual use of wine of the types being studied are all that are required. As the need for accuracy increases, the number of wines and/or attributes per sample should be reduced to limit taster fatigue. In addition, professional panelists need to be tested for an adequate degree of acuity, trained in the use of precise descriptive language, and be assessed for consistent and reliable scoring.

Although sensory evaluation can assist in the production of better wine, there remains a gap in correlating sensory attributes with wine chemistry. New techniques, such as gas chromatography, combined with olfactometry (Acree, 1997), aroma extract dilution analysis (AEDA) (Grosch, 2001), and atmospheric pressure chemical ionization mass spectroscopy (APCI-MS) (Linforth et al., 1998) are starting to tackle the thorny issue correlating wine chemistry with the perception of sensory quality.

SELECTION AND TRAINING OF TASTERS

Basic Requirements

Except winemakers, negotiants, or wine critics, for whom tasting is part of their job description, participation in a tasting panel is usually voluntary. Thus, the most crucial requirement for membership may be motivation. It provides the desire required for both effective learning and consistent attendance. Without a sincere interest in wine, it is unlikely that members will retain the dedication needed for the periodic bouts of arduous concentration that may extend over days, weeks, or months. Good health is also a criterion, not only because medical issues can interfere with attendance or sensory acuity, but also because many medications distort perception (Doty and Bromley, 2004). Examples are diuretics disrupting ion channels (essential for receptor cell activation), antimicrobial agents (suppressing cytochrome P450-dependent enzymes in the nasal mucus), and chemotherapeutic agents (affecting receptor-cell regeneration).

For consistency, retaining a common nucleus of members is desirable. Therefore, the selection process ideally should include more candidates than required. In addition, experience suggests that no more than 60% of potential candidates may possess the necessary sensory skills. If a high degree of discrimination is required, an even larger number of candidates may need to be tested. Tact is often required in rejecting candidates, especially with those eager to participate, or those who feel their experience merits inclusion. For long-term employees of a wine store, sensory lab, or winery, this could be particularly embarrassing. In such a situation, selection as part of a backup group, or forming several panels with different (simpler or less significant) tasks could be part of a diplomatic solution.

To maintain enthusiasm and attendance, panel members need to have clear indications of the significance placed on their work. Providing a comfortable, dedicated, tasting room is a major sign of importance. Training sessions or periodic demonstration of how their work is used can be valuable. Feedback on effectiveness of individual panel members (presented privately) can be both encouraging as well as an effective incentive to improve (Desor and Beauchamp, 1974). Finally, depending on legal issues and appropriateness, presentation of vouchers for wine purchase is a tangible expression of the appreciation placed on panelist participation.

Besides interest and dedication, panelists must be consistent in their assessment, or soon develop this property. Consistency can be assessed using one of several statistical tools. For example, insignificant variance in ranking repeat samples, or term use, can serve as indicators of consistency. Analysis of variance (ANOVA) can also indicate whether panel members are individually, or collectively, using a score sheet uniformly. Although tasters show fluctuations in sensory skill, this should be minimal, if subtle differences need to be detected. Because tasters are often being used as semianalytic instruments, their responses to sensory features should be highly uniform. Whether this property is equally desirable when assessing wine quality (an integrative property) is a moot point. If the experimenter wants to distinguish minute differences, then minimal intermember variation is essential. In contrast, if the intent is to know whether consumers might differentiate among particular wines, panel variation should be representative of the target group. Regrettably, the degree of sensory variation in any target group is rarely known.

Professional tasters are expected to possess an extensive knowledge of accepted norms. This typically means a working knowledge of varietal, regional, and stylistic archetypes, an ability to discriminate between these, and consistent use of an agreed-upon lexicon (Bende and Nordin, 1997; Ballester et al., 2008). Excellent sensory acuity is clearly a prerequisite, combined with a well-developed odor memory. However, these attributes, even among trained individuals, such as Masters of Wine or sommeliers, should not be taken for granted. Several studies have shown these individuals to be less skilled at distinguishing among varietal or regional wines than normally thought (Winton et al., 1975; Noble et al., 1984; Morrot, 2004). In addition, they are often unable to recognize wines based on their own or collective descriptions of the wines sampled (Lawless, 1985; Hughson and Boakes, 2002; Lehrer, 2009). However, this is not a universal finding (Gawel, 1997). In addition, professional tasters should be able to describe wine independent of personal preferences, using properties ideally based on training and experience, and they should have a frank and analytic mind.

Although enthusiasm, experience, and an extensive knowledge of wine are desirable, these aspects alone cannot be used in lieu of testing skill (Frøst and Noble, 2002). Unknown sensory idiosyncrasies (see Fig. 5.2) could compromise group results. Each sensory attribute that needs to be investigated is an independent entity, and must be assessed for adequacy in each potential panel member. In most instances, adequate acuity, consistency, and an excellent odor memory are the principal sensory requisites of a superior taster.

As noted, sensitivity to particular characteristics often influences panelist suitability. For example, in descriptive sensory analysis, high acuity to particular attributes is typically required. In such instances, there is no need or intention for the panel to represent consumers. Usually, though, panel members need only average sensitivity, and this may be preferable. For example, a panel composed of individuals hypersensitive to bitterness and astringency may unduly downgrade young red wines, but be required if the effect of skin contact on these attributes is under investigation. Typically, the ability to learn and consistently use detailed sensory terms is important. However, for quality evaluation, the appropriate use of integrative or holistic terms, such as complexity, development, balance, body, and finish, may be the most important critical properties.

Because the features preferred for trained panelists typically differ from those of most consumers, panel rankings rarely reflect consumer preference (Williams et al., 1982). In their study, flavor in-mouth was highly correlated with overall ranking (0.91) by all groups, with little importance being placed on aroma (0.40) or color (0.54) by the public (Table 5.1). In contrast, ranking of the wines was highly correlated with aroma for both experts and trained panelists. Color was seemingly significantly correlated with ranking by experts only when the specific wine style was known in advance (and thus could have been distorted due to experience-based models of the style). Among different consumer age groups, aroma was least correlated with ranking for the youngest group, but of significant interest to those between 35 and 44 (Table 5.2). Drinking habits also significantly influenced the relative importance of color, aroma, and flavor to ranking (Table 5.3). As one might expect, aroma was more important in the ranking of frequent wine consumers, and slightly more so for those consuming more expensive wines. Those who had greater experience with wine ranked red wines with a brownish tinge higher, whereas those with less experience preferred those with a more intense red color. Another finding was that consumers often thought the wine was too acidic and insufficiently sweet, whereas experts considered the samples too sweet.

Other studies have also noted the proportional significance of gustatory sensations to consumers (Solomon, 1997), there being a distinct preference for wines with minimal astringency, bitterness, or sourness, as well as perceptible sweetness and/or pétillance. Thus, if taste panels are used in marketing studies, careful selection of members showing similar preferences to the target audience is required if the results are likely to be of relevance. Consumer-based studies are discussed at greater length in Chapter 6, Qualitative Wine Assessment.

Identification of Potential Wine Panelists

Because of the considerable time and expense involved in training, it is judicious to do an initial screening of potential panelists for the requisite skills. No known series of tests can infallibly identify the sensory potential

TABLE 5.1　Correlation between overall acceptability and hedonic scores for general wine attributes

	General public	Master of wine		Trained panelists
		As commercial product	As wine of specific type	
Appearance	—	—	—	−0.01
Color	0.54	0.56	0.74	0.39
Aroma	0.40	0.96	0.95	0.78
Flavor by mouth	0.91	0.92	0.89	0.96

From Williams et al., 1982, reproduced by permission.

TABLE 5.2　Correlation between overall acceptability and hedonic scores for general wine attributes within population subsets

	Age			
	18–24	25–34	35–44	45–64
Color	0.40	0.43	0.45	0.59
Aroma	0.17	0.32	0.73	0.46
Mouth flavor	0.90	0.82	0.90	0.82

From Williams et al., 1982, reproduced by permission.

TABLE 5.3　Correlation between overall acceptability and hedonic scores for general wine attributes within wine price categories and frequency of wine consumption

	Red wine: Drinking habits			
	<£ Per bottle		>£ Per bottle	
	Frequent drinkers	Infrequent drinkers	Frequent drinkers	Infrequent drinkers
Color	0.42	0.57	0.56	0.48
Aroma	0.63	0.36	0.73	0.47
Mouth flavor	0.88	0.94	0.84	0.96

From Williams et al., 1982, reproduced by permission.

of a candidate. This reflects as much the variety of skills that may be required (sensory acuity, olfactory memory, consistency, objectivity), as it does the efficacy of the testing procedure. Possibly, a finely tuned ability to discriminate among wines is the most useful indicator of potential. Unfortunately, discrimination is probably based as much on odor memory (experience) as on sensory acuity. Thus, discriminatory skill may not be readily recognizable in novice candidates. Nevertheless, indications of tasting capacity, such as sensory acuity (low detection thresholds), odor and flavor recognition, term use and recall, scoring consistency, as well as existent ability to discriminate, recognize, and identify wines can be measured. These measures can act as predictors of the potential skill, and what knowledge or skills need to be gained or perfected.

Other potential qualities are more difficult to quantify. These include the ability to recognize and rank wine based on traditional standards. Although these facilities are inherently subjective, they are based on experience (and thus, can potentially be acquired). These skills typically demand both deductive and inductive reasoning. For example, detection of a wine's origin and quality usually entails assessment of attributes both present and absent. In general, it appears that exceptional olfactory memory, rather than phenomenal perceptual acuteness, is what distinguishes a superior taster.

Examples of screening tests that might be useful are noted in the following section. The skills measured indicate aptitude for sensory analysis, as well as demonstrate individual strengths and weaknesses. The proficiency required in any test will naturally depend on the skills required by the researcher. Sensory assessment is markedly different from assessing quality in manufactured goods. In the latter, quality standards are usually precise and objectively measurable.

Ideally, testing should be conducted over several weeks. This reduces stress on the candidates and, therefore, improves the likelihood of valid assessment (Ough et al., 1964). However, in commercial situations, multiple wines may need to be sampled in quick succession. Thus, concentrating the testing within a short period may provide the stress that highlights those best able to perform adequately under such conditions.

In preparing the tests, it is obvious that wine samples should be free of fault (unless that is intended), and truly representative of the variety or type. Repute or previous appropriateness is no substitute for sampling by the researcher/instructor prior to use. Excellence in representing the attributes desired, not price, should be a sole condition for sample selection.

Testing and Training

In the past, wine evaluation was conducted primarily by winemakers or wine merchants. Their training tended to focus on recognizing accepted stylistic, regional, or varietal norms. This established a frame of reference for judging acceptance, in accordance with traditional attributes. A problem with this approach is that it accentuated the importance of accepted archetypes (Helson, 1964). However, such paradigms may become inappropriate, shifting as technological innovation permits improvements or consumer preferences change. Although personal opinion may be acceptable in many parts of the wine trade, it is becoming less pervasive in modern commerce. Individual biases and sensory deficiencies can limit sales in an evolving marketplace. To offset such inherent limitations, most critical wine evaluations are now conducted by panels. This has required the training and qualification of more professionals than generated by the older, informal, in-house, experience approach. It has also led to more focused and standardized training programs. In contrast, for research purposes, training is typically designed to achieve particular standards for a specific experiment, with required skills often varying from project to project.

As noted, extraordinary olfactory acuity is usually not required. In addition, initial skill in recognizing odors is typically of little significance, as this feature typically reflects experience (Cain, 1979), not potential. Training (repeat exposure) can enhance discriminatory skills, but potentially only to the types of wines used, without affecting broader discriminatory abilities (see Owen and Machamer, 1979). Similarly, odorant training increased panel sensitivity, but only to the odorant used (Dalton et al., 2002; Fig. 5.3), there being little transfer of improved acuity to other odorants. For example, repeat exposure to androstenone, was correlated with increased sensitivity to and nerve activation in the olfactory patches (Wang et al., 2004). In addition, enhanced perception may also show a cerebral

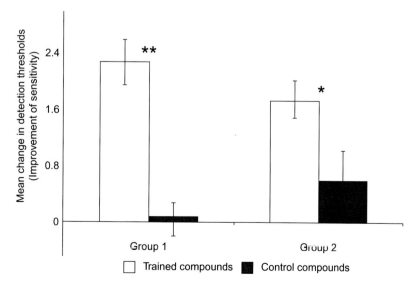

FIGURE 5.3 Changes in the detection threshold with two groups for compounds for which they had (trained compounds) or had not been trained (control compounds). For Group 1, the trained compound was linalool; the control was diacetyl. For Group 2, the trained and control compounds were reversed. Significant differences (Duncan test) were *$p < 0.05$; ** $p < 0.01$. *Reprinted from Tempere, S., Cuzange, E., Bougeant, J.C., de Revel, G., Sicard, G., 2012. Explicit sensory training improves the olfactory sensitivity of wine experts. Chem. Percept. 5, 205–213, with kind permission of Springer Science + Business.*

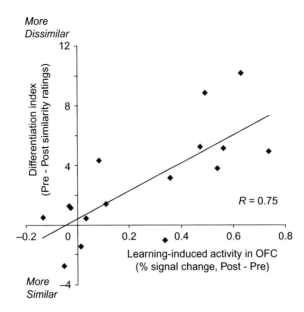

FIGURE 5.4   Neural plasticity is illustrated by regression analysis of enhanced orbitofrontal cortex responses to learning-induced perceptual odor expertise (indexed via behavioral ratings of odor quality similarity: post- minus preexposure). Each point (◆) represents a different subject. *Reprinted from Li, W., Luxenberg, E., Parrish, T., Gottfried, J.A., 2006. Learning to smell the roses: Experience-dependent neural plasticity in human piriform and orbitofrontal cortices. Neuron 52, 1097–1108. with permission from Elsevier.*

component, reacting differently to familiar versus unfamiliar odors (Wilson, 2003). Training also enhances response in the orbitofrontal cortices (Fig. 5.4), as well as inducing structural reorganization (Delon-Martin et al., 2013).

Even exercises requesting subjects to imagine specific odors can enhance olfactory performance, both in detection and identification (Tempere et al., 2014). Such enhanced odor sensitivity was detected only in experts, not novices or undergraduate enology students and, then, only for some odorants (Fig. 5.5). Surprisingly, a related experiment requesting subjects to visualize the objects associated with odorants resulted in deterioration of detection and identification, but only for experts. Thus, not all forms of training seem to be equally beneficial, or even desirable.

Odorant training can also have unexpected influences. For example, although training is often associated with improved sensitivity (reduction in the detection threshold), it can also lower the perceived intensity of the compounds. As with sensitivity training, an associated reduction in perceived intensity applies only to those odorants used during the training exercises (Fig. 5.6). This effect does not appear to apply to trigeminal stimulants, however, e.g., carbon dioxide (Livermore and Hummel, 2004).

What is unclear is the long-term stability of odor training. Odor memory fades (Lawless, 1978), but seemingly at a rate the inverse of familiarity (Kärneküll et al., 2015). Thus, it is important for odor samples used in training to be present during analysis sessions. Just the visibility of appropriate terms can facilitate identification (Frank et al., 2011). Measures of a panelist's ability to learn seem more indicative of a member's potential than their initial ability to recognition odors (Stahl and Einstein, 1973).

Instructions during training (and subsequent assessment) affect whether sensations are integrated into multimodal attributes (e.g., vanilla associated with sweetness), or retain their individual modal identities (Prescott et al., 2004). Thus, training may concentrate on individual aromatic attributes, as for sensory analysis, on sensory integration, where training is designed to concentrate on regional, stylistic, or varietal characteristics, or some other interest of the researcher.

Humans have the ability to differentiate thousands of odorants, compared side by side, but often have remarkable difficulty identifying even common odors out of context. This supports the common significance of crossmodal integration among modalities in odor memory (e.g., Gottfried and Dolan, 2003). Improvement in specific identification is enhanced, however, with feedback on misidentifications (Cain, 1979), which is the essence of good teaching.

Of particular significance is the difficulty people experience in identifying components in odor mixtures (Fig. 3.24), notably more than two. This could compromise experiments where panelists are trained with pure compounds, and then expected to recognize them within the context of a wine. An exception to this generality appears to involve off-odors. This may relate to their perceived intensity being sufficiently intense to dominate or suppress other wine fragrances (Takeuchi et al., 2013), even in minute amounts.

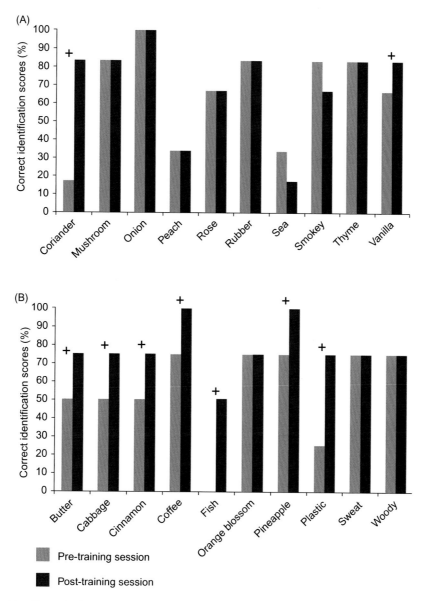

FIGURE 5.5   Odor-specific effects of olfactory mental imagery training on experts' correct identification scores of odorants before and after training. " + " indicates odorants for which there was significant enhancement over the training period. *From Tempere, S., Hamtat, M.L., Bougeant, J.C., de Revel, G., Sicard, G., 2014. Learning odors: The impact of visual and olfactory mental imagery training on odor perception. J. Sens. Stud. 29, 435–449, reproduced with permission of John Wiley and Sons, copyright.*

How odor samples are prepared can also have unexpected effects. For example, training with a mixture of odorants can modify the subsequent perceived quality of a sample's components (Stevenson, 2001). This tendency, the generation of a holistic odor quality for a mixture, is limited if the odorants are familiar and already have distinct and recognized odor qualities. Learning combinations of odors before recognizing their components makes later differentiation of the components more difficult.

In most instances, training's primary role is in enlarging a trainee's odor-memory base. Where of value, training may also focus on minimizing the influence of multimodal interactions (e.g., taste–taste or taste–odor interaction). Although training tends to concentrate on olfactory perception, central to most sensory assessments, it may also extend to taste, mouth-feel, and visual attributes. Where necessary, it becomes part of the evaluation and selection of panel members.

Once experimentation has commenced, additional training exercises may be designed to refresh and reinforce odor memories. This can be particularly useful where terms do not possess visual and other contextual attributes

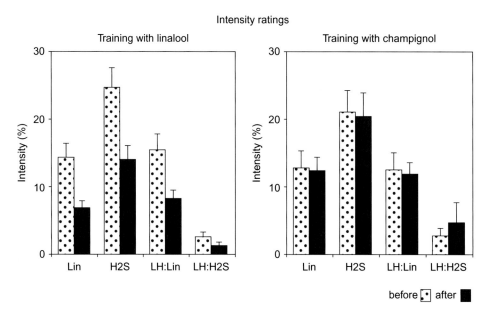

FIGURE 5.6   Intensity ratings (mean, standard error of the mean) for linalool and H_2S, when presented as single stimuli (Lin, H_2S) or assessed individually but as a binary mixture of both linalool and H_2S (LH). *Reprinted from Livermore, A., Hummel, T., 2004. The influence of training on chemosensory event-related potentials and interactions between the olfactory and trigeminal systems. Chem. Senses 29, 41–51, by permission of Oxford University Press.*

(e.g., chemicals versus familiar objects). Terms possessing rich semantic and emotional elements typically generate memories that are far easier to retain and recognize (Jehl et al., 1997).

Beyond learning to recognize individual sensory attributes, panel members may need to become adept at recognizing the properties that traditionally characterize wine quality. Thus, training may involve extensive exposure to a selection of varietal, regional, and stylistic wine types (paradigm learning). This is essential in evaluative assessments. For this, samples should represent not only archetypical examples of the type, but also the range that can ordinarily be expected to be found in commerce.

Depending on the importance of the abilities required, training continues until candidates show the level of proficiency and consistency deemed necessary, or they are eliminated. Training usually also includes the development of an appropriate descriptive language and its consistent use. Lexicon development may entail panel discussion during training on the terms most appropriate for the purpose(s) of the experiment. Alternatively, the researchers may supply both the descriptors and samples to be used in training and in the experiment.

As noted, candidates without the desired consistency are typically eliminated. Natural variation is inappropriate for the fine distinction often desired. Alternatively, the effects of variation may be minimized by statistical procedures such as Procrustes analysis (see below). Nonetheless, there is a risk in its use, as the assumptions on which the procedure is based may be invalid. Although the collection of precise data on a wine's attributes, obtainable with an expert panel, is necessary to understand the sources of sensory variation among wines, it is often equally important to understand the extent and distribution of consumer sensory variation, to appreciate the data's potential commercial relevance. As with training, the appropriate form of data analysis often depends on the intent of the task. For further discussion of these issues, see Stevens (1996).

With the availability of computerized data accumulation and analysis, trainees can receive rapid feedback on their abilities and deficiencies, facilitating learning and hopefully enhance motivation if needed. It could also permit trainees to learn their relative status in comparison with other members (if deemed valuable). An example of how rapid feedback is used is provided by Findlay et al. (2007). Rapid feedback can also help the experimenter/trainer to see where training may be deficient and needs improving. Cain (1979) considers that successful odor identification requires three elements: commonality, prolonged odor-name association, and supplemental clues.

For economy and convenience, fragrances commonly considered to resemble grape varietal aromas may be partially reproduced in odor samples (see Appendix 5.1). These samples have the advantage of being continuously available for reference. Standards may be prepared in a neutral or artificial wine base, stored under paraffin in sample bottles (Williams, 1975, 1978), or in a cyclodextrin (Reineccius et al., 2003) solution. Because pure compounds

TABLE 5.4 Organization of samples for a taste–mouth-feel test

Sample (solution)	Amount (per 750 mL, Water or wine)	Sensations
Sugar	15 g sucrose	Sweet
Acid	2 g tartaric acid	Sour
Bitter	10 mg quinine sulfate	bitter
Astringent	1 g tannic acid	Astringent woody
Alcohol	48 mL ethanol	Sweet, hot, body, alcoholic odor

may contain trace impurities, which can modify odor quality, it may be necessary to conduct initial purification before use (Meilgaard et al., 1982). Although sample preparation is often necessary for training, most samples have comparatively short shelf-lives, and need to be replaced frequently. This, in its own right, creates additional problems, notably when samples are derived from natural or commercial products. Because such sources are not always identical, replacement samples are unlikely to be as identical as desired.

In addition to recognizing wine odors, identification of off-odors should be a vital component of training. In the past, faulty samples were obtained from wineries, but samples prepared in the laboratory can be standardized and available as needed (see Appendix 5.2). In addition, prepared samples have the advantage of being producible in any wine or at intensity level desired.

Basic Selection Tests

The tests described below have been used in selecting panelists. However, their applicability will depend on the tasks required. Thus, they are provided as examples of tests that could be used.

Taste Recognition

As noted, fine discrimination of taste sensations is seldom of critical importance in wine assessment. This does not imply insignificance, but merely a lack of clearly defined standards. In addition, mixtures of tastants and odorants (as in wine) can significantly modify the perception of individual gustatory modalities. Thus, the tests mentioned are primarily for the benefit of the participants, to give them an opportunity to discover any taste and mouth-feel idiosyncrasies of which they may be unaware. This can help panelists gain insight into human sensory variability, as well as how matrix (water versus wine) can influence perception.

Taste Acuity

For the initial assessment, samples are prepared in water. This simplifies detection and identification, each chemical being isolated from the complex chemistry of wine. Subsequent samples are prepared in both white and red wine. Table 5.4 provides an example of preparing such a series.

Dissolve the samples in the base solution about an hour in advance. For the samples prepared in wine, provide participants with a sample of the base wine, to permit comparison with adjusted samples.

Present the adjusted samples, identified only by number or colored stickers (glasses containing ~30 mL), in random order. For each sample, participants should note all detected taste/touch sensations (Appendix 5.3), and record the intensity of the most dominant sensation along a supplied line scale.

Attention should be paid to where and how quickly each sensation occurs most prominently in the mouth, especially with the aqueous samples. This permits individual characterization of sensitivity throughout the mouth. After having sampled the tastants in water, subsequent trials examine the ability to recognize the same taste and touch stimuli in the context of a white and red wine. Detection often differs significantly from those in water.

Fig. 5.7 illustrates the diversity of perceived intensity ratings of the dominant tastant that can be expressed by participants with a series of aqueous samples as prepared as in Table 5.4. Table 5.5 illustrates the diversity of perceptions that can be elicited by the same samples. The imagined perceptions noted in the water samples may reflect either lingering or crossover sensations from the previous sample, or the belief that certain sensations could or should have been present. Participants were not instructed to expect a water control. Fig. 5.8 demonstrates how the matrix (water or wine) can influence response to a bitter tastant.

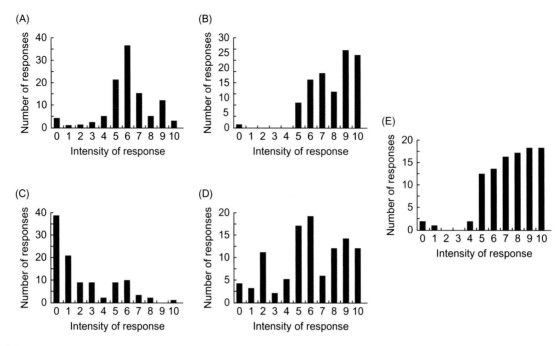

FIGURE 5.7   Intensity response of 105 people to several taste solutions: (A) sucrose (15 g/L), (B) tartaric acid (2 g/L), (C) quinine sulfate (10 mg/L), (D) tannic acid (1 g/L), and (E) ethanol (48 mL/L).

TABLE 5.5   The variation in responses of 27 people to the dominant taste sensations in water solutions of: sucrose (15 g/L), tartaric acid (2 g/L), quinine sulfate (10 mg/L), tannic acid (1 g/L), and ethanol (48 mL/L)

Solution	Percent of responses							
	Sweet	Sour	Bitter	Astringent	Alcoholic	Dry	Salty	Nothing
Sucrose	94	6	0	0	3	0	0	0
Tartaric acid	3	47	17	12	3	0	22	0
Quinine	0	15	40	15	0	4	0	26
Tannic acid	0	16	25	47	0	7	7	0
Ethanol	7	6	12	1	74	0	1	0
Water	7	15	7	14	0	0	0	57

Relative Sensitivity (Sweetness)

For this test, dissolve 2.25, 4.5, 9, and 18 g sucrose in separate 750 mL wine samples. Prepare a 0.5% pectin solution or have unsalted crackers as a palate cleanser. Participants should familiarize themselves with the base wine in advance. It should be keep handy for reference during the test.

Each participant receives five randomly numbered samples to rank in order of relative sweetness. The test is repeated with different wines. Table 5.6 presents the setup for three sessions with different wines.

Similar sensitivity tests for sour, bitter, and astringent compounds can be conducted using citric or tartaric acids for sourness (5, 10, 20, 40 g/L), quinine or caffeine for bitterness (2.5, 5, 10, 20 mg/L), and tannic acid for bitterness and astringency (5, 10, 20, 40 g/L). Alum (aluminum potassium sulfate) can be used as a substitute for tannic acid if astringency is desired without any accompanying bitterness (e.g., 2.5, 5, 10, 20 g/L).

Although a common aspect of training, sensitivity to tastants can be eliminated if time is limited. Only rarely are tasters required to assess individual taste and mouth-feel modalities. In addition, oral sensations are, to varying degrees, influenced by other wine constituents. This is particularly significant relative to the perception of sweetness and sourness. For example, perceived sweetness in dry wines is typically a reflection of the wine's aromatics influencing how the orbitofrontal cortex interprets sensory impulses. The experience of how the matrix of a wine can influence its basic oral attributes can, by itself, be adequate justification for the tests.

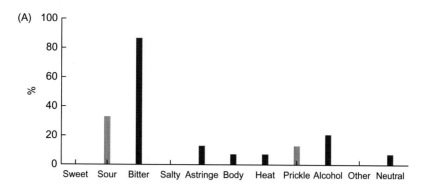

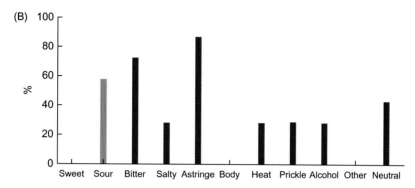

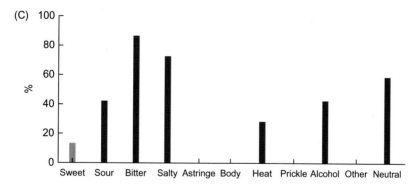

FIGURE 5.8   Variation of responses of 15 people to the presence of quinine sulfate (10 mg/L) in (A) water, (B) Cabernet Sauvignon wine, (C) Sémillon/Chardonnay wine.

TABLE 5.6   Sweetness sensitivity test

Session 1 Chardonnay	Session 2 Valpolicella	Session 3 Riesling
#5[a] (A[b])	#5 (E)	#1 (E)
#3 (B)	#3 (D)	#4 (D)
#2 (D)	#4 (C)	#2 (C)
#1 (C)	#2 (B)	#3 (A)
#4 (E)	#1 (A)	#4 (B)

[a]# 1–5 Label identifying the ample.
[b]A, control (0 g/l sucrose); B, 3 g/l; C, 6 g/l; D, 12 g/l; E, 24 g/l.

Threshold Assessment

Because of the time and labor involved in threshold measurement, the specific recording of individual threshold values is seldom conducted, unless there is some rationale for selecting panelists based on their sensitivity to particular compounds. Threshold measurements are conducted more to determine the detection and/or recognition

TABLE 5.7   Hypothetical responses to an A-not-A test for determining the detection threshold for glucose

		Sample (% glucose)					
	Control	0.2%	0.3%	0.4%	0.5%	0.75	0.9%
Correct responses[a]	3	4	3	4	5	5	5
Percent correct	50	66.7	50	66.7	83.3	83.3	83.3

[a]Total number of responses = 6.

thresholds of particular compounds than of individual panel members, though this could be useful, depending on the purpose of the tasting.

Threshold detection has been primarily used in studies investigating the relative significance of particular compounds to a wine's varietal aroma. Typically, only those compounds occurring at or above their threshold value significantly affect the aroma. Threshold determination is also important in assessing the concentration at which off-odors may begin to affect wine rejection. Although detection is known to be affected by the type of wine in which an off-odor occurs, perception of other attributes can also be influenced at below-threshold values. For example, trichloranisole (TCA) can suppress general olfactory sensitivity at concentrations below detection values (Takeuchi et al., 2013). Subthreshold influences on perception have also been detected using fMRI (Hummel et al., 2013).

Threshold determination can also be influenced by issues such as marked individual variation (Fig. 3.19), bimodal response curves (Lundström et al., 2003), and the influence of the wine matrix (Table 3.3). Due to these issues, whether the mean or median value is the more appropriate statistic is almost academic (e.g., Pineau et al., 2007). For more discussion on the complexities of threshold assessment see Chapter 3, Olfactory Sensations, Bi and Ennis (1998), Walker et al. (2003), and Lawless (2010).

A simple procedure for determining thresholds is as follows. A series of concentrations of the compound in question is prepared in water (or some other appropriate medium). The number of samples should cover a wide range of concentrations. In assessing taste sensitivity, for example, concentrations of 0, 1, 2, 3, 4, 5, 6, and 7 g/L glucose, or 0.03, 0.07, 0.10, and 0.15 g/L tartaric acid are often adequate. If more precision were desired, once the rough threshold range has been established, additional sets could more precisely bracket the threshold value. For example, if the rough threshold were between 0.4% and 0.5% glucose, concentrations such as 0.40%, 0.42%, 0.44%, 0.46%, 0.48%, and 0.50% might prove appropriate.

Each sample concentration is paired with a control solution. The samples are arranged and marked so that the control sample is tasted first, followed by the tastant or another control sample. Participants note whether the second sample is the same or different from the control. This type of procedure is called A-not-A test. Each concentration pair should appear at random at least six times, but ideally more. Chance alone should produce about 50% correct responses at below threshold. Detection of a legitimate differentiation among samples is normally considered to have occurred when a correct response rate of >75% (50% above chance) occurs. The samples should be presented in random order to avoid panelists anticipating an increase or decrease in concentration. Having the panelists do the test slowly, and in a random sequence, helps minimize adaptation or fatigue. In the example given in Table 5.7, the participant would have a detection threshold of between 0.4% and 0.5% glucose.

Odor-Recognition Tests

Fragrance (Aroma and Bouquet)

The fragrance test measures the identification of several odors often viewed as being characteristic of wine aromas. Appendix 5.1 gives an example of formulas for sample preparation. Because learning to recognize is often as much a component of the test as measuring recognition, samples should be presented for familiarization before each test. Participants should be encouraged to take notes of the aromatic features they think may help them recognize the samples, and the base wine (control).

Cover the mouth of each glass with tightfitting covers (e.g., 60-mm plastic petri dish bottoms, Plate 5.1). This permits vigorous swirling. The samples should only be smelt, tasting being unnecessary. By obviating tasting, preparation (number of samples required) and cost can be minimized.

The samples should be presented in black, preferably ISO, glasses (Plates 5.2, 5.3), which can be purchased from several sites (e.g., Amazon.fr and wineware.co.uk). This negates any visual clues as to origin (the color or cloudiness of some samples is unavoidably affected). Substitute some odor samples with control samples in each test session. This helps to minimize any tendency for identification to be aided by a process of elimination.

PLATE 5.1   Preparation of samples for a sensory analysis experiment. Note the small petri-dish covers over the tops of each clear ISO wine-tasting glass. Cool Climate Oenology and Viticulture Institute (CCOVI) Tasting Lab, Brock University. *Photo courtesy R. S. Jackson.*

PLATE 5.2   Tasting setup using black ISO wine-tasting glasses. Cool Climate Oenology and Viticulture Institute (CCOVI) Tasting Lab, Brock University. *Photo courtesy R. S. Jackson.*

The answer sheet lists all samples as well as provides space for recording an unknown number of controls. Descriptor terms are provided because identification is being assessed, not word recall. For simplicity, participants mark the sample number opposite the appropriate term, or record it as a control.

After the test, encourage participants to reassess misidentified samples to assist learning. Three training-testing sessions may be sufficient to judge ability to learn, but are unlikely to be adequate to train a panel to use terms accurately and consistently.

Fig. 5.9 illustrates a set of data from this type of test. When participants misidentify samples, they frequently (but not consistently) choose a related term if present (e.g., bell pepper for herbaceous). Because odor memories are often organized in experience-related categories, this may explain why exposure to phenylethyl alcohol (possessing a floral odor) enhanced subsequent differentiation between other floral odors (Li et al., 2006). This influence did not extend to unrelated odorants.

Off-Odors: Basic Test

The off-odors test is a variant of the previous test, designed to teach and assess the ability to recognize several characteristic odor faults. Appendix 5.2 gives an example of sample preparation. Training sessions with the samples

PLATE 5.3   Tasting session. Cool Climate Oenology and Viticulture Institute (CCOVI) Tasting Lab, Brock University. *Photo courtesy R. S. Jackson.*

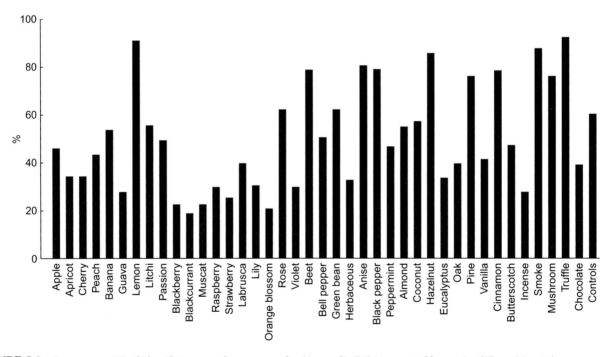

FIGURE 5.9   Success rate (%) of identifying several aroma samples (Appendix 5.1) in a neutral base wine (15 participants).

should be held a few hours prior to the test. During training, participants should be encouraged to record their impressions, as a recognition aid later on.

The answer sheet lists all off-odors, plus space for several controls. Participants smell each sample and mark the sample number opposite the most appropriate term, or as a control. The samples are covered when not in use to minimize odor contamination of the surroundings. Again, the samples should be presented in black ISO wine-tasting glasses or equivalents. Appearance may give clues as to the fault.

Three training-testing sessions are usually adequate to screen people for their ability to detect off-odors, and initiate learning. Additional training sessions should be provided to stabilize their odor memory, as well as periodically to maintain proficiency. Fig. 5.10 illustrates an example of responses from 15 participants. When samples

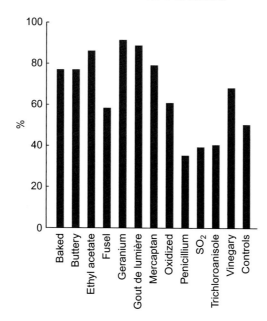

FIGURE 5.10   Success rate (%) of identifying several off-odors (Appendix 5.2) in a neutral base wine (15 participants).

are misidentified, they may be selectively interchanged, e.g., baked and oxidized; mercaptan and *goût de lumière*; guaiacol and TCA; and ethyl acetate and acetic acid, fusel alcohols, or plastic.

Off-Odors in Different Wines

The previous test presents off-odors at a single concentration and in a relatively neutral-flavored wine. Because off-odor detection can be markedly affected by the wine in which they occur (e.g., Martineau et al., 1995; Mazzoleni and Maggi, 2007), a more realistic assessment is provided by presenting faults in two (or more) base wines, and at two (or more) concentrations (closer to those that might occur naturally).

In the test design given in Appendix 5.4, only a selection of the more important and easily prepared off-odors are used. Participants should be provided with an opportunity to smell the unmodified (control) wines for familiarization before performing the test.

As per standard, arrange the faulty and control samples randomly. List all off-odor names on the answer sheet, even though only some are presented. Leave space for an undisclosed number of controls. Participants smell each sample and mark the number of the sample opposite the appropriate fault, or in the area set aside for controls.

Fig. 3.25 illustrates results from this type of test. Off-odors tend to be more readily detected in white than red wines, presumably because of the less-fragrant context of white wines. Intriguingly, in this example, almost half the control samples were identified as having an off-odor, even though participants sampled the base wines before the test. This is likely an example of how expectation can influence perception.

Discrimination Tests

Varietal Dilution

The test measures discrimination among subtle differences in wine samples. Wines possessing a varietally distinctive aroma are diluted with neutral-flavored wine of similar color. If appropriate wines of similar color are unavailable, the samples should be presented in black ISO glasses, tested under red light, or artificially colored. The dilution series can be at any level desired, but dilutions of 4, 8, 16, and 32% seem to provide a reasonable range for discrimination.

Five sets of three glasses at each dilution are the minimum requirements. Pour diluted (or undiluted) wine into two of the three glasses. The remaining glass holds the undiluted (or diluted) sample. Thus, each set has one different sample, but not consistently the diluted or undiluted. Position the sets at random. This arrangement is termed the triangle test. Table 5.8 illustrates an example of a setup for this type. Other test procedures, such as the pair and

TABLE 5.8   Example of a setup for a varietal dilution test

Dilution fraction (%)[a]	Green (g)	Yellow (y)	Purple	Most different[b] sample	Arrangement sequence of samples		
					#1[c]	#2	#3
4	$_c$d	x	c	y	1	15	13
4	x	c	c	g	8	18	9
4	c	c	x	p	13	5	19
4	x	c	x	y	6	12	18
4	c	c	x	p	19	1	8
8	c	x	c	y	12	20	4
8	c	x	x	g	17	13	14
8	x	c	x	y	5	4	15
8	x	c	c	g	18	7	1
8	c	x	c	y	9	3	7
16	c	c	x	p	11	11	2
16	x	c	x	y	2	16	10
16	c	x	x	g	10	9	5
16	c	c	x	p	16	2	3
16	c	x	c	y	14	6	6
32	c	x	x	g	4	19	12
32	c	c	x	p	7	10	20
32	c	x	c	y	15	14	17
32	c	x	x	g	3	17	11
32	x	c	x	y	20	8	16

[a]Needs 4 bottles of each wine plus a bottle for dilution; 4 empty bottles for diluted samples to dilute samples: 4% = 384 mL wine + 16 mL diluting wine; 8% = 368 mL wine + 32 mL diluting wine; 16% = 336 mL wine + 64 mL diluting wine; 32% = 272 mL wine + 128 mL diluting wine.

[b]The base of each glass is marked with a colored sticker (g, green; p, purple; y, yellow). The participants identify the most different sample by noting the color of the sample on the test sheet.

[c]Sessions #1: Cabernet Sauvignon, #2: Zinfandel, #3: Chardonnay.

[d]c, control (undiluted) sample; x, diluted sample.

duo-trio tests, are equally applicable, but are less economic in wine use. In addition, the triangle test tends to be more rigorous, requiring increased concentration.

It may be simplest for the participants to remain standing, pass by each set of glasses, remove the covers, and smell each sample. They record the number (or color patch) of the *most different* sample on the answer form. Because the similar samples may have come from separate bottles, they may not be as identical as one might desire. This approach limits the problem that adaptation may result in identical samples not being perceived as duplicates. If participants are not certain which is the *most different*, they are required to guess. The statistical test assumes that some correct responses will be guesses.

Probability tables provided in Appendix 5.5 indicate the level at which participants begin to distinguish differences among samples. Under the conditions used in this example (five replicates), the participants must correctly identify four out of the five replicates to indicate successful differentiation at any dilution level.

Although the test is designed to distinguish the sensory skills of individuals, group results can also be informative. Some individuals may be able to detect all levels of dilution at above chance, but personal experience indicates that successful differentiation by groups may not exceed 60 percent, even at the highest level of dilution employed.

Varietal Differentiation

This test assesses the ability of participants to distinguish between examples of varietal wines. As in previous tests, the triangle procedure is used.

TABLE 5.9 Example of the arrangement for samples in a wine differentiation test

	Green (g)	Yellow (y)	Purple (p)	Most different[a] sample	Position#
Sangiovese	1	2	1	y	1
	2	1	1	g	5
(1, Melini Chianti;	2	2	1	p	7
2, Ruffino Chianti)	1	2	1	y	11
	1	2	2	g	15
	1	1	2	p	18
	1	2	2	g	21
	2	2	1	p	23
	1	2	1	y	26
	1	2	2	y	28
Cabernet Sauvignon	2	1	2	y	2
	1	2	2	g	4
(1, Santa Rita;	1	1	2	p	8
2, Santa Carolina)	2	1	1	g	10
	2	1	1	g	13
	2	1	2	y	17
	1	2	1	y	20
	1	1	2	p	24
	2	1	2	y	27
	1	1	2	p	29
Pinot noir	1	2	2	g	3
	2	1	2	y	6
(1, Drouhin;	2	1	2	y	9
2, Pedauque)	1	2	1	y	12
	2	1	1	g	14
	1	1	2	p	16
	2	2	1	p	19
	1	1	2	p	22
	2	1	2	y	25
	1	2	1	y	30

[a]The base of each glass is marked with a colored sticker. The participants identify the most different sample on the test sheet.

Choose distinctive pairs of varietal wines (ones that the instructor or researcher at least can distinguish blind). For each pair, prepare 10 sets of 3 glasses. Two glasses contain one of the wines, whereas the third contains the other. Use black ISO glasses or red illumination if the wines are recognizably different in color. Alternatively, stating that the color of the wines has been altered should eliminate color from influencing their responses. Random positioning of the sets minimizes the likelihood of identical pairs occurring in sequence. Table 5.9 gives an example using three varietal-wine pairs. With 10 replicates of each set of wines, the participant must obtain 7 correct responses to signify identification above chance ($p = 0.05$) (see Appendix 5.5).

Participants assess each pair of the wines before the test. This minimizes prior experience from being a major determinant in differentiation. This is especially important if, as usual, the participants are asked to identify the varietal origin of each of the 30 sets of wine.

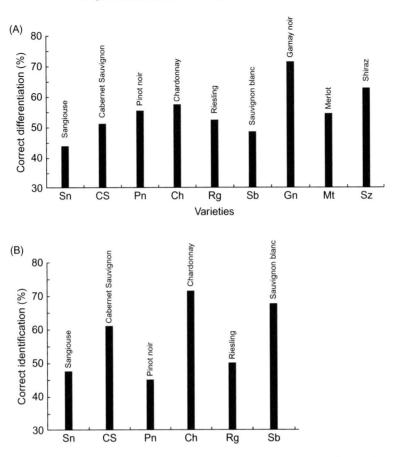

FIGURE 5.11   The success of several groups of participants in differentiating paired sets of several varietal wines (A) and recognizing their varietal origin (B), when the pairs were randomly intermixed with two separate pairs of varietal red or white wines (B).

The test is particularly valuable in assessing tasting ability, as the participants are required to recognize the subtle differences that distinguish similar wines. Where participants are expected to also distinguish among the three sets of cultivars, it is especially important that the sample wines chosen do demonstrate the features that aromatically distinguish them. Although wines are not usually assessed only by smell (as in this test), data from Aubry et al. (1999) suggest that it can be almost as discriminatory as under standard tasting conditions.

Fig. 5.11 provides an illustration of the type of group results one might expect. For example, participants in the particular situation provided had more difficulty separating the two Beaujolais (Gamay noir) than recognizing that they were Beaujolais. In contrast, the group found it easier to separate the two Chardonnays than to recognize that they were Chardonnay wines. Individually, participants varied from recognizing 33% to 90% of the wine's varietal origin, and from 33% to 70% in differentiating between sample pairs.

Short-Term Wine Memory

The wine-memory test is particularly significant as it measures the ability to recognize wines sampled previously. This skill is essential for fair assessment of wines compared, as usual, in groups.

In preparation for the test, each participant samples a set of five wines. Each wine should be sufficiently unique to permit the participants a fair chance at subsequent differentiation. Participants assess each wine for odor, taste, and flavor, using a standard detailed score card. Each wine is identified by its varietal, regional, or stylistic origin, whichever seems appropriate. Participants are told that they will be expected to identify the wines as to type in the subsequent test. Whether the participants are permitted to retain any notes they take in this preparatory phase is up to the designer. My experience is that notes may be useful, if only as a psychological crutch.

For the test, samples may be presented in black ISO glasses, or under dim red lighting, to eliminate color as a distinguishing criterion. Sample glasses are distinguished by number, letter, or colored stickers. However, in this instance, instead of five wines being presented, seven appear. Participants are told that among the seven are the five wines sampled earlier. They are also informed that among the seven glasses are two that either are repeats of ones sampled previously, or contain wines different from those previously sampled. Participants sample each wine, identifying them as one of the previously sampled wines (by name, if sampled in the preparatory phase), or as a new sample. Although deceptively simple, experience has shown this to be one of the most challenging of screening tests.

Taster Training

In most of the screening tests noted in the preceding section, training is incorporated into the testing procedure. In most cases, though, further specific training follows panelist selection. Part of this may include sessions on how to taste wines critically. For experienced individuals, this step may be unnecessary, but can be useful as a refresher. Part of training usually involves practice in the use of evaluation sheets (or forms available with sensory computer software). If scaling is used, developing experience in how it should be used is equally essential.

A critical component of many training sessions involves practice in the correct and consistent recognition of a set of standards. In academic research this usually involves specified odor (and occasionally taste) representatives. Determining individual threshold values may be useful when assessing these compounds, but is not inherently indicative of discriminatory ability. Occasionally, training may be preceded by sessions in which the panelists generate, and subsequently select, the vocabulary to be used during sensory analysis. This may be particularly useful with descriptive sensory analysis. It they generate the terms to be used (those that seem appropriate to them), it may prevent panelists from expecting or imagining the presence of descriptive terms, simply because they were provided by the instructor. An even more insidious problem can arise if panelists, able to recognize odors in isolated form, are unable to identify them in the matrix of wines. In complex odor mixtures, such as wine, individual compounds may become associated in memory with other odorants, generating qualitatively different sensations that may interfere with identifying their individual constituents.

Despite the time and effort involved in training, the benefits of training are clear. An example is its efficacy, relative to term use, is illustrated in Fig. 5.12.

In contrast to training for sensory analysis, most training in the wine industry is typically focused on recognizing representative examples of varietal, regional, and stylistic types of wines or quality grades. The extent and duration of such training depend on the task. No specific recommendations are possible as training is too dependent on what the group or individuals are expected to do.

Before training begins, candidates may be first assessed on their ability to discriminate between a range of wines. Although improved discrimination is frequently observed during training, it may be judicious to eliminate those for whom training may be ineffectual. The use of 30–40 discrimination trials, spread over several days, is suggested as usually adequate for the first round of member selection (Stone et al., 2012).

Assessing Taster and Panel Accuracy and Consistency

Critical to the effective function of any panel is its accuracy and consistency. At its simplest, this may involve comparing the results of each taster with the group average. ANOVA can provide additional information. Typically, consistency and score stability are the measures used. Consistency refers to the ability of individual panel members to repeatedly generate similar assessments over time, whereas accuracy refers to an individual's ability to assess wines or attributes similarly to other panelists. Although many procedures have been suggested to assess features such as consistency, accuracy, reliability, and discrimination, no consensus has arisen as to the most appropriate statistics to be used (Alvelos and Cabral, 2007; Hodgson, 2009; Hyldig, 2010; Stone et al., 2012; Tomic et al., 2013). Some references provide working examples, such as Vaamonde et al. (2000), King et al. (2001), and Bi (2003). Deterioration in skill may indicate the need for retraining, declining interest, or another factor that needs to be investigated and corrected where possible. In such investigations, it is important to distinguish between panel versus sample variation.

Because of the additional effort involved in making these assessments, they are seldom performed on a regular basis. However, with the introduction of computer-based, automated, data entry, there is little reason not to conduct regular member assessment. More difficult, though, is deciding on what constitutes a minimally acceptable level of performance. Here, statistics are of no avail.

Consistency in all aspects of critical assessment is essential. In descriptive sensory analysis, near-homogeneous use of descriptive terms is obligatory. Continuous monitoring of term use during training can be analyzed to determine if there is low taster x term variance. For evaluation-type tasting, random repeat tasting of a few wines over several weeks or months can provide similar data (e.g., Gawel and Godden, 2008). In the latter assessment, it is

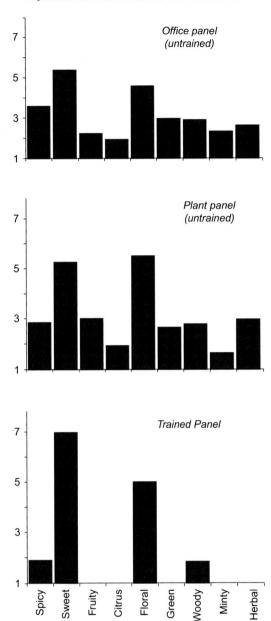

FIGURE 5.12   Mean profiles of data for a trained panel and two untrained panels. *Reproduced from Lawless, H. T., 1999. Descriptive analysis of complex odors: reality, model or illusion? Food Qual. Pref. 10, 325–332, with permission from Elsevier.*

important that the wines given repeatedly not be easily recognized as such. If wines are particularly distinctive, tasters will soon detect this, realize why the same wine(s) are appearing frequently, and potentially modulate their scores accordingly. Variance that is insignificant, or small values obtained from measuring standard deviation or tests of skewness, suggests taster consistency.

With computer statistical packages, use of ANOVA has become the standard means for assessing aspects of consistency. Because these packages incorporate F-distribution and t-distribution tables, there is little value in reproducing them here. If required, they can be found in any modern set of statistical tables. Even without dedicated statistical software, spreadsheet programs in Office-type software packages can be readily set up to perform analyses of variance.

Table 5.10A presents hypothetical tasting results from which data are used to assess taster consistency. In the example, scoring data on two different wines, randomly presented twice during five separate tastings are compiled. From the ANOVA table (Table 5.10B), the least-significant difference (LSD) can be derived from the formula:

$$\text{LSD} = t_{\pm} \sqrt{2v/n} \qquad\qquad (5.1)$$

TABLE 5.10 (A) Scores given six wines (four identical) on five different occasions by a single taster. (B) Analysis of variance (ANOVA) table

A

Testing occasion	Wines				Total
	A_1	A_2	B_1	B_2	
1	9	10	6	5	30
2	10	8	7	6	31
3	7	9	5	7	28
4	8	9	6	5	28
5	9	8	7	6	30
Total	43	44	31	29	
Mean	8.6	8.8	6.2	5.8	

B

Source	SS	df	ms	F	$F^a_{.05}$	$F_{.01}$	$F_{.001}$
Total	50.55	19					
Wines	36.95	3	12.32	12.52	3.49	5.95	10.8
Replicates	1.79	4	0.45	0.46	3.26	5.41	9.36
Error	11.8	12	0.98				

$G = \sum$ Totals = (43 + 44 + 31 + 29) = (30 + 31 + 28 + 28 + 30) = 147 $C = G^2/n = (147)^2/20 = 1080.45$ Total SS $= \sum$ (individual scores)$^2 - C = (9^2 + 10^2 + 7^2 + 8^2 + \ldots\ldots 6^2) - C = 1131 - 1080.45 = 50.55$ Wine SS $= \sum$ (wine total)$^2/n - C = (43^2 + 44^2 + 31^2 + 29^2)/5 - C = 1117.4 - 1080.45 = 36.95$ Replicates SS $= \sum$ (replicate totals)$^2/n - C = (30^2 + 31^2 + 28^2 + 28^2 + 30^2/5 - C = 1082.25 - 1080.45 = 1.8$ Error SS $= \sum$(individual scores)$^2 - \sum$ (wine total)$^2/n - \sum$ (replicate totals)$^2/n = 50.55 - 36.95 - 1.8 = 11.8$df (degrees of freedom): Total (# scores − 1) = (20−1) = 19; Wine (# wines − 1) = (4−1) = 3; Replicate (# replicates − 1) = (5−1) = 4; Error (total df − Wine df − Replicates df; = 19 − 3 − 4 = 12ms.- Wines (Wine SS ÷ Wine df) = 36.95/3 = 12.32; Replicates (Replicate SS ÷ Replicate df) = 1.79/4 = 0.45; Error (Error SS ÷ error df) = 11.8/12 = 0.98 F; Wines (Wine ms ÷ error ms) = 12.32/0.98 = 12.52; Replicates (Replicates ms ÷ Error ms) = 0.46/0.98 = 0.46

[a]F-distribution values are available in all standard statistical tables.

where t_α is the t value with error degrees of freedom (available from standard statistics charts), at a specific significance level (α); v is the error variance (ms); and n is the number of scores on which each mean is based. At a 0.1% level of significance, the formula for data from Table 5.10 becomes:

$$LSD = 2.91\sqrt{2(0.983)/5} = 1.916 \tag{5.2}$$

For significance, the difference between any two means must exceed the calculated LSD value (1.916). The difference between the mean scores for both wines (A and B) indicates that there is no significant difference between the rankings of either wine. For wine A, the mean difference was 0.2 (8.8–8.6), and for wine B, the mean difference was 0.4 (6.2–5.8). Both values are well below the LSD values needed for significance, i.e., 1.916. Equally, the results show that the two wines were well distinguished by the taster. All combinations of the mean differences between replicates of wines A and B were greater than the LSD value (1.916):

$$A_1 - B_1 = 8.6 - 6.2 = 2.4 \tag{5.3}$$

$$A_1 - B_2 = 8.6 - 5.8 = 2.8 \tag{5.4}$$

$$A_2 - B_1 = 8.8 - 6.2 = 2.6 \tag{5.5}$$

$$A_2 - B_2 = 8.8 - 5.8 = 3.0 \tag{5.6}$$

Similar results are obtained directly from the ANOVA table (Table 5.10B). The F value from the two wines ($F = 12.52$) shows the wines were differentiated by the tasters at the 0.1% level of significance, while the replicate scores from the same wines were insignificant ($F = 0.46 < F_{.05} = 3.26$).

Score Variability

High levels of agreement among panelists is essential if wines of similar character are to be differentiated. However, agreement may also indicate lack of skill, or an inadequate reflection of human variability. For example, inexperienced tasters may show lower score variability than experienced tasters, possibly because experience permits confidence in use of the full range of a marking scheme. Tasters with consistent, but differing, views on wine quality will also increase score variability. Acceptance of this variability may, however, result in a reduction in likelihood of differentiation. As noted, absence of accepted standards for wine quality makes it difficult to decide whether score variability results from divergence in perceptive ability, experience, concepts of quality, or other factors.

In the absence of clear and definitive quality standards, the best that can be done is to measure panel-score variability. If tasters have been shown to be consistent in the past, then significant variance among panelists' scores for single wines probably reflects differences in perception. However, significant differences among score averages for several wines probably indicate real differences among the wines. Table 5.11A provides data on the ranking of four wines assessed by five tasters.

In the example, ANOVA shows that panelists clearly demonstrated the ability to distinguish among the four wines (Table 5.11B). The calculated F value (21.2) is greater than F statistics up to the 0.1% level of significance (10.8). However, no significant difference appeared among the scores of the five tasters; the calculated F value (1.17) is less than the $F_{.05}$ statistic (3.26). This indicates group-scoring similarity (at least for those wines on that particular occasion).

ANOVA can also suggest whether a panel possesses a common concept of quality. Table 5.12A provides data where the wines cannot be considered significantly different, but the individual scores show significant difference. The ANOVA table (Table 5.12B) generates an F value of 2.23 for variance among wines. Because this is less than the $F_{.05}$ statistic (2.78), this indicates that no significant difference ($p < 0.05$) was detected among the wines. In contrast, the calculated F value of 3.35 for variance among tasters is greater than the $F_{.05}$ statistic for tasters (2.51). This suggests that panel members probably did not have a common view of wine quality. Removing data generated by panelists either unfamiliar with that type of wine, inconsistent in scoring, or possessing aberrant views of wine quality might show that the wines were differentiable, and the remaining panelists consistent in their perception of quality (at least for those wines).

TABLE 5.11 (A) Data on the score for four wines tasted by five panelists of 4 wines. (B) Analysis of variance (ANOVA) table

A

Tasters	Wines				Sum	Mean
	W_1	W_2	W_3	W_4		
1	15	9	15	12	51	12.8
2	16	10	12	13	51	12.8
3	18	10	13	11	52	13
4	19	11	14	12	56	14
5	17	12	13	15	57	14.3
Sum	85	52	67	63		
Mean	17	10.4	13.4	12.6		

B

Source	SS	df	ms	F	$F_{.05}$	$F_{.01}$	$F_{.001}$
Total	142.55	19					
Wines	112.95	3	37.65	21.21	3.49	5.95	10.8
Tasters	8.30	4	2.08	1.17	3.26	5.41	9.36
Error	21.30	12	1.77				

TABLE 5.12 (A) Scoring results of seven panelists for five Riesling wines. (B) Analysis of variance (ANOVA) table

A

Tasters	Wines					Sum	Mean
	W_1	W_2	W_3	W_4	W_5		
1	3	5	4	7	7	26	5.2
2	5	6	5	6	9	31	6.2
3	9	3	2	4	5	23	4.6
4	7	2	8	8	6	31	6.2
5	6	5	4	5	6	26	5.2
6	7	8	5	8	7	35	7
7	3	4	7	8	9	31	6.2
Sum	40	33	35	46	49	203	
Mean	5.7	4.71	5	6.6	7		

B

Source	SS	df	ms	F	F.05
Total	56.6	34			
Wines	9.6	4	2.38	2.23	2.78
Tasters	21.5	6	3.58	3.35	2.51
Error	25.7	24	1.07		

Assessment of panelist consistency, both of individuals and groups, should be based on many tastings. Individuals have off days and views of quality can vary among wines, being more consistent for some wines than others.

SUMMARY

Current evidence suggests that becoming a good taster requires both training and experience; both are essential. One cannot substitute one for the other. Although exceptionally acute sensitivity is not always essential, it can be a desirable attribute. What appears more essential is the ability to develop an extensive memory for the olfactory and oral attributes that distinguish wines. Additional properties are the ability to consistently describe and rank wines, relative to standard norms, development of a standard lexicon of wine descriptive terms, recognition of subtle differences between similar wines, assessment of wine attributes independent of personal preferences, and being healthy and dedicated to attendance at tastings; these properties are often condensed to consistency, concentration, and commitment. A further property that many tasters desire, but is usually not necessary for panel membership, is the ability to identify wine provenance. This usually requires identifying both attributes the wine expresses as well those it does not.

What is regrettable is evidence on the best ways to develop these characteristics. Clearly motivation is required, but how is this to be accomplished and maintained? Neither do we know the meaning of motivation at a molecular nor neuronal level, nor do we understand what environmental stimuli are required for its activation. It is also uncertain how memories are established and sustained. Are there fundamental differences between incidental learning (e.g., the association of color, texture, and flavor everyone develops as part of one's upbringing) and attentive, declarative learning (e.g., language acquisition and event-driven memory), or are their apparent differences just an illusion and simply conceptually designated zones along a spectrum? It seems that the more emotionally significant the situation, the better the encoded memory trace. Were these fundamental aspects better known, presumably we could provide better and more effective learning experiences.

Few people are likely to be inherently blessed with, or develop, all of these desirable traits, or express them consistently. For this reason, if for no other, a panel is required.

PRETASTING ORGANIZATION

Tasting Area

Indirect north lighting has often been considered ideal for assessing wine color, possibly because of was thought to be uniform. However, the supposed uniformity is only illusory (Fig. 5.13). In addition, indirect north lighting is rarely possible without forethought in designing a tasting room (e.g., Plate 5.4). In practice, most tasting rooms possess only artificial lighting, with little if any natural illumination.

Of artificial light sources, fluorescent tubes provide more uniform illumination. Daylight tubes may be preferable, but cool white fluorescent tubes seem adequate. The mind is amazingly adept at adjusting color perception (a cerebral construct anyway) within a fairly wide range of spectral and intensity changes (Brou et al., 1986). Thus, tasters quickly adapt, usually without realizing it, to the color characteristics of most forms of standard illumination. Within typical ranges, the intensity and diffuse nature of fluorescent illumination are more important than its spectral quality. Alternatively, recessed or track halogen or LED lights can provide individual, brilliant, full-spectrum (or low-intensity red illumination for color distortion) at each station. Wine-color distortion can be further minimized by having tasting rooms and booths in uniform, neutral (white or off-white) colors, with counters or tabletops matte white.

When the influence of wine color on perception needs to be avoided, discrimination based solely on gustatory and/or olfactory sensations can be achieved with black ISO glasses. Alternatively, blue- or red-colored glasses or low-intensity red light (Plates 5.5 and 5.7) may be used. Where low-intensity red light is used, and samples are supplied through a port in each booth, light from the prep room should not be permitted to illuminate the samples during the transfer. Although low-intensity red light distorts color perception, whether it actually negates potential color bias has not been demonstrated. That colored light can modify wine perception is clear (Spence et al., 2014), but in the study, light intensity was within a standard range. However, as red light markedly affected the perception of taste, with less significant effects on intensity and liking, the results raise questions as to whether even pale

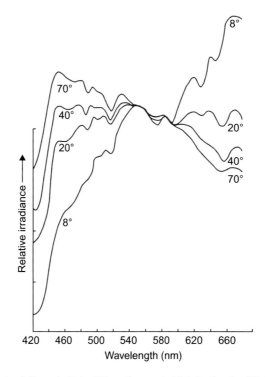

FIGURE 5.13   Daily spectral variability in diffuse skylight. When the sun is high in the sky (70 degrees above the horizon), the peak radiation is in the blue section. This shifts as the sun begins to set to a point where it is in the red by half an hour before sunset (8 degrees above the horizon). *Adapted from Henderson, S.T., 1977. Daylight and its Spectrum. Wiley, New York (Henderson, 1977), reproduced by permission.*

PLATE 5.4   Research sensory analysis laboratory with natural north lighting in the Institut des Sciences de la Vigne et du Vin, Villenave-d'Ornon, France. *Photo courtesy R. S. Jackson.*

red lighting might unknowingly prejudice perception in ways that have hitherto been unexpected, or is it really effective as assumed. Experiments in perception psychology have demonstrated that models of reality developed during infancy and later can be so strong that our brain overpowers the information sent by our senses, even when we realize the perceived "reality" is illogical or in error. For an example, look at https://www.youtube.com/watch?v=G-lN8vWm3m0 for an example of the McGurk effect (McGurk and MacDonald, 1976).

Where back-and-forth comparison of samples is not required, avoiding color-induced distortion of perception may be achieved by presenting the wines one at a time (Stone et al., 2012). Of course, this assumes that any color differences among the wines are not so obvious as to be easily remembered.

Under most tasting situations, wine color may not be a significant factor, thus, maintaining normal lighting conditions is preferable. Color is a normal component of a wine's gestalt, i.e., how wine is recorded in memory (Morrot et al., 2001; Österbauer et al., 2005). This is especially important if the results are intended to have any relevance to the marketplace.

Tasting rooms should ideally be air conditioned, or at the least, well ventilated. This is not only for the comfort of the panel, but also to prevent odor accumulation and adaptation. Positive air pressure in the tasting room also limits extraneous odors from entering from surrounding areas. This can be especially important where the tasting room is close to a wine cellar or storage area. Tasters become adapted to mild background odors, but if these odors are stale or possess off-odors, they may compromise the fairness of the tasting.

When air conditioning or adequate ventilation is not available, air purifiers can assist in providing an aromatically neutral tasting environment. Another aspect of reducing odor contamination is to have all wine glasses covered when not being sampled. This also applies to cuspidors. It hopefully goes without saying that use of personal perfumes, colognes, etc., is prohibited. Tasting rooms equally need to be free of annoying sounds that could disrupt concentration.

Tasting stations should be physically isolated to prevent taster interaction. Research has confirmed the widely held belief that comments can modify perception (Herz and von Clef, 2001; Herz, 2003). Thus, in deference to Shakespeare:

A rose by any other name *may not* smell as sweet.

Ideally, the tasting room should be designed specifically for the purpose, with separate tasting booths (Figs. 5.14 and 5.15; Plates 5.2–5.6). The back of each booth ideally opens onto a preparation room (Plate 5.7). Glasses should also ideally be stored close to the student or research tasting room (Plate 5.8). Alternate lighting sources permit the use of white or low-intensity red lighting. A port permits samples to be easily presented and removed. Various closures have been used, from simple sliding openings to counterbalanced guillotine models. For ease of maintenance, the closure should be produced from some opaque, readily cleaned material. An opening about 20 cm high and 25 cm wide is usually adequate. Other models are discussed in sensory evaluation texts noted in the Suggested Reading

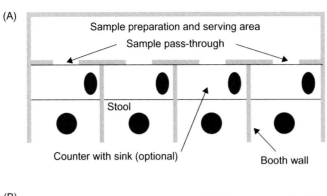

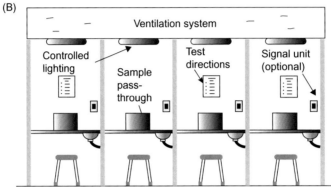

FIGURE 5.14 Diagram of a sensory-evaluation booth (not drawn to scale). *From Stone, H., Bleibaum, R., Thomas, H.A., 2012. Sensory Evaluation Practices. 4th ed. Academic Press, Orlando, FL, reproduced by permission.*

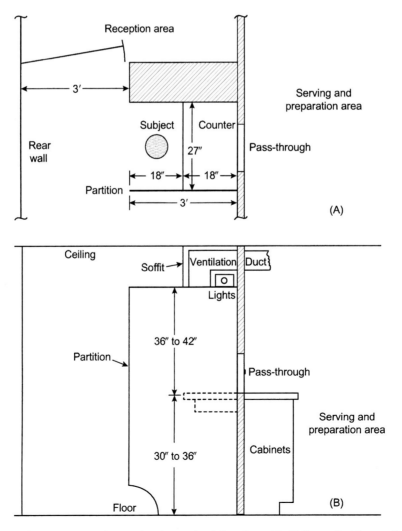

FIGURE 5.15 Design details for a sensory-evaluation booth. *Reprinted from Stone, H., Bleibaum, R., Thomas, H.A., 2012. Sensory Evaluation Practices. 4th ed. Academic Press, Orlando, FL, with permission from Elsevier.*

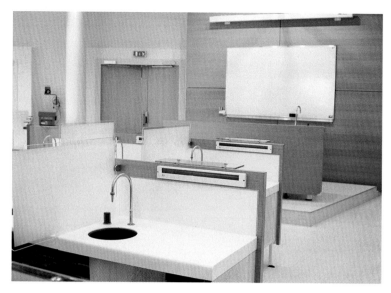

PLATE 5.5   Sensory analysis classroom in the Institut des Sciences de la Vigne et du Vin, Villenave-d'Ornon, France. Note the availability of red light illumination. *Photo courtesy R. S. Jackson.*

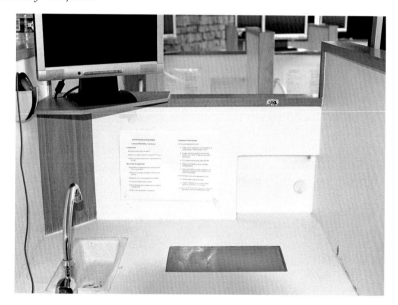

PLATE 5.6   Tasting station in the sensory laboratory, Institut Universitaire de la Vigne et du Vin Jules Guyot, Dijon, Burgundy, France. *Photo courtesy R. S. Jackson.*

PLATE 5.7   Prep area for organizing and presenting samples behind tasting cubicles. Note the presence of ceiling red lighting is present if desired. Cool Climate Oenology and Viticulture Institute (CCOVI) Tasting Lab, Brock University. *Photo courtesy R. S. Jackson.*

PLATE 5.8   Prep area of the tasting lab showing the arrangement of black ISO wine-tasting glasses, ready for use in the adjacent sensory laboratory, Institut Universitaire de la Vigne et du Vin Jules Guyot, Dijon, Burgundy, France. *Photo courtesy R. S. Jackson.*

at the end of the chapter. Where considered of value (frequent presentation of a series of samples), some means of easy communication between panelists and servers should be included. Plates 5.5 and 5.6 illustrate alternative student teaching/research rooms. Plate 7.9 shows a tasting-room setup in Cognac.

Where the tasting room is multipurpose, folding partitions can be constructed as tasting booth dividers. Partitions are often constructed from 3/8-in. white melamine (which is easily cleaned). The sections should be about 1 m × 1 m (allowing about 50 cm to extend over the edge of the counter or table). Two long hinges permit the panels to be conveniently folded for storage. If tasting stations cannot be compartmentalized, separate and random presentation of the samples to each taster can reduce the likelihood of significant taster interaction, especially if the tasters know that the samples have been randomized.

Chairs or stools should be adjustable as well as comfortable. Panel members may be spending hours concentrating on the difficult tasks of identification, differentiation, or evaluation. Every effort should be made to make the working environment as pleasant and comfortable as possible, for example, adding an adjoining pleasant waiting/discussion room.

Rooms specifically designed for tasting should possess dentist-like cuspidors at each station (Plates 5.2–5.7). They are both hygienic and minimize odor buildup. Because many tasting environments are less well equipped, 1-liter opaque plastic buckets can act as substitutes. The buckets need a cover and should be voided frequently to limit odor contamination of the surroundings.

Pitchers of water to rinse glasses (if reused during a tasting) and samples of unsalted crackers or other palate cleansers should be available at each tasting station. Several studies on the efficacy of palate cleansers have reached diverging conclusions, some suggesting unsalted crackers, 1% pectin, or 0.55% carboxymethyl cellulose solutions as preferable (see Chapter 1). Water has repeatedly been shown to be an ineffective palate cleanser.

Computer terminals positioned at each booth or tasting station greatly facilitate data collection. Rapid analysis permits almost instantaneous feedback during training, while samples are still present for reassessment. The presence of touch screen monitors, showing a form for collecting the desired data, frees the tasting surface of clutter. If notation demands more than just checking items, a keypad installed on a slide rail under the counter is an option. It eliminates problems that frequently arise from illegible handwriting.

The preparation area should possess a series of refrigerated units so that wines can be brought to, and maintained at any desired temperature. Cool-temperature laboratory incubators store more wine, over a wider range of temperatures, and are far less expensive than commercial, refrigerated wine-storage units. Refrigeration is also useful for storing tasting reference standards. There also needs to be ample cupboard space for glass storage as well as other equipment such as microbalances, beakers, and graduated cylinders. The construction material should be neutral in odor, so as not to contaminate the glasses during storage. Industrial dishwashers are indispensable for the rapid, efficient, and sanitary cleaning of the large amount of glassware involved in most tastings. Odorless cleaners, combined with proper rinsing, provide glassware that is crystal clear, as well as odor- and detergent-free.

There also needs to be ample counter space for sample preparation and facilitating transfer of wines to panelists. In addition, an adequate supply of odorless, tasteless water is needed. In large research centers, this is usually supplied by a double distiller. Where this is not possible, the ready availability of under-counter reverse-osmosis units can usually supply an adequate supply of flavorless water.

Number of Samples

There is no generally accepted number of samples appropriate for a tasting. If the samples are similar, only a few wines should be assessed together. For accurate evaluation, the taster must be able to simultaneously remember the attributes of each sample, which is no easy task. Also, if description as well as rating are involved, the more detailed the tasting, the fewer the number of wines that can be adequately evaluated concurrently.

If the samples are markedly different, relatively large numbers of wines can often be evaluated together. Nevertheless, six wines tend to be the limit that can be jointly compared adequately. In contrast, wine tasters in commercial competitions are often expected to sample more than 30 wines within a relatively short period. Obviously, these assessments can be only quick and simple, even when sampled in groups. Serious consideration of the duration of the wine's fragrance and its development is not feasible. Under such conditions, wines are judged on a simple, overall, immediate impression. Although rapid, it approximates how most consumers assess wine. Nonetheless, it does injustice to the potential of the wine to show its more highly esteemed qualities. If the reader considers my view of how most consumers assess wine unduly harsh, I suggest they take furtive glances at how most people taste wine in restaurants and at wine tastings.

Replicates

In most tastings, replicates are unavailable, primarily for economic reasons. If the tasters are skilled and consistent, there may be little need for replication. However, if taster consistency is unknown, some repeats should be incorporated to assess taster reliability and consistency. Also, replicates may be required to replace faulty samples. If the samples are not checked in advance, the presence of a faulty sample may be hinted at by detecting an atypical variation in assessments, or from comments provided by the panelists.

Temperature

White wines are typically served at cool temperatures, whereas red wines are served at room temperature. Rosé wines are served at a temperature somewhere in-between. Besides these general preferences, there is little precise agreement among authorities. Fig. 4.7 illustrates some of the effects serving temperature can have on the perception of a wine's characteristics.

The recommended cool temperature for white wines can vary from 8 to 12°C, with the upper end being more frequent. Anywhere within this range is usually adequate. What is more important is that all the wines be served at the same temperature. As with other aspects of tasting, it is essential that wines be sampled under identical conditions, or at least as close to identical as possible. Sweet (dessert) white wines are typically viewed as showing best at the lower end of the standard temperature range, whereas dry white wines are more preferred at the upper end of the scale. Dry white wines can even show well at 20°C. This should not be too surprising, as wine soon reaches and surpasses 20°C in the mouth. Some aspects of what is termed wine development may be partially associated with increased volatility as the wine warms.

Red wines are generally recommended to be assessed at between 18 and 22°C. This range enhances fragrance and diminishes perceived bitterness and astringency. Only light, fruity red wines, such as carbonic maceration wines (e.g., Beaujolais), are taken somewhat cooler, at between 15 and 18°C. Rosé wines are typically served at about 15°C.

Sparkling wines are best served at between 4 and 8°C. This range enhances the expression of the toasty aspect so desired in most dry sparkling wines. In addition, it retards the release of carbon dioxide and extends the duration of the effervescence it generates (Fig. 5.16). Cold temperatures also enhance the prickling sensation of sparkling wines (see Green, 1992). Finally, the cool sensation gives the wine part of its refreshing sensation. Unfortunately, cold temperatures can also enhance the metallic perception occasionally found in some sparkling wines.

Most fortified wines, such as sherries, are sampled cool to cold. This mellows their intense bouquet. It also decreases the burning influence of their high alcohol content and the sweetness of cream sherries. In contrast, ports are typically served at a temperature near the lower range for red wines.

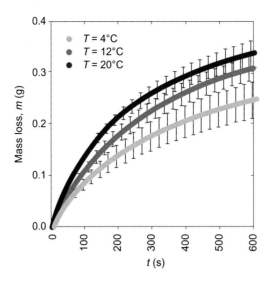

FIGURE 5.16   Effect of temperature on the loss of carbon dioxide (mass loss) from champagne following pouring into flutes. Vertical lines represent standard deviations from six successive pourings. *Reprinted with permission from Liger-Belair, G., Villaume, S., Cilindre, C., Jeandet, P., 2009. Kinetics of CO2 fluxes outgassing from champagne glasses in tasting conditions: The role of temperature. J. Agric. Food Chem.* **57**, *1997–2003. Copyright 2009, American Chemical Society.*

Temperature preferences have often been thought to reflect habituation or cultural norms, but recommendations also appear to reflect legitimate physicochemical effects on sensory perception. Consumers reported preferring a mild red wine at room temperature (20–25°C) or hot (80–85°C), but only the latter when blindfolded (Zellner et al., 1988). The wine was not appreciated at a cold (5°C) temperature. In a study on aroma intensity at several temperatures, the aroma was enhanced slightly at warmer temperatures with a white wine (Fig. 4.7A), but more with a red wine (Fig. 4.7B). The effect of temperature on red wine aroma was similar (being greater at 22°C), with the wine's sour, bitter, and astringent characteristics being enhanced at cooler temperatures (Ross et al., 2012). These results correspond to what might be expected from other studies on the effects of temperature on taste and volatility.

Cool temperatures have often, but not consistently, been found to reduce sensitivity to sugars more than acids, possibly explaining why dessert wines appear more balanced at cooler temperatures. Coolness can also generate a pleasant freshness. This gives rosé wines part of their summer appeal as sipping wines. In contrast, reduced bitterness and astringency help to explain why red wines are typically served at or above 18°C. As far as fragrance, warmer temperatures favor the escape of aromatics from the wine. This applies not only from the main body of wine, but especially from the thin film of wine that adheres to the sides of a glass after swirling. Although cool wine temperatures may initially limit the bonding and reactivity of tastants and volatility of aromatics in the mouth, rapid warming likely quickly reduces these effects.

Wines should be brought to the desired temperature several hours before tasting, to avoid any last minute rush. In most commercial situations, temperature control is maintained in special, temperature-controlled units. These units are usually subdivided to maintain wines at different temperatures. Alternatively, wines can be stored in adjustable cool-temperature incubators, a refrigerator, or a cold room until the desired temperature is reached, or immersed in water at the desired temperature. Fig. 5.17 provides examples of the rates of temperature change in a 750 mL bottle of wine in air versus water. Temperature adjustment is about five times faster in water than air, with equilibration often being reached within minutes rather than hours. Once the desired temperature has been reached, the wine can be kept within an acceptable range for a short period in well-insulated containers.

Because a wine's temperature begins to warm upon pouring, it is often advisable to present the wine at the lower limit of suitable temperatures. Thus, the wine will remain within a desirable range for a longer period. This is particularly important if features such as duration and development are to be assessed over 20–30 min. Unfortunately, most tastings occur so rapidly that these noteworthy properties are missed.

Cork Removal

No corkscrew is fully adequate in all situations. For general use, those with a helical coil are preferable. The waiter's corkscrew is a classic example, but may require more force that many people prefer. Correspondingly,

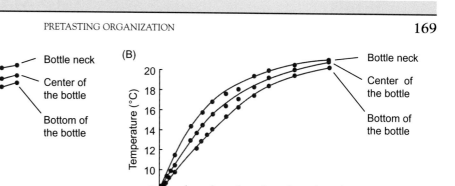

FIGURE 5.17 Distribution and change in the temperature of wine in a 750 mL bottle placed (A) at 21°C in air or (B) in water at 20°C. *From Gyllensköld, H., 1967. Att Temperera Vin. Wahlström and Widstrand, Stockholm, 1967.*

PLATE 5.9 Double-action corkscrew, original model of the Screwpull and foil cutter. *Photo courtesy of Le Creuset, New York, NY.*

double-action models are available and simpler to use (Plate 5.9). They both sink the screw into the cork, and then remove it. Several even easier-to-use lever models are available for both restaurant and home use (Plate 5.10). Fig. 5.18 illustrates the force typically required to remove several types of corks.

Regardless of design, most corkscrews have difficulty removing old corks. With time, natural cork closures lose their resiliency and tend to tear apart on removal. In this situation, a two-prong, U-shaped device, occasionally called the Ah-So (Fig. 6.2A), or a hand pump, connected to a long hollow needle, may prove invaluable, but only if the cork is still firmly attached to the sides of the bottle neck. In the case that the cork is inadvertently pushed into the bottle, there are also devices that can extract the ill-fated cork from the wine (e.g., Fig. 6.2B).

Decanting and Pouring

Decanting is valuable when sampling old wines that may have developed sediment, a situation rarely associated with technical sensory analysis (regrettably). Most modern wines are sufficiently stabilized to avoid developing a sediment. However, even without the issue of sediment, decanting prior to tasting does permit the early detection of off-odors or otherwise aberrant samples, permitting the option of finding an alternate, appropriate version before the panelists commence tasting.

PLATE 5.10   Double-action corkscrews, lever model of the Screwpull. *photo courtesy of Le Creuset, New York, NY.*

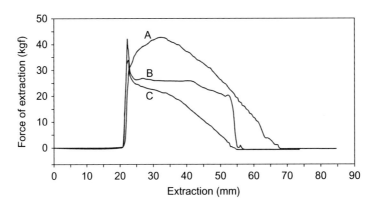

FIGURE 5.18   Typical extraction force profile of (A) natural cork, (B) injected synthetic, and (C) coextruded synthetic closures. The text was conducted at a speed of 5mm/s. *Reprinted from Giunchi, A., Versari, A., Parpinello, G. P., Galassi, S., 2008. Analysis of mechanical properties of cork stoppers and synthetic closures used in wine bottling. J. Food Engin. 88, 576–580, with permission from Elsevier.*

During pouring, and subsequently, aromatics begin to escape from the wine into the surrounding air. This is of less concern if the wine is comparatively young (it possessing more aromatic compounds than an older version). Nonetheless, decanters or filled glasses should be covered immediately, and sampling commence shortly thereafter. However, with very old wines, their feeble fragrance rapidly dissipates. Thus, they need to be assessed almost immediately upon decanting/pouring.

Changes in the equilibrium between gaseous, dissolved volatile, and loosely bound, nonvolatile states of aromatic compounds in a wine probably explain anecdotes concerning the benefits of decanting/aerating. Whatever the explanation, it is more interesting for the taster to experience this phenomenon in the glass, rather than for it to occur unnoticed in a decanter or other container.

Sample Volume

The volume of each sample should be identical. Depending on the purpose of the tasting, an adequate volume can range from 35 to 70mL. Where only simple evaluation is required, volumes less than the 35 mL range may be fully adequate. If a more detailed or prolonged assessment is desired, volumes in the 50 to 60mL range are more appropriate. With volumes in the 50 mL range, a 750 mL bottle can serve 12–14 tasters.

Dispensers

Wine dispensers are usually refrigerated units holding various numbers of bottles. They may possess separate compartments for keeping both red and white wines at separate temperatures (Plate 5.11). Each bottle is separately connected to a gas cylinder containing an inert gas, usually nitrogen. The pressure in the cylinder supplies the force

PLATE 5.11   Eight-bottle wine dispenser with refrigerated compartment for white wines. *photo courtesy of WineKeeper, Santa Barbara, CA.*

PLATE 5.12   Spigot of wine dispenser. *photo courtesy of WineKeeper, Santa Barbara, CA.*

required to dispense wine when the spigot is activated (Plate 5.12). Originally designed for commercial wine bars, these dispensers also have particular value in training sessions. Wines can be economically sampled over several days or weeks. If refrigeration is less important, almost any number of samples can be supplied with individual spigots connected to a common gas cylinder (Plate 9.4). This relatively inexpensive option is especially useful where odor references are kept available over several weeks.

One of the limitations of any system that dispenses small quantities of wine over an extended period is the progressive loss of aromatics from the wine into the increasing headspace volume above the wine. Another issue relates to the spigot retaining a small amount of wine. Over time, this becomes infested with acetic acid bacteria, noticeably contaminating the first few millimeters of wine poured. Thus, spigots need periodic disinfection.

Representative Samples

With relatively young wines, any randomly selected bottle is likely to be representative of the production. However, as a wine ages, unequal aging conditions are likely to progressively generate sensory differences among individual bottles. To neutralize such differences, if more than one bottle is available for tasting, all samples should be blended before serving.

When samples from cooperage are to be assessed, when checking development or for faults, they should be placed in sealable containers and transported directly to a tasting room. Cellars are notoriously poor areas in which to assess samples, due to ambient odors. If assessment must be delayed for any expended period, the samples should be cooled and the containers flushed with nitrogen or carbon dioxide before being sealed. This minimizes the possibility of oxidative or other changes before testing.

Within-tank (or barrel) variation can develop due to physicochemical differences throughout the volume. Bottom samples are likely to be cloudy and potentially be tainted with hydrogen sulfide or mercaptans, due to the low redox potential that can develop in any accumulated lees. In addition, regions next to oak will possess higher concentrations of wood extractives. Thus, samples taken from the middle of the tank are more likely to be representative of the volume as a whole. Several liters may have to be drained before obtaining a representative sample.

Between-barrel variation can be greater than that which exists between large cooperages. These variations can develop due to barrel manufacture, conditioning, or prior use, as well as nonuniform conditions throughout the cellar. Samples from barrels in several regions throughout the cellar are usually needed to achieve a representative sample of the wine. Frequently, though, a fully representative sample is not required, individual barrel samples being taken to check for faults.

Glasses

Glasses for wine assessment should possess specific characteristics, notably being crystal clear and colorless. Without these properties, any accurate assessment of the wine's visual attributes will be seriously compromised. This, of course, assumes that color is unlikely to bias the taster's assessment, in which case black glasses are the simplest solution.

The bowl should be wider at the base than the top, and possess sufficient volume to permit vigorous swirling of between 35 and 60 mL of wine. In addition, the stem should be adequate for convenient holding and swirling. These features have been incorporated into the International Standard Organization (ISO) wine-tasting glass (Fig. 1.2). Its tulip shape facilitates viewing and vigorous swirling, as well as concentrating aromatics in the headspace above the wine. The latter is particularly useful in detecting subtle fragrances.

Most ISO glasses are made of slim crystal. Although this enhances elegance, it may make regular cleaning in commercial dishwashers risky. In addition, it is essential to minimize scratching and avoid the buildup of a whitish, film-like patina on the glass. Thus, less expensive, thicker-glass versions (such as those available from Durand or Libbey) may be preferable. Alternate versions are illustrated in Plate 5.13. Where they are used solely for in-house tastings, a circular line etched around the glass denotes an appropriate fill level. This is of particular value where tasters fill their own glasses. Most stores that do engraving will etch glasses for a nominal fee.

As noted, where the biasing influence of color needs to be avoided, black or colored glasses are preferable. Examples of their need include training to recognize standard odor samples (they often differ in color and clarity), or where wines are being assessed for the presence of oxidative odors (differences in color could significantly influence a taster's perception). There may be other producers of black ISO glasses, but one is Verrerie de la Marne (France). Riedel in Austria produces a black tasting glass, but of a larger volume than the typical ISO glass. Blue glasses are available from Libbey, and possibly other producers. Standard wine glasses can be painted black, but have the distinct disadvantage that they need to be hand-washed to avoid removing the paint. An alternative technique, of untried efficacy, is to indicate to the panelists that the wines may have had their color distorted, and thus they should disregard the wine's color.

Although ISO glasses have become standard for assessing wines critically, only comparatively recently has the influence of glass shape on wine assessment been seriously investigated. These studies often require elaborate precautions to avoid participants detecting the glass shape (Plate 5.14). Examples of studies on glass shape can be found in Delwiche and Pelchat (2002), Fischer (2000), Hummel et al. (2003), and Russell et al. (2005). The ISO shape has been found fully adequate in all studies, and usually preferred. ISO glasses have also been shown to be superior for color discrimination (Cliff, 2001).

PLATE 5.13   Wine-tasting glasses: left, Royal Leerdam Wine Taster #9309RL – 229 mL, 7 ¾ oz (ISO model); and right, Citation All Purpose Wine #8470 – 229 mL, 7 ¾ oz. *Photo courtesy Libbey Inc., Toledo, OH.*

PLATE 5.14   Setup for assessing the effect of glass shape on the sensory attributes of wine. *Photo courtesy of Dr. J. Delwicke, Ohio State University.*

Unfortunately, none of the studies have addressed the physicochemical reasons for any differences detected. The sources are most likely to be a complex function of several factors, notably:

surface area of the wine in the glass (πr_w^2)
surface area of the wine adhering to the sides of the glass following swirling ($2\pi r_s\, dh_1$)
volume of headspace gas above the wine ($\pi r_s^2 h_2$)
diameter at the glass mouth (πr_m^2)

where r_w = radius of the wine surface in the glass; r_s = variable radius of sides of glass covered by wine following swirling; r_m = radius of the mouth of the glass; h_1 = height from the surface of the wine (meniscus) to the top of wine film adhering to the sides of the glass; h_2 = height from surface of the wine to the opening of the glass; π = 3.14159.

The first two equations relate to the escape of volatiles from the wine into the headspace. The headspace volume sets a limit on the equilibrium between aromatics in the wine and the headspace. The surface area of the mouth regulates the escape of volatiles from the headspace into the surrounding air. What the equations do not address is the dynamics of how different aromatics escape from the wine.

The surface area of the wine film coating the sides of the glass changes rapidly, decreasing as gravity draws it down to the main volume of wine, following swirling, and increasing again as a result of repeat swirling. The evaporation of ethanol (and other volatiles) changes the dynamics of their partial pressures, and therefore volatilization. Those compounds with higher partial pressures are the most immediately affected. They tend to volatilize more rapidly than they are supplied from the wine to a static surface. Those with lower partial pressures build up in concentration in the headspace more slowly, but more consistently. Changes in surface tension also undoubtedly affect volatility. As compounds escape, the dissociation constants between dissolved volatiles and their weakly bound forms in the wine change. As compounds in the headspace above the wine dissipate into the surrounding air, the equilibrium between the composition of the headspace volatiles and their dissolved forms in the wine continue to change. The latter is primarily influenced by the surface area of the mouth. For a critical measurement of glass shape on volatilization, continuous assessment of headspace gas composition is required.

In contrast to most wines, sparkling wines are usually assessed in flutes (Plate 5.15). The shape permits a detailed analysis of the wine's effervescence. This includes the size, persistence, and nature of the chain of bubbles; the mounding of bubbles on the wine's surface (*mousse*); and the ring of bubbles around the edge of the glass (*cordon de mousse*). Colored flutes, such as the blue versions available from Libbey, can be used where color differences between the samples might unduly influence their assessment. Where effervescence is not a critical part of the assessment (as in formulating the *assemblage cuvée* before the second fermentation in sparkling wine production), regular ISO glasses are fully adequate, and preferable.

Cognac has also traditionally had its own preferred glass shape, with a short stem and an enlarged bowl. However, those in Cognac prefer a small (150 mL) version of the ISO wine-tasting glass (Plate 7.8). Their view is that the large-bowled glass is only necessary for lower-grade brandies, needing the extra volume to accumulate its lower aromatic content.

Industrial dishwashers not only effectively clean but also sterilize. Extensive rinsing removes detergent and odor residues. Once glasses have been cleaned and dried, they should be stored upright in a dust- and odor-free environment. The upright position helps limit aromatic contaminants collecting on the insides of the glass. This is especially critical if the glassware is stored in cardboard boxes (a poor idea to start with). Hanging stemware upside down may be acceptable in restaurants, where the glasses are constantly being reused, but is ill-advised in cabinetry.

No matter the type of glass used, it is a useful habit to smell the bowl of any glasses to be used before filling. This will confirm that the glass itself is not contaminated with odors, which could be mistaken as coming from the wine. Although uncommon, improperly cleaned or rinsed glasses have spoiled the impression of many a good wine.

PLATE 5.15 Sparkling wine flutes: left, Royal Leerdam Allure Flute #9100RL – 214 mL, 7 ¼ oz; right, Citation Flute # 8495, 185 mL, 6 ¼ oz. *Photo courtesy Libbey Inc., Toledo, OH.*

Number of Tasters

Up to a point, the larger the number of tasters, the greater the probability of obtaining valid data (Lawless and Heymann, 2010), or the data being more representative of the clientele for whom a panel may be intended to represent. In practice, though, as few tasters as possible are used, for convenience as well as to minimize costs. The number of participants chosen typically reflects the nature of the experiment, the skill and consistency of the participants, and the degree of precision and confidence in the results needed or desired. Repeat tasting is a partial alternative to improve statistical reliability.

Typically, a nucleus of 15–20 trained tasters is appropriate. This should provide at least 12 tasters for any tasting. In some ways, it is an advantage that wines vary so such from vintage to vintage. Rough approximations may be all that is necessary. This is not so for most beverages (see Schindler, 1992).

Because tasters are occasionally out of form, continuous monitoring of taster function is advisable. However, designing an appropriate test for this purpose depends on the task desired. Panel members should be encouraged to self-acknowledge when they are not feeling up to par, and excuse themselves from tastings.

For quality-control work in small wineries, the number of tasters may be very low (the winemaker). Nevertheless, it is far better that several people be involved. Daily variation in perception is often too marked to leave critical decisions up to one or two people, no matter how experienced (Fig. 5.19). If at all feasible, only when immediate

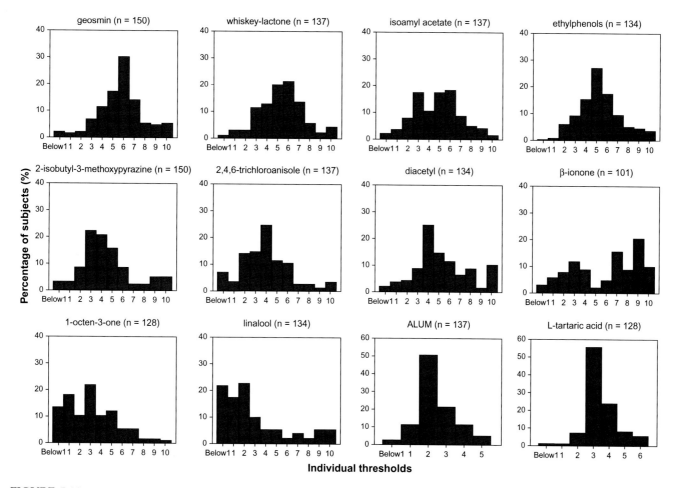

FIGURE 5.19 Frequency distribution of experts' individual detection thresholds for 10 odorants and 2 oral stimuli. The number of subjects tested is indicated in brackets. The dilution levels are indicated on the x-axis. Individual detection threshold labeled *below 1* indicates subjects with the lowest sensitivity. In contrast, depending on the compound, 10, 5, or 6 represent subjects showing the highest sensitivity. Concentration ranges for odorants in µg/l were geosmin (1–1000), whiskey lactone (2.47–2471), isoamyl acetate (1.59–1591), ethylphenols (1.66–166 and 0.19–19.3 for 4-ethyl phenol and 4-ethylguaiacol, respectively), 2-isobutyl-3-methoxypyrazine (0.05–46), 2,4,6-trichloroanisole (0.1–104.4), diacetyl (0.2–198.4), β-ionone (1.91–1911), 1-octen-3-one (0.11–108.5), linalool (0.11–112.8), tartaric acid (0.03–0.8 g/l), alum (3–600 mg/l). *Reprinted from Tempere, S., Cuzange, E., Malik, J., Cougeant, J. C., de Revel, G., Sicard, G., 2011. The training level of experts influences their detection thresholds for key wine compounds. Chem. Percept. 4, 99–115, with kind permission of Springer Science + Business reproduced with permission.*

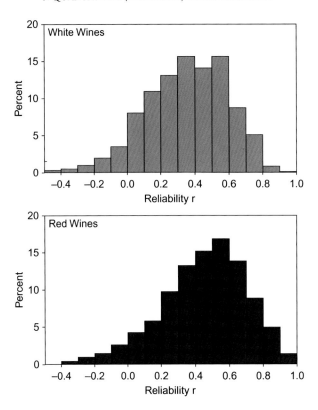

FIGURE 5.20 Distribution of reliability of individual assessors, as measured by Pearson's correlation coefficient, applied to scores given to the same wine on duplicate presentations ($n = 571$). *From Gawel, Goodman, 2008, Aust. J. Grape Wine Res., reproduced with permission from John Wiley and Sons, copyright.*

attention to some problem is required, and there is no time to convene a panel, should the decision be left to single individual. The days of the lone expert still lingers in the wine industry, notably in the guise of "flying winemakers," but it is an anachronism elsewhere in the food industry, and justly so.

With small groups (≤5), valid statistical analysis is unlikely. Thus, tasters should be known to be sufficiently acute for the task and be very consistent. It is also important that assessments be done individually, not by consensus. Consensus too often reflects the views of the most dominant member (Myers and Lamm, 1975).

Quality-control programs should retain samples that represent the range of acceptability desired. These should be continuously available as standards for the panel members. Clearly, the panel needs to be well trained, appropriately selected, and of adequate size (>10) to provide for adequate statistical analysis. In a direct investigation of trained assessors, at least 8, and preferably 10, panelists seemed sufficient (Heymann et al., 2012), or 11 (Silva et al., 2014). Further evidence of sensory idiosyncrasies, indicating the need for a coterie of panelists, is provided in Figs. 5.20 and 5.21. In addition, provision for assessing the relevance of the data obtained may be desirable to ensure that procedural issues and panel performance are adequate. As noted, members showing the diversity of a particular subcategory of consumers may be desirable in wine evaluation, but unacceptable for descriptive sensory analysis.

TASTING DESIGN

How a tasting is designed obviously needs to reflect its purpose. This can vary from descriptive sensory analysis at one end of the spectrum to consumer preference studies and simple ranking at the other. Most of what is discussed in this chapter relates to analytical wine assessment. Consumer preference studies and related tasting situations are primarily dealt with in Chapter 6, Qualitative Wine Assessment.

Information Provided

What information is provided is one of the more important decisions concerning any assessment. By indicating what is expected one can bias the data obtained (Lawless and Clark, 1992), but without direction the desired data

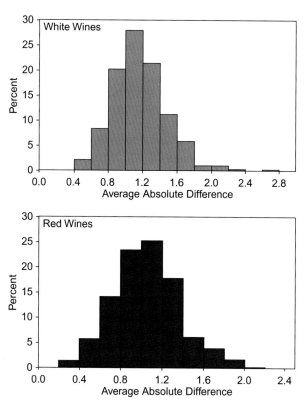

FIGURE 5.21 Distribution of the average absolute difference between scores given to the same wine on duplicate presentations ($n = 571$). *From Gawel, Goodman, 2008, Aust. J. Grape Wine Res., reproduced with permission from John Wiley and Sons, copyright.*

may not be generated. The act of measuring has to be so designed to minimize the likelihood that it may significantly modify what is to be measured (Lawless and Schlegel, 1984). For example, asking the panel to check for wine faults is likely to result in more comments relative to perceived faults than would have been obtained were the issue of faults not raised in the introduction to the tasting. Equally, requesting the panel to sort samples relative to their expression of certain attributes can exaggerate differences in those very attributes, to the exclusion of other, possibly more significant features. Such skewing of data may be desired, but is likely a distortion of reality (assuming there is one "reality"). This is similar to the dilemma researchers have with statistical analysis, steering between acceptance of a false hypothesis (Type I error) and rejection of a valid hypothesis (Type II error).

Where the procedure is new, or its intent different from previous tasting sessions, providing necessary instructions both verbally and in print (at the tasting station) is a judicious move. Tastings should not be an exercise to assess panel members' ability to remember instructions. It is far better to provide adequate details both verbally and in print than to have to repeat the experiment unnecessarily.

When wines are to be assessed critically, their identity usually should to be withheld. Precise knowledge of the wine's varietal, stylistic, or regional origin, price, or vinery name could bias assessment, either positively or negatively. For example, in a study in Bordeaux, enology students were presented a wine for assessment, but permitted to observe what appeared to the empty bottle nearby. When the wine was presumed to be a *vin de table*, term use was skewed toward faults, in contrast to predominantly positive comments when the empty bottle was of a *grand cru classé* (Brochet and Morrot, 1999). In a similar mode, Lange et al. (2002) found that knowledgeable consumers in the Champagne region seldom distinguished between champagne grades in a blind-tasting situation. However, when the bottles (and labels) were in clear evidence, ranking and estimated monetary value coincided with traditional views. Contextual effects on perception have also been directly observed in modified brain activity (McClure et al., 2004). Indicated price (either high or low) has also been demonstrated to have a marked effect on both reported appreciation and the degree of activation of parts of the brain associated with pleasure (Plassmann et al., 2008). The connection between perceived quality and price seems to be a common consumer phenomenon (Ariely, 2010).

Preparing Samples

As usual, where concealing the samples' origins is necessary, prepouring in marked glasses (or carafes) assures anonymity. This also has the advantage that the wines can be checked for appropriateness (e.g., freedom from faults), or decanted to avoid any sediment in the bottle clouding the samples. If black glasses are used, the potential for any biasing influence from the wine's appearance can be minimized if not avoided. To minimize aroma loss or modification, pouring should occur comparatively shortly before tasting and the tops of each glass (or carafe) should be covered (Wollen et al., 2016).

In most tasting situations, this degree of concealment of the wines' origins is likely to be unnecessary. Simply covering the bottles with a paper bag may be sufficient. Although hiding the label, bottle color and neck design remains evident. Both can give clues to potential origin. Residual corroded material on the neck may also provide hints as to wine age. Bottle shape may also suggest wine origin. Having the wine poured by nonparticipants helps limit access to this information.

Sources of Perceptive Error

Without appropriate precaution, sequence errors can invalidate tasting results. Sequence errors distort perception based on the order in which samples are presented. A common example is if all tasters sample the wines in the same order. In such a situation, the first in a series of red wines is often ranked more highly than would be expected by chance. This may result from the removal of tannins by precipitation with saliva proteins, making the first wine appears smoother and more balanced. An analogous sequence error can result from taste adaption, notably sweetness—the first in a series appearing comparatively sweeter. Equally, a wine tasted after a faulty sample will probably be perceived better than it would have had it followed a faultless wine (contrast error). Similar effects can occur whenever markedly different wines are presented in the same sequence. Grouping wines by category is a standard technique to minimize the occurrence of this form of sequence error. An illustration of sequence error is provided in Fig. 5.22, where the most marked differences in perception occurred when wines of lower alcohol contents were sampled before those of higher content. An opposite (convergence) effect can occur when several wines in a tasting are very similar, their differences being minimized in relation to other wines.

The effect of sequence errors may be partially avoided by allowing sufficient time (at least 2 min) between samples, and adequate palate cleansing, but this has not been conclusively demonstrated. In addition, incidental randomization of sampling during repeat tasting may negate sequence errors. However, these procedures may not be practical, or possible. Therefore, presenting wines to each taster in a different sequence is the simplest means of avoiding group-generated sequence errors. One method of achieving a randomized design is through the use of the Latin square (Table 5.13), or its modifications. This method works well for most situations where panelists taste all samples. Where this is not possible, an incomplete block design or other treatments may be required (e.g., Lawless and Heymann, 2010).

Other sources of perceptive errors include expectation and anticipation. Examples of an expectation error could result if wine color unduly influenced the perception of flavor intensity or varietal identification. Anticipation errors can arise if panelists are instructed to check for particular attribute(s), leading to exaggeration of their presence. Stimulus errors are similar, but derived from prior knowledge about the source of the wines. A central tendency error relates to the tendency of panelists, at least initially, to preferentially use midrange values of a scale, thereby minimizing actual differences among the samples. A halo effect occurs where one assessment affects a second, for example preference tests being followed by assessment of the wines' attributes. The comments generated from the assessment may be adjusted to justify their previous preference ranking. Leniency error is more a problem with consumer studies, where participants may provide views based on how they feel they should, or are expected to, respond. For example, if a panel is "honored" with the presentation of a prestigious wine, which might happen to be faulty, the fault is often politely ignored, assaulting the truth. This has occurred at tastings that I attended where the wines in question showed obvious *Brett* or ethyl acetate off-odors.

Even numbers used to label samples can generate subliminal (anchoring) bias (Furnham and Boo, 2011). To minimize this, each wine can be assigned a randomly generated three-or-more-digit code. Such codes may be obtained from several Internet sites, such as http://www.mrs.umn.edu/~sungurea/introstat/public/instruction/ranbox/randomnumbersII.html or http://warms.vba.va.gov/admin20/m20_2/Appc.doc.

Marking glasses is usually done with a marking pen directly on the glass. However, small colored coding labels, such as Avery #579x, are equally as effective, avoiding any potential numerical anchoring bias.

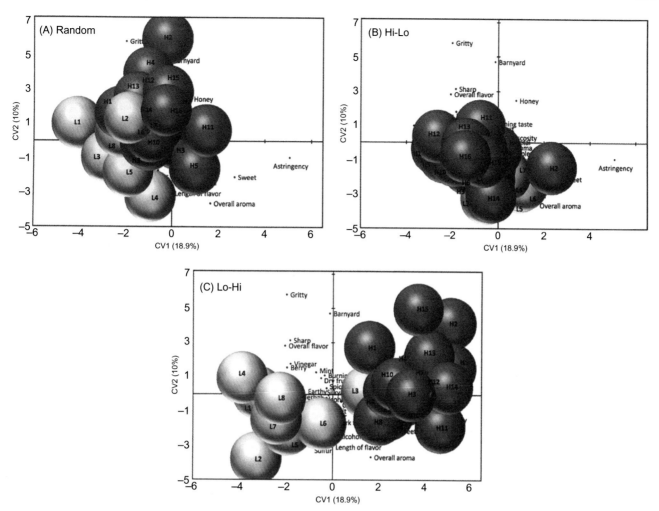

FIGURE 5.22 Canonical variate analysis (CVA) by group of wine interactions for three descriptive analysis groups with different tasting orders: (A) randomized (*Random*); (B) *Hi-Lo*, high-alcohol wines (greater than 14% v/v) before low-alcohol wines (less than 14% v/v) in each session, and (C) *Lo-Hi*, low-alcohol wines before high-alcohol wines in each session, for aroma, taste, and mouth-feel attributes from a formal descriptive analysis of 24 U.S. Cabernet Sauvignon wines. *White circles*, low-alcohol wines (less than 14% v/v, L1- L8); *black circles*, high-alcohol wines (greater than 14% v/v, H1–H16); numbers represent increasing alcohol concentrations. Circles represent 95 percent confidence intervals (n = 3 replicates), where circles that overlap are not significantly different from one another. *Reprinted from King, E.S., Dunn, R.L., Heymann, H., 2013. The influence of alcohol on the sensory perception of red wines. Food Qual. Pref. 28, 235–143, with permission from Elsevier.*

TABLE 5.13 Example of use of a Latin square design for randomizing the sequence order of six wines for six tasters

Taster	Order of testing					
	1st	2nd	3rd	4th	5th	6th
A	1	3	6	4	2	5
B	2	4	1	5	3	6
C	3	5	2	6	4	1
D	4	6	3	1	5	2
E	5	1	4	2	6	3
F	6	2	5	3	1	4

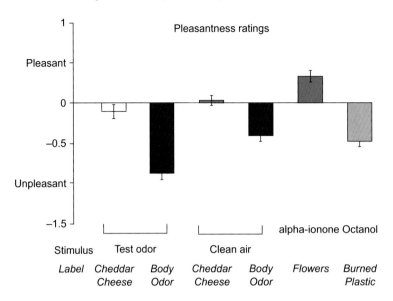

FIGURE 5.23   Subjective pleasantness ratings to labeled odors. The means ± SEM across subjects are shown. The corresponding stimulus and label to each bar are listed in the lower part of the figure. Note that the test odor and clean air were paired in different trials with a label of either "cheddar cheese" or "body odor." *Reproduced from de Araujo, I.E., Rolls, E.T., Velazco, M.I., Margot, C., Cayeux, I., 2005. Cognitive modulation of olfactory processing. Neuron 46, 671–679, with permission from Elsevier.*

Other problems can be avoided (or arise) depending on the information provided. Stating that identical samples could be present might condition tasters not to exaggerate differences. However, the same information may induce tasters to ignore legitimate differences. Thus, as noted, it is essential that much forethought go into the design of any critical tasting.

The potential influence of observing facial reactions is well known. The more difficult the assessment, the more likely tasters may be swayed by suggestion. Correspondingly, cubicles are used to isolate panelists. This also helps prevent verbal communication between tasters. Although mentioning odor-related terms (but not other words) can directly affect sensory perception (González et al., 2006), the selective use of terms during a tasting is likely to be prejudicial. For example, in one psychological experiment, mentioning the terms "cheese" or "body odor," while exposing subjects to isovaleric acid (possessing a sweaty odor) or cheese flavor, elicited markedly different hedonic responses from the participants (Fig. 5.23). Differences were also observed in the areas of the brain activated. These studies suggest intimate connections exist between language, olfactory, and emotional centers in the brain. Expectation is also a significant influencer in flavor perception (Pohl et al., 2003). Panel members should be made aware of these influences, so that they can attempt to guard against their influence.

The standard recommendations of assessing white wines before red wines, dry before sweet, and young before old have much logic, but again have not been experimentally assessed as to their actual relevance. If conditions permit, each subsequent set of wines sampled should be different. This hopefully will sustain panel interest, while minimizing sensory fatigue.

Typically, only similar wines are tasted together, e.g., stylistic, regional, or varietal wines. Such arrangements are necessary if assessment of typicity is intended. Unfortunately, tasting within narrow categories does little to encourage change or improvement. Cross-regional and cross-varietal comparisons are excellent means of identifying where advances could, or should, be made.

Tasting similar wines tends to force panelists to search for differences. This is desirable when differentiation is desired. However, it also tends to exaggerate differences that otherwise go unnoticed. This is termed the range effect (Lawless and Malone, 1986). Thus, comparing equivalent wines alone generates a greater score range than might occur if the wines were sampled with dissimilar wines. This suggests that the ranking of wines is relative and cannot be correlated to the same wines ranked under different circumstances.

Other sources of error, at least when sensory analysis is used to approximate consumer perception, relate to the conditions essential for critical assessment that can modify perception. Part of this concerns the artificiality of the laboratory setting. It is antithetical to how consumers taste wines, where the context of the tasting can be as or more important than the sensory characteristics of the wine. Another aspect of the divide between home and laboratory tasting is wine expectoration. As a consequence, the duration and characteristics of the after-smell are modified (Déléris et al., 2014). However, the degree to which expectoration significantly modifies wine assessment is currently unknown.

Timing

Tastings are commonly held in the late morning, when people are supposedly at their most alert. Where this is inconvenient or impossible, tastings are commonly scheduled in the late afternoon or mid-evening. This usually eliminates problems that might arise following teeth cleaning. Most toothpastes contain flavorants (such as menthol, thymol, methyl salicylate, eucalyptol, or cinnamaldehyde), or surfactants (such as sodium lauryl sulfate). These can disrupt olfactory and gustatory sensations, respectively. Usually an hour is sufficient to avoid these sensory distortions (Allison and Chambers, 2005).

Organizing tastings 2–3 hours after eating also seems to be correlated with more acute sensory skills. Reduced acuity, associated with satiety, partially results from suppression of central nervous system activity (see Rolls, 1995). It may also involve direct suppression of receptor sensitivity. For example, the secretion of leptin that elicits a feeling of satiety after eating suppresses the recognition thresholds of sweet tastants (Nakamura et al., 2008). Conversely, the "hunger" hormone, ghrelin, secreted by the intestinal tract, enhances olfactory responsiveness (Loch et al, 2015).

WINE TERMINOLOGY

The popular language of wine is often colorful, poetic, and evocative, seemingly appropriate for sensory inputs that traverse the amygdala before reaching the cognitive centers of the brain (Figs. 3.9, 3.11, 3.38). Thus, although wine language can be rich and redolent, wine memories often encode only an impressionistic glimpse of a wine's most distinctive aspects. Without recording these perceptions in words, they soon morph into one another. As noted in Fig. 5.23, even the terms used can distort perception, as well as potentially affect sensory processing. The degree to which this occurs probably relates to personal experience and the emotional loading associated with images the terms or events evoke. For example, initially indistinguishable odor enantiomers (structurally mirror images of one another) become distinguishable when associated with aversive (threatening) experiences (Li et al., 2008) as also can isotopic (deuterated) differences (Gane et al., 2013).

Regrettably, the terms commonly employed in popular wine writing poorly articulate the sensory perceptions they are supposed to describe. Odors, unlike tastes, are usually described relationally, in terms of concrete objects or experiences, rather than symbolically. Thus, the terms usually involve the addition of suffixes, such as -*like* (e.g., rose-like), -*y* (e.g., corky), -*ic* (e.g., metallic), -*ful* (e.g., flavorful), and -*ous* (e.g., harmonious). Verb participles may also be used in describing wine qualities, e.g., "balanced" or "refreshing." Occasionally allusions to music may be conjured up, to what value one can only wonder (Spence and Wang, 2015). The use of chemical names would be precise, but are only of relevance to those with a chemic background. In addition, most naturally occurring odors are chemically complex, making chemical notation either impossible or nonsensically complex. Finally, humans are ill adept at distinguishing the individual components of a mixture (e.g., wine). Thus, there seems little choice but to use comparative representations. The terms generated may assist in communication, but may be more illusory than real. For example, a wine that may be thought to possess a redcurrant-like odor may possess compounds bearing little relationship to those found in redcurrants. It is somewhat analogous to the yellow appearance of an object that may possess no yellow pigments.

The difficulties of assessing the validity of descriptive terms are discussed by Wise et al. (2000). They point out significant liabilities and limitations to their use, notably that they are often highly context-, experience-, and culture-sensitive, and therefore tend to be individualistic, not wine specific. Requiring people to make precise statements about wine may actually defeat the purpose. Oronasal responses are largely implicit, encoded subconsciously.

Nonetheless, terms can attain precise meaning and significance when developed, and used repeatedly by a group. Even with consistent use, though, descriptors rarely give adequate expression to individual differences in perception. Is one person's iris-, rose-, or tulip-like the same as another's? Anyone familiar with these horticultural plants will know that different cultivars often have clearly distinguishable aromas. The act of verbal description may be so unfamiliar as to invalidate its use when dealing with consumer preferences (Köster, 2003). Worst of all, most terms do not permit precise designation of qualitative differences. Without confidence in qualitative measurement, the ability to understand the origins of any differences identified is severely limited. For example, when tasters note a blackcurrant or violet aroma in Cabernet Sauvignon wines, are people referring to the specific aroma of the named object, or is it simply training or tradition that has taught them to consider these descriptors as appropriate?

Although of dubious value, many people seem to desire descriptors to assist in recognizing cultivar wines. In that regard, examples of frequently used descriptors for the aroma of several cultivar wines are noted in Tables 7.2 and 7.3. These, and the terms used in odor reminders (aroma wheels and charts) are only potential resemblances, and may primarily reflect more the fragrant experiences and perceptions of their originators than the wines. Once published, such lists seem to take on a life of their own, almost developing predestined authority, in spite of few if any having been assessed for their appropriateness or usefulness. Possibly it is a reflection of "something is better than nothing," regardless of how inappropriate the expressions may be.

Despite the credence given aroma aids, aromatic compounds cannot be neatly categorized relative to their perceptive qualities. Some esters do smell like fruit, others like fats; some terpenes resemble the odors of flowers, but not all; some thiols have fruit-like aromas, others definitely not; and moldy odors can belong to a wide range of chemical groups, as can those of flowers. The arrangement in aroma wheels or charts is a crutch, albeit maybe a useful one, which satisfies the need to organize. The chemistry of aromatics has yet to be directly correlated with perceptive quality, and likely never will. Perceptive quality is a cerebral construct, associated with experience; change the experience and the descriptive nature of the odor quality may change. There is no equivalent direct correlation between primary color perception and parts of the electromagnetic spectrum, or sound with audio wavelength. Odor is seemingly infinitely receptive to new qualitative associations.

Sensory researchers need to consider the option of limiting term use to a selected range, at least until there is a consensus among the panel members. Even then, there is dubious merit in restricting panelists to a select list of terms. With a few exceptions (e.g., pepper in Shiraz and litchi in Gewürztraminer), the presence of black cherry, raspberry, iris, violet, or truffle odors have not been demonstrated definitively to occur in wine. Even the association of particular descriptors with specific chemicals is fraught with problems. Diacetyl may be described as being buttery, but not to others (I being one of them). In addition, the odor quality of many compounds alters as their concentration changes, or in the presence of other compounds. These potential problems can be minimized if sufficient time is taken for panel members to come to agreement on term use. For example, in spite of some members not considering a term appropriate, buttery can be agreed upon as the descriptor for whatever perception diacetyl engenders, thereby avoiding the specter of sensory "dumping." Lawless (1999) presents additional views on these issues.

In descriptive sensory analysis, distinction among similar samples is the usual intent, not a complete description of the wine's sensory attributes. In this case, many of the problems just noted do not arise, or can be sidestepped. Terms selected usually refer only to attributes considered discriminatory, and do not overlap. They are typically represented by physical or chemical reference standards. They are used in lexicon training and available during assessment. In addition, panelists are selected for their ability to accurately and consistently use the terms. That laboratory tasting conditions do not reflect consumer experience is not a concern. The intent is discrimination, with no inherent aim to reflect preference or quality.

In normal usage, flavor impressions are funneled through the language of odor. This often involves categories of conceptually related objects, such as fruity, flowery, herbal (Fig. 5.24). However, even experienced tasters differ in their interpretation of these terms (Ishii et al., 1997). Descriptions may also include terms for which precise representations do not exist, for example allusions to size or shape (e.g., big, round), power (e.g., robust, weak), or weight (e.g., heavy, light, watery). Even broad emotional reactions, such as pleasant, unctuous, vivacious, scintillating can be envisioned. Such terms may have value in enunciating emotional or preference responses to wine, but are typically so ill defined and personal as to be of no value in critical wine assessment.

Training and experience for wine panels is associated with an expanded lexicon, employing terms involving samples representing specific, reproducible sensations. In contrast, metaphoric illusions (e.g., heavenly, voluptuous, feminine, nervous) are assiduously avoided as being too nebulous or indefinable. Nonetheless, the terms used often reflect the cultural, geographic and specific upbringing of their users; truffle is commonly identified in wines from the Rhône or northern Italy (where truffles grow indigenously), whereas violet is detected in Bordeaux wines by critics in that region. It is probably exposure to the terms used by wine writers that explains the preponderant use of fruit and flower descriptors by aficionados versus teetotalers in describing wines. Those with little experience with wine tend to use terms such as pungent, sharp, vinegary, or alcoholic (Fig. 5.25).

Terms used in the popular press may also employ expressions such as jammy, harmonious, and complex when referring to highly regarded wines, whereas unbalanced, distasteful, astringent, or simple are used for wines that are disliked. Only rarely are specific wine constituents sufficiently distinct to be individually recognizable, usually in a negative sense (e.g., mercaptans, methoxypyrazines, hydrogen sulfide).

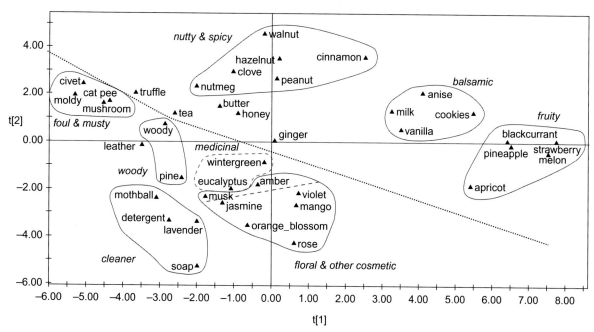

FIGURE 5.24   First two principal components (t[2] vs t[1]) of common, everyday odors, grouped related to perceived relatedness. The *x*-axis (t[1]) related compounds on a scale of relative pleasantness. The dotted line separates odors from edible and inedible related substances. *From Zarzo, 2008, J. Sens. Stud., reproduced with permission of John Wiley and Sons, copyright.*

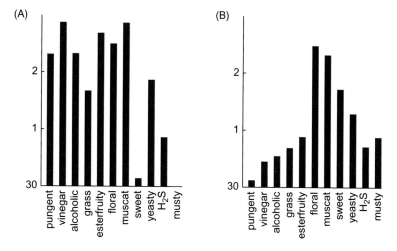

FIGURE 5.25   The relative choice of terms used to describe wine by wine abstainers (A) and wine consumers (B). *From Duerr, 1984, reproduced by permission.*

Several descriptors (cherry, peach, melon, apple) reflect the color of the wine (in this instance, white wines). In contrast, red wine flavors are often associated with red or dark-colored objects (Morrot et al., 2001). These connections appear to mirror the integration and interpretation of visual, olfactory, and taste sensations in the orbitofrontal cortex (Fig. 3.11), and the prejudicial influence of visual sensations.

Many of the commonly used terms, especially when ranking wines, denote holistic, prototypic, or integrated aspects of perception, as well as their intensity or development (Lehrer, 1975, 1983; Brochet and Dubourdieu, 2001). Terms such as balance, dynamism, complexity, development, body, finish, and memorability integrate multiple, often independent sensations. Such abstract terms reflect perceived attributes, although they are impossible to define precisely or represent with physicochemical examples. Thus, evaluation of their appropriate application by tasters is fraught with difficulty.

The difference between how terms are used, based on experience, appears to partially reflect which hemisphere in the brain is primarily active. For example, holistic expressions appear to arise from the selective activation of the right hemisphere (Herz et al., 1999; Savic and Gulyas, 2000). This region typically deals with integrated aspects of expression. In contrast, sommeliers showed more rapid and focused activity in various areas of the brain implicated in higher-level cognitive processes and the left hemisphere than control subjects (Castriota-Scanderbeg et al., 2005; Pazart et al., 2014). Verbalization of knowledge, and most analytic processes, tend to be concentrated in the left hemisphere, at least in right-handed people (Deppe et al., 2000; Knecht et al., 2000). However, odor memory appears to require both hemispheres for optimal recognition (Dade et al., 2002).

Poor verbalization of sensory perception probably reflects a lack of training. Current research on brain plasticity is not inconsistent with this view. Either way, only a small portion of the brain is set aside for processing olfactory information in comparison with vision; this may also partially explain why humans have such a limited odor lexicon. Without a detailed and unique odor lexicon, people are forced to describe their vinous impressions imprecisely in terms used for familiar objects, experiences, or the emotional responses they engender. The specific terms used probably also reflect genetic individuality.

Although our limited ability to enunciate odors precisely may be discouraging, it should not be unexpected. Humans are very adept at recognizing faces, but would be hard pressed to describe them verbally. We equally learn the visual patterns that distinguish groups of people. Thus, what is more valuable than naming odors is to harness our inherent ability to identify patterns, and apply it to distinguishing different wines. But, even more valuable to the consumer is concentrating on the vinous characteristics they enjoy, and hopefully understanding why. From an evolutionary viewpoint, identifying odors and tastes that gave pleasure, and were not toxic, was far more important than being able to verbally detail their specific flavor constituents.

Despite the difficulty of expressing odors in abstract terms, and the on-the-tip-of-the-nose phenomenon (Lawless and Engen, 1977), words can have an amazing influence on perception. Thus, language can play a variety of roles in wine tasting other than description. Wine columnists often use language as much to entertain as to inform. In a home setting, comments typically express subjective holistic feelings, or are used to seek reassurance. Under other conditions, comments may be used to complement the host or donator, or simply function to lubricate social exchange. Regrettably, "winespeak" may also be used to impress or cajole. Only with long and concerted training is it possible to approach the semblance of precision in describing a wine's sensory attributes.

For critical analysis, communication needs to be as precise and accurate as possible, ideally expressing the impressions of the taster in a clear, unambiguous manner. Terms chosen should be readily distinguishable from one another (minimum overlap), be represented by stable chemical standards, facilitate wine differentiation, and preferably permit subsequent wine recognition. Meilgaard et al. (1979) were the first to develop a standardized terminology (for beer). A related system was subsequently generated for wine (Noble et al., 1987). Regrettably, neither has been studied relative to their effectiveness, appropriateness, or term overlap. Subsequently, lexicons for sparkling wine (Noble and Howe, 1990; Duteurtre, 2010), brandy (Jolly and Hattingh, 2001), and other beverages have been developed. Gawel et al. (2000) proposed a set of wine-specific mouth-feel terms. The latter seems to have even more problems than those for olfaction (King et al., 2003), partially due to the difficulty in clearly defining gradations of particular sensations, or because their perception may be expressed in several semidistinct, overlapping modalities. Despite these difficulties, prolonged training is reported to generate panelists who can use the terms consistently (Gawel et al., 2001).

Most collections of terms have been presented in the form of wheels, with concentric tiers intended to provide greater precision for each category and subcategory. In contrast, the versions provided in Figs. 1.3 and 1.4 are prepared in chart form. This was intended to avoid having to rotate the sheet (or one's head) to view various portions of the wheel. As noted above, none have any scientifically established relevance. They are presented only to provide a framework for training in the use of selected terms (ideally possessing formulae for preparing samples representing each term), and/or to encourage consumers to focus on the most complex and fascinating aspect of wine.

As one would suspect (or at least hope), training does improve term use. For familiar odors, several attempts may be needed to achieve recognition in the absence of visual clues (Cain, 1979). With new odors, concentration on verbal memorization (rather than their sensory attributes) seems initially to retard odor memory development (Melcher and Schooler, 1996; Köster, 2005). In addition, recognition and familiarity with complex, naturally occurring odors (mixtures of several aromatic compounds) can complicate recognition of their individual components (Case et al., 2004). Such long-established (encoded) associations are often stable and resistant to recoding (Lawless and Engen, 1977; Stevenson et al., 2003). They have usually formed incidentally (without intentional learning), and are associated with particular experiences. This may explain why some trainees fail to establish the desired skills and need to be eliminated.

Although selection for consistency promotes the likelihood of more significant statistical results, they may be consumer irrelevant, due to data reflecting only the views of a select subset of individuals (the panel). That wine consumers fall into different groups is self-evident, but only recently have the factors affecting how this influences wine purchasing habits (Goodman et al., 2006), and by whom (Hughson et al., 2004), attracted attention. As previously noted, which approach to panel selection (consistency or representativeness) is preferable depends on the intent of the tasting.

Although the development of a common lexicon is important for technical tastings, odor training also tends to selectively direct attention to the attributes denoted by the terms (Deliza and MacFie, 1996). Selective focusing is essential when searching for faults, or characteristics distinctive of varietal or regional wines, but may induce exaggeration of their presence. In addition, searching for particular descriptors may result in equally important, but unnamed fragrances and flavors being ignored or misrepresented. Whether selective attention has the same effect during wine tastings as in the famous "gorilla" experiment in psychology is unknown, but is potentially relevant. When viewers were requested to count some fast-moving activity in the field of view, as a person dressed as a gorilla costume sauntering through the scene, most people failed to notice the gorilla (see https://www.youtube.com/watch?v=IGQmdoK_ZfY).

Although experts use descriptive language more precisely, it may not significantly improve their ability to identify wine origin (Lehrer, 1983). Various studies have investigated the differences between expert and novice tasters (Lawless, 1984; Solomon, 1990; Bende and Nordin, 1997; Hughson and Boakes, 2002; Parr et al., 2002, 2004; Ballester et al., 2008; Lehrer, 2009), or trained and untrained (but experienced) tasters (Gawel, 1997). The issue is complicated by having no objective measure of the differences among the wines assessed, or for what constitutes an expert, there often being considerable difference in experience, training, or both in the participants from study to study. In addition, there is typically wide variation in the sensitivity of experts to different wine constituents (Tempere et al., 2011), unless the panel members are selected specifically for uniformity. Nonetheless, ability to differentiate specific attributes appears to increase with both experience and training, training enhancing sensory and verbal skills, and the ability to use scalar methods. This often occurs, however, without any accompanying overall improvement in general sensory acuity.

In research investigating wine language, participants are often divided into novices and experts. While novices are often consumers with little knowledge of, or experience with wine, experts generally fall into two relatively distinct categories. One category involves those having expensive, experience-based expertise, while the other includes those having been specifically trained in sensory analysis (training-based expertise). This is similar to the distinction proposed by the ISO (1992). Experience-based experts include sommeliers, the majority of wine writers, and many winemakers. Training-based experts include members of panels used in descriptive sensory analysis and the majority of university-trained winemakers. Experience-based experts tend to have extensive practice in assessing the relative quality of varietal, regional, and stylistic wine types. This form of training tends to produce individuals who focus on the features that distinguish wines, aspects that may not necessarily be obvious to the novice. This seems equivalent to how botanists view trees, in contrast to how most individuals distinguish broad categories (e.g., conifers, palms, broad-leaved trees). Each taster tends to use a comparatively limited descriptive lexicon, but an extensive range of emotive expressions that are idiosyncratic. There is little commonality in the terminology used by such experts. In contrast, training-based experts tend to use a standard (panel-based) lexicon. These usually refer to specific reference samples with which they have been trained. Emotive and anthropomorphic language is eschewed as an anathema. Training-based experts may or may not have extensive experience in recognizing the features that may characterize varietal, regional, or stylistically distinctive wines.

Each type of expert has their rightful place in wine tasting, a phenomenon that includes a host of different situations, each with its own particular modus operandi. Thus, no form of expertise is inherently preferable. Their respective values depend on what is desired. Nonetheless, when one is looking at studies on wine language usage, it is important to discern the category of taster involved. Without acknowledging this feature, comparing studies could easily lead to the impression that their conclusions are as often in conflict as in agreement.

Simple taste and touch sensations are easier to differentiate than odors. This partially results from their limited number of modalities, and each possessing relatively distinct receptors. Gustatory compounds may interact synergistically or suppressively, but always retain their individual sensory qualities. In contrast, olfactory compounds often interact in complex ways, generating an almost infinite number of distinct sensory qualities. This should be expected since olfactory receptors often respond differentially to a range of compounds, and any individual compound may variously activate several different receptors. Only a comparatively few olfactory compounds generate qualities that can be grouped together and accepted as similar, independent of cultural heritage. These are largely compounds that trigger nasal trigeminal nerve endings, being categorized abstractly, e.g., pungent, acrid, or putrid.

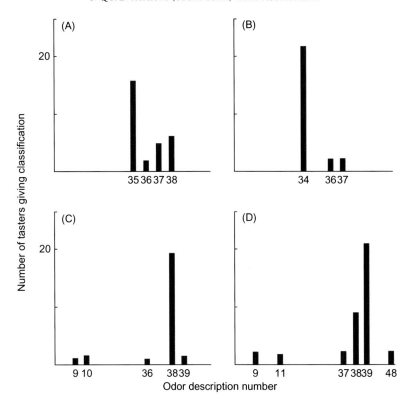

FIGURE 5.26   Percent allocation of four fragrances, (A) banana, (B) strawberry, (C) peardrop, and (D) apple, to particular descriptors: 9, fusel alcohol-like; 10, ethyl acetate-like; 11, acetaldehyde-like; 34, strawberry-like; 35, banana-like; 36, pineapple-like; 37, pear-like; 38, peardrop-like; 39, apple-like; 48, spicy. *From Williams, A.A., 1975. The development of a vocabulary and profile assessment method for evaluating the flavour contribution of cider and perry aroma constituents. J. Sci. Food Agric. **26**, 567–582. Copyright Society of Chemical Industry. Reproduced with permission. Permission is granted by John Wiley & Sons Ltd on behalf of the SCI.*

Although a universal, precise language for wine would be desirable, no such system has yet evolved, nor is it likely to. Human variability in experience and olfactory acuity is too great. As noted, people even have difficulty correctly naming common household odors, notably in the absence of visual clues with which they are usually associated. Figs. 5.19 and 5.26 illustrate the extent of human variability in sensory acuity. It may be that the value of naming wine fragrances develops so late in life that it is easier to relate them to olfactory experiences we already recognize and can name. In addition, the ability to identify compounds is inconsistent, varying from day to day (Ishii et al., 1997). Wine-judging experience can even complicate the use of preset categories, due to the difficulty of learning new terms for odors already fixed in memory. Thus, the categories and terms found in aroma compendia may reflect only an illusionary reality. Even experts can have considerable difficulty in recognizing wines from verbal descriptions, generated either by others or themselves (Lawless, 1984; Duerr, 1988). Thus, that wine terminology is often idiosyncratic, and reflects familial, cultural, and geographic contexts, is almost inevitable. In addition, the chemical diversity in wine aromas, and the lack of childhood odor training, makes it almost unavoidable that the terms commonly used will be individualistic and largely reflect personal experience.

WINE EVALUATION

Score Forms

The diversity of wine tasting events is reflected in their diversity of score forms. They span the spectrum from purely academic investigations, such as sensory psychophysiology, to assessing the consistency of a proprietary blend, to recommending the "best buy" in some category.

Despite their diversity, tasting forms can be grouped roughly into two categories: synthetic and analytic. In synthetic versions, wines are assessed holistically and/or hedonically, and the wines' attributes assessed collectively

Ranking: Name of wine

First:
Second: _____
Third: _____
Fourth: _____
Fifth: _____
Sixth: _____

FIGURE 5.27   Hedonic score card (hierarchical ranking).

on qualitative scales, such as balance, complexity, development, and duration. Alternatively, or in addition, wines may be assessed relative to their expression of some established paradigm, such as varietal, regional, or stylistic characteristics, or a combination thereof. In analytic versions, individual visual, gustatory, mouth-feel, and olfactory attributes tend to be assessed separately. These are usually rated as to their relative intensities and qualities. As noted earlier, accurately assessing individual sensory attributes can be fraught with difficulty, notably due to integration of most sensory qualities. Training minimizes such influences, but is likely not to eliminate them.

For preference rating, sheets recording ordinal divisions may be sufficient (Fig. 5.27). For this function, they may be as effective as detailed assessments (Lawless et al., 1997).

In contrast, most forms used to assess wine rank individual attributes (Fig. 5.28), or provide ample space for detailed descriptions. The categories may focus on the important traits that give particular wines their distinctive characteristics (Fig. 5.29), or may emphasize integrated impressions (Fig. 5.30). Detailed score sheets are particularly useful in identifying the specific strengths and weaknesses of each sample. Most forms are, however, deficient in adequately rating intensity characteristics or gauging the negative influences of faults. They also require the taster to give a quantitative estimate to the importance of individual and/or integrated groupings of visual, gustatory, and olfactive sensations. Furthermore, they assume that ranking wines can be fairly assessed by a sum of their component parts. A place for adding points for exceptionally qualities in one or more categories is seldom provided, except where "fudge" points are permitted to adjust the score to a desired value. Statistical analysis is also complicated by the variable number of marks permitted for each category, and assigned points are restricted to whole numbers. This is particularly limiting when the score range for particular attributes is limited (two or less). There are also considerable difficulties in consistency of use. Preparing adequately representative standards for "good" versus "poor" indications that may appear on a form is essentially impossible. Panelists may have differing concepts of what these categories mean, and use personal criteria to integrate and quantify complex qualitative judgments. Use of such sheets effectively and consistently, if possible, presumably requires considerable training and extensive practice. Without this, the data obtained are suspect. They are certainly not of a quality on which specific wine production or purchase decisions should be made.

To offset these limitations, tasting forms designed specifically for descriptive sensory analysis have been created (e.g., Figs. 5.49 and 5.50). They permit assessment of the intensity of individual attributes. The features chosen for the sheet usually reflect those that most effectively permit discrimination among the types of wines being assessed. By avoiding the integration of sensory attributes, the source of the quality attributes of each sample is highlighted. This is particularly valuable if decisions are to be made on future vineyard or winery actions. Limitations to the use of descriptive sensory analysis (see below) include the time and expense involved in both panel training as well as data collection and analysis; issues such as data dumping and halo effects; and limitation of assessment to attributes for which adequate reference standards can be prepared. Another limitation is absence of any means of recording changes in the intensity of attributes over the course of the tasting. This issue is currently being addressed with techniques such as temporal dominance of sensations (TDS); see below.

Few tasting forms have been studied to determine how quickly they come to be used effectively. Thus, reliance on the data derived partially depends on how accurately and consistently tasters use the forms. Generally, the more detailed the form, the longer it takes for the form to be used consistently (Ough and Winton, 1976). If the score sheet is too complex, features may be disregarded, and rankings revert to preexisting quality concepts (Amerine and Roessler, 1983). Conversely, insufficient choice may result in dumping, or the use of unrelated terms to register important perceptions not represented on the sheet (Lawless and Clark, 1992; Clark and Lawless, 1994). This can be avoided only by selectively choosing terms that adequately express the important impressions detected by the tasters (Schifferstein, 1996).

Possibly the most widely used scoring system is that based on the University of California (UC) Davis score card (a modification is illustrated in Fig. 5.28). It was developed as a tool to identify wine production defects. Thus, it is

Wine: _____   Date: _____

Feature	Description	Score
Appearance and color		
0	POOR - Dull or slightly off-color	
1	GOOD - Bright with characteristic color	
2	SUPERIOR - Brilliant with characteristic color	
Aroma and bouquet		
0	FAULTY - Clear expression of an off-odor	
1	OFF CHARACTER - Marginal expression of an off-odor	
2	ACCEPTABLE - No characteristic varietal-regional-stylistic fragrance or aged bouquet	
3	PLEASANT- Mild varietal-regional-stylistic fragrance or aged bouquet	
4	GOOD - Standard presence of a varietal-regional-stylistic fragrance or aged bouquet	
5	SUPERIOR - Varietal-regional-stylistic fragrance or aged bouquet distinct and complex	
6	EXCEPTIONAL - Varietal-regional-stylistic fragrance or aged bouquet rich, complex, refined	
Acidity		
0	POOR - Acidity either too high (sharp) or too low (flat)	
1	GOOD - Acidity appropriate for the wine style	
Balance		
0	POOR - Acid/sweetness ratio inharmonious; excessively bitter and astringent	
1	GOOD - Acid/sweetness ratio adequate; moderate bitterness and astringency	
2	EXCEPTIONAL- Acid/sweetness balance invigorating; smooth mouth-feel	
Body		
0	POOR - Watery or excessively alcoholic	
1	GOOD - Typical feeling of weight (substance) in mouth	
Flavor		
0	FAULTY - Off-tastes or off-odors so marked as to make the wine distinctively unpleasant	
1	POOR - Absence of varietal, regional, or stylistic flavor characteristics in the mouth	
2	GOOD - Presence of typical varietal, reginal, or stylistic flavor characterstics	
3	EXCEPTIONAL - Superior expression of varietal, regional, or stylistic flavor characteristics	
Finish		
0	POOR - Little lingering flavor in the mouth, excessive astringency and bitterness	
1	GOOD - Moderate lingering flavor in the mouth, pleasant aftertaste	
2	EXCEPTIONAL - Prolonged flavor in the mouth (>10 to 15 s), delicate and refined aftertaste	
Overall quality		
0	UNACCEPTABLE - Distinctly off-character	
1	GOOD - Acceptable representation of traditional aspects of the wine type	
2	SUPERIOR - Clearly better than the majority of the wines of the type	
3	EXCEPTIONAL - So nearly perfect in all sensory qualities as to be memorable	

FIGURE 5.28   General score sheet based on the Davis 20-point model.

Wine: _____   Date: _____

Feature	Description	Score
Appearance and color		
0	POOR - Dull or slightly off-color	
1	GOOD - Bright with characteristic color	
2	SUPERIOR - Brilliant with rich color	
Effervescence		
0	POOR - Few, large bubbles in loose chains, short duration	
1	GOOD - Many mid-sized bubbles in long chains, long duration, no *mousse**	
2	SUPERIOR - Many fine bubbles in long continuous chains, long duration, *mousse* present	
3	EXCELLENT - Multiple long chains of fine, compact bubbles, *mousse* traits fully developed	
Aroma and bouquet		
0	FAULTY - Clear expression of an off-odor	
1	OFF-CHARACTER - Marginal expression of an off-odor	
2	STANDARD - Mild varietal fragrances and process bouquet	
3	SUPERIOR - Subtle varietal fragrances with complex, toasty, process bouquet	
4	EXCELLENT - Complex, subtly rich aromas and refined toasty bouquet, long duration	
Acidity		
0	POOR - Acidity either too high (sharp) or too low (flat)	
1	GOOD - Acidity fresh and invigorating	
Balance		
0	POOR - Watery, acid/sweetness inharmonious, overly bitter, metallic tasting	
1	GOOD - Standard Acid/sweetness balance, smooth mouth-feel, no metallic sensation	
2	SUPERIOR - Fresh dynamic acid/sweetness balance, rich mouth-feel, harmonious	
Flavor		
0	FAULTY - Off-tastes or off-odors so marked as to make the wine distinctively unpleasant	
1	POOR - Absence of traditional flavor characteristics, soapy effervescence	
2	GOOD - Presence of traditional subtle flavors, prickling effervescence sensation	
3	EXCEPTIONAL - Rich traditional flavors and vibrant mouth-feel from the effervescence	
Finish		
0	POOR - Little lingering flavor in the mouth, astringency and bitterness	
1	GOOD - Moderate lingering flavor in the mouth, pleasant aftertaste	
2	EXCEPTIONAL - Prolonged flavor in the mouth (>10 to 15 s), delicate and refined aftertaste	
Overall quality		
0	UNACCEPTABLE - Distinctly off-character	
1	GOOD - Standard representation of traditional aspects of type	
2	SUPERIOR - Clearly better than the majority of sparkling wines	
3	EXCEPTIONAL - to read Nearly perfect in all sensory perceptions, truly remarkable	

* collection of fine bubbles on the wine's surface in the center of the glass and as a ring around the outer edges of the glass
† attributes derived from treatments given after fermentation is complete.

FIGURE 5.29 Modified Davis-like score sheet designed to be appropriate for sparkling wines.

OCCASION

| commission n° | sample n° | vintage | name of wine | | presentation category |

| date | time | | test |

		EXCELLENT	VERY GOOD	GOOD	FAIR	UNSATISFACTORY	POOR	NEGATIVE	NON CORRESPONDENCE	EXCESS	LACK	IMBALANCE	DEFECT	NATURE OF DEFECTS
SIGHT	LIMPIDITY	6	5	4	3	2	1	0	■	■		■		
	COLOUR HUE	6	5	4	3	2	1	0		■	■	■		biological ☐
	COLOUR INTENSITY	6	5	4	3	2	1	0				■		
BOUQUET	GENUINENESS	6	5	4	3	2	1	0	■	■		■		
	INTENSITY	8	7	6	5	4	2	0		■		■		chemical physical ☐
	REFINEMENT	8	7	6	5	4	2	0		■		■		
	HARMONY	8	7	6	5	4	2	0	■	■	■			
TASTE FLAVOUR	GENUINENESS	6	5	4	3	2	1	0	■	■		■		
	INTENSITY	8	7	6	5	4	2	0		■		■		accidental ☐
	BODY	8	7	6	5	4	2	0				■		
	HARMONY	8	7	6	5	4	2	0	■	■	■			
	PERSISTENCE	8	7	6	5	4	2	0	■	■		■		congenital ☐
	AFTER TASTE	6	5	4	3	2	1	0		■	■			
	OVERALL JUDGMENT	8	7	6	5	4	2	0		■	■			
partial TOTALS	tens / units							TOTAL						

remarks

member/s of committee   signature/s

FIGURE 5.30 Sensorial analysis tasting sheet for wine-judging competitions. *From Anonymous., 1994. OIV standard for international wine competitions. Bull. O.I.V. 67, 558–597 (Anonymous, 1994), reproduced by permission.*

not fully applicable to wines of equal or high quality (Winiarski et al., 1996). In addition, the card may rate aspects that are inappropriate to certain wines (e.g., astringency in white wines), or lack features central to a particular style (e.g., effervescence in sparkling wines). Evaluation forms designed for sparkling wines incorporate a special section on effervescence attributes (Fig. 5.29; Anonymous, 1994). Specific forms may be designed for particular varietal or regional wines, or when wines are rated on integrated quality attributes (Figs. 5.30 and 5.31). An alternative to numerical ranking is letter-grading, similar to that often used in universities for student grades (Fig. 5.32). Whatever the choice, it should be as simple as possible, while consistent with adequate precision to achieve the goal(s) for which the assessment was designed.

The total point range used should be no greater than that which can be used effectively and consistently. In decimal systems, permitting half-points increases breadth in the midrange (central tendency); most tasters avoid the extremes of any scale (Ough and Baker, 1961). Ends of a range tend to be used only when fixed to specific quality designations.

Assuming that wines show a normal distribution of quality, scores should show a normal distribution. While possible, scores often tend to be skewed to the right (Fig. 5.33), reflecting the infrequent appearance of faulty wines. When tasters make more use of the lower end of the range, scoring distribution tends to be abnormal (Fig. 5.34). Tasters showing atypical distributions using the Davis 20-point card demonstrated standard distributions when using fixed-point scales (Ough and Winton, 1976).

The relative weightings given the appearance, odor, taste/flavor, and typicity in several professionally generated score sheets (Table 5.14) show considerable uniformity in their views of significance for these attributes. However, this similarity may be fortuitous, being based on similar views of the originators, not research into the relative importance of these attributes to wine professionals (or consumers). It is also a moot point whether quality can be quantified adequately by summing the values given to particular attributes. In addition, experienced judges may not use any of these sheets as intended, first assigning an overall numerical score, then going back to assign marks to individual categories to reach what subjectively seems to be an appropriate total.

When time for wine assessment is limited, as in many competitions, only a holistic ranking is requested, based on the wine's immediate impression, sidestepping any specific weighting relative to specific attributes. This avoids

Wine: _____ Date: _____

Feature	Description	Score
Appearance and color		
0	POOR - Dull or slightly off-color	
1	GOOD - Bright with characteristic color	
2	SUPERIOR - Brilliant with characteristic color	
Aroma and bouquet		
0	FAULTY - Clear expression of an off-odor	
1	OFF CHARACTER - Marginal expression of an off-odor	
2	ACCEPTABLE - Absence of characteristic varietal-regional stylistic fragrance or bouquet	
3	GOOD - Mild to standard varietal-regional-stylistic fragrance or aged bouquet	
4	SUPERIOR - Varietal-regional-stylistic fragrance or aged bouquet distinct and complex	
5	EXCEPTIONAL - Rich, complex traditional fragrance or refined lingering aged bouquet	
Taste and flavor		
0	FAULTY - Off-tastes or off-odors so marked as to make the wine distinctively unpleasant	
1	POOR - Absence of varietal, regional or stylistic taste and flavor characteristics	
2	GOOD - Presence of distinctive varietal, regional, or stylistic taste and flavor characteristics	
3	EXCEPTIONAL - Superior varietal, regional or stylistic taste and flavor characteristics	
Balance		
0	POOR - Acid/sweetness ratio inharmonious; excessively bitter and astringent	
1	GOOD - Acid/sweetness ratio adequate; moderate bitterness and astringency	
2	EXCEPTIONAL - Acid/sweetness balance invigorating; smooth mouth feel	
Development/duration		
0	POOR - Fragrance simple, does not develop, of short duration	
1	STANDARD - Fragrance typical, develops in complexity, does not fade during tasting	
2	SUPERIOR - Fragrance improves in intensity and/or character, lasts throughout tasting	
3	EXCEPTIONAL - Rich fragrance, improves in intensity and character, long lasting	
Finish		
0	POOR - Little lingering flavor in the mouth, excessive astringency and bitterness	
1	GOOD - Moderate lingering flavor in the mouth, fresh aftertaste	
2	EXCEPTIONAL - Prolonged flavor in the mouth (>10 to 15 s), subtle refined after-sensations	
Overall quality		
0	UNACCEPTABLE - Distinctly off-character	
1	GOOD - Acceptable representation of traditional aspects of the type	
2	SUPERIOR - Clearly better than the majority of the wine type	
3	EXCEPTIONAL - Memorable experience	

FIGURE 5.31 Modified Davis-like score sheet to better reflect assessing aesthetic wine quality.

Letter-grading score sheet

Wine #	Verbal descriptional:	Letter grade
___		
		___
___		
		___
___		
		___
___		
		___

Rank	Subcategories*			Characteristics	Proportion(%)**
A	A⁺	A	A⁻	exceptional fine for its class	5
B	B⁺	B	B⁻	clearly better than average for class	25
C	C⁺	C	C⁻	average for class	40
D		D		below average	25
D		E		faulty	5

* The A, B, and C categories are subdivided into grouping of below average, average and superior representation with each grade level. The subdivisions are more likely to be skewed to the left, with more in the '–' level than in the '+' grouping since excellence is a rare phenomenon in any aspect of human endeavor. If desired, each of the categories could be substituted by a number to achieve a numerical comparison of a series of wines.

**Assuming a random quality distribution among wines.

FIGURE 5.32   Letter-grade wine score sheet.

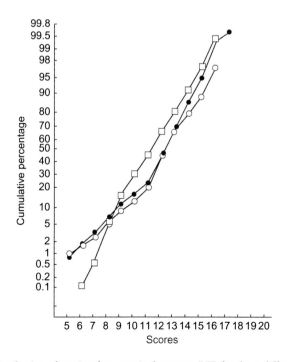

FIGURE 5.33   Cumulative percent distribution of scoring for a particular taster (XII) for three different years using the Davis 20-point score card: ○, 1964; ●, 1965; □, 1968. *From Ough, C.S. and Winton, W.A., 1976. An evaluation of the Davis wine-score card and individual expert panel members. Am. J. Enol. Vitic. 27, 136–144, reproduced by permission.*

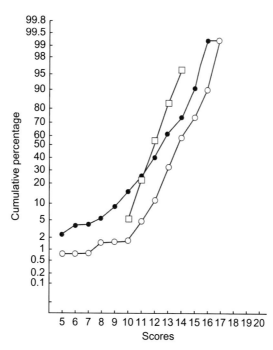

FIGURE 5.34 Cumulative percent distribution of scoring for a particular taster (XI) for three different years using the Davis 20-point score card: ○, 1964; ●, 1965; □, 1968. *From Ough, C.S. and Winton, W.A., 1976. An evaluation of the Davis wine-score card and individual expert panel members. Am. J. Enol. Vitic. 27, 136–144, reproduced by permission.*

TABLE 5.14 Comparison of score sheets showing the percentage weighting given for each sensory category

Score sheet	Appearance	Smell	Taste/flavor	Overall/typicity
Associazione Enotecnici Italiani (1975)	16	32	36	12
Davis score card (1983)	20	30	40	10
O.I.V. tasting sheet (1994)	18	30	44	8

the problem that many forms have, i.e., possessing no option to significantly downgrade wines with a serious fault, or upgrade due to exceptional quality. An example of the holistic approach is the 100-point ranking system popular in consumer publications. However, there is no evidence that wines can be accurately distinguished to this degree of precision. That the portion of the scale most commonly recorded ranges from 80 to 100 is an indication of a lack of effective use. Parr et al. (2006) have demonstrated that 100-point systems are no more discriminative than 20-point systems.

Other ranking schemes permit tasters to assign any number they want. In this instance, there is no upper limit and the taster's ratings are not constrained. Procrustes analysis is typically used to analyze the data. Procrustes analysis is based on the dubious assumption that, although people may use different terms or scales, the attributes are perceived essentially identically (e.g., one person's 10 can be legitimately equated to another's 17). Procrustes procedures search for common trends, and shrink, expand, or rotate the values to develop the best data matching. However, if people's responses differ fundamentally, and not just by degree, then Procrustes statistics is based on a false premise, and could lead to invalid conclusions. In addition, default settings in the messaging of the data involved in Procrustes statistics could remove extremes that have significant behavioral relevance in consumer studies. This can be limited by dividing participants into more homogenous subgroups, and analyzing the data from each subgroup separately. With sensory panels, there is usually sufficient training and selection to ensure that members are comparatively similar in their acuity, and fully understand the tasting procedure to permit the valid application of statistical analysis.

If both ranking and detailed sensory analyses are desired, independent scoring systems should be used. Not only does this simplify assessment, but it can avoid the halo effect, a situation where one assessment prejudices another (McBride and Finlay, 1990; Lawless and Clark, 1992). For example, astringent bitter wines may be scored poorly on overall quality and drinkability, but be rated more leniently when the wine's sensory aspects are assessed and scored independently.

Many studies have shown that the type of task (analytic or synthetic) demanded of panelists can modify significantly the results derived (see Prescott, 2004). Thus, the experimenter must have a clear idea of what is desired, and how the tasting procedure might influence the validity of interpretations based on the data derived.

Statistical Analysis

Statistical analysis is used primarily to establish the degree of confidence to which experimental data have veracity (not due to happenstance). Ideally, it provides an objective means of avoiding accepting hypotheses when false, or rejecting them when valid (Type I and Type II statistical errors). Statistics also can be used to investigate large amounts of data, and to isolate those factors that explain the data's variation. For example, descriptive sensory analysis is often used to assess if, and in what manner, or to what degree, changes in vineyard and winery practice affect the resultant wines. Measuring the statistical significance of detectable differences is important, but more valuable is determining the sensory attributes that lead to significance, and their relative sensory importance. Sensory techniques such as descriptive analysis (DA) is often used to rank the perceived intensities of particular attributes, whereas time-intensity (TI) analysis is used to explore their temporal dynamics. Another procedure, temporal dominance of sensations (TDS), can help avoid potential exaggeration of the relative importance of individual traits. Ultimately, the goal is to correlate wine chemistry (and the sources of any relevant differences) with wine flavor, so that grape growers and winemakers have guideposts by which to direct their efforts.

Regrettably, most statistical procedures are designed for assessing commercial products with well-defined properties, such as cookies, chocolate bars, or soft drinks. Their traits can be assessed quickly. In contrast, wine attributes are complex in character, vary with the cultivar and style, and often develop in the glass, demanding assessment over about 20–30 min. The most desired characteristic may develop only after several to many years in-bottle, and can vary markedly from vintage to vintage, or provenance. Additional complicating factors include the inherent variability of human sensory perception. It is with this latter issue that statistics can be of particular value, helping to distinguish between panel- and wine-based variation.

Various statistical procedures, such as principle component analysis, are valuable when direct comparison of the complex data is difficult and confusing. Statistical analytic programs can assist in identifying the factors that led to detectable differences. Although imperfect, when used with caution, and interpreted within the limitations of such tests, statistical analysis can provide guidance not easily obtained otherwise. Confirmation of interpretations requires further experimentation. Unfortunately, this is too often lacking. This is one of the major problems with time-based funding, leading to research being conducted within the residency period of graduate students or postdoctoral fellows.

Other problems that can affect the validity or applicability of interpretations are inherent in the experimental design. Wine assessed in a laboratory setting is very different from that at the dinner table. Also, data derived from panels, selected for consistency in specific sensory skills, may bear little relevance to real-life situations. Statistical significance does not necessarily imply practical applicability. Thus, wines distinguishable under experimental conditions may be irrelevant to the conditions under which consumers consume wine. For example, that attendees at a dairy convention did not notice that only margarine was served at meals is indicative for the importance of context. In addition, there are multiple instances where correlation may be highly significant but causally unrelated. Humorous examples are given by Vigen (2015), such as the almost-perfect match of margarine consumption and the divorce rate in Maine between 2000 and 2009, or the number of stay-at-home dads and Walt Disney Company revenue between 2000 and 2013.

Increasing the number of panelists can augment statistical discrimination, but not linearly (Fig. 5.35). Thus, the benefit of a larger panel, or more trials, relative to statistical discrimination becomes increasing less efficient in the time and effort spent. Although significant p values (<0.05 and <0.01) are standard, there is nothing sacrosanct about these values, and they have often been misused (Wasserstein and Lazar, 2016). It depends on the degree of confidence required in the data's interpretation. In addition, statistical analysis must never take precedence over common sense. As noted by Lovell (2013) and Nuzzo (2014), the size (degree) of the effect is equally and possibly more important to relevance. This can be especially applicable to sensory analysis, where variation between individuals is often marked and differences between wines are limited.

Another fundamental problem with interpretation, especially for nonstatisticians, is the difficulty of intuitively confirming the veracity of the numbers, especially associations derived from complicated analysis. It is far too easy to be dazzled by the impressive graphical presentation of data derived from the now-readily-available computer

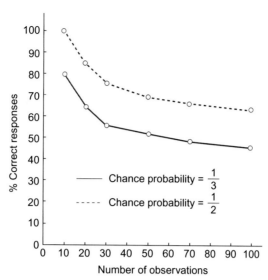

FIGURE 5.35 Plot of the percent correct responses necessary for significance at p < 0.01 for one-tailed tests when chance probabilities are triangle tests (1/3) or paired-difference tests (1/2). *Reprinted from Amerine, M.A., Pangborn, R.M., Roessler, E.B., 1965. Principles of Sensory Evaluation of Foods. Academic Press, New York, NY., with permission from Elsevier.*

statistical packages. Conclusions can be no better than the validity of data on which they are based, derived from the use of a proper experimental design (Leek and Peng, 2015), avoidance of psychological errors, adequate replication, use of competent (and appropriate) tasters, and conducted with due concern for the assumptions on which the statistical analysis is based. Otherwise, the comment popularized over a century ago by Mark Twain can be as valid today as when first penned to paper:

> There are three types of lies: lies, damned lies, and statistics.

Except for those with extensive statistical training, it is advisable to seek professional advice in choosing an appropriate tool for data analysis, and with assistance in its proper interpretation. New statistical approaches are constantly being developed and introduced. For most researchers, it may not be readily apparent which might be the most appropriate. Thus, it is preferable to work with a statistician who has a good working knowledge of sensory evaluation procedures. It is important that statisticians be well aware of the limitations of dealing with humans as sensory instruments. Without access to a statistician, Qannari and Schlich (2006) and Findlay and Hasted (2010) provide good introductions to statistics relative to sensory analysis, provide practical advice, and present case studies and worked examples. Other examples dealing specifically with wine ranking can be found in Quandt (2006) and Cao (2014).

Because an adequate discussion and illustration of multivariate procedures are the prerogative of statistical textbooks, only examples of simple statistical tests are provided below. More complex statistical tests are noted only relative to their potential use in wine analysis, and where their use might be justified, e.g., in visualizing patterns and associations among variables (Fig. 5.36).

As noted at the beginning of the chapter, such statistical rigor is often unnecessary in everyday winery practice, or in wine competitions. In most instances, simple statistical tests adequately determine if wines can be differentiated. The results can be obtained rapidly, and without complex computation necessitating calculators or computers. Only in research, or where critical production decisions are involved, is establishing statistical significance clearly warranted.

At its simplest, statistical analysis can suggest that a particular group of individuals can detect differences among a set of wines. In addition, the data can often be used to identify those wines that are distinguishable, or whether panelists have scored the wines similarly (see Tables 5.10, 5.11, and 5.12). Taster consistency is particularly important because inclusion of data from panelists with poor scoring skills can potentially nullify the discrimination of skilled panelists. Establishment of a detectable difference does not confirm that there is an equivalent preference difference, any more than distinguishing between apples and oranges implies that a preference exists. Equally, clear preferences may exist, but the scores be identical.

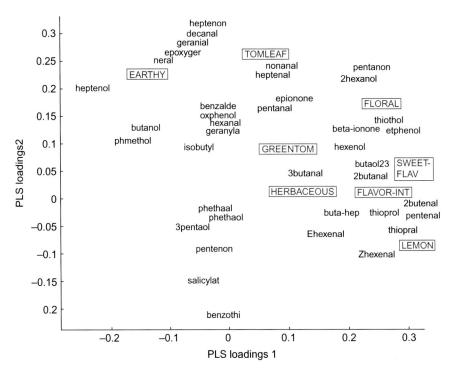

FIGURE 5.36   Illustration of the correlation between 8 sensory attributes (*FLAVOR-INT*, flavor intensity; *GREENTOM*, green tomato; *SWEET-FLAV*, sweet flavor; *TOMLEAF*, tomato leaf) from 20 tomato varieties, with aromatics isolated (noted in shortened names) from these tomatoes, based on loading of the first to partial least-squares (PLS) components. *Reprinted from Qannari, E.M., Schlich, P., 2006. Matching sensory and instrumental data. In: Flavour in Food. (A. Voilley and P. Etiévant, eds.) Woodhouse Publ. Inc., Cambridge, UK. pp. 98–116, reproduced with permission for Elsevier.*

Simple Tests

Table 5.15A presents hypothetical results from a ranking of six wines by five tasters. A cursory look at the data suggests that differences exist among the wines. However, statistical analysis of the data (Appendices 5.6 and 5.7) indicates that distinction is not accepted at a 5% level. For significance, rank total range would have had to have fallen outside the range of 9–26. The actual rank total range was from 9 to 24. Lack of differentiation results from the excessive variability among the individual rankings.

As noted above, increasing the number of tasters can improve the probability of detecting statistical differences. For example, if the results from 10 tasters were as noted in Table 5.15B (where the scoring pattern of the new five tasters replicated that of the first five), the score range now becomes 18–48. From Appendix 5.6, differentiation (at the 5% level) is indicated only when the total rank range exceeds 22–48. Thus, with 10 tasters, Wine A is recognized as being significantly different from the others. In the same manner, were there 15 tasters of the same skill used, both Wines A and E would be considered distinguishable (the highest- and lowest-ranked wines). Even having a few inconsistent tasters, or possessing different concepts of quality, can suggest that the wines are indistinguishable.

Table 5.15 illustrates one of the limitations of ordinal ranking. Wines may be differentiated, but the degree of their apparent differentiation remains undefined. This is improved slightly by ranking based on accumulated scores.

Table 5.16A presents data from a situation where five wines are ranked by seven tasters. For differentiation among wines, the *range* of total scores (26) must be greater than the product of the upper statistic from Appendix 5.8 (0.58) and the *sum* of score ranges (24); in this instance, 13.9. Because the range of total scores (26) is greater than the product (13.9), one can conclude that the tasters could distinguish among the wines. Which wines were distinguishable can be determined with the second (lower) statistic in Appendix 5.8 (0.54). It is multiplied by the sum of score ranges, producing the product 13. When the difference between the total scores of any pair of wines is greater than the calculated product (13), the wines may be considered significantly different; if less, they are considered indistinguishable. Differentiation among the wines (from Table 5.16B) can be visualized by joining wines (by lines) considered to be indistinguishable:

E   B   A   C   D

TABLE 5.15 Ordinal (hierarchical) ranking of six wines by (A) 5 and (B) 10 panelists

A. Five tasters

	Wines					
Taster	**A**	**B**	**C**	**D**	**E**	**F**
1	1	3	6	4	5	2
2	2	1	4	5	3	6
3	3	6	2	5	4	1
4	1	4	5	3	6	2
5	2	5	3	4	6	1
Rank total	9	19	20	21	24	12
Overall rank	6	4	3	2	1	5

B. Ten tasters

	Wines					
Taster	**A**	**B**	**C**	**D**	**E**	**F**
1	1	3	6	4	5	2
2	2	1	4	5	3	6
3	3	6	2	5	4	1
4	1	4	5	3	6	2
5	2	5	3	4	6	1
6	1	3	6	4	5	2
7	2	1	4	5	3	6
8	3	6	2	5	4	1
9	1	4	5	3	6	2
10	2	5	3	4	6	1
Rank total	18	38	40	42	48	24
Overall rank	8	4	3	2	1	5

To determine if differences exist between tasters, the sum of taster score ranges (37) is multiplied by the upper statistic (0.58). Because the product (20) exceeds the difference between the range of their score totals (11), the tasters appear to be assessing the wines similarly. This obviates the need for a second calculation to determine which tasters scored differently.

Table 5.16C shows the ANOVA table using the same data. It confirms significant differences between wines, but no differentiation between tasters.

Multivariate Techniques

Although simple statistics can be useful in assessing if wines are distinguishable, they provide no information on why. This is the data grape growers or winemakers need to improve their wines. For this, their sensory attributes need to be evaluated during ranking, or poorer and better wines need subsequently to be assessed critically. Such data is typically subjected to analysis of variance (ANOVA), or multivariate analysis of variance (MANOVA) (Figs. 5.37 and 5.38). In so doing, the likely interaction among various factors involved in ranking can be assessed. ANOVA evaluates different factors (dependent variables) one at a time, whereas MANOVA simultaneously measures the significance of their interaction. For example, the intensity of several independent attributes may be measured when performing descriptive sensory analysis. MANOVA would assess for the presence of any statistical differences between the wines based on the assessed attributes. Were a significant F-value is obtained, performing individual ANOVAs of the data could indicate the origin of the statistical difference detected by MANOVA.

TABLE 5.16 (A)   Data from five wines scored by seven panelists. (B) Difference among the sum of scores for pairs of wines. (C) Analysis of variance (ANOVA).

A. Wine-scoring data

Taster	Wines					Total	Range
	A	B	C	D	E		
1	14	20	12	13	14	**73**	8
2	17	15	14	13	18	**77**	5
3	18	18	15	13	17	**82**	5
4	15	18	17	14	18	**82**	4
5	**12**	16	15	12	19	**74**	7
6	13	14	13	14	17	**71**	4
7	15	17	13	14	16	**75**	4
Mean	15	16.9	14.1	13.3	17		
Total	104	118	99	93	119		
Range	6	6	5	2	5		

B. Difference among the sum of scores for pairs of wines

	Wine A	Wine B	Wine C	Wine D	Wine E
Wine A	—				
Wine B	14[a]	—			
Wine C	6	19[a]	—		
Wine D	12	25[a]	6	—	
Wine E	14[a]	1	**10**	25[a]	—

C. Analysis of variance (ANOVA)

Source	SS	df	Ms	F	$F_{.05}$	$F_{.01}$	$F_{.001}$
Total	164.2	34					
wines	76.2	4	19.0	6.74	2.78	4.22	6.59[b]
Tasters	20.2	6	3.4	1.19	2.51		
Error	67.8	24	2.8				

Sum of score ranges: (Wines) = (6+ 6+ 5+ 2+ 5) = 24 (Tasters) = (8+ 5+ 5+ 4+ 7+ 4+ 4) = 37. Range of total scores (Wines) = (119–93) = 26 (Tasters) = (82–71) = 11 Appendix 5.8 provides two statistics for 5 wines and 7 tasters: 0.58 (upper) and 0.54 (lower). For significance, differences among wine totals must exceed sum of wine score ranges (24) multiplied by the upper statistic (0.58) = 13.9. For significance, differences among taster totals must exceed sum of taster score ranges (37) multiplied by the upper statistic (0.58) = 20.0.
For significance, differences between pairs of samples differences among totals must exceed 24 (0.54) = 13. Table 5.16B represents the difference in total scores between pairs of wines.
[a]*Significance at the 5% level.*
[b]*Significance at the 0.1% level.*

Without a significant result from MANOVA, performing separate ANOVAs for all the attributes would be unwarranted. In addition, MANOVA assesses whether combinations of attributes may discriminate among wines, even if single attributes do not. Thus, it detects aspects potentially missed by assessing the significance of individual attributes in isolation. Various forms of these analyses are available, depending on the conditions and assumptions that apply in particular cases. A brief introduction to multivariate methods in grape and wine analysis is given in Cozzolino et al. (2010), with worked examples of the technique in Stone et al. (2012).

Data analysis from subsequent tastings of identical wines or their attributes can generate means, standard deviations, and probability values for the rating of each panel member. This is one means of assessing their ability to distinguish and consistently rate wines or their attributes. This could indicate either undesirable variability among

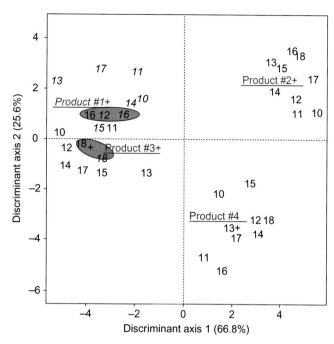

FIGURE 5.37   Example of ANOVA-analyzed data on panel differentiation between four products. The products are well separated, with #1 and #3 being more alike than #2 and #4. In addition, features about individual panelists can be detected. For example, Panelist 16 considered product #3 closer to product #1, whereas Panelist 18 found #1 more related to #3. *Reprinted from Stone, H., Bleibaum, R., Thomas, H.A., 2012. Sensory Evaluation Practices. 4th ed., Academic Press, London, UK, with permission from Elsevier.*

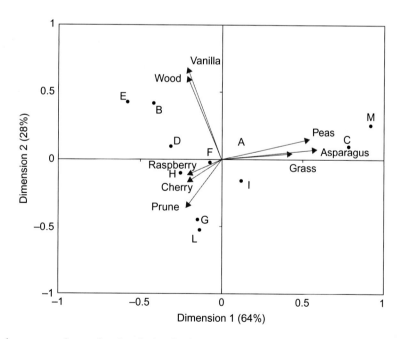

FIGURE 5.38   Example of a perceptual map showing the bi-plot (scores and loadings) from principal component analysis (PCA) of sensory descriptive data (eight aroma attributes) from 11 wines (A–M). The first two dimensions account for approximately 92% of the variation (64% and 28%). The first dimension, from left to right, primarily discriminates wines M and C from the rest by contrasting herbaceous or vegetative attributes (asparagus, pea, and grass) with fruity (raspberry, cherry, dried prune) and wood (vanilla and wood) attributes. The second dimension, from the bottom to the top, indicates that wines E and B are separated from the rest by contrasting the wood notes with dried prune. *Reprinted from Monteleone, E. (2012). Sensory methods from product development and their application in the alcohol beverage industry. In: Alcoholic Beverages. Sensory Evaluation and Consumer Research. (J. Piggott, ed.) pp. 66–100. Woodhouse Publ. Ltd., Cambridge, UK, with permission from Elsevier.*

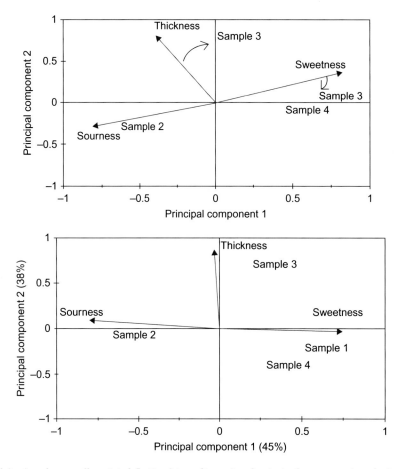

FIGURE 5.39   Unrotated (top) and manually rotated (bottom) two-dimensional principal component analysis of hypothetical data. Arrows with open arrowheads indicate direction of rotation. PC1 explains 45% of the variance in the data set. This PC is composed of a contrast between sweetness and sourness. PC2 explains an additional 38% of the variance in the data. It is primarily a function of bitterness. Sample 1 is sweet but not sour and mildly bitter; Sample 4 is also sweet and not sour, but is slightly less bitter than Sample 1. Sample 4 is as bitter as Sample 1, but is more sour and less sweet. Sample 3 is similar in sourness and sweetness to Sample 4, but is more bitter in character. *Modified from Sensory Evaluation of Food: Principles and Practices. 1998, Lawless, H. T., and Heymann, H., with kind permission of Springer Science and Business Media.*

panelists (and the need for additional training), unusual variability between samples (if they were from different bottles), or differences in sample preparation, storage, etc., that need to be addressed.

Other statistical tests, such as discriminant analysis, may be used to separate wines into groups that are significantly different. Discriminant analysis involves an investigation of the variance of data-point means (e.g., wine versus attribute), based on the amount of disagreement between panelists. As usually occurs in descriptive sensory analysis, multiple attributes are involved in wine discrimination. Occasionally, some of these attributes may be correlated; e.g., the influence of some aromatic compounds on the perception of sweetness. To investigate this feature, and reduce redundancy where various attributes are affected by the same properties, one may perform principal component analysis. It transforms the data into new variables called principal components (see Borgognone et al. (2001) for worked examples). These components maximize the variation, and related attributes are grouped together. The first principal component corresponds to those attributes that generate the maximum amount of variance in the data. Subsequent principal components successively account for smaller amounts of variation distinguishing the wines. When discrimination is based primarily on a few components, the results may be visually represented on a two- or three-dimensional linear graph. In other words, it maps the related attributes (components) spatially. In the hypothetical example represented in Fig. 5.39, the samples were differentiated primarily on their sweetness, sourness and bitterness, with sweetness and sourness being related, but opposed, attributes. Occasionally, attributes may group together along a common principal component axis, occurring on the same (+/-) side of the graph. Their location close together suggests related discriminability, possibly based on similar or related stimuli (for example bitterness and astringency). Those components showing the highest loading (values closer to the maximum of

+ or −1) are the most significant in discrimination. Although of academic interest, such analyses generally do not provide information of practical value to either grape growers or winemakers.

Even more problems arise when analyzing the data from consumers. Because training alters perception, its use is inappropriate in such studies. In this instance, and some others, free-choice profiling is used, where tasters use their own scales and terms. For analysis, Procrustes statistics is the method of choice. It searches for common patterns, adjusting for idiosyncratic term or scale use, standardizes ("messages") the data based on this analysis, then searches for the sources of variation, similar to that generated by principal component analysis. The data generated are graphed similarly. The technique has also been used to compare data from different panels, or where sensory data are compared with instrumental analysis. Where there is ample justification to believe that the panelists are perceiving and ranking attributes similarly (despite divergent term use or scaling), Procrustes analysis is appropriate. However, with consumer groups, possessing diverse likes and dislikes, and different perceptive acuities, its use is suspect. In the latter instance, Procrustes analysis may generate spurious correlations. As always, the researcher must guard against mistaking correlation for causation. Even when there is a causal relationship, there is the question of which is the cause and which the effect.

An alternative procedure, considered more appropriate for consumer data, is preference mapping (Yenket et al., 2011; MacFie and Piggott, 2012). It is particularly applicable when looking at those attributes or wines distinguished by groups of tasters. In this regard, it helps identify those features that may be affecting the preferences of various target groups. The principal drawback of preference mapping is the condensation involved in the procedure. As a consequence, important facets of the data may be overlooked. In addition, if the groupings plot close together, it may signify only that the participants were unable to differentiate among the samples, may be making choices based on attributes not noted on the score sheets, or are inconsistent in their assessment. This method is best combined with other statistical methods, such as ANOVA of individual attributes or wines.

Canonical variate analysis (CVA) permits investigation of the differences between two or more sets of data, with respect to several variables simultaneously (Darlington et al., 1973; Klecka, 1980; Peltier et al., 2015). Linear combinations of differences, which include all variables, are termed canonical variates. Variable importance is represented by the greater weighting it receives in the presentation. Chi-square approximation assesses statistical significance of each dimension. For details see Thompson (1984).

Because these techniques require sophisticated calculation, computers have permitted their current level of use. Direct electronic incorporation of data, as generated by a panel, further eases analysis of large sets of data. This has developed to the point where computer programs specifically designed for the food and beverage industries are commercially available (e.g., Compusense *five*, Fizz Network). They can be adjusted to suit the needs of the user.

For detailed discussions of sensory statistics, see texts such as Bi (2006), Bower (2013), and Meullenet et al. (2007) (in Suggested Readings).

Applicability of Sensory Analytic Results

Training and selection help produce panels that are not only more consistent but also more likely to generate reliable data. Regrettably, training can also lead to increasing halo effects, such as wine-color biasing perception, as well as constricting the range of natural sensory acuity. Limiting panel variance to within parameters set by the experimenter is necessary to enhance a panel's discriminatory ability, and thereby improve the confidence limits of any statistical analysis. Although the influence of natural variation may be adjusted for with Procrustes statistics, it is based on a potentially false assumption that individual members perceive and respond similarly to identical sensations, differing only in degree, rather than kind. As with thresholds, perception may vary in bi- or multimodal ways. Furthermore, in certain circumstances (Köster, 2003), the laboratory tasting conditions required for wine assessment may invalidate the relevance of results. Wine consumed with a meal can appear vastly different than when sampled in the laboratory or in a competitive tasting. Equally, wine perception changes depending on the accompanying food, or with the same food on different days, or under different circumstances (home versus restaurant). Because environmental and psychological influences can so markedly influence perception and appreciation, wines can be impartially compared only under the admittedly contrived conditions of the laboratory. Finally, the data are comparative, not absolute.

Although situations under which wines are usually consumed differ markedly from those under which wines are assessed critically, they are necessary to avoid psychological influences on sensory perception. They clearly provide a better basis for making winemaking or purchasing decisions than those derived from a single or small group of individuals under unknown and potentially prejudicial conditions. Despite the limitations and artificiality of sensory analysis, it is hoped that conclusions derived from such tastings have relevance to real-life tasting situations. To date, the degree to which this hope is justified has yet to be established.

SENSORY ANALYSIS

Sensory analysis can involve a range of techniques, each designed to study particular characteristics of wines, how they are perceived, and how they relate to features such as the wine's chemical nature or varietal, regional, and stylistic origin. That is, they are primarily research tools. These procedures include techniques such as discrimination testing, descriptive sensory analysis, time–intensity analysis, charm analysis, and TDS.

Discrimination Testing

Of analytic sensory procedures, discrimination testing may be the most relevant to winery operations. It assesses whether two (or a few) wines can be differentiated based on one or more attributes. Its use applies to situations in which brands are intended to remain consistent from year to year, or should be distinctively different is some aspect(s). This applies particularly to large wineries and most sparkling wine, sherry, and port producers. In most such situations, the final wine is blended to produce a brand-named product. Because of variation in supply of the base wines, the blend usually needs to be changed periodically. Thus, the producer needs some reliable means of assessing that the new formulation is perceptively identical to the existing blend. For this task, the winery needs a panel that is both consistent and fairly representative of the target purchaser. A panel composed of professionals may be too critical, demanding more similarity than commercially necessary. Where more than two wines need to be assessed, other procedures such as ranking or scaling may be more appropriate.

In preparing a formulation, an initial assessment by winery staff can determine if a more in-depth assessment merits use of a larger group. It all depends on how crucial it is that the new and old blends be sensorially indistinguishable. Conversely, discrimination may be used to confirm that a particular blend that is to be labeled differently is detectably distinct. However, in most such situations, the wines will be graded to determine not only if they are detectably different, but also to what degree. This could be useful in determining pricing.

Alternatively, discrimination testing can be used with groups of wines (e.g., Tables 5.11 and 5.12). The degree to which panel members can distinguish different pairs is a relatively unbiased measure of the degree to which the wines diverge. It avoids problems associated with semantic ambiguity, subjective term use in sensory profiling, or differences in how panelists scale criteria. Fig. 5.40 provides another example of the importance of training in influencing the scale-range use by panelists. Discrimination testing is also applicable to research questions, e.g., measuring the differences in a homologous series of aromatic compounds (Fig. 5.41). The technique is also preliminary to deciding whether the more complex procedure of descriptive analysis is merited to investigate the origin of any differences.

Different discrimination methodologies are available, but all assess if samples are distinguishable or not (Ennis et al., 2014). This determination is often more difficult than might appear at first. For example, one sample in a blend may affect several attributes of the final product, inducing both synergistic and antagonistic influences, and compounding sensory differences between panel members. The most commonly used discrimination procedures are the triangle and duo-trio tests, whereas the simplest is the paired-comparison test. The latter is particularly appropriate when one of the samples is likely to possess a lingering effect, such as an off-odor.

When employing the paired-difference test, the panel may be simultaneously provided with pairs an equal number of times. Panelists may only be asked which is stronger is some aspect (AB and BA sequences), or may be required to indicate whether the samples are similar or not, in which case identical pairs (AA and BB) are also included. Panelists tend to anticipate a difference, whether evident or not, and this must be accounted for in the statistics. In the case of an unsuspected off-odor, it is judicious to request that any potential faults be noted. Although this may increase the likelihood of falsely detecting a taint, releasing wine from the winery without confirming that no fault was detectable could lead to expensive recalls. The minimum number of correct answers considered necessary to confirm differences using paired-difference is given in Appendix 5.9.

Another variant-difference test involves comparing pairs, one of which does not contain the compound (or any additive) whose sensory significance is under study. The test can be used in the assessment of threshold values.

A somewhat similar test, called A-Not-A, involves odor memory. The first sample is presented, removed, and replaced by a second. Panelists are asked whether the second sample is the same or different from the first. This test can be particularly useful in studying different blend formulations. It is particular applicable to wine blends since it is unlikely that consumers will specifically compare samples from different bottles. This discrimination-testing variant resembles slightly more natural tasting conditions.

Another variant is termed the duo-trio test. In it, panelists are simultaneously presented three samples, one of which is a reference wine. In most instances, the reference sample will be the same in all comparisons. The taster is

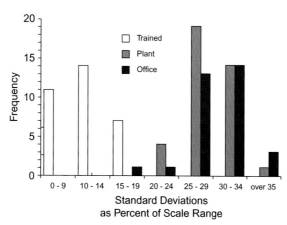

FIGURE 5.40 Standard deviations as a percent of scale range in fragrance descriptive analysis. Data from a trained panel and two consumer model panels (untrained). Ratings were made across 10 commercial air freshener fragrances on 9 attribute scales. *Reproduced from Lawless, H.T., 1999. Descriptive analysis of complex odors: reality, model or illusion? Food Qual. Pref. 10, 325–332, with permission from Elsevier.*

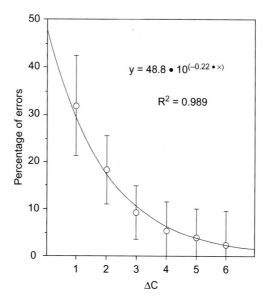

FIGURE 5.41 Discrimination performance (mean ± SD) of 20 subjects as a function of differences in carbon-chain length. ΔC1 corresponds to the discrimination of ketones which differ by only one carbon atom, and ΔC2– Δ C6 to the discrimination of ketones, which differ by 2–6 carbon atoms, respectively. *Reproduced from Laska, M., Hübener, F., 2001. Olfactory discrimination ability for homologous series of aliphatic ketones and acetic esters. Behav. Brain Res. 119, 193–201, with permission from Elsevier.*

requested to indicate which of two test samples is more similar to the reference wine. The order of the test samples (AB/BA) is randomized. Appendix 5.9 indicates the minimum number of correct answers considered appropriate to confirm a detectable difference. The duo-trio test is more discriminatory because the panelists have a reference against which to compare the test samples.

The third major variant, the triangle test, is probably the most commonly used (Kunert and Meyers, 1999; Carbonell et al., 2007). It also presents panelists with three samples, two of which are identical. The samples are grouped so that the two wines are represented an equal number of times. In addition, the sequence order (AAB, ABA, BAA, BBA, BAB, and ABB) is randomized, with each arrangement occurring equally. The participants are requested to indicate which of the coded samples is the odd sample. Again, this can be far from easy, depending on the participants accurately remembering the attributes of the first two samples, before comparing their similarities or differences with that of the third sample. According to Frijters (1980), differentiation is improved if participants are requested to indicate which two samples are the "most similar" rather than which is different. Appendix 5.5 indicates the minimum number of correct answers needed to support differentiation. Providing the panel with a particular attribute to assess in making their comparisons may enhance the value of the test. In this instance, the

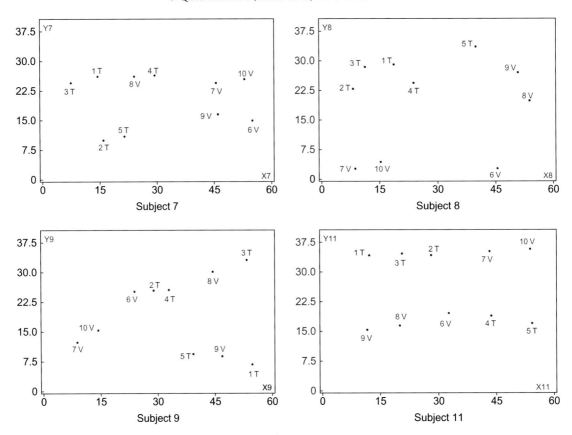

FIGURE 5.42   Example of the results of four panelists using projective mapping to position 12 white Loire wines relative to perceived degrees of similarity/dissimilarity. The sheet was 60 × 40 cm. *T*, Touraine (Sauvignon blanc); *V*, Vouvray (Chenin blanc). *Reprinted from Pagès, J., 2005. Collection and analysis of perceived product inter-distances using multiple factor analysis: Application to the study of 10 white wines from the Loire Valley. Food Qual. Pref. 16, 642–649, with permission from Elsevier.*

decision is not which sample is different, but which is the more distinctive in the noted attribute. In this situation, the triangle test is termed a 3-AFC (three-alternative forced choice).

Although individual comparisons are comparatively simple, the procedure becomes increasing complex and time consuming as the number of pairwise comparisons expands. For example, a comparison of 10 wines would require 45 pair combinations. One solution is projective mapping (napping) (Pagès et al., 2010; Dehlholm et al., 2012; MacFie and Piggott, 2012). It can rapidly generate data to guide more in-depth studies, if deemed warranted.

In projective mapping, panelists are requested to arrange the samples on a sheet of paper, based on their degree of perceived similarity and dissimilarity (Fig. 5.42). Various paper shapes (rectangular, square, circular, and oval) have been used, but there is no clear evidence suggesting one shape is more appropriate or useful than another (Louw et al., 2015). The two-dimensional positioning of the samples is collated and analyzed, and the data then used to represent the sensory relatedness of the samples (Fig. 5.43). Panelists use whatever criteria they consider appropriate in positioning the samples, or may be given instructions on what attributes to use. Up to 10–12 samples have been assessed together without appearing to result in panelist confusion (Pagès, 2005).

When panelists are given a list of attributes from which to make their judgments, this risks the psychological error of dumping, where unrecognized attributes affect how panelists arrange samples. Conversely, free-choice positioning permits each taster to combine any attributes they consider important, in a manner they see fit, within the constraints of two dimensions. Future developments in 3D computing may enable more complex positioning, but still will not solve problems involved with the multidimensional nature of sensory perception. This limitation may be partially offset if the panelists note what attributes they used in making their decisions. There are several statistical packages designed for this situation.

As with other sensory studies, the appropriateness of any group of tasters (consumers, trained professionals, industry experts) will depend on the information desired. The number of panelists required depends partially on their level of experience with the procedure and the product being analyzed. For consumer studies with diverse products, Vidal et al. (2014) recommend 50 participants. Examples of the use of projective mapping with wines

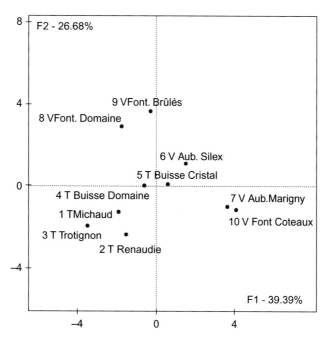

FIGURE 5.43   Representation of the collated data from the use of projective mapping to assess the relative similarity of 10 white Loire wines. *T*, Touraine (Sauvignon blanc); *V*, Vouvray (Chenin blanc). *Reprinted from Pagès, J., 2005. Collection and analysis of perceived product inter-distances using multiple factor analysis: Application to the study of 10 white wines from the Loire Valley. Food Qual. Pref. 16, 642–649, with permission from Elsevier.*

Preference Ranking

Check the box that best characterizes your overall response to each sample.

Sample number _____

    ☐   Like extremely
    ☐   Like very much
    ☐   Like moderately
    ☐   Like slightly
    ☐   Neither like or dislike
    ☐   Dislike slightly
    ☐   Dislike moderately
    ☐   Dislike very much
    ☐   Dislike extremely

FIGURE 5.44   An example of a 9-point hedonic ranking form.

involve an investigation of the effect of serving temperature on the sensory properties of red wine (Ross et al., 2012), comparing the aroma perception of naive and experienced assessors (Torri et al., 2013), and the effect of alcohol content on the perception of brandy (Louw et al., 2014).

For more details on discrimination testing, and its requirements, limitations, and misappropriation, see O'Mahony and Rousseau (2002) and Stone et al. (2012). The latter contains useful examples.

Scalar Ranking

Although differentiation tests determine whether two or more wines are distinguishable, they do not clearly denote ranking. For this, score sheets are commonly used. However, most are of little use in sensory analysis, providing little or no information on the reasons why wines were ordered in the manner they were, or the magnitude of the differences in preference. Thus, more informative scaling techniques may be chosen, e.g., the 9-point preference scale (Fig. 5.44).

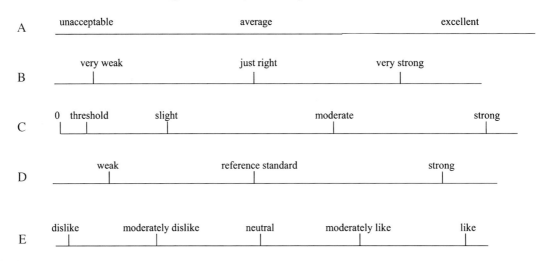

FIGURE 5.45   Examples of types of labeled intensity (magnitude) scales.

The 9-point hedonic scale presents a balanced set of values, four on each side of a neutral or central point. It is commonly used to measure the degree to which a product is liked or disliked in consumer evaluations. Although there is a tendency for people to avoid the ends of a scale, the interval between the categories is represented as being equal. If so used, it permits the results to be converted numerically (1 through 9) for parametric statistical analysis. Other than basic instruction on how to proceed, no training is required. Where desired, demographic and wine-usage data can be simultaneously collected. This could permit analysis of the data based on consumer subcategories. Suggested improvements in its design and value are noted in Lim et al. (2009), with comments (Prescott, 2009).

The 9-point hedonic scale can also be used to derive more specific information concerning particular attributes. When so used, training is typically required. Otherwise, tasters may interpret the attributes in different ways. For example, consumers are unlikely to have a clear and common understanding of the meaning of terms such as viscosity or body. Astringency is another term with various shades of meaning and may be used dissimilarly. Depending on the property, a range of anchor points may be designated. Although it would be useful to understand the origin of consumer preferences, scaling is rarely employed in such studies, due to the training required and its potential to modify preferences. In addition, consumer perception in a test situation is unlikely to accurately reflect natural (home or restaurant) sampling conditions. As always, it is important to recognize the potential limitations to any method. Consumer testing procedures are discussed in greater detail in Chapter 6, Qualitative Wine Assessment.

Alternative measurement techniques employ scaled lines (Fig. 5.45), where positions along the line can be converted into numbers for statistical analysis. Line ends may be designated by terms such as *none* or *very weak* versus *very evident* or *intense*. Alternatively, intensities may be scaled relative to some central point or standard, considered normal or optimal.

The principal disadvantage of scaling is the time and concentration it takes, especially if several attributes are independently scaled. Additional issues can arise if the attributes change in quality as well as intensity over the expected range, such as astringency or oxidation, both being multimodal perceptions. It is important not to incorporate or present tasks that demand measuring perceptive intensity with attributes possessing pronounced hedonic responses, e.g., mercaptans. Otherwise, subjective responses are likely to distort intensity measurements. Finally, it is important to realize that the values derived from interval scales are relative and not likely to represent absolute differences in the attributes scored. In other words, a value of 5 on a scale of 9 is unlikely to be twice that of a value of 2.5.

Except for specific purposes, such as basic research, scaling assessments are rarely used in daily winery practice. The time and effort involved in training panel members is seldom justified. Training usually involves reference samples for each of the anchor points. Otherwise, the use of procedures such as Procrustes analysis becomes essential. While training is beneficial, it is unlikely to overcome contextual effects, such as contrast between samples and changes in the range of intensities from test to test. Where feasible, context effects may be minimized by repeat sample presentation and providing wines in random order.

The use of scalar measurements has been greatly facilitated by the introduction of touchscreen monitors. Panelists simply slide a bar along the appropriate line with their finger to indicate intensity or magnitude. Lines for distinct attributes are usually positioned above each other. Besides being easier to use, computers can automatically digitize, collate, and perform statistical procedures deemed applicable.

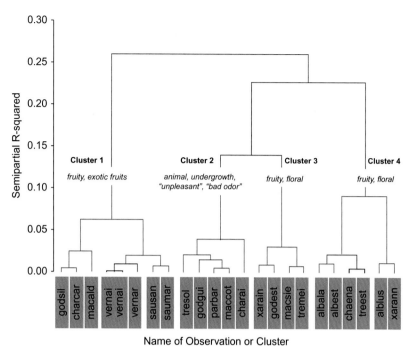

FIGURE 5.46   Composition of the clusters derived from the sorting task of varietal Spanish wines (first 3 letters) and representative producers (last 3 letters). *From Campo et al. 2008, Aust. J. Grape Wine Res., reproduced with permission from John Wiley and Sons, copyright.*

Although line scales are often used in sensory analysis, some researchers question the legitimacy of their use. For example, Engen and Pfaffmann (1959, 1960) found that people were surprisingly poor at accurately identifying the odor intensity of various samples. Suggested solutions have been to substitute measures of odor relatedness for odor intensity (Schutz, 1964).

Sorting Analysis

Sorting analysis in another discrimination test used to investigate preexisting conceptual models for varietal (Fig. 5.46), regional, or stylistic wines. Panelists are required to sort wines into two or more groups, based on similarity. The technique has been used to investigate varietal typicality as well as sensory attributes (Parr et al., 2010; Ballester et al., 2008; Perrin and Pagès, 2009). Because matching archetypic models to geographical typicity depends on the extensive experience, with considerable overlap with adjacent regions being frequent (Maitre et al., 2010), the results are less than convincing. Without understanding the chemical origins of any verifiable regional typicity, the results are of little use in directing the activities of either grape grower or winemaker, or providing confidence to the consumer.

Descriptive Sensory Analysis

Descriptive sensory analysis attempts to quantitatively describe the distinctive sensory attributes of a food or beverage. It evolved as an offshoot of producing a complete flavor profile. However, developing a flavor profile is time consuming in terms of panel training (often months), and usually involves features unnecessary for differentiation, or detecting changes in production procedure.

Relative to wine, descriptive sensory analysis has primarily been used as a research tool and in quality control. Surprisingly, it has seen little application, in print, to developing wines designed for particular consumer subgroups. The concept of wine being manufactured, as most processed foods, is anathema to the popular view of wines as natural, spontaneously generated beverages, with artisan cellar masters functioning only as Nature's handmaiden. The artisanal approach certainly has its rightful place, and produces some of the world's finest wines. However, this approach also produces much of the world's mundane wines. Designed wines are increasingly required to supply consistent-quality wine with defined features to a burgeoning audience of savvy, sophisticated consumers.

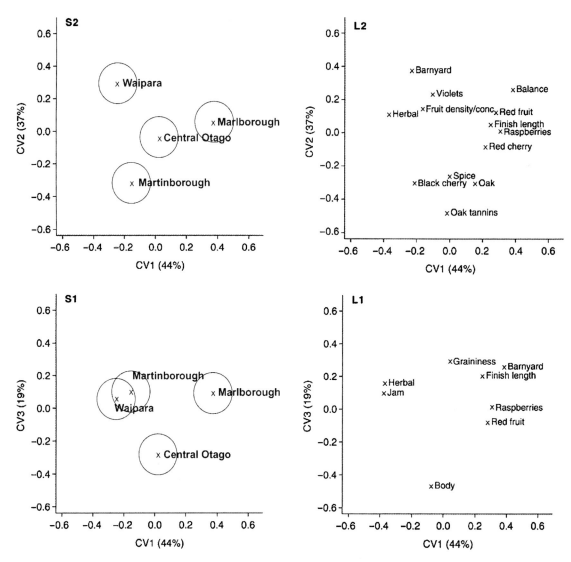

FIGURE 5.47   Separation of Pinot noir wines by region (Central Otago, Marlborough, Martinborough, and Waipara) using canonical variate analysis (CVA). *From Tomasino et al., 2013, reproduced with permission 12013 American Society for Enology and Viticulture. AJEV 64:357–363.*

In addition, fewer aficionados are inclined to pay high prices for wines that need decades to express their finest qualities. Thus far, though, descriptive sensory analysis has been largely applied to studying or validating features that distinguish wines made from particular varieties, produced in specific regions (typicity), or in distinctive styles (Williams et al., 1982; Guinard and Cliff, 1987; McCloskey et al., 1996; Figs. 5.47 and 5.48). Descriptive sensory analysis has also been used in investigating the climatic features of a region that influence wine production (Falcetti and Scienza, 1992). When studying the effects of production techniques on sensory attributes, it is essential that all other factors be as equivalent as possible.

In combination with chemical analysis, descriptive sensory analysis can facilitate the identification of compounds responsible for specific sensations. The procedure can also clarify those attributes or chemicals that may be critical to quality perception. Finally, the technique can be combined with preference studies in exploring commonly held views on consumer partialities. For example, Williams et al. (1982) confirmed the relative importance of sweetness, low acidity, and minimal bitterness and astringency to infrequent wine drinkers.

In most forms of descriptive analysis, members of the panel are trained in the use of a specific set of sensory terms. Where a complete sensory profile is desired, standards for all pertinent sensory modalities are required. For wine, this would include visual, gustatory, haptic, and olfactory traits. Typically, though, only selected aspects of the aroma or flavor are investigated. The terms need not adequately describe the wine's flavor profile. Only those features that differentiate samples are typically required. Thus, defining those attributes is a prerequisite to conducting

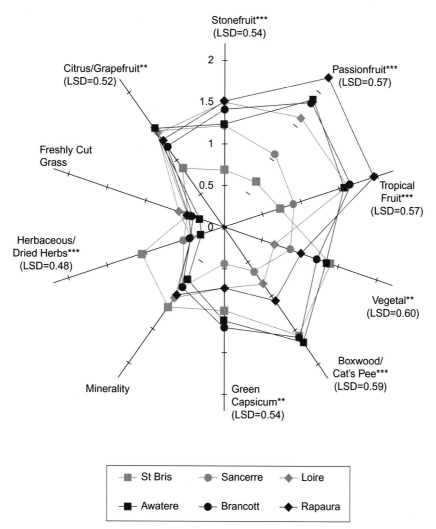

FIGURE 5.48 Aroma profiles of wines as a function of country of origin (France versus New Zealand) and subregion (France: Saint Bris, Sancerre, Loire; New Zealand: Awatere, Brancott, Rapaura). *Reprinted from Parr, W.V., Valentin, D., Green, J.A., Dacremont, C., 2010. Evaluation of French and New Zealand Sauvignon wine by experienced French wine assessors. Food Qual. Pref. 21, 55–64, with permission from Elsevier.*

the main study. Quantification involves scoring the perceived intensity of each attribute. Correspondingly, panelists need to learn not only how to use the terms consistently and uniformly, but also become proficient in the use of intensity scales.

Because descriptive analysis is one of the best and most-used tools in critically assessing wine, Lawless (1999) felt it necessary to drawn attention to the assumptions upon which its appropriate use are based:

- that odor quality can be analyzed and measured as a set of independent olfactory attributes
- that each descriptive term be unrelated and perceived separately
- that individual attributes vary in perceived intensity relative to stimulus concentration

Lawless argued that it is better to ask panelists to rate terms on their degree of appropriateness rather than intensity. This appears to be what panelists frequently do, regardless of instructions. In addition, descriptive wine terms only loosely resemble the olfactory characteristics named, and may overlap. Thus, the aroma of a Chardonnay wine may be described in terms of resemblances to apples, melons, and peaches. However, these terms rarely do more justice to the actual impression of the Chardonnay wine than does the verbal description of a musical composition adequately depict its sonic expression or appeal. Descriptors are our attempt to find a "best fit" for the characteristics of a wine. Regrettably, in the process, the mind may generate illusions (artifacts) that do not exist.

These concerns do not negate the potential value of descriptive sensory analysis in differentiating or characterizing wines, but instead give a reason for caution about overinterpreting the data. The spider plots and other visual representations of data clarify associations between attributes, but do not adequately represent our perceptions of wines, any more than maps do justice to the geographic diversity of a region.

There are several variants of descriptive sensory analysis. Only the three most common examples are outlined below: quantitative descriptive analysis, spectrum analysis, and free-choice profiling. For the first two, screening of potential panelists is category specific, as is subsequent training. In contrast, panel members for free-choice profiling seldom undergo sensory screening, nor are measures of sensory variability determined. For a detailed discussion of these techniques and worked examples, see Stone et al. (2012).

Development of the descriptive lexicon for the first two procedures requires considerable attention and partially differentiates quantitative descriptive analysis from spectrum analysis. In addition, sensory features tend to be grouped into related (i.e., commonly accepted) categories, definitions developed for each attribute, reference samples prepared, and members trained in their use and how to score intensity (or degree of appropriateness).

Terms should not overlap or be redundant, and should constitute primary attributes (e.g., apple, litchi, oak, mushroom) versus collective/holistic terms (e.g., balanced, perfumed, rich), permit clear enunciation of the differences that distinguish the wines, adequately note the attributes that may group wines by type, and ideally be represented by standards (Civille and Lawless, 1986).

For quantitative descriptive and spectrum analyses, reference standards permit panel members to refer to them when required for clarification or confirmation. Although reference standards are preferred, absence of an acceptable reference does not necessarily exclude its use. Where present, standards should demonstrate only the property intended, without confusing tasters with additional attributes. They should also be sufficiently subtle to avoid sensory interaction or fatigue. It is equally important that standards remain stable over the experimental period. If they need to be replaced, the panel should be questioned about acceptability. When not in use, standards are usually refrigerated and stored in the absence of oxygen. Using nitrogen as a headspace gas helps, as does incorporation of a small amount of antioxidant, such as sulfur dioxide (~ 30 mg/L). Incorporation of cyclodextrin is an additional measure extending the useful shelf-life of standards.

Reference standards are equally essential when new panel members are incorporated. They permit new members to gain the experience necessary for appropriate term usage. Reference standards are particularly important in training members for spectrum analysis, where members are not involved in term development.

With quantitative descriptive analysis, members work toward a consensus on descriptors that adequately represent the wine's distinctive attributes (McDonnell et al., 2001; Lawless and Civille, 2013). The duration of training is often proportional to the degree of differentiation required (Chambers et al., 2004). Each sample is analyzed separately, attribute by attribute, and on several occasions. The data are usually analyzed with ANOVA.

Training occurs under the supervision of a leader who both guides and arbitrates the discussion, attempting to avoid polarizing around initial opinions (Lamm and Myers, 1978), or supporting the views of one or more dominant individuals. The leader does not directly participate in term development or in the tastings; otherwise, prior knowledge might bias the features chosen and the terms used. The leader's responsibility is limited to encouraging fruitful discussion, organizing the tests, and analyzing the results. Although a particular term is attached to a particular standard, a range of descriptive terms may be employed initially to encourage members to search for the central attributes of the descriptor. For example, benzaldehyde may be described as having elements of cherry pits, wild cherry, and almond oil. Homa and Cultice (1984) have noted that panel feedback and diverse descriptive components facilitate learning. Training is usually shorter if members develop their own terms, rather than use unfamiliar technical terms or chemical names. Subsequently, panelists assess wines using these descriptors to determine their efficacy (lack of confusion or redundancy), and establish which terms are the most discriminatory. This typically involves analysis of tasting results after several trials using identical wines. Only then can the number of attributes be reduced to those actually required. Analysis also permits members' use of terms to be measured. At this stage, inconsistent tasters should be removed from the panel. Earlier elimination is inappropriate because the members are still in the training phase.

Table 5.17 illustrates one method for studying term efficacy. Significant judge–wine (J x W) interaction (among mint/eucalyptus, earthy, and berry-by-mouth ratings) indicates an unacceptable level of variation in term usage. These attributes were subsequently eliminated. High levels of individual judge–wine interaction can indicate tasters who are using terms inconsistently.

Another procedure involves correlation matrices. It can highlight the occurrence of term redundancy. For example, Table 5.18 illustrates that attributes such as fresh berry and berry jam, leather and smoke, and astringency and bitterness showed highly significant correlation. The latter is certainly not unexpected as both astringency and bitterness are elicited by similar, and often identical, compounds. However, in this instance, significant correlation does not indicate term redundancy because both refer to distinct sensory phenomena. Data from sensory analysis must be interpreted critically and only in association with other information.

TABLE 5.17 Analyses of variance of attribute ratings (seven judges) for Pinot noir wines: Degrees of freedom (df), F-ratios, and error mean squares (MSE)

	F Ratios						
	Judges (J)	Reps (R)	Wines (W)	J R	J W	R W	MSE
Red color	30.74***	0.02	61.17***	0.90	1.12	1.97**	94.87
Fresh berry	37.30***	0.08	2.61***	0.59	1.17	1.15	198.73
Berry jam	70.34***	2.20	2.90***	0.28	1.32*	0.85	150.17
Cherry	71.91***	0.35	2.71***	0.39	1.42*	0.83	130.65
Prune	72.46***	0.23	1.39	0.66	1.21	1.38	147.57
Spicy	130.64***	1.17	1.78*	1.82	1.28	0.99	89.32
Mint/eucalyptus	121.27***	0.90	2.32***	0.55	2.10***	1.40	105.02
Earthy	121.47***	1.33	5.28***	1.03	1.64***	1.93**	125.67
Leather	56.59***	0.63	2.39***	0.98	1.17	0.85	181.54
Vegetal	103.28***	0.01	4.74***	2.64*	1.38*	0.77	130.74
Smoke/tar	110.38***	0.07	4.02***	2.81*	1.19	1.22	129.90
Berry by mouth	36.05***	3.67	2.21**	1.79	1.55**	0.92	171.75
Bitterness	111.21***	5.72*	3.83***	2.22*	1.19	1.06	134.94
Astringency	128.78***	0.29	5.05***	1.45	1.24	1.65	146.91
df	6	1	27	6	162	27	162

From Guinard and Cliff, 1987, reproduced by permission.
*, **, ***, significant at $p < 0.05$, $p < 0.01$, and $p < 0.001$, respectively.

TABLE 5.18 Correlation matrix among the descriptive terms (df = 26) used for differentiation Pinot noir wines

Term	1	2	3	4	5	6	7	8	9	10	11
1. Red color	1.00										
2. Fresh berry	−0.05	1.00									
3. Berry jam	0.01	0.60***	1.00								
4. Cherry	0.13	0.71***	0.42*	1.00							
5. Prune	−0.22	−0.22	0.04	−0.19	1.00						
6. Spicy	0.32	0.29	0.27	0.47*	−0.25	1.00					
7. Leather	0.18	−0.50**	−0.45*	−0.28	0.14	0.16	1.00				
8. Vegetal	−0.46*	−0.57**	−0.27	−0.61***	0.51**	−0.54**	0.11	1.00			
9. Smoker/tar	0.46*	−0.66***	−0.41*	−0.35	0.04	0.20	0.70***	0.15	1.00		
10. Bitterness	0.09	−0.14	−0.14	−0.08	0.35	0.04	0.35	0.10	0.27	1.00	
11. Astringency	0.57**	0.00	0.00	−0.02	−0.04	0.23	0.23	−0.27	0.36	0.61***	1.00

From Guinard and Cliff, 1987, reproduced by permission.
*, **, ***, significant at $p < 0.005$, $p < 0.01$, and $p < 0.001$, respectively.

Spectrum analysis provides panelists with preselected attributes to measure, reference standards, and established intensity scales. In situations where the discriminatory attributes are known, and the wines relatively uniform, training time can be reduced. Spectrum analysis is also appropriate where the experimenter has interest in only specific wine attributes. However, this runs the risk that panel members may expropriate terms to represent properties not present in the designated terminology (dumping). Training in term use is similar to that of quantitative descriptive analysis.

Despite training and selection, both quantitative descriptive and spectrum methods may experience concerns associated with inconsistent term use. In addition, some researchers have pondered about problems associated with polarization during term development, idiosyncratic term use, and even whether correct sets of descriptors are possible (Solomon, 1991). In an attempt to offset these concerns, free-choice profiling may be used. In it, panelists are allowed to use their own terminology. Although training is minimal, time is still required for tasters to gain experience with the wines, develop their own lexicon, and become familiar with intensity-scale use. Similar numbers of terms make analysis simpler, but are not obligatory. It is assumed (correctly or not) that panel members experience identical sensations, even though their term and intensity-scale use differ.

As noted previously, one of the concerns associated with free-choice profiling arises from the complex mathematical adjustment of the data involved. In generalized Procrustes analysis (Oreskovich et al., 1991; Dijksterhuis, 1996), algorithms involving rotation, stretching, and shrinking of the data search for common trends. Thus, the possibility exists that relationships may be generated where none exist (Stone et al., 2012), or legitimate variation eliminated (Huitson, 1989). In addition, relating the generated axes to specific sensory attributes can be difficult. This often requires combination of the data with information derived from sensory profiling or chemical analyses. An alternative procedure involves global Chi-square analysis (Symoneaux et al., 2012).

Typically, free-choice profiling is used with consumer panels, permitting some tentative conclusions to be drawn. Ideally, these should be subjected to additional assessments, to verify the conclusions. In addition, free-choice profiling circumvents issues associated with the need to develop a descriptive lexicon, and the connected potential for sensory modification during training. Free-choice profiling has also been proposed as a simple procedure for use by professional tasters (enologists, winemakers) in determining the typicity of regional or varietal wines, or in the preparatory stage of term generation for conventional sensory studies (Laurence et al., 2013).

A variant technique, termed the Flash profile (Delarue and Sieffermann, 2004), has the advantage of being less time consuming than some other similar procedures, with participants ranking products on an ordinal scale. Although they still use their own terms, participants are encouraged to use specific sensory attributes rather than hedonic concepts. It is particularly useful at the beginning of a project to select the most appropriate terms for conventional descriptive analysis. Recently it has been combined with napping to increase the value of both techniques (Liu et al., 2016).

An alternative approach to offset consumer idiosyncrasy is by analyzing the ranking or scores for pairs of wines for each participant, rather than averaging the data. The differences can then be studied statistically for significance with techniques such as ANOVA.

All analytic procedures attempt to minimize panel variation, either by member selection or with multidimensional mathematical models such as Procrustes analysis. The consequential disadvantage is the enhanced likelihood that the data may be irrelevant to groups with different perceptions.

An essential element in most descriptive analyses is intensity scoring. In quantitative descriptive analysis, panelists note intensity on a line scale. Traditionally, this involves a line 15 cm long, with two vertical lines placed 1.27 cm from each end. Above each vertical line is an expression denoting the intensity and direction. Examples of polar ends are always provided. In spectrum analysis, several types of scales may be employed. Specific numerical anchors are often provided for particular sensory characteristics. In free-choice profiling, a single type of line scale is used. In all procedures, intensity indications are converted to numerical values for analysis.

Examples of tasting sheets applicable to descriptive sensory analysis are given in Figs. 5.49 and 5.50. Fig. 5.49 illustrates the use of intensity scales for several sensory attributes. Fig. 5.50 involves a simplification for categorizing large numbers of wines, more than can easily be assessed by traditional sensory analysis procedures. In the latter instance, panel members select up to five terms, from a list of the 10 attributes that most distinguish individual wines. The frequency of use (a measure of appropriateness) is used to characterize the wines.

All methods incorporate replicate trials, with three to five replicates generally being considered minimally adequate. Most panels consist of between 10 and 20 members.

Data generated from these procedures are often visualized as polar (radar, spider, cobweb) plots. The distance from the center represents the mean intensity value for each attribute (Fig. 5.48). Additional information can be shown if correlation coefficients are used to define the angles between the lines connecting the intensity values (Fig. 5.51). This presentation works well for up to five comparisons (especially when presented in different colors), but becomes visually cluttered when more attributes are compared. In such situations, statistical methods are used to reduce the number of points involved in each comparison. Multivariate analysis becomes invaluable, and only legitimate, when dealing with large data sets (Meilgaard et al., 2006). Each sensory attribute can be considered as a point in multidimensional space, the coordinates of which are the magnitude of the component attributes. Multivariate analysis helps highlight the most discriminant attributes.

Polar plots have become particularly popular means of representing data from descriptive analysis. However, caution should be used in a simplistic interpretation of the data. The data represent sensory attributes that the panelists used to *distinguish* among the wines, not the attributes that necessarily *characterize* the wines. In that regard

Wine: _____   Date: _____

Attribute	Relative intensity scale*

Flavor

Low   High

Berry

Blackcurrant

Green bean

Herbaceous

Black pepper

Bell pepper

Tannic

Oaky

Vanilla

Leather

Cigar box

Taste/Mouth-feel

Sourness

Bitterness

Astringency

Body

Balance

* Place a vertical line across the horizontal line at the point which best illustrates how you rate the intensity of the attribute.

FIGURE 5.49   Example of a descriptive sensory analysis tasting form adjusted for Bordeaux wines.

they are similar to floral diagrams used in plant taxonomy (Ijiri et al., 2005). The latter highlight the distinguishing features of a flower, but certainly do not represent the holistic impression of the flower. In addition, the specific set of terms, and their physical reference standards, are not available to the viewer of the polar plot. Thus, any use in understanding consumer perception would be highly dubious, as consumers rarely analyze wine or assess the relative intensities of distinct flavor attributes.

Another analytic procedure, termed principal component analysis, has been frequently used to visualize correlations between multiple attributes. Fig. 5.52 provides an example where peach, floral, and citrus notes of Chardonnay

Name: _____   Date: _____

Wine # : _____   Attributes:* _____ , _____ , _____ , _____ , _____

Wine # : _____   Attributes: _____ , _____ , _____ , _____ , _____

Wine # : _____   Attributes: _____ , _____ , _____ , _____ , _____

Wine # : _____   Attributes: _____ , _____ , _____ , _____ , _____

Wine # : _____   Attributes: _____ , _____ , _____ , _____ , _____

Wine # : _____   Attributes: _____ , _____ , _____ , _____ , _____

Wine # : _____   Attributes: _____ , _____ , _____ , _____ , _____

Wine # : _____   Attributes: _____ , _____ , _____ , _____ , _____

Wine # : _____   Attributes: _____ , _____ , _____ , _____ , _____

Wine # : _____   Attributes: _____ , _____ , _____ , _____ , _____

Wine # : _____   Attributes: _____ , _____ , _____ , _____ , _____

Wine # : _____   Attributes: _____ , _____ , _____ , _____ , _____

Wine # : _____   Attributes: _____ , _____ , _____ , _____ , _____

Wine # : _____   Attributes: _____ , _____ , _____ , _____ , _____

* Describe each wine sample with the use of two to five of the set of listed terms, examples are given for two grape varieties:

     Chardonnay wines: apple, citrus, muscat, fruit, buttery, honey, caramel, oak, herbaceous, neutral
     Semillion wines: floral, lime, pineapple, honey, nutty, grassy, toasty, tobacco, smoky, oak

FIGURE 5.50   Multiwine descriptive analysis score card.

wines are shown to be closely correlated (perceived similarly), but negatively correlated (rarely associated) with pepper notes (first principal component). The second principal component illustrates an independent relationship between sweet and vanilla notes, which is inversely correlated with the presence of bitterness. These associations can be represented in conventional histograms, but are less obvious and do not present quantitative indicators of correlation. Combination of average data with that of individual wines can demonstrate whether particular wines cluster with specific flavor characteristics (e.g., Carneros Pinot noir, Fig. 5.53). Principal component analysis can also be used to suggest which attributes are insufficiently discriminatory and might be eliminated from subsequent studies. Combining principal component analysis with ANOVA is illustrated in Luciano and Naes (2009).

While generating academically interesting information, the procedures can be taxing on the panel. Thus, without clear indication of the relevance of the data to improving wine production or other practical ends, maintenance of member enthusiasm may wane and fatigue may set in. Both can result in qualitative data deterioration. On the other hand, if too many attributes are eliminated, panelists may receive a negative message, suggesting that the considerable effort expended in preparing the attribute list was unwarranted. Researchers need good personnel skills as much as anyone.

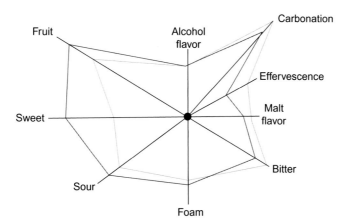

FIGURE 5.51 Quantitative descriptive analysis of two competitive products: the distance from the center is the mean value for that attribute, and the angles between the outer lines are derived from the correlation coefficients. *From Stone, H., Sidel, J., Oliver, S., Woolsey, A., Singleton, R.C., 1974. Sensory evaluation by quantitative descriptive analysis. Food Technol., Nov. 24–34, reproduced by permission.*

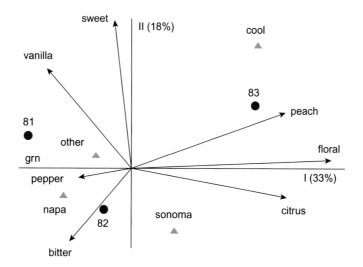

FIGURE 5.52 Principal component analysis of Chardonnay wines. Projection of sensory data on principal components I and II. Attribute loadings (vectors) and mean factor scores for wines from 1981, 1982, and 1983 (●) and from Napa Valley, Sonoma Valley, cool regions, and other locations (▲). *Reproduced from Noble, A.C., 1988. Analysis of wine sensory properties. In Wine Analysis (H. F. Linskens and J. F. Jackson, eds.), pp. 9–28. copyright Springer-Verlag.*

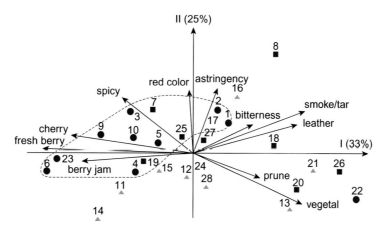

FIGURE 5.53 Principal component analysis of the mean ratings of 28 Pinot noir wines from three regions: Carneros (●), Napa (■), and Sonoma (▲). *From Guinard and Cliff, 1987, reproduced by permission.*

Cluster analysis is another statistical tool that can be applied to highlight quantitative relationships between sensory factors. These often resemble the branches of a tree (Fig. 5.46). Connections between the branches indicate the degree of correlation. Nonetheless, it has seen little use in wine analysis.

To date, descriptive sensory analysis has been used primarily to distinguish or characterize groups of wines. This limitation is regrettable, especially in relation to the time and effort involved. For example, investigation in the reasons of the associations noted in Fig. 5.53 could have been particularly informative, due to term redundancy, related chemicals, clonal origin, or a unifying winemaking procedure? Equally revealing would have been investigation into why certain attributes were inversely correlated (Fig. 5.52).

The potential of descriptive sensory analysis is illustrated in the work of Williams et al. (1982) and Williams (1984), where preference data were correlated with particular aromatic and sapid substances. Relating wine chemistry to consumer preferences could lead to production changes intended to appeal to the consumer and/or accentuate varietal or stylistic characteristics. Designing wine may not fit the romantic image of winemaking, as some idealistic expression of the estate, region, or producer, but could lead to the production of more flavorful, enjoyable wines. It can also be highly remunerative. The success of "wine coolers" is a clear example. Finally, it could also help keep financial wolves from the cellar door.

Another incidental use of sensory analysis is in training winemakers. Winemakers must develop astute sensory skills. Preparing blends requires the evaluation of each wine's sensory properties and how they interact on combination. While blending is still an art, it is based on understanding how different attributes of wines can combine to enhance their positive qualities and minimize any deficiencies. The intense concentration and scrutiny required in descriptive sensory analysis provide a wonderful learning experience for both young and seasoned winemakers.

Time-Intensity Analysis and Temporal Dominance of Sensation

One of the deficiencies in the methods mentioned above is the absence of any indication of the temporal dynamics of the sensations assessed. Time-intensity (TI) analysis partially offsets this deficiency by assessing changes in the intensity of various attributes over time, notably gustatory sensations (Lawless and Clark, 1992; Dijksterhuis and Piggott, 2001), but also retronasal odor (Fig. 3.35). The technique has panelists record how the perceived intensity of particular sensations changes over the course of the tasting. The use of computers has significantly facilitated the recording of data, with vertical positioning of the cursor on the monitor indicating intensity as screen moves to the left at a constant rate.

As illustrated in Fig. 5.54, four basic aspects of the dynamics of perception have been distinguished. These include the time taken to reach maximum intensity, maximal perceived intensity, duration of the perception; and the temporal dynamics of perception. These features vary from person to person, but remain relatively constant for the individual.

The technique has been particularly useful in quantifying the dynamics of individual and group reactions to particular tastants, either in water or wine-like solutions. Thus, it has considerable use in investigating the kinetics of

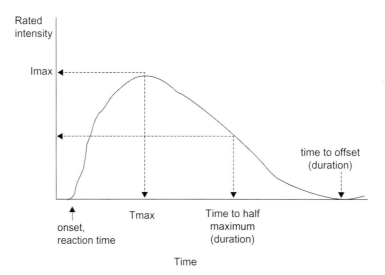

FIGURE 5.54   Time-intensity (TI) scaling (I_{max}, maximum perceived intensity; T_{max}, time taken to reach maximum intensity). *From Lawless and Clark, 1992, reproduced by permission.*

oral sensations. For example, TI studies have identified clear differences between sweet, sour, bitter, and astringent perceptions (at least under controlled laboratory conditions). The technique has been used extensively in designing foods to have specific taste profiles.

The use of TI methods with wine has been more problematic. The chemistry of wine often modifies gustatory sensations in complex and unsuspected ways (Chapter 4, Oral Sensations (Taste and Mouth-Feel)). Regardless, TI studies can provide insights into how various blends can affect the sensory perception of particular consumer groups. A delay in the perception of bitterness, and adjustment in the detection of the various qualities of astringency could make a red wine more appealing to the majority of consumers. TI methods can also reveal details on how a wine's gustatory sensations are modified when taken with food. For example, total taste intensity and sweetness were reduced in combination with various constituents that activate oral trigeminal nerves (Lawless et al., 1996). This can have relevance to how particular foods pair with wines (see Chapter 9, Wine and Food Combination).

In response to the multiple qualitative attributes of some tastants, and their lingering aftertastes, several modifications of TI tests have been developed. Examples include techniques designed to investigate two or more attributes simultaneously, consecutively, or primarily on aftertaste characteristics (Obst et al., 2014).

Generally, concentration has limited effect on the time taken to reach maximum perceived intensity, although it does affect duration. Because wines are normally expectorated shortly after sampling, this tradition must be modified to study adaptation. In so doing, it has been shown that increasing concentration can delay the onset of adaption (Fig. 5.55). The timing of expectoration must also be changed to study the effect of repeat exposure on features such as astringency (Fig. 5.56). To date, the procedure has been used little to investigate the interaction between sapid substances, or the joint influences of sapid and olfactory compounds on flavor perception. Details on panel training for TI analysis are given in Peyvieux and Dijksterhuis (2001).

Although providing valuable insights into the nature of perception (Lawless and Heymann, 2010), TI does not describe how various oral sensations change dynamically over the tasting experience or the finish. As noted in Chapter 1, Introduction, a crude form of TI recording was advocated by Vandyke Price (1975), visualizing dynamic changes in wine flavor during the course of tasting (Fig. 1.7). Its closest professional equivalents are techniques such as temporal order of sensations (Duizer et al., 1997) and progressive profiling (DeRovira, 1996).

An alternative view on sensory dynamics is citation frequency (Campo et al., 2010). It differs in that it is not the intensity of the attributes that are noted but the frequency with which they are mentioned by the panel (Fig. 5.57). Citation frequency appears more suited to situations where achieving a better representation of a complex aroma is desired.

An increasingly popular technique, based partially on citation frequency, is termed temporal dominance of sensation (TDS) (Pineau et al., 2009). It assesses the sensory dynamics of a range of attributes, more or less simultaneously, while the food or wine is in the mouth. Panelists are instructed to use a cursor to note (on a computer screen) the most dominant (evident) sensation at any point in time, score its intensity (on a bar associated with each attribute), and continue to record any changes in these perceptions over the duration of oral retention. For wines, this often varies from 1 to 1.5 min. The most dominant perception at any time need not be the most intense. Thus, the method

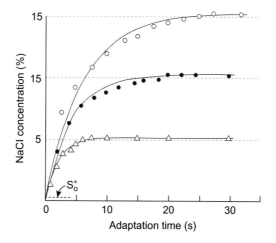

FIGURE 5.55 Adaptation during 30 s of the threshold of perception under stimulation with 5% (△), 10% (●), and 15% (○) sodium chloride solutions. The unadapted threshold S_o^* is 0.24%. The measuring points are from Hahn (1934). *From Overbosch, P. (1986). A theoretical model for perceived intensity in human taste and smell as a function of time. Chem. Senses 11, 315-329. Reproduced by permission of Oxford University Press.*

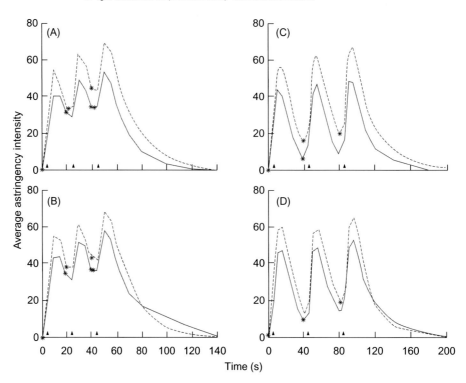

FIGURE 5.56   Average time-intensity (TI) curves for astringency in wine with 0 (——) and 500 (– – –) mg/L of added tannic acid upon three successive ingestions: (A) 8-mL samples, 20s between ingestions; (B) 15-mL samples, 20s between ingestions; (C) 8-mL samples, 40s between ingestions; and (D) 15-mL samples, 40s between ingestions. Sample uptake and swallowing are indicated by a star and an arrow, respectively (n = 24). *From Guinard et al., 1986, reproduced by permission.*

reflects the reality that perception often shifts from attribute to attribute, as well as in intensity, during the course of a tasting. Tastings can be repeated. To avoid the sequence of the attributes listed influencing the results, their order is randomized for each panelist and tasting session. The data are subsequently collated, analyzed, and prepared for graphical representation. The results present the perception of the group for all noted attributes over time (Fig. 5.58). Bars can be added to indicate the level of statistically significant. A worked example of statistical analysis of TDS data is provided in Meyners and Pineau (2010).

The simultaneous assessment of multiple attributes minimizes the potential disadvantage of focusing on a single attribute, that of exaggerating its significance or presence, or result in other attributes inducing halo-dumping. This is particularly relevant to wine, where many attributes are complex (e.g., astringency) or may influence the expression of other modalities. One of the strongest advantages of TDS is that it records features of great importance to aficionados, i.e., flavor development and finish.

Not surprisingly, TDS has shown marked potential for illustrating the complexities of a wine's sensory dynamics (Meillon et al., 2010), and the influences of wine-production techniques on these properties (Sokolowsky et al., 2015). Fig. 5.59 illustrates the complex effects of alcohol removal (A–D), and sugar addition (E), on the perception of individual attributes, as well as their integration (complexity). Figs. 5.60 and 5.61 illustrate how TDS can complement standard intensity assessments, in this case, how different skin-contact times influenced the attributes of a Gewürztraminer wine. Thus, TDS appears to have tremendous potential for clarifying the effects of various modifications in vineyard and winery procedures, which are issues of great practical significance.

TDS also has the potential of representing properties, such as development and duration, if sampling and assessment occur over a sufficiently extended period. Each assessment could be put in sequence to illustrate the changes over the course of tasting. TDS also could be used to detail changes in a wine's finish, by continuing to monitor flavor changes after the wine has been expectorated.

Because of the expanding use of TDS, and its applicability to practical vini-viticulture, there is even more importance to being sure that the panel is performing both accurately and consistently. For research on this aspect, the reader is directed to Meyners (2011) and Lepage et al. (2014).

Comparisons of TI and TDS assessments can be found in Le Révérend et al. (2008) and Pineau et al. (2012). They found that for investigating the temporal dynamics of individual attributes, TI was superior, whereas TDS was better at providing a perception of the whole product as a function of time.

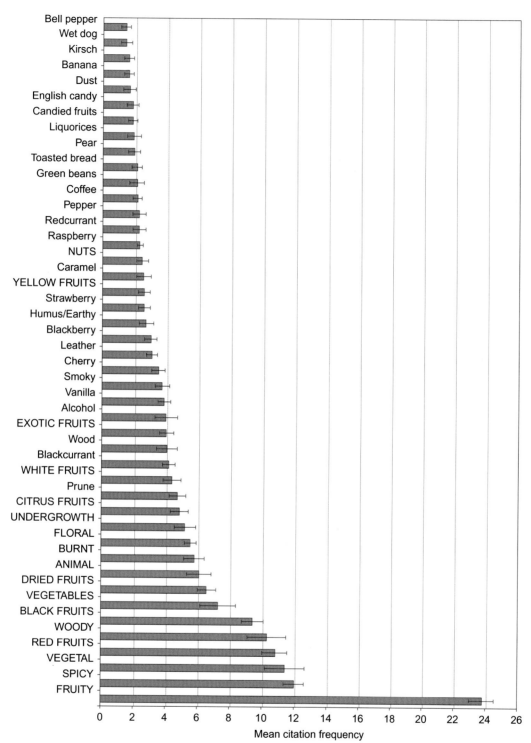

FIGURE 5.57 Use of citation frequency means from odor attributes to study wines. Error bars are calculated as $s/(n)^{1/2}$; (s) standard deviation; (n) number of samples. *Reprinted from Campo, E., Ballester, J., Langois, J., Dacremont, C., Valentin, D., 2010. Comparison of conventional descriptive analysis and a citation frequency-based descriptive method of odor profiling: An application to Burgundy Pinot noir wines. Food Qual. Pref. 21, 44–55, with permission from Elsevier.*

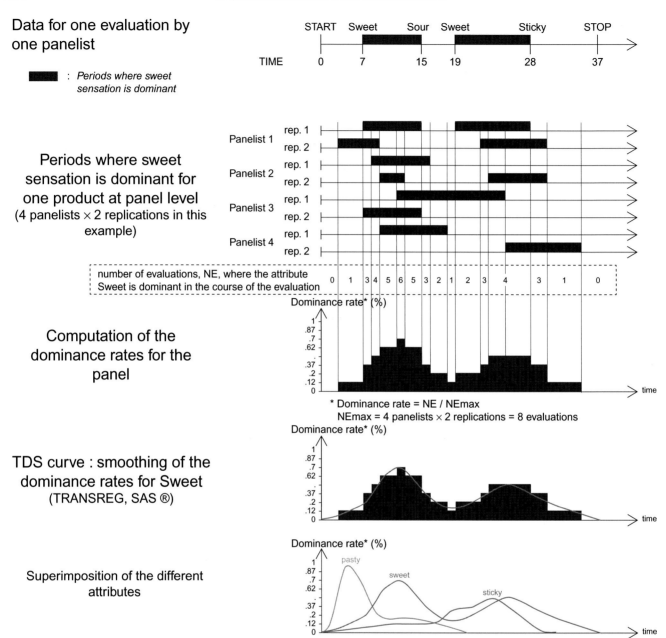

FIGURE 5.58 The methodology of computing temporal dominance of sensations (TDS) curves. *Reprinted from Pineau, N., Schlich, P., Cordelle, S., Mathonnière, C., Issanchou, S., Imbert, A., Rogeaux, M., Etiévant, P., Köster, E., 2009. Temporal Dominance of Sensations: Construction of the TDS curves and comparison with time-intensity. Food Qual. Pref. 20, 450–455, with permission from Elsevier.*

CHEMICAL INDICATORS OF WINE QUALITY AND CHARACTER

Gas Chromatography–Olfactometry (GC-O) Analysis

None of the abovementioned techniques directly assesses the chemical origins of the sensory phenomena they measure. One technique designed to look into this issue, relative to olfactory compounds, is gas chromatography–olfactometry (GC-O) (Debonneville et al., 2002; Plutowska and Wardencki, 2008; d'Acampora Zellner et al., 2008). As the name suggests, the technique combines the advantages of odorant separation and quantitative analysis (via gas chromatography) and their qualitative identification (by humans). Thus, it directly associates chemical presence with potential sensory significance. As aromatic compounds are separated by the gas chromatograph, a

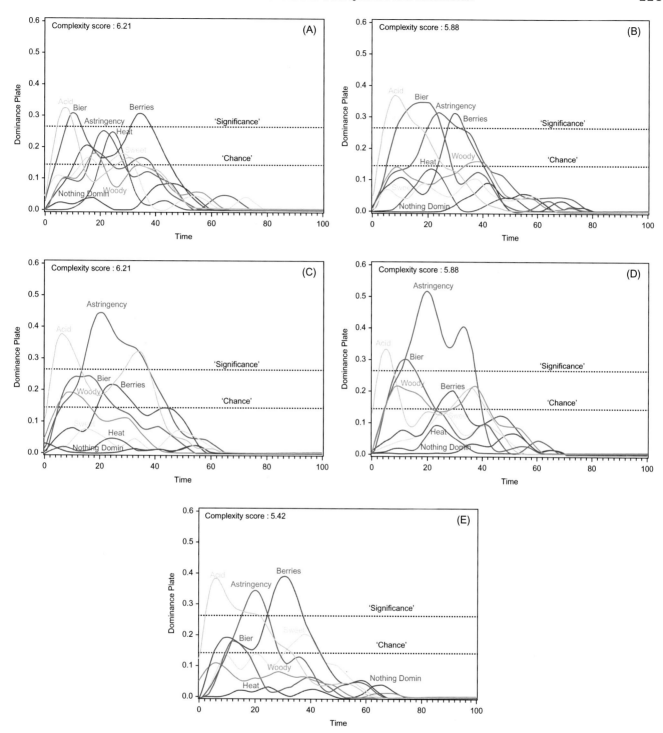

FIGURE 5.59 The effect of alcohol reduction on multiple sensory attributes of a Shiraz wine using temporal dominance of sensations (TDS). (A) unmodified wine (13.5% alcohol) and dealcoholized wines: (B) 11.5%, (C) 9.5%, (D) 7.9%, (E) 7.9% with 8.44 g sugar. *Reprinted from Meillon, S., Viala, D., Medel, M., Urbano, C., Guillot, G., and Schlich, P., 2010. Impact of partial alcohol reduction in Syrah wine on perceived complexity and temporality of sensations and link with preference. Food Qual. Pref. 21, 732–740, with permission from Elsevier.*

taster sniffs the compounds as they exit the device (Fig. 5.62), and suggests an appropriate descriptor name. The terms may come from a list supplied by the researcher or volunteered by the participants. This information is then correlated with the rates at which the compounds were eluded (retention time, *RT*; or indices, *RI*), assessed for their concentration, and identified using procedures such as mass spectroscopy and flame ionization detection

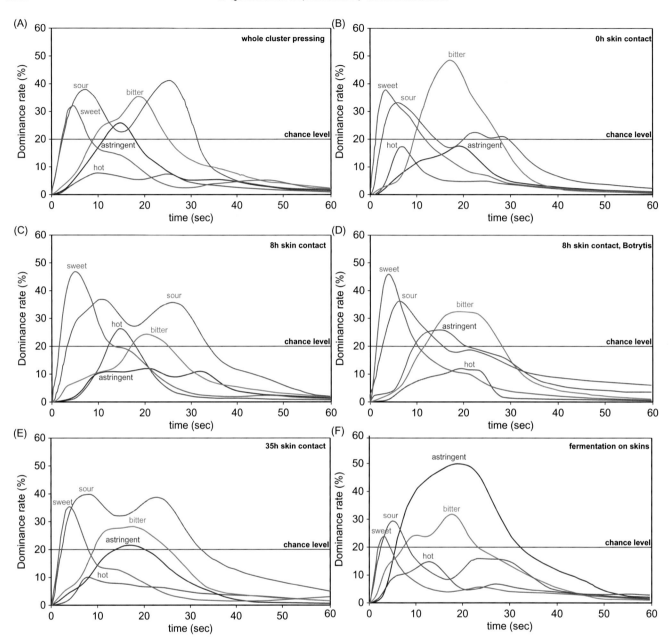

FIGURE 5.60 The effect of different skin-contact procedures as shown by temporal dominance of sensations (TDS) for 2010 Gewürztraminer wines from juice following (A) whole-cluster pressing, (B) 0 h skin contact, (C) 8 h skin contact, (D) 8 h skin contact of berries 30% infected by *Botrytis cinerea*, (E) 35 h skin contact, and (F) fermentation on skins (13 × 3 replicates). *Reprinted from Sokolowsky, M., Rosenberger, A., Fischer, U., 2015. Sensory impact of skin contact on white wines characterized by descriptive analysis, time–intensity analysis and temporal dominance of sensations analysis. Food Qual. Pref. 39, 285–297, with permission from Elsevier.*

(Fig. 5.63). By repeating the process with various assessors on different occasions, or with different instruments, odorant chemistry can be correlated with qualitative descriptors. In this manner, potentially significant aromatics can be identified from the hundreds of odorants typically present in wine. These can subsequently be subjected to a range of techniques designed to investigate their potential impact in generating the olfactory attributes that characterize the particular wine.

Several dilution techniques address the issue of the nonlinear relation between chemical composition and perceived intensity (a function of Stevens' law). This can involve measuring a compound's olfactory potency, or the dilution point at which its presence is no longer detected. Termed aroma extract dilution analysis (AEDA), it provides a hierarchal list of the relative potency of the compounds detected. A variant procedure, termed combined

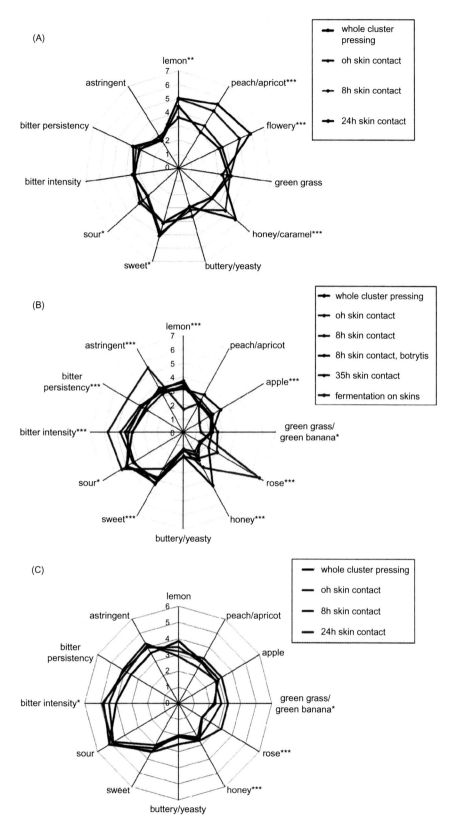

FIGURE 5.61 Aromatic and in-mouth attribute mean scores from descriptive analysis of wines varying in skin-contact time: (A) 2009 Gewürztraminer, (B) 2010 Gewürztraminer, and (C) Riesling (16 or 17 × 3 replications, respectively). Levels of significance: *p < 0.05; **p < 0.01; ***p < 0.001). *Reprinted from Sokolowsky, M., Rosenberger, A., Fischer, U., 2015. Sensory impact of skin contact on white wines characterized by descriptive analysis, time–intensity analysis and temporal dominance of sensations analysis. Food Qual. Pref. 39, 285–297, with permission from Elsevier.*

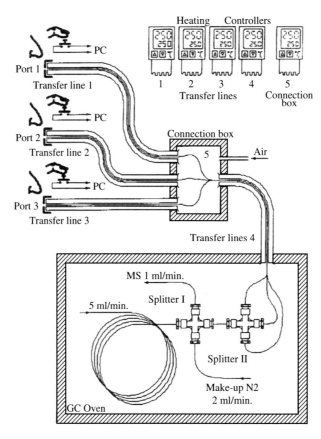

FIGURE 5.62   Scheme of multisniffing hardware for gas chromatography–olfactometry. *Reprinted with permission from Debonneville, S., Orsier, B., Flament, I., Chaintreau, A., 2002. Improved hardware and software for quick gas chromatography-olfactometry using Charm and GC-"SNIF" analysis. Anal. Chem. 74, 2345–2351. Copyright 2002, American Chemical Society.*

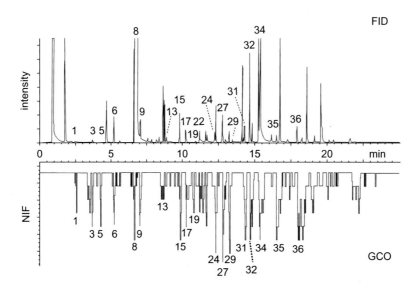

FIGURE 5.63   Example of a comparison of the aroma profiles of a Croatian Rhine Riesling wine obtained by gas chromatography with flame-ionization detector (FID) (upper) and olfactometry detector (GC-O) (lower), using the nasal impact frequency (NIF) detection method. *Reprinted from Komes, D., Ulrich, D., Lovric, T., 2006. Characterization of odor-active compounds in Croatian Rhine Riesling wine, subregion Zagorje. Eur. Food Res. Technol. 222, 1–7, with kind permission of Springer Science + Business).*

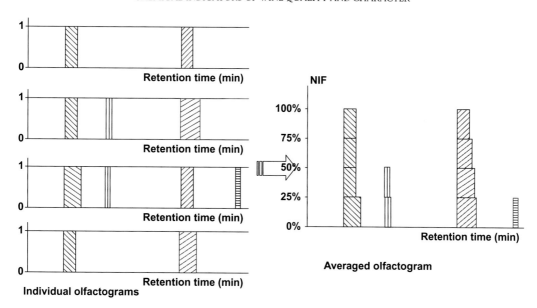

FIGURE 5.64 Scheme of olfactogram forming in detection frequency methods, when four evaluators participate in the experiment. *Reprinted from Plutowaska, B., Wardencki, W., 2008. Application of gas chromatography-olfactometry (GC–O) in analysis and quality assessment of alcoholic beverages – A review. Food Chem. 107, 449–463, with permission from Elsevier.*

hedonic aroma response measurement (CHARM), measures how long the odor is detected. Dilutions are presented in a random order to avoid expectation becoming part of a panelist's response. In contrast, detection frequency methods, such as nasal impact frequency (NIF) (Pollien et al., 1997), use only one dilution level, where the peak detection of several panelists (minimum of 6, preferably 8–10) are summed to determine the percentage detection (Fig. 5.64). Data analysis is based on detection frequency versus successive dilutions. The procedure is considered independent of assumptions based on the interaction between concentration and perceived intensity, and individual variances in response. A third set of techniques employs time-intensity methods, for example finger-span cross-modality matching (FSCM). It has been used to analyze and compare the olfactory characteristics of wines (Etiévant et al., 1999; Bernet et al., 2002). Its advantage is that it avoids extensive training to achieve an adequate level of consistency in term use. The procedure generates scaled-intensity measures of specific aroma compounds as they leave the chromatograph, potentially useful in indicating chemical differences between samples.

These procedures could also be used to analyze the abilities of panelists, and select those with a desired degree of sensory acuteness (Vene et al., 2013, Fig. 5.65).

Wine Color

There has been a surprisingly good correlation between color density measurements and the assessed quality of certain red wines (Fig. 8.12). Although fascinating, Somers (1975) did not assess whether similar results were obtained when the same wines were tasted blind. Thus, learned associations between wine color and quality could explain this correlation, rather than the flavor attributes of the wines. The influence of wine color on quality perception was first demonstrated in a study on rosé wines (André et al., 1970). This prejudicial influence has been repeatedly shown subsequently, most markedly by Morrot et al. (2001) (Fig. 3.33). Color may be more influential in affecting wine evaluation than generally appreciated.

The association between color and perceived quality is not unexpected, being as it is correlated with grape maturity, flavor development, and the duration of maceration (anthocyanin, tannin, and flavor extraction). However, it is equally clear that dark red wine is not consistently highly regarded, e.g., the so-called black wines of Cahors. In wine competitions, it is usual to group wines by category, thereby partially offsetting color differences among wines based on the cultivars used, the wine's age, its region of origin, or stylistic nature. Thus, the judges typically will know in advance what to expect, relative to color, for the wines being assessed.

The closest chemical association between the color of a young red wine and its quality appears to be based on the proportion of colored anthocyanins (notably ionized flavylium forms). Similar color-quality associations have

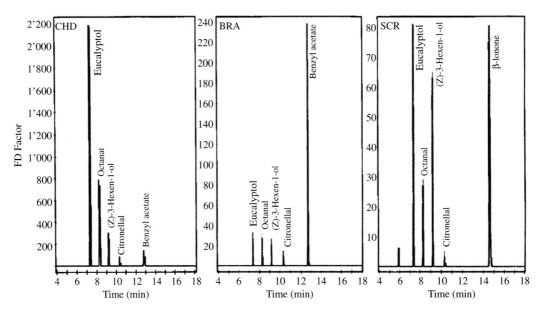

FIGURE 5.65 Example of the variation in sensitivity among three panelists shown by combined hedonic aroma response measurement (CHARM) analysis. *Reprinted with permission from Debonneville, S., Orsier, B., Flament, I., Chaintreau, A., 2002. Improved hardware and software for quick gas chromatography-olfactometry using Charm and GC-"SNIF" analysis. Anal. Chem. 74, 2345–2351. Copyright 2002, American Chemical Society.*

been obtained with visible near-infrared spectroscopy (Dambergs et al., 2002; Cozzolino et al., 2008). They support the contention that anthocyanin content is a major predictor of perceived red wine quality (Dambergs et al., 2002; Fig. 5.66). Spectroscopy has been less successful in predicting the quality of white wines. This may relate to the quality of white wines being more associated with its fragrance, and inversely with phenolic content.

Were spectroscopic techniques ever to be perfected, and sufficiently simplified to function as an effective indicator of quality, they could greatly assist wineries as well as wine competition organizers. However, it seems unlikely that any single instrument will ever be able to quantify the multiplicity of features involved in actual sensory wine quality. Nonetheless, even partial success could facilitate the initial assessment of quality, highlighting the best samples in a competition worthy of detailed evaluation. The result would be a marked savings in time and reduction in judge fatigue. Analytic equipment has many advantages over human assessors: it is objective, accurate, rapid, reproducible, available at any time desired, does not suffer from exhaustion or adaptation, possesses no experience-based biases, and is not influenced by psychological factors.

Electronic Noses

Electronic noses (e-noses) are one of the more interesting developments in objective sensory analysis (Martí et al., 2005; Röck et al., 2008; Korotcenkov, 2013). Some are already produced in handheld models. Most detector systems possess a collection of sensors composed of an electrical conductor (such as carbon black) homogeneously embedded in a nonconducting absorptive polymer. Each sensor possesses a distinctive polymer that differentially absorbs a range of aromatic compounds. Each disc possesses a pair of electrical contacts bridged by a composite film (such as alumina). When volatile compounds pass over the sensors, absorption causes the disc to swell (Fig. 5.67A). Swelling separates the electrically conductive particles, resulting in a rise in electrical resistance. The speed and degree of resistance recorded from each sensor generate a unique fingerprint (smellprint) of the compounds in a sample (Fig. 5.67B). The response pattern generated by known odorants is used in interpreting the output from the detectors. As the number of sensors increases, so does the instrument's potential discriminating power. Commercial e-nose systems, such as Cyranose 320 (Plate 5.16), possess 32 distinct sensors. After each sample, as the aromatics absorbed escape into and are vented by a flushing gas, the sensors return their original size and sensitivity.

Electronic (e-) tongues are similar in concept, but the receptors respond to chemicals in solution rather than in air. They have been used experimentally to discriminate among white wines (Pigani et al., 2008), monitor maturation in oak cooperage (Parra et al., 2006), estimate port aging (Rudnitskaya et al., 2007), quantify the total phenolic content of wines (Cetó et al., 2012), and measure metallic content (da Costa et al., 2014). Although useful, there is significant effort involved before the data can be interpreted in a meaningful way. It is not as yet "off-the-shelf" technology.

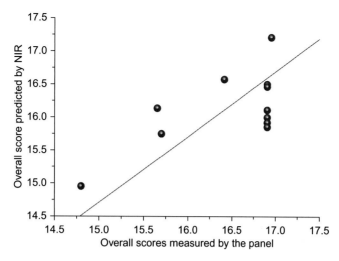

FIGURE 5.66 Near-infrared predicted scores versus reference value for commercial Shiraz wine in validation of wine-show judges. *With kind permission from Springer Science + Business Media: Cozzolino, D., Cowey, G., Lattey, K.A., Godden, P., Cynkar, W.U., Dambergs, R.G., Janik, L., Gishen, M., 2008. Relationship between wine scores and visible–near-infrared spectra of Australian red wines. Anal. Bioanal. Chem. 391, 975–981.*

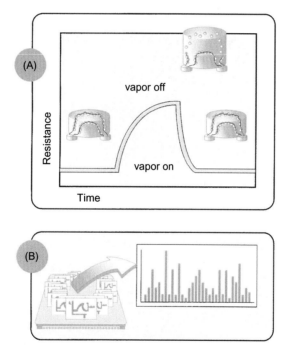

FIGURE 5.67 Most electronic noses consist of a series of sensors that, on absorbing a compound, start to swell, increasing the electrical resistance, until current flow stops (A). The different absorption properties of the sensors generate patterns that are often unique to a particular aromatic object or substance (B). *Courtesy of Cyrano Sciences, Inc., Pasadena, CA.*

Electrical resistance data from each sensor are adjusted (smoothed) for background electrical noise before being analyzed. The data can generate a histogram but, because of the complexity of the pattern and overlaps, are typically subjected to one or more statistical recognition tests. These tests usually include principal component analysis and mathematical algorithms. The patterns generated (Fig. 5.68) resemble the principal component relationships derived from descriptive sensory analysis. Computer software supplied with the device conducts the mathematical analyses automatically.

When the data are combined with computer neural networks (Fu et al., 2007), recognizable odor patterns can be derived. However, reference samples need to be chosen carefully to represent the full range of characteristics

PLATE 5.16   The Cyranose 320 electronic nose. *photo courtesy of Cyrano Sciences, Inc., Pasadena, CA.*

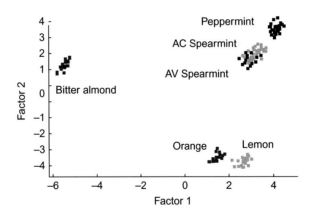

FIGURE 5.68   Principal factor analysis of different essential oils using the Cyranose 320 electronic nose. *courtesy of Cyrano Sciences, Inc., Pasadena, CA.*

found in the wines tested. Similar to sensory descriptive analysis, but unlike traditional analytic techniques, separation, identification, and quantification of individual volatile compounds are not obtained. This potential drawback may be reduced with the introduction of genetically engineered olfactory proteins (Wu and Lo, 2000), or peptide receptor-based detectors (e.g., to estimate bitterness and astringency from the wine's phenolic content (Umali et al., 2015). Such improvements in receptor specificity would dramatically increase sensitivity and reduce "noise" from nontarget chemicals.

A different form of e-nose technology is based on flash chromatography. Unlike traditional gas chromatography, samples can be analyzed in a few seconds, rather than hours. In the zNose (Plate 5.17), a sample is initially preconcentrated in a Tenax trap, before heat-vaporizing the aromatics into a stream of helium gas. The gas passes through a heated, 1-m long capillary column to separate the constituent volatiles. When the chemicals emerge from the column, they are directed onto a quartz surface acoustic wave (SAW) detector. As the aromatics absorb and desorb from the detector, the frequency of sound emitted from the acoustic detector changes. The frequency change is recorded every 20 msec, with the degree of change indicating concentration. Because the acoustic impulse can be correlated

PLATE 5.17   z-Nose sampling machine. *photo courtesy of Electronic Sensor Technology, Newbury Park, CA.*

with retention time on the capillary column, identification of the compounds is possible. This involves reference standards, as is typical with traditional gas chromatography. What is especially appealing is that the data can be quickly presented in a manner reminiscent of polar plots. As such, they visually represent aspects of the wine's aromatic nature. Because the presence or concentration of most wine aromatics is not unique to particular wines or cultivars, the instrument can be instructed to record only the emissions from selected compounds. This presents a clearer visual representation of the distinctive aromatic character of the wine. Adjustment of features such as preconcentration, coil temperature, and acoustic sensor temperature can influence both sensitivity and chemical resolution.

Present z-Nose models primarily resolve and quantify hydrocarbons. Because wines contain many hydrocarbons, their dominant presence could mask the cooccurrence of significant impact compounds. For example caprylate and caproate might mask the presence of the off-odor caused by 2,4,6-TCA (Staples, 2000). In addition, certain impact compounds, specific to particular cultivars, cannot currently be resolved with flash chromatography. This drawback may disappear as improvements in carrier systems are developed. Nevertheless, current models can readily distinguish between several varietal wines (Fig. 5.69).

E-nose technology is already becoming routinely used in food quality control (Peris and Escuder-Gilabert, 2009). Its speed, accuracy, economy, portability, and applicability are its primary benefits, but the ability to correlate the data with panel-developed standards permits e-noses to replace panels under some conditions. E-nose technology is also finding useful applications in disease detection, the cosmetic and pharmaceutical industries, as well as criminal, security, and military investigations.

The potential of electronic noses to quickly measure important varietal fragrances (Lozano et al., 2005; Fig. 5.70) could facilitate the timing of grape harvests to achieve predetermined, desired flavor characteristics. Fig. 5.71 illustrates the potential difference in aromatic complexity that can exist between fully and artificially ripened fruit. Nonetheless, there is considerable difference between the sensitivity of analytic equipment and the human nose, analytic instruments being better with some constituents, whereas the nose is superior with others. Even more importantly, there is still a huge gap between our knowledge of a wine's chemistry and those attributes that humans consider important in differentiating between wines. Thus, as in art, paint pigments do not a great work define.

Nonetheless, e-nose technology may supply a useful quantitative quality-control tool. Potential uses include discriminating among oak-barrel toasting levels (Chatonnet and Dubourdieu, 1999), monitoring aroma production during fermentation (Pinheiro et al., 2002), and detecting contamination with 2,4,6-TCA or *Brett* off odors (4-ethylphenol and 4-ethylguaiacol) (Cynkar et al., 2007). In addition, e-nose technology has shown some success in differentiating among wines from a single winery (García et al., 2006), and wines from the same cultivar but different regions (Buratti et al., 2004). Furthermore, some terms used to describe wine attributes correlate with e-nose measurements (Lozano et al., 2007).

Although significant advances are occurring rapidly in this field, there are still significant hurdles in using electronic nose technology in routine wine analysis. Wines are extremely complex chemically, consisting of hundreds of components, ranging from ng/L to g/L concentrations, and spanning a vast range of polarities, solubilities,

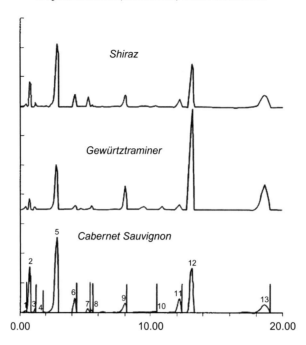

FIG. 5.69   Example of the differentiation possible between varietal wines using z-nose, a chromatographic form of electronic nose technology. *courtesy of Electronic Sensor Technology, Newbury Park, CA.*

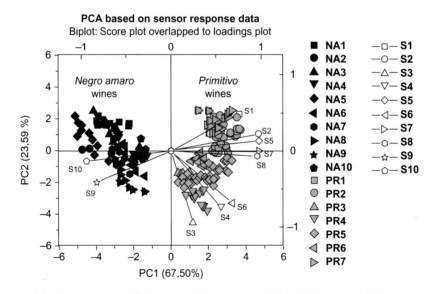

FIGURE 5.70   Illustration of the distinction possible between 10 Negro amaro (*NA*) and 7 Primitivo (*PR*) wines using electronic-based detection systems. Sensors S1–S8 were more correlated to Primitivo versus sensors S9–S10, which were more correlated with Negro amaro. *Reprinted from Capone, S., Tufariello, M., Francioso, L., Montagna, G., Casino, F., Leone, A., Siciliano, P., 2013. Aroma analysis by GC/MS and electronic nose dedicated to Negroamaro and Primitivo typical Italian Apulian wines. Sens. Actuat. B 179, 259–269, with permission from Elsevier.*

volatilities, and pHs. Often extraction and concentration are required to detect the compounds, leading to the possibilities of degradation and artifact production.

Humans will always be needed to assess the sensory significance of taste, mouth-feel, and olfactory sensations, as well as the cerebral constructs of flavor, complexity, balance, and body. After all, we are the consumers, not machines. Nonetheless, electronic noses are likely to join panels in many situations in which people have been used in lieu of objective olfactory instruments (Fig. 5.72). In addition, panels of trained tasters will coexist with instrumental analysis as the chemical nature of sensory perception is clarified. Once this goal is achieved (if ever), the

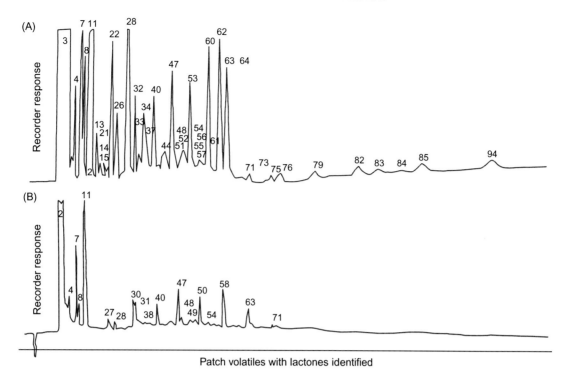

FIGURE 5.71 Illustration of the significantly increased aromatic complexity of tree-ripened (A) versus artificially ripened (B) peaches. *From Do et al., 1969, reproduced by permission.*

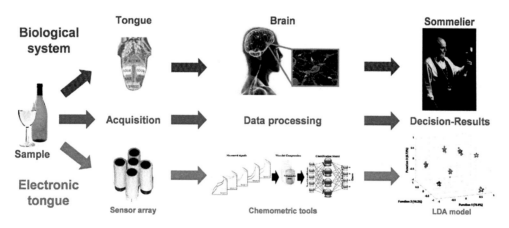

FIGURE 5.72 Comparison of the recognition process of a sample by the biological (top) and the biomimetric system (bottom). *Reprinted from Cetó, X., González-Calabuig, A., Capdevila, J., Puig-Pujol, A., and del Valle, M., 2015. Instrumental measurement of wine sensory descriptors using a voltammetric electronic tongue. Sensor Actuator B 207, 1053–1059, with permission from Elsevier.*

ease, efficiency, reliability, and accuracy of analytic equipment may replace the expense and imprecision of human panels for objective assessments in many routine areas of sensory evaluation.

OCCUPATIONAL HAZARDS OF WINE TASTING

It is atypical to consider occupational hazards in relation to wine tasting. If any were acknowledged, it would probably have been concern about triggering latent alcoholism. While legitimate, the practice of expectoration should minimize such a likelihood. Scholten (1987) reported that expectoration avoided a significant rise in blood alcohol level during wine tasting. Despite this, it may be judicious for women, pregnant, or desirous of becoming so, to excuse themselves from extended tastings, as well as anyone with a family history of alcoholism.

Headache induction is another well-known potential hazard of wine tasting, however, usually associated with excess consumption, not tasting. For some people, though, consumption of even small amounts of white or red wines, or their combination, can initiate a headache. Currently, there seems no established explanation for the phenomenon. However, for those who find young red wines potent activators, headache activation may relate to the concentration of small phenolics. Unlike their polymers (tannins) that accumulate as the wine ages, monomeric phenolics traverse the intestinal lining and enter the blood easily. Typically, absorbed phenolics are rapidly detoxified (o-methylated or sulfated) by plasma enzymes. However, some phenolics suppress the action of platelet phenolsulfotransferase (PST) (Jones et al., 1995). Regrettably, the level of PST tends to decline with age, as does so much else. Low levels of platelet-bound PST are apparently correlated with migraine susceptibility (Alam et al., 1997). Reduced PST also limits the sulfation (detoxification) of a variety of endogenous and xenobiotic (foreign) compounds, including phenolics and biogenic amines. Without inactivation, biogenic amines (found in low concentrations in most wines) may activate the liberation of 5-hydroxytryptamine (5-HT, or serotonin), an important brain neurotransmitter. 5-HT also induces the dilation of small blood vessels in the brain. It can provoke intracranial pain (stretching of the meninges that cover the brain), which may incite a migraine (Pattichis et al., 1995). In addition, if phenolics absorbed from wine are not inactivated rapidly in the blood, they are likely to be oxidized to o-quinones. If these traverse the blood–brain barrier, o-quinones could inhibit the action of catechol-O-methyltransferase (COMT). This limits the breakdown of the neurotransmitter dopamine, and the availability of µ-opioid (painkilling) receptors. Consequently, the perception of pain associated with cerebral blood vessel dilation may be exacerbated.

Also, potentially related to wine-induced headaches is the release of E-prostaglandins. These compounds are involved in cerebral vessel dilation. This may explain the action of prostaglandin-synthesis inhibitors, such as acetylsalicylic acid, acetaminophen, and ibuprofen in wine-induced headache prevention. For maximal benefit, any of these prostaglandin-synthesis inhibitors should be taken in advance of tasting (Kaufman, 1992). Personally, they limit both headache induction and the facial flushing associated with assessing multiple wines over a short period. Small phenolics also prolong the action of potent hormones and nerve transmitters, such as histamine, serotonin, dopamine, adrenalin, and noradrenaline. These could also influence the severity of headaches (and various allergies and sensitivities).

Biogenic amines, such as histamine and tyramine, have often been suggested as an activator of headache induction. Nonetheless, double-bind tests with self-professed sensitive individuals have not supported this contention (Masyczek and Ough, 1983). In addition, the levels of biogenic amines in wines are usually lower than those considered sufficient to activate a migraine. Nonetheless, alcohol can suppress the action of diamine oxidase, an important intestinal enzyme that inactivates histamine and other biogenic amines (Jarisch and Wantke, 1996). However, this does not match the correlation between histamine content and those products most frequently associated with migraine attacks, i.e., spirits and sparkling wines (Nicolodi and Sicuteri, 1999). Both are lower in histamine content that other wines or alcohol-containing beverages.

The sulfite content of wines has often been noted as a potential culprit in wine-related headaches. However, as yet, no scientifically validated evidence of this has been presented. That sulfites can cause asthma in sensitive individuals is clear, but is not directly involved in headache induction. With the continuing reduction in the use of sulfur dioxide, its involvement as an occupational hazard for wine tasters is diminishing.

Potentially the principal occupational hazard for the professional wine taster is dental erosion (Mandel, 2005; Fig. 5.73). This results from the frequent and extended exposure to wine acids. Dissolving calcium (Fig. 5.74) can lead to enamel softening (Lupi-Pegurier et al., 2003) and dentine erosion that can affect both tooth shape and size. Cupping, a depression in the enamel (exposing dentine at the tip of molar cusps), is a frequent clinical sign. Erosion can also contribute to severe root abrasion at the gum line. Protection is partially achieved by rinsing the mouth with an alkaline mouthwash after tasting, application of a fluoride gel (such as acidulated phosphate fluoride, or APF) or Tooth Mousse, and refraining from tooth brushing for at least one hour after tasting (see Ranjitkar et al., 2012). The delay in brushing permits minerals in the saliva to rebind to enamel. The problem is not known to affect consumers who take wine with meals. Food and saliva secretion limit, if not prevent, demineralization of tooth enamel and dentine.

Nonetheless, wine may also contribute to oral health. Proanthocyanidins in wine can limit the adherence and biofilm-forming activity of caries-inducing *Streptococcus mutans* (Daglia et al., 2010). Gibbons (2013) provides a fascinating insight into the association of this bacterium with changes in human diet, which resulted from a switch from a hunter–gather to an agricultural lifestyle.

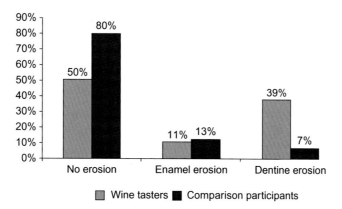

FIGURE 5.73 Frequency and severity grade of dental erosive wear for wine tasters and a comparison group. *Reprinted from Mulic, A., Tveit, A. B., Hove, L. H., Skaarf, A. B., 2011. Dental erosive wear among Norwegians wine tasters. Acta Odontol. Scand. 69, 21–26, reproduced by permission from Taylor & Francis Ltd, www.tandfonline.com.*

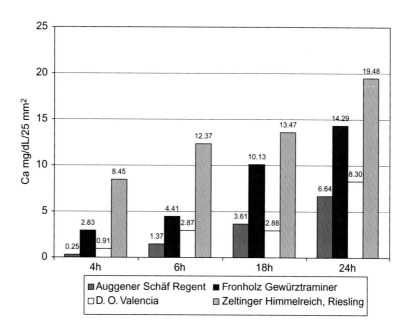

FIGURE 5.74 Time-dependent calcium release (mg dL^{-1} 25 mm^{-2}) from enamel samples after separate incubation in 3 mL of wine. Red: Auggener Schäf Regent (pH 4.02), D.O. Valencia (pH 3.44). White: Fronholz Gewürztraminer (pH 3.65), Zeltinger Himmelreich Riesling (pH 2.99) for 4, 6, 18, and 24 hours. *Reprinted from Willershausen, B., Callaway, A., Azrak, B., Kloß, C., Schulz-Dobrick, B., 2009, Prolonged in vitro exposure to white wines enhances the erosive damage on human permanent teeth compared with red wines. Nutr. Res. 29, 558–567, with permission from Elsevier.*

Another complex situation that may relate to wine tasting involves the potential for interaction between wine constituents and certain medications (Adams, 1995; Fraser, 1997; and Weathermon and Crabb, 1999). However, since wine is not consumed during professional tasting sessions, this is unlikely to be a significant issue.

Finally, a potential occupational "hazard" is a result of changed perspective, where insipid, characterless wines are no longer acceptable, despite price or repute. Daily wine must, at least, not insult one's sensibilities and be both interesting and flavorful. Thus, the vast majority of red, and many white, wines may become no longer acceptable, and much effort now becomes required to locate good wines at a modest price. In addition, other than being appreciated at wine outlets for your above-average wine purchases, you may become an annoyance due to your return of faulty wines.

POSTSCRIPT

Much of the discussion in this chapter has dealt with the complexities involved with selecting, training, and using panels of assessors in various sensory tests. This involves not only issues relating to inherent human variability, but also the effects of training in enhancing an existing gap between their percepts of wine and those of the majority of consumers. At one end of the spectrum are panels intended to function as analytic equipment. They investigate the sensory attributes that distinguish particular types of wines, the features that may appeal to consumers, the sensory effects of grape-growing and winemaking procedures, and the nature of the psychophysiology of sensory perception. In the middle of the spectrum are panels (or individuals) whose function is to assess the subjective quality of wines and their origins. These typically consist of wine "experts" who may also provide purchase guidance. Wine aficionados are their consumer equivalents. At the other end of the spectrum are panels of consumers used to study what features they prefer, assess purchase intentions, and investigate those intrinsic and extrinsic factors involved in purchase. The more commercial-/consumer-oriented aspects of wine assessment are the principal topics of Chapter 6.

Ideally, the primary function of sensorially evaluating wine is to discover those chemical attributes that generate wines that people will savor. Ultimately, this should guide grape growers and winemakers to produce better (or at least better appreciated) wines. However, because consumers are seldom interested in wine chemistry (usually negatively so), what they are concerned with is the psychological pleasure wines can donate. For most consumers, this is simply sensory pleasure. For others, however, it is imbued with extraneous aspects associated with culture, status, and, yes, sophistication or elitism. These latter factors are largely beyond the control of winemakers and grape growers. Nonetheless, technical advances in grape production and winemaking suggest a bright future for achieving fine quality, affordable wine for all.

Suggested Readings

Adams, W.L., 1995. Interactions between alcohol and other drugs. Int. J. Addict. 30, 1903–1923.

Bi, J., 2006. Sensory Discrimination Tests and Measurements: Statistical Principles, Procedures and Tables. Blackwell, Ames, IA.

Bower, J.A., 2013.) Statistical Methods for Food Science: Introductory Procedures for the Food Practitioner. Wiley Blackwell, West Sussex, UK.

Buglass, A.J., Caven-Quantrill, D.J., 2013. Instrumental assessment of the sensory quality of wine. In: Kilcast, D. (Ed.), Instrumental Assessment of the Food Sensory Quality. A Practical Guide, Woodhead Publ. Ltd., Cambridge, U.K, pp. 466–546.

De Vos, E., 2010. Selection and management of staff for sensory quality control. In: Kilcast, D. (Ed.), Sensory Analysis for Food and Beverage Quality Control. A Practical Guide, Woodhead Publ. Ltd., Cambridge, U.K, pp. 17–36.

Dijksterhuis, G., 1995. Multivariate data analysis in sensory and consumer science: An overview of developments. Trend Food Sci. Technol. 6, 206–211.

Dijksterhuis, G.B., Piggott, J.R., 2001. Dynamic methods of sensory analysis. Trends Food Sci. Technol. 11, 284–290.

Durier, C., Monod, H., Bruetschy, A., 1997. Design and analysis of factorial sensory experiments with carry-over effects. Food Qual. Pref. 8, 141–149.

Earthy, P.J., MacFie, H.J.H., Hederley, D., 1997. Effect of question order on sensory perception and preference in central location trials. J. Sens. Stud. 12, 215–238.

Köster, E.P., 2003. The psychology of food choice: Some often encountered fallacies. Food Qual. Pref. 14, 359–373.

Köster, E.P., 2005. Does olfactory memory depend on remembering odors? Chem. Senses 30, i236–i237.

ISO, 2006. Sensory analysis—Methodology—Initiation and training of assessors in the detection and recognition of odours. #5496. International Standards Organization, Geneva, Switzerland.

ISO, 2007. Sensory analysis—General guidance for the design of test rooms. #8589. International Standards Organization, Geneva, Switzerland.

ISO, 2008a. Sensory analysis—General guidance for the selection, training and monitoring of assessors—Part 2: Expert sensory assessors. # 8586. International Standards Organization, Geneva, Switzerland.

ISO, 2008b. Sensory analysis—Vocabulary. #5492. International Standards Organization, Geneva, Switzerland.

Lawless, L.J.R., Civille, G.V., 2013. Developing lexicons: A review. J. Sens. Stud. 28, 270–281.

Lawless, H.T., Heymann, H., 2010. Sensory Evaluation of Food: Principles and Practices, 2e. Springer, New York.

Lehrer, A., 2009. Wine and Conversation, second ed. Oxford University Press, Oxford, UK.

Leland, J.V., Scheiberle, P., Buettner, A., Acree, T.E. (Eds.), 2001. Gas Chromatography-Olfactometry. The State of the Art. ACS Symposium Series, No. 782, Oxford University Press, Oxford, UK.

Meilgaard, M.C., Civille, G.V., Carr, T.C., 2006.). Sensory Evaluation Techniques, fourth ed. CRC Press, Boca Raton, FL.

Meullenet, J.-F., Heymann, H., Xiong, R., Findlay, C., 2007. Multivariate and Probabilistic Analyses of Sensory Science Problems. IFT Press (Blackwell Publishing, Ames, IA.

Oreskovich, D.C., Klein, B.P., Sutherland, J.W., 1991. Procrustes analysis and its applications to free-choice and other sensory profiling. In: Lawless, H.T., Klein, B.P. (Eds.) Sensory Science Theory and Application in Foods, Dekker, New York, pp. 353–393.

Pinheiro, C., Rodrigues, C.M., Schäfer, T., Crespo, J.G., 2002. Monitoring the aroma production during wine-must fermentation with an electronic nose. Biotechnol. Bioengin. 77, 632–640.

Rantz, J.M. (ed.). (2000). Sensory Symposium. In Proc. ASEV 50th Anniv. Ann. Meeting, Seattle, WA., June 19–23, 2000. pp. 3-8. American Society for Enology and Viticulture, Davis, CA.

Stevenson, R.J., Boakes, R.A., 2008. A mnemonic theory of odor perception. Psychol. Rev. 110, 340–364.

Stone, H., Bleibaum, R., Thomas, H.A., 2012. *Sensory Evaluation Practices*, fourth ed. Academic Press, London, UK.

Website

Projective Mapping: http://www.sensorysociety.org/knowledge/sspwiki/Pages/Projective Mapping.aspx

A Guide to Analyze Sensory Evaluation Test Data Using Spss Software. https://www.youtube.com/watch?v=Ka2TJKoXU_E

References

Acree, T.E., 1997. GC/O Olfactometry. Anal. Chem. 69, 170A–175A.

Alam, Z., Coombes, N., Waring, R.H., Williams, A.C., Steventon, G.B., 1997. Platelet sulphotransferase activity, plasma sulfate levels, and sulphation capacity in patients with migraine and tension headache. Cephalalgia 17, 761–764.

Allison, A.-M.A., Chambers, D.H., 2005. Effects of residual toothpaste flavor on flavor profiles of common foods and beverages. J. Sens. Stud 20, 167–186.

Alvelos, H., Cabral, J.A.S., 2007. Modelling and monitoring the decision process of wine tasting. Food Qual. Pref. 18, 51–57.

Amerine, M.A., Roessler, E.B., 1983. Wines, Their Sensory Evaluation, second ed Freeman, San Francisco, CA.

Amerine, M.A., Pangborn, R.M., Roessler, E.B., 1965. Principles of Sensory Evaluation of Foods. Academic Press, New York, NY.

André, P., Aubert, S., Pelisse, C., 1970. Contribution aux études sur les vins rosés meridionaux. I. La couleur. Influence sur la degustation. Ann. Technol. Agric. 19, 323–340.

Anonymous, 1994. OIV standard for international wine competitions. Bull. O.I.V. 67, 558–597.

Aqueveque, C., 2015. The influence of experts' positive word-of-mouth on a wine's perceived quality and value: The moderator role of consumers' expertise. J. Wine Res. 26, 181–191.

Areni, C.S., Kim, D., 1993. The influence of background music on shopping behavior. Classical versus top-forty music in a wine store. Adv. Consum. Res. 20, 336–340.

Ariely, D., 2010. Predictably Irrational: The Hidden Forces That Shape Our Decisions. Harper Collins, New York, NY.

Aubry, V., Sauvageot, F., Etiévant, P., Issanchou, S., 1999. Sensory analysis of Burgundy Pinot noir wines. Comparison of orthonasal and retronasal profiling. J. Sensory Stud. 14, 97–117.

Ballester, J., Patris, B., Symoneaux, R., Valentin, D., 2008. Conceptual vs. perceptual wine spaces: Does expertise matter? Food Qual. Pref. 19, 267–276.

Bécue-Bertaut, M., Álvarez-Esteban, R., Pagès, J., 2008. Rating of products through scores and free-text assertions: Comparing and combining both. Food Qual. Pref. 19, 122–134.

Bende, M., Nordin, S., 1997. Perceptual learning in olfaction: Professional wine tasters versus controls. Physiol. Behav. 62, 1065–1070.

Bernet, C., Dirninger, N., Claudel, P., Etiévant, P., Schaeffer, A., 2002. Application of Finger Span Cross Modality Matching Method (FSCM) by naive accessors for olfactometric discrimination of Gewürztraminer wines. Lebensm.-Wiss. u -Technol. 35, 244–253.

Bi, J., 2003. Agreement and reliability assessments for performance of sensory descriptive panel. J. Sens. Stud. 18, 61–76.

Bi, J., Ennis, D.M., 1998. Sensory thresholds: Concepts and methods. J. Sens. Stud. 13, 133–148.

Borgognone, M.G., Bussi, J., Hough, G., 2001. Principal component analysis in sensory analysis: Covariance or correlation matrix? Food Qual. Pref. 12, 323–326.

Bower, J.A., 2013. Statistical Methods for Food Science: Introductory Procedures for the Food Practitioner. Wiley Blackwell, West Sussex, UK.

Brochet, F., Dubourdieu, D., 2001. Wine descriptive language supports cognitive specificity of chemical senses. Brain Lang. 77, 187–196.

Brochet, F., Morrot, G., 1999. Influence du contexte sur la perception du vin. Implications cognitives et méthodologiques. J. Int. Sci. Vigne Vin 33, 187–192.

Buratti, S., Benedetti, S., Scampicchio, M., Pangerod, E.C., 2004. Characterization and classification of Italian Barbera wines by using an electronic nose and an amperometric electronic tongue. Anal. Chim. Acta 525, 133–139.

Cain, W.S., 1979. To know with the nose: Keys to odor identification. Science 203, 467–469.

Campo, E., Do, B.V., Ferreira, V., Valentin, D., 2008. Aroma properties of young Spanish monovarietal white wines: A study using sorting task list of terms and frequency of citation. Aust. J. Grape Wine Res. 14, 104–115.

Campo, E., Ballester, J., Langois, J., Dacremont, C., Valentin, D., 2010. Comparison of conventional descriptive analysis and a citation frequency-based descriptive method of odor profiling: An application to Burgundy Pinot noir wines. Food Qual. Pref. 21, 44–55.

Capone, S., Tufariello, M., Francioso, L., Montagna, G., Casino, F., Leone, A., et al., 2013. Aroma analysis by GC/MS and electronic nose dedicated to *Negroamaro* and *Primitivo* typical Italian Apulian wines. Sens. Actuat. B 179, 259–269.

Carbonell, L., Carbonell, I., Izquierdo, L., 2007. Triangle tests. Number of discriminators estimated by Bayes' rule. Food Qual. Pref. 18, 117–120.

Case, T.I., Stevenson, R.J., Dempsey, R.A., 2004. Reduced discriminability following perceptual learning with odors. Perception 33, 113–119.

Castriota-Scanderbeg, A., Hagberg, G.E., Cerasa, A., Committeri, G., Galati, G., Patria, F., et al., 2005. The appreciation of wine by sommeliers: A functional magnetic resonance study of sensory integration. Neuroimage 25, 570–578.

Cetó, X., Céspedes, F., del Valle, M., 2012. BioElectronic Tongue for the quantification of total polyphenol content in wine. Talanta 99, 544–551.

Cetó, X., González-Calabuig, A., Capdevila, J., Puig-Pujol, A., del Valle, M., 2015. Instrumental measurement of wine sensory descriptors using a voltammetric electronic tongue. Sensor Actuator B 207, 1053–1059.

Chambers, D.H., Allison, A.-M., Chambers IV, E., 2004. Training effects on performance of descriptive panelists. J. Sens. Stud. 19, 486–499.

Chatonnet, P., Dubourdieu, D., 1999. Using electronic odor sensors to discriminate among oak barrel toasting levels. J. Agric. Food Chem. 47, 4319–4322.

Civille, G.V., Lawless, H.T., 1986. The importance of language in describing perceptions. J. Sensory Stud. 1, 203–215.

Clark, C.C., Lawless, H.T., 1994. Limiting response alternatives in time-intensity scaling. An examination of the halo-dumping effect. Chem. Senses 19, 583–594.

Cliff, M.A., 2001. Influence of wine glass shape on perceived aroma and colour intensity in wines. J. Wine Res. 12, 39–46.

Cozzolino, D., Cowey, G., Lattey, K.A., Godden, P., Cynkar, W.U., Dambergs, R.G., et al., 2008. Relationship between wine scores and visible–near-infrared spectra of Australian red wines. Anal. Bioanal. Chem. 391, 975–981.

Cozzolino, D., Cynkar, W.U., Shah, N., Dambergs, R.G., Smith, P.A., 2010. A brief introduction to multivariate methods in grape and wine analysis. Int. J. Wine Res. 1, 123–130.

Cynkar, W., Cozzolino, D., Dambergs, B., Janik, L., Gishen, M., 2007. Feasibility study on the use of a head space mass spectrometry electronic nose (MS e-nose) to monitor red wine spoilage induced by Brettanomyces yeast. Sens. Actuat. B 124, 167–171.

d'Acampora Zellner, B. d'A., Dugo, P., Dugo, G., Mondello, L., 2008. Gas-chromatography-olfactometry in food flavour analysis. J. Chromatograph. A 1186, 123–143.

da Costa, A.M.S., Delgadillo, I., Rudnitskaya, A., 2014. Detection of copper, lead, cadmium and iron in wine using electronic tongue sensory system. Talanta 129, 63–71.

Dade, L.A., Zatorre, R.J., Jones-Gotman, M., 2002. Olfactory learning: Convergent findings from lesion and brain imaging studies in humans. Brain 125, 86–101.

Daglia, M., Stauder, M., Papetti, A., Signoretto, C., Giusto, G., Canepari, P., et al., 2010. Isolation of red wine components with anti-adhesion and anti-biofilm activity against Streptococcus mutans. Food Chem. 119, 1182–1188.

Dalton, P., Doolittle, N., Breslin, P.A.S., 2002. Gender-specific induction of enhanced sensitivity to odors. Nature Neurosci. 5, 199–200.

Darlington, R.B., Weinberg, S.L., Walberg, H.J., 1973. Canonical variate analysis and related techniques. Rev. Educ. Res. 43, 433–454.

de Araujo, I.E., Rolls, E.T., Velazco, M.I., Margot, C., Cayeux, I., 2005. Cognitive modulation of olfactory processing. Neuron 46, 671–679.

Debonneville, S., Orsier, B., Flament, I., Chaintreau, A., 2002. Improved hardware and software for quick gas chromatography-olfactometry using Charm and GC-"SNIF" analysis. Anal. Chem. 74, 2345–2351.

Delarue, J., Sieffermann, J.M., 2004. Sensory mapping using flash profile. Comparison with a conventional descriptive method for the evaluation of the flavour of fruit dairy products. Food Qual. Pref. 15, 383–392.

Déléris, I., Saint-Eve, A., Lieben, P., Cypriani, M.-L., Jacquet, N., Brunerie, P., et al., 2014. Impact of swallowing on the dynamics of aroma release and perception during the consumption of alcoholic beverages. In: Ferreira, V., Lopez, R. (Eds.), Flavour Science. Proceedings from XIII Weurman Flavour Research Symposium. Academic Press, London, UK, pp. 533–537.

Deliza, R., MacFie, H.J.H., 1996. The generation of sensory expectation by external cues and its effect on sensory perception and hedonic ratings. A review. J. Sens. Stud. 11, 103–128.

Delon-Martin, C., Plailly, J., Fonlupt, P., Veyrac, A., Royet, J.-P., 2013. Perfumers' expertise induces structural reorganization in olfactory regions. NeuroImage 68, 55–62.

Delwiche, J.F., Pelchat, M.L., 2002. Influence of glass shape on wine aroma. J. Sens. Stud. 17, 19–28.

Deppe, M., Knecht, S., Lohmann, H., Fleischer, H., Heindel, W., Ringelstein, E.B., et al., 2000. Assessment of hemispheric language lateralization: A comparison between fMRI and fTCD. J. Cereb. Blood Flow Metab. 20, 263–268.

DeRovira, D., 1996. The dynamic flavour profile method. Food Technol 50, 55–60.

Desor, J.A., Beauchamp, G.K., 1974. The human capacity to transmit olfactory information. Percep. Psychophys. 16, 551–556.

Dijksterhuis, G., 1996. Procrustes analysis in sensory research In: Naes, T. Risvik, E. (Eds.), Multivariate Analysis of Data in Sensory Science. Data Handling in Science and Technology, Vol. 16. Elsevier Science, Amsterdam, pp. 185–220.

Do, J.Y., Salunkhe, D.K., Olson, L.E., 1969. Isolation, identification and comparison of the volatiles of peach fruit as related to harvest maturity and artificial ripening. J. Food Sci. 34, 618–621.

Doty, R.L., Bromley, S.M., 2004. Effects of drugs on olfaction and taste. Otolaryngol. Clin. N. Am. 37, 1229–1254.

Duerr, P., 1984 Sensory analysis as a research tool. In: Proc. Alko Symp. Flavour Res. Alcoholic Beverages. Helsinki 1984. Nykänen, L., and Lehtonen, P. Foundation Biotech. Indust. Ferm. 3, 313–322.

Duerr, P., 1988 Wine description by expert and consumers. pp. 342–343. In: Proceeding of the Second International Symposium for Cool Climate Viticulture and Oenology, Auckland, N.Z., eds. Smart, R.E., Thornton, S.B., Rodriguez, S.B., and Young, J.E., N.Z. Soc. Vitic. Oenol.

Duizer, L.M., Bloom, K., Findlay, C.J., 1997. Dual-attribute time-intensity sensory evaluation: A new method from temporal measurement of sensory perceptions. Food Qual. Pref. 8, 261–269.

Duteurtre, B., 2010. Le Champagne: de la tradition à l science. Lavoisier Tec & Doc, Paris, France.

Engen, T., Pfaffmann, C., 1959. Absolute judgements of odor intensity. J. Expt. Psychol. 58, 23–26.

Engen, T., Pfaffmann, C., 1960. Absolute judgements of odor quality. J. Expt. Psychol. 58, 214–219.

Ennis, J.M., Rousseau, B., Ennis, D.M., 2014. Sensory difference tests as measurement instruments: A review of recent advances. J. Sens. Stud. 29, 89–102.

Etiévant, P.X., Callement, G., Langlois, D., Issanchou, S., Coquibus, N., 1999. Odor intensity evaluation in gas chromatography-olfactometry by finger span method. J. Agric. Food Chem. 47, 1673–1680.

Falcetti, M., Scienza, A., 1992. Utilisation de l'analyse sensorielle comme instrument d'évaluation des choix viticoles. Application pour déterminer les sites aptes à la culture du cépage Chardonnay pour la production de vins mousseux en Trentin. J. Int. Sci. Vigne Vin 26 (13–24), 49–50.

Findlay, C., Hasted, A., 2010. Statistical approaches to sensory quality control. In: Kilcast, D. (Ed.), Sensory Analysis for Food and Beverage Quality Control. Woodhouse Publ, Oxford, UK, pp. 118–140.

Findlay, C.J., Castura, J.C., Lesschaeve, I., 2007. Feedback calibration: A training method for descriptive panels. Food Qual. Pref. 18, 321–328.

Fischer, U., (2000). Practical applications of sensory research: Effect of glass shape, yeast strain, and terroir on wine flavor. In Proc. ASEV 50th Anniv. Ann. Meeting, Seattle, WA., June 19–23, 2000. pp. 3–8. American Society for Enology and Viticulture, Davis, CA.

Frank, R.A., Rybalsky, K., Brearton, M., Mannea, E., 2011. Odor recognition memory as a function of odor-naming performance. Chem. Senses 36, 29–41.

Fraser, A.G., 1997. Pharmacokinetic interactions between alcohol and other drugs. Clin. Pharmacokinet. 33, 79–90.

Frijters, J.E.R., 1980. Three-stimulus procedures in olfactory psychophysics: An experimental comparison of Thurstone-Ura and three-alternative forced-choice models of signal detection theory. Percep. Psychophys. 28, 390–397.

Frøst, M.B., Noble, A.C., 2002. Preliminary study of the effect of knowledge and sensory expertise on liking for red wines. Am. J. Enol. Vitic. 53, 275–284.

Fu, J., Li, G., Qui, Y., Freeman, W.J., 2007. A pattern recognition method for electronic noses based on an olfactory neural network. Sensors Actuators B 125, 487–497.

Furia, T.E. Bellance, N. (Eds.), FENAROLI's Handbook of Flavor Ingredients, second ed In: Vols. 1 and 2. CRC Press, Cleveland, OH.

Furnham, A., Boo, H.C., 2011. A literature review of the anchoring effect. J. Socio-Econom. 40, 35–42.

Gane, S., Georganakis, D., Maniati, K., Vamvakias, M., Ragoussis, N., Skoulakis, E.M.C., et al., 2013. Molecular vibration-sensing component in human olfaction. PLos One 8, e55780.

García, M., Aleixandre, M., Gutiérrez, J., Horrillo, M.C., 2006. Electronic nose for wine discrimination. Sensors Actuators B 113, 911–916.

Gawel, R., 1997. The use of language by trained and untrained experienced wine tasters. J. Sens. Stud. 12, 267–284.

Gawel, R., Godden, P.W., 2008. Evaluation of the consistency of wine quality assessments from expert wine tasters. Aust. J. Grape Wine Res. 14, 1–8.

Gawel, R., Oberholster, A., Francis, I.L., 2000. A "Mouth-feel Wheel": Terminology for communicating the mouth-feel characteristics of red wine. Aust. J. Grape Wine Res. 6, 203–207.

Gawel, R., Iland, P.G., Francis, I.L., 2001. Characterizing the astringency of red wine: A case study. Food Qual. Pref. 12, 83–94.

Gibbons, A., 2013. How sweet it is: Genes show how bacteria colonized human teeth. Science 339, 896–897.

Giunchi, A., Versari, A., Parpinello, G.P., Galassi, S., 2008. Analysis of mechanical properties of cork stoppers and synthetic closures used in wine bottling. J. Food Engin. 88, 576–580.

González, J., Barros-Loscertales, A., Pulvermuller, F., Meseguer, V., Sanjuán, A., Belloch, V., et al., 2006. Reading cinnamon activates olfactory brain regions. NeuroImage 32, 906–912.

Goodman, S., Lockshin, L., Cohen, E., 2006. What influences consumer selection in the retail store? Aust. NZ Grapegrower Winemaker 515, 61–63.

Gottfried, J.A., Dolan, R.J., 2003. The nose smells what the eye sees: Crossmodal visual facilitation of human olfactory perception. Neuron 39, 375–386.

Green, B.G., 1992. The effects of temperature and concentration on the perceived intensity and quality of carbonation. Chem. Senses 17, 435–450.

Grosch, W., 2001. Evaluation of the key odorants of foods by dilution experiments, aroma models and omission. Chem. Senses 26, 533–545.

Guinard, J., Cliff, M., 1987. Descriptive analysis of Pinot noir wines from Carneros, Napa, and Sonoma. Am. J. Enol. Vitic. 38, 211–215.

Guinard, J.-X., Pangborn, R.M., Lewis, M.J., 1986. The time–course of astringency in wine upon repeated ingestion. Am. J. Enol. Vitic. 37, 184–189.

Gyllensköld, H., 1967. Att Temperera Vin. Wahlström and Widstrand, Stockholm.

Hahn, H., 1934. Die Adaptation des Geschmackssinnes. Z. Sinnesphysiol. 65, 105–145.

Heath, H.B., 1981. Source Book of Flavor. AVI, Westport, CT.

Helson, H.H., 1964. Adaptation-Level Theory. Harper & Row, New York, NY.

Henderson, S.T., 1977. Daylight and its Spectrum. Wiley, New York.

Herz, R.S., 2003. The effect of verbal context on olfactory perception. J. Exp. Psychol. Gen. 132, 595–606.

Herz, R.S., von Clef, J., 2001. The influence of verbal labeling on the perception of odors: Evidence for olfactory illusions? Perception 30, 381–391.

Herz, R.S., McCall, C., Cahill, L., 1999. Hemispheric lateralization in the processing of odor pleasantness vs odor names. Chem. Senses 24, 691–695.

Heymann, H., Machado, B., Torri, L., Robinson, A.L., 2012. How many judges should one use for sensory descriptive analysis? J. Sens. Stud. 27, 111–122.

Hodgson, M.D., Langridge, J.P., Linforth, R.S.T., Taylor, A.J., 2005. Aroma release and delivery following the consumption of beverages. J. Agric. Food Chem. 53, 1700–1706.

Hodgson, R.T., 2009. How expert are "expert" wine judges? J Wine Econ. 4, 233–241.

Homa, A., Cultice, J., 1984. Role of feedback, category size, and stimulus distortion on the acquisition and utilization of ill-defined categories. J. Exp. Psychol.: Learning, Memory and Cognition 10, 83–93.

Hughson, A., Ashman, H., de la Huerga, V., Moskowitz, H., 2004. Mind-sets of the wine consumer. J. Sens. Stud. 19, 85–105.

Hughson, A.L., Boakes, R.A., 2002. The knowing nose: The role of knowledge in wine expertise. Food Qual. Pref. 13, 463–472.

Huitson, A., 1989. Problems with Procrustes analysis. J. Appl. Stat. 16, 39–45.

Hummel, T., Delwiche, J.F., Schmidt, C., Hüttenbrink, K.-B., 2003. Effects of the form of glasses on the perception of wine flavors: A study in untrained subjects. Appetite 41, 197–202.

Hummel, T., Olgun, S., Gerber, J., Huchel, U., Frasnelli, J., 2013. Brain responses to odor mixtures with sub-threshold components. Front. Psychol. 4 (786), 8. http://dx.doi.org/10.3389/fpsyg.2013.00786.

Hyldig, G., 2010. Proficiency testing of sensory panels. In: Kilcast, D. (Ed.), Sensory Analysis for Food and Beverage Quality Control. A Practical Guide. Woodhead Publ. Ltd, Cambridge, U.K, pp. 37–48.

Ijiri, T., Owada, S., Okabe, M., Igarashi, T., 2005. Floral diagrams and inflorescences: Interactive flower modeling using botanical structural constraints. ACM Trans. Graph. 24, 720–726.

Ishii, R., Kemp, S.E., Gilbert, A.N., O'Mahony, M., 1997. Variation in sensory conceptual structure: An investigation involving the sorting of odor stimuli. J. Sensory Stud. 12, 195–214.

ISO, 1992. Sensory Analysis, Vocabulary, 1st ed International Standards Organization #5492, Geneva, Switzerland.

Jarisch, R., Wantke, F., 1996. Wine and headache. Intl. Arch. Allergy Immunol. 110, 7–12.

Jehl, C., Royet, J.-P., Holley, A., 1997. Role ov verbal encoding in short- and long-term odor recognition. Percept. Psychosphys. 59, 100–110.

Jolly, N.P., Hattingh, S., 2001. A brandy aroma wheel for South African brandy. S. A. J. Enol. Vitic. 22, 16–21.

Jones, A.L., Roberts, R.C., Colvin, D.W., Rubin, G.L., Coughtrie, M.W.H., 1995. Reduced platelet phenolsulphotransferase activity towards dopamine and 5-hydroxytryptamine in migraine. Eur. J. Clin. Pharmacol. 49, 109–114.

Kahan, G., Cooper, D., Papavasiliou, A., Kramer, A., 1973. Expanded tables for determining significance of differences for ranked data. Food Technol. 27, 64–69.

Kärneküll, S.C., Jönsson, F.U., Willander, J., Sikström, S., Larsson, M., 2015. Long-term memory for odors: Influences of familiarity and identification across 64 days. Chem. Senses 40, 259–267.

Kaufman, H.S., 1992. The red wine headache and prostaglandin synthetase inhibitors: A blind controlled study. J. Wine Res. 3, 43–46.

King, E.S., Dunn, R.L., Heymann, H., 2013. The influence of alcohol on the sensory perception of red wines. Food Qual. Pref. 28 235–143.

King, M.C., Hall, J., Cliff, M.A., 2001. A comparison of methods for evaluation the performance of a trained sensory panel. J. Sens. Stud. 16, 567–581.

King, M.C., Cliff, M.A., Hall, J., 2003. Effectiveness of the 'Mouth-feel Wheel' for the evaluating of astringent subqualities in British Columbia red wines. J. Wine Res. 14, 67–78.

Klecka, W.R., 1980. Discriminant Analysis. Sage University Paper #19. Sage Publ, Newbury Park, CA.

Knecht, S., Drager, B., Deppe, M., Bobe, L., Lohmann, H., Floel, A., et al., 2000. Handedness and hemispheric language dominance in healthy humans. Brain. 123, 2512–2518.

Komes, D., Ulrich, D., Lovric, T., 2006. Characterization of odor-active compounds in Croatian Rhine Riesling wine, subregion Zagorje. Eur. Food Res. Technol. 222, 1–7.

Korotcenkov, G. (Ed.), 2013. Handbook of Gas Sensor Materials: Properties, Advantages and Shortcomings, Vols. 1&II. Springer, New York, NY.

Köster, E.P., 2003. The psychology of food choice: Some often encountered fallacies. Food Qual. Pref. 14, 359–373.

Köster, E.P., 2005. Does Olfactory Memory Depend on Remembering Odors? Chem. Senses 30, i236–i237.

Kunert, J., Meyners, M., 1999. On the triangle test with replications. Food Qual. Pref. 10, 477–482.

Kurtz, T.E., Link, T.E., Tukey, R.F., Wallace, D.L., 1965. Short-cut multiple comparisons for balanced single and double classifications. Technometrics 7, 95–165.

Lamm, H., Myers, D.M., 1978. Group-induced polarization of attitudes and behavior. Adv. Exp. Soc. Psychol. 11, 145–195.

Landon, S., Smith, C.E., 1998. Quality expectations, reputation and price. South. Econom. J. 364, 628–647.

Lange, C., Martin, C., Chabanet, C., Combris, P., Issanchou, S., 2002. Impact of the information provided to consumers on their willingness to pay for Champagne: Comparison with hedonic scores. Food Qual. Pref. 13, 597–608.

Laska, M., Hübener, F., 2001. Olfactory discrimination ability for homologous series of aliphatic ketones and acetic esters. Behav. Brain Res. 119, 193–201.

Lattey, K.A., Bramley, B.R., Francis, I.L., 2010. Consumer acceptability, sensory properties and expert quality judgements of Australian Cabernet Sauvignon and Shiraz wines. Aust. J. Grape Wine Res. 16, 189–202.

Lawless, H.T., 1984. Flavor description of white wine by "expert" and nonexpert wine consumers. J. Food Sci. 49, 120–123.

Lawless, H.T., 1985. Psychological perspectives on wine tasting and recognition of volatile flavours. In: Birch, G.G., Lindley, M.G. (Eds.), Alcoholic Beverages. Elsevier, London, pp. 97–113.

Lawless, H.T., 1999. Descriptive analysis of complex odors: Reality, model or illusion? Food Qual. Pref. 10, 325–332.

Lawless, H.T., 2010. A simple alternative analysis for threshold data determined by ascending forced-choice methods of limits. J. Sens. Stud. 25, 332–346.

Lawless, H., Engen, T., 1977. Associations of odors: Interference, mnemonics and verbal labeling. J. Expt. Psycho. Hum. Learn. Mem. 3, 52–59.

Lawless, H.T., Clark, C.C., 1992. Physiological biases in time-intensity scaling. Food Technol. 46, 81–90.

Lawless, H.T., Heymann, H., 2010. Sensory Evaluation of Food: Principles and Practices, 2nd ed Springer, New York.

Lawless, H.T., Malone, J.G., 1986. The discriminative efficiency of common scaling methods. J. Sens. Stud. 1, 85–96.

Lawless, H.T., Schlegel, M.P., 1984. Direct and indirect scaling of taste-odor mixtures. J. Food Sci. 49, 44–46.

Lawless, H.T., Tuorila, H., Jouppila, K., Virtanen, P., Horne, J., 1996. Effects of guar gum and microcrystalline cellulose on sensory and thermal properties of a high fat model food system. J. Texture Stud. 27, 493–516.

Lawless, H.T., Liu, Y.-F., Goldwyn, C., 1997. Evaluation of wine quality using a small-panel hedonic scaling method. J. Sens. Stud. 12, 317–332.

Lee, K.-Y.M., Paterson, A., Piggott, J.R., 2001. Origins of flavour in whiskies and a revised flavour wheel: A review. J. Inst. Brew. 107, 287–313.

Leek, J.T., Peng, R.D., 2015. What is the question? Science 347, 1314–1315.

Lehrer, A., 1975. Talking about wine. Language 51, 901–923.

Lehrer, A., 1983. Wine and Conversation. Indiana Univ. Press, Bloomington, ID.

Lehrer, A., 2009. Wine and Conversation, second ed Oxford University Press, Oxford, UK.

Lepage, M., Neville, T., Rytz, A., Schlich, P., Martin, N., Pineau, N., 2014. Panel performance for Temporal Dominance of Sensations. Food Qual. Pref. 38, 24–29.

Le Révérend, F.M., Hidrio, C., Fernandes, A., Aubry, V., 2008. Comparison between temporal dominance of sensations and time intensity results. Food Qual. Pref. 19, 174–178.

Li, W., Luxenberg, E., Parrish, T., Gottfried, J.A., 2006. Learning to smell the roses: Experience-dependent neural plasticity in human piriform and orbitofrontal cortices. Neuron 52, 1097–1108.

Li, W., Howard, J.D., Parrish, T.B., Gottfried, J.A., 2008. Aversive learning enhances perceptual and cortical discrimination of indiscriminable odor cues. Science 319, 1842–1845.

Liger-Belair, G., Villaume, S., Cilindre, C., Jeandet, P., 2009. Kinetics of CO_2 fluxes outgassing from champagne glasses in tasting conditions: The role of temperature. J. Agric. Food Chem. 57, 1997–2003.

Lim, J., Wood, A., Green, B.G., 2009. Derivation and evaluation of a labeled hedonic scale. Chem. Senses 34, 739–751.

Liu, J., Grønbeck, S., Di Monaco, R., Giacalone, D., Bredie, W.L.P., 2016. Performance of Flash Profile and Napping with and without training for describing small sensory differences in a model wine. Food Qual. Pref. 48, 41–49.

Livermore, A., Hummel, T., 2004. The influence of training on chemosensory event-related potentials and interactions between the olfactory and trigeminal systems. Chem. Senses 29, 41–51.

Loch, D., Breer, H., Strotmann, J., 2015. Endocrine modulation of olfactory responsiveness: Effects of the orexigenic hormone ghrelin. Chem. Senses 40, 469–479.

Louw, L., Oelosfe, S., Naes, T., Lambrechts, M., van Rensburg, P., Niedwoudt, H., 2014. Trained sensory panellist's response to product alcohol content in the projective mapping task: Observation on alcohol content, product complexity and prior knowledge. Food Qual. Pref. 34, 37–44.

Louw, L., Oelosfe, S., Naes, T., Lambrechts, M., van Rensburg, P., Niedwoudt, H., 2015. The effect of tasting sheet shape on product configurations and panellist's performance in sensory projective mapping of brandy products. Food Qual. Pref. 40, 132–136.

Lovell, D.P., 2013. Biological importance and statistical significance. J. Agric. Food Chem. 61, 8340–8348.

Lozano, J., Santos, J.P., Horrillo, M.C., 2005. Classification of white wine aromas with an electronic nose. Talanta 67, 610–616.

Lozano, J., Santos, J.P., Arroya, T., Aznar, M., Cabellos, J.M., Gil, M., et al., 2007. Correlating e-nose responses to wine sensorial descriptors and gas chromatography-mass spectrometry profiles using partial least squares regression analysis. Sensors Actuators B 127, 267–276.

Luciano, G., Naes, T., 2009. Interpreting sensory data by combining principal component analysis and analysis of variance. Food Qual. Pref. 20, 167–175.

Lundström, J.N., Hummel, T., Olsson, M.J., 2003. Individual differences in sensitivity to the odor of 4,16-androstadien-3-one. Chem. Senses 28, 643–650.

Lupi-Pegurier, L., Muller, M., Leforestier, E., Bernard, M.F., Bolla, M., 2003. In vitro action of Bordeaux red wine on the microhardness of human dental enamel. Arch. Oral Biol. 48, 141–145.

MacFie, H.J.H., Piggott, J.R., 2012. Preference mapping: Principles and potential applications to alcoholic beverages. In: Piggott, J. (Ed.), Alcoholic Beverages. Sensory Evaluation and Consumer Research. Woodhouse Publ. Ltd., Cambridge, UK, pp. 436–476.

Maitre, I., Symoneaux, R., Jourjon, F., Mehinagic, E., 2010. Sensory typicality of wines: How scientists have recently dealt with this subject. Food Qual. Pref. 21, 726–731.

Mandel, L., 2005. Dental erosion due to wine consumption. J. Am. Dent. Assoc. 136, 71–75.

Martí, M.P., Boqué, R., Busto, O., Guasch, J., 2005. Electronic noses in the quality control of alcoholic beverages. Trends Anal. Chem. 24, 57–66.

Martineau, B., Acree, T.E., Henick-Kling, T., 1995. Effect of wine type on the detection threshold of diacetyl. Food Res. Inst. 28, 139–144.

Masyczek, R., Ough, C.S., 1983. The "red wine reaction" syndrome. Am. J. Enol. Vitic. 32, 260–264.

Mazzoleni, V., Maggi, L., 2007. Effect of wine style on the perception of 2,4,6-trichloroanisole, a compound related to cork taint in wine. Food Res. Int. 40, 694–699.

McBride, R.L., Finlay, D.C., 1990. Perceptual integration of tertiary taste mixtures. Percept. Psychophys. 48, 326–336.

McCloskey, L.P., Sylvan, M., Arrhenius, S.P., 1996. Descriptive analysis for wine quality experts determining appellations by Chardonnay wine aroma. J. Sensory Stud. 11, 49–67.

McClure, S.M., Li, J., Tomlin, D., Cypert, K.S., Montague, L.M., Montague, P.R., 2004. Neural correlates of behavioral preference for culturally familiar drinks. Neuron 44, 379–387.

McDonnell, E., Hulin-Bertaud, S., Sheehan, E.M., Delahunty, C.M., 2001. Development and learning process of a sensory vocabulary for the odor evaluation of selected distilled beverages using descriptive analysis. J. Sens. Stud. 16, 425–445.

McGurk, H., MacDonald, J., 1976. Hearing lips and seeing speech. Nature 264, 746–748.

Meilgaard, M.C., 1988. Beer flavor terminology — A case study In: Moskowitz, H. (Ed.), Applied Sensory Analysis of Foods, Vol. 1. CRC Press, Boca Raton, Florida, pp. 73–87.

Meilgaard, M.C., Dalgliesh, C.E., Clapperton, J.F., 1979. Progress towards an international system for beer flavour terminology. Am. Soc. Brew. Chem. 37, 42–52.

Meilgaard, M.C., Reid, D.C., Wyborski, K.A., 1982. Reference standards for beer flavor terminology system. J. Am. Soc. Brew. Chem. 40, 119–128.

Meilgaard, M.C., Civille, G.V., Carr, T.C., 2006. Sensory Evaluation Techniques, fourth ed CRC Press, Boca Raton, FL.

Meillon, S., Viala, D., Medel, M., Urbano, C., Guillot, G., Schlich, P., 2010. Impact of partial alcohol reduction in Syrah wine on perceived complexity and temporality of sensations and link with preference. Food Qual. Pref. 21, 732–740.

Melcher, J.M., Schooler, J.W., 1996. The misremembrance of wines past: Verbal and perceptual expertise differentially mediate verbal overshadowing of taste memory. J. Memory Lang. 35, 231–245.

Meyners, M., 2011. Panel and panelist agreement for product comparisons in studies of Temporal Dominance of Sensations. Food Qual. Pref. 22, 365–370.

Meyners, M., Pineau, N., 2010. Statistical inference for temporal dominance of sensations data using randomization tests. Food Qual. Pref. 21, 805–814.

Monteleone, E., 2012. Sensory methods from product development and their application in the alcohol beverage industry. In: Piggott, J. (Ed.), Alcoholic Beverages. Sensory Evaluation and Consumer Research. Woodhouse Publ. Ltd., Cambridge, UK, pp. 66–100.

Morrot, G., 2004. Cognition et vin. Rev. Oenologues 111, 11–15.

Morrot, G., Brochet, F., Dubourdieu, D., 2001. The color of odors. Brain Lang. 79, 309–320.

Myers, D.G., Lamm, H., 1975. The polarizing effect of group discussion. Am. Sci. 63, 297–303.

Nakamura, Y., Sanematsu, K., Ohta, R., Shirosaki, S., Koyano, K., Nonaka, K., et al., 2008. Diurnal variation of human sweet taste recognition thresholds is correlated with plasma leptin levels. Diabetes 57, 2661–2665.

Nicolodi, M., Sicuteri, F., 1999. Wine and migraine: Compatibility or incompatibility? Drugs Exp. Clin. Res. 25, 147–153.

Noble, A.C., 1988. Analysis of wine sensory properties. In: Linskens, H.F., Jackson, J.F. (Eds.), Wine Analysis. Springer-Verlag, Berlin, pp. 9–28.

Noble, A.C., Howe, P., 1990. Sparkling Wine Aroma Wheel. The Wordmill, Healdsburg, CA.

Noble, A.C., Williams, A.A., Langron, S.P., 1984. Descriptive analysis and quality ratings of 1976 wines from four Bordeaux communes. J. Sci. Food Agric. 35, 88–98.

Noble, A.C., Arnold, R.A., Buechsenstein, J., Leach, E.J., Schmidt, J.O., Stern, P.M., 1987. Modification of a standardized system of wine aroma terminology. Am. J. Enol. Vitic. 36, 143–146.

North, A., Hargreaves, D., McKendrick, J., 1999. The influence of in store music on wine selections. J. Appl. Psychol. 84, 271–276.

Nuzzo, R., 2014. Statistical errors. Nature 506, 150–152.

O'Mahony, M., Rousseau, B., 2002. Discrimination testing: A few ideas, old and new. Food Qual. Pref. 14, 157–164.

Obst, K., Paetz, S., Ley, J.P., Engel, K.-H., 2014. Multiple time-intensity profiling (mTIP) as an advanced evaluation tool for complex tastants. In: Ferreira, V., Lopez, R. (Eds.), Flavour Science. Proceedings from XIII Weurman Flavour Research Symposium. Academic Press, London, UK, pp. 45–49.

Oreskovich, D.C., Klein, B.P., Sutherland, J.W., 1991. Procrustes analysis and its applications to free-choice and other sensory profiling. In: Lawless, H.T., Klein, B.P. (Eds.), Sensory Science Theory and Application in Foods. Marcel Dekker, New York, pp. 353–393.

Österbauer, R.A., Matthews, P.M., Jenkinson, M., Beckmann, C.F., Hansen, P.C., Calvert, G.A., 2005. Color of scents: Chromatic stimuli modulate odor responses in the human brain. J. Neurophysiol. 93, 3434–3441.

Ough, C.S., Baker, G.A., 1961. Small panel sensory evaluation of wines by scoring. Hilgardia 30, 587–619.

Ough, C.S., Winton, W.A., 1976. An evaluation of the Davis wine-score card and individual expert panel members. Am. J. Enol. Vitic. 27, 136–144.

Ough, C.S., Singleton, V.L., Amerine, M.A., Baker, G.A., 1964. A comparison of normal and stressed-time conditions on scoring of quantity and quality attributes. J. Food Sci. 29, 506–519.

Overbosch, P., 1986. A theoretical model for perceived intensity in human taste and smell as a function of time. Chem. Senses 11, 315–329.

Owen, D.H., Machamer, P.K., 1979. Bias-free improvement in wine discrimination. Perception 8, 199–209.

Pagès, J., 2005. Collection and analysis of perceived product inter-distances using multiple factor analysis: Application to the study of 10 white wines from the Loire Valley. Food Qual. Pref. 16, 642–649.

Pagès, J., Cadoret, M., Lê, S., 2010. The sorted napping: A new holistic approach in sensory evaluation. J. Sens. Stud. 25, 637–658.

Parker, M., Pollnitz, A.P., Cozzolino, D., Francis, I.L., Herderich, M.J., 2007. Identification and quantification of a marker compound for 'pepper' aroma and flavor in Shiraz grape berries by combination of chemometrics and gas chromatography mass spectrometry. J. Agric. Food Chem. 55, 5948–5955.

Parr, W.V., Heatherbell, D., White, K.G., 2002. Demystifying wine expertise: Olfactory threshold, perceptual skill, and semantic memory in expert and novice wine judges. Chem. Senses 27, 747–755.

Parr, W.V., White, K.G., Heatherbell, D., 2004. Exploring the nature of wine expertise: What underlies wine expert's olfactory recognition memory advantage? Food Qual. Pref. 15, 411–420.

Parr, W.V., Green, J.A., White, K.G., 2006. Wine judging, context and New Zealand Sauvignon blanc. Rev. Eur. Psychol. Appl. 56, 231–238.

Parr, W.V., Valentin, D., Green, J.A., Dacremont, C., 2010. Evaluation of French and New Zealand Sauvignon wine by experienced French wine assessors. Food Qual. Pref. 21, 55–64.

Parra, V., Arrieta, A.A., Fernández-Escudero, J.A., Íñiguez, M., de Saja, J.A., Rodríguez-Méndez, M.L., 2006. Monitoring of the ageing of red wines in oak barrels by means of a hybrid electronic tongue. Anal. Chim. Acta 563, 229–237.

Pattichis, K., Louca, L.L., Jarman, J., Sandler, M., Glover, V., 1995. 5-Hydroxytryptamine release from platelets by different red wines: Implications for migraine. Eur. J. Pharmacol. 292, 173–177.

Pazart, L., Comte, A., Magnin, E., Millot, J.-L., Moulin, T., 2014. An fMRI study on the influence of sommeliers' expertise on the integration of flavor. Front. Behav. Neurosci. 8, 15. #385.

Peltier, C., Visalli, M., Schlich, P., 2015. Canonical variate analysis of sensory profiling data. J. Sens. Stud. 30, 316–328.

Peris, M., Escuder-Gilabert, L., 2009. A 21st centurey technique for food control: Electronic noses. Anal. Chim. Acta. 638, 1–15.

Perrin, L., Pagès, J., 2009. A methodology for the analysis of sensory typicality judgements. J. Sens. Stud .24, 749–773.

Perrin, L., Symoneaux, R., Maître, I., Asselin, C., Jourjon, F., Pagès, J., 2008. Comparison of three sensory methods for use with the Napping procedure: Case of ten wines from Loire valley. Food Qual. Pref. 19, 1–11.

Peyvieux, C., Dijksterhuis, G., 2001. Training a sensory panel for TI: A case study. Food Qual. Pref. 12, 19–28.

Pigani, L., Foca, G., Ionescu, K., Martina, V., Ulrici, A., Terzi, F., et al., 2008. Amperometric sensors based on poly(3,4-ethylenedioxythiophene)-modified electrodes: Discrimination of white wines. Anal. Chim. Acta 614, 213–222.

Pineau, B., Barbe, J.-C., van Leeuwen, C., Dubourdieu, D., 2007. Which impact for β-damascenone on red wines aroma? J. Agric. Food Chem. 55, 5214–5219.

Pineau, N., Schlich, P., Cordelle, S., Mathonnière, C., Issanchou, S., Imbert, A., et al., 2009. Temporal Dominance of Sensations: Construction of the TDS curves and comparison with time-intensity. Food Qual. Pref. 20, 450–455.

Pinheiro, C., Rodrigues, C.M., Schäfer, T., Crespo, J.G., 2002. Monitoring the aroma production during wine-must fermentation with an electronic nose. Biotechnol. Bioengin. 77, 632–640.

Plassmann, H., O'Doherty, J., Shiv, B., Rangel, A., 2008. Marketing actions can modulate neural representations of experienced pleasantness. PNAS 105, 1050–1054.

Plutowaska, B., Wardencki, W., 2008. Application of gas chromatography-olfactometry (GC–O) in analysis and quality assessment of alcoholic beverages – A review. Food Chem. 107, 449–463.

Pohl, R.F., Schwarz, S., Sczesny, S., Stahlberg, D., 2003. Hindsight bias in gustatory judgements. Expt. Psychol. 50, 107–115.

Pollien, P., Ott, A., Montigon, F., Baumgartner, M., Muñoz-Box, R., Chaintreau, A., 1997. Hyphenated headspace-gas chromatography-sniffing technique: Screening of impact ordoarants and quantitative aromagram comparisons. J. Agric. Food Chem. 45, 2630–2637.

Prescott, J., 2004. Psychological processes in flavour perception. In: Taylor, A.J., Roberts, D. (Eds.), Flavour Perception. Blackwell Publishing, London, pp. 256–278.

Prescott, J., 2009. Rating a new hedonic scale: A commentary on "Derivation and evaluation of a labeled hedonic scale" by Lim, Wood and Green. Chem. Senses 34, 735–737.

Prescott, J., Johnstone, V., Francis, J., 2004. Odor-taste interactions: Effects of attentional strategies during exposure. Chem. Senses 29, 331–340.

Qannari, E.M., Schlich, P., 2006. Matching sensory and instrumental data. In: Voilley, A., Etiévant, P. (Eds.), Flavour in Food. Woodhouse Publ. Inc., Cambridge, UK, pp. 98–116.

Quandt, R.E., 2006. Measurement and inference in wine tasting. J. Wine Econ. 1, 7–30.

Ranjitkar, S., Smales, R., Lekkas, D., 2012. Prevention of tooth erosion and sensitivity in wine tasters. Wine Vitic. J 27 (1), 34–37.

Reineccius, T.A., Reineccius, G.A., Peppard, T.L., 2003. Flavor release from cyclodextrin complexes: Comparison of alpha, beta, and gamma types. J. Food Sci. 68, 1–6.

Röck, F., Barsan, N., Weimar, U., 2008. Electronic nose: Current status and future trends. Chem. Rev. 108, 705–725.

Roessler, E.B., Pangborn, R.M., Sidel, J.L., Stone, H., 1978. Expanded statistical tables for estimating significance in prepared-preference, paired-difference, duo-trio and triangle tests. J. Food Sci. 43, 940–943.

Rolls, E.T., 1995. Central taste anatomy and neurophysiology. In: Doty, R.L. (Ed.), Handbook of Olfaction and Gustation. Marcel Dekker, New York, pp. 549–573.

Ross, C.F., Weller, K.M., Alldredge, J.R., 2012. Impact of serving temperature on sensory properties of red wine as evaluated using projective mapping by a trained panel. J. Sens. Stud. 27, 463–470.

Rudnitskaya, A., Delgadillo, I., Legin, A., Rocha, S.M., Costa, A.-M., Simões, T., 2007. Prediction of the Port wine age using an electronic tongue. Chemomet. Intelligent Lab. Syst. 88, 125–131.

Russell, K., Zivanovic, S., Morris, W.C., Penfield, M., Weiss, J., 2005. The effect of glass shape on the concentration of polyphenolic compounds and perception of Merlot wine. J. Food Qual. 28, 377–385.

Sáenz-Navajas, M.-P., Tao, Y.-S., Dizy, M., Ferreira, V., Fernández-Zurbano, P., 2010. Relationship between nonvolatile composition and sensory properties of premium Spanish red wines and their correlation to quality perception. J. Agric. Food Chem. 58, 12407–12416.

Savic, I., Gulyas, B., 2000. PET shows that odors are processed both ipsilaterally and contralaterally to the stimulated nostril. Brain Imaging 11, 2861–2866.

Schifferstein, H.J.N., 1996. Cognitive factors affecting taster intensity judgements. Food Qual. Pref. 7, 167–175.

Scholten, P., 1987. How much do judges absorb? Wines Vines 69 (3), 23–24.

Schindler, R.M., 1992. The real lesson of New Coke; the value of focus groups for predicting the effects of social influence. Market. Res. 4, 22–27.

Schutz, H.G., 1964. A matching-standards method for characterizing odor qualities. Ann. N.Y. Acad. Sci. 116, 517–526.

Silva, R.C.S.N., Minim, V.P.R., Silva, A.N., Minim, L.A., 2014. Number of judges necessary for descriptive sensory texts. Food Qual. Pref. 31, 22–27.

Sokolowsky, M., Rosenberger, A., Fischer, U., 2015. Sensory impact of skin contact on white wines characterized by descriptive analysis, time–intensity analysis and temporal dominance of sensations analysis. Food Qual. Pref. 39, 285–297.

Solomon, G.E.A., 1990. Psychology of novice and expert wine talk. Am. J. Psychol. 103, 495–517.

Solomon, G.E.A., 1991. Language and categorization in wine expertise. In: Lawless, H.T., Klein, B.P. (Eds.), Sensory Science Theory and Applications in Foods. Marcel Dekker, New York, pp. 269–294.

Solomon, G.E.A., 1997. Conceptual change and wine expertise. J. Learn. Sci. 6, 41–60.

Somers, T.C., 1975. In search of quality for red wines. Food Technol. Australia 27, 49–56.

Spence, C., 2011. Wine and music. World Fine Wine 31, 96–104.

Spence, C., Wang, Q.J., 2015. Wine and music (I): On the crossmodal matching of wine and music. Flavour 4, 34. (14 pp.)

Spence, C., Richards, L., Kjellin, E., Huhnt, A.-M., Daskal, V., Scheybeler, A., et al., 2013. Looking for crossmodal correspondences between classical music and fine wine. Flavor 2, 29. (13 pp.)

Spence, C., Velasco, C., Knoeferle, K., 2014. A large sample study on the influence of the multisensory environment on the wine drinking experience. Flavor 3, 8. (12 pp.).

Stahl, W.H., Einstein, M.A., 1973. Sensory testing methods In: Snell, F.D. Ettre, L.S. (Eds.), Encyclopedia of Industrial Chemical Analysis, Vol. 17. John Wiley, New York, pp. 608–644.

Staples, E.J. (2000). Detecting 2,4,6 TCA in corks and wine using the zNose™. http://www.estcal.com/tech_papers/papers/Wine/TCA_in_Wine_Body.pdf.

Stevens, D.A., 1996. Individual differences in taste perception. Food Chem. 56, 303–311.

Stevenson, R.J., 2001. Associative learning and odor quality perception: How sniffing an odor mixture can alter the smell of its parts. Learn Motiv. 32, 154–177.

Stevenson, R.J., Case, T.I., Boakes, R.A., 2003. Smelling what was there: Acquired olfactory percepts are resistant to further modification. Learn. Motivat. 34, 185–202.

Stone, H., Bleibaum, R., Thomas, H.A., 2012. Sensory Evaluation Practices, fourth ed Academic Press, Orlando, FL.

Stone, H., Sidel, J., Oliver, S., Woolsey, A., Singleton, R.C., 1974. Sensory evaluation by quantitative descriptive analysis. Food Technol., **Nov.** 24–34.

Symoneaux, R., Galmarini, M.C., Mehinagic, E., 2012. Comment analysis of consumer's likes and dislikes as an alternative tool to preference mapping. A case study on apples. Food Qual. Pref. 24, 59–66.

Takeuchi, H., Kato, H., Kurahashi, T., 2013. 2,4,6-Trichloroanisole is a potent suppressor of olfactory signal transduction. PNAS 110, 16235–16240.

Tempere, S., Cuzange, E., Malik, J., Cougeant, J.C., de Revel, G., Sicard, G., 2011. The training level of experts influences their detection thresholds for key wine compounds. Chem. Percept. 4, 99–115.

Tempere, S., Cuzange, E., Bougeant, J.C., de Revel, G., Sicard, G., 2012. Explicit sensory training improves the olfactory sensitivity of wine experts. Chem Percept. 5, 205–213.

Tempere, S., Hamtat, M.L., Bougeant, J.C., de Revel, G., Sicard, G., 2014. Learning odors: The impact of visual and olfactory mental imagery training on odor perception. J. Sens. Stud. 29, 435–449.

Thompson, B., 1984. Canonical Correlation Analysis: Uses and Interpretation. Sage University Paper #47. Sage Publ, Newbury Park, CA.

Tomasino, E., Harrison, R., Sedcole, R., Frost, A., 2013. Regional differentiation of New Zealand Pinot noir wine by wine professionals using Canonical Variate Analysis. Am. J. Enol. Vitic. 64, 357–363.

Tomic, O., Forde, C., Delahunty, C., Næs, T., 2013. Performance indices in descriptive sensoyr analysis – A complementary screening tool for assessor and panel performance. Food Qual. Pref. 28, 122–133.

Torri, L., Dinnella, C., Recchia, A., Naes, T., Tuorila, H., Monteleone, E., 2013. Projective mapping for interpreting wine aroma differences as perceived by naive and experienced assessors. Food Qual. Pref. 29, 6–15.

Umali, A.P., Ghanem, E., Hopfer, H., Hussain, A., Kao, Y.-t, Zabanal, L.G., et al., 2015. Grape and wine sensory attributes correlate with pattern-based discrimination of Cabernet Sauvignon wines by a peptidic sensor array. Tetrahedron 71, 3095–3099.

Vaamonde, A., Sánchez, P., Vilariño, F., 2000. Discrepancies and consistencies in the subjective ratings of wine-tasting committees. J. Food Qual. 23, 363–372.

Vandyke Price, P.J., 1975. The Taste of Wine. Random House, New York, NY.

Vene, K., Seisonen, S., Koppel, K., Leitner, E., Paalme, T., 2013. A method for GC-olfactometry panel training. Chem. Percept. 6, 179–189.

Vidal, L., Cadena, R.S., Antúnez, L., Giménez, A., Varela, P., Ares, G., 2014. Stability of sample configurations from projective mapping: How many consumers are necessary? Food Qual. Pref. 34, 79–87.

Vigen, T., 2015. Spurious Correlations. Hachette Books.

Walker, J.C., Hall, S.B., Walker, D.B., Kendal-Reed, M.S., Hood, A.F., Niu, X.-F., 2003. Human odor detectability: New methodology used to determine threshold and variation. Chem. Senses 28, 817–826.

Wang, L., Chen, L., Jacob, T., 2004. Evidence for peripheral plasticity in human odour response. J. Physiol. 55, 236–244.

Wasserstein, R.L., Lazar, N.A., 2016. The ASA' statement on p-values: Context, process, and purpose. Am. Statistician 70, 129–133.

Weathermon, R., Crabb, D.W., 1999. Alcohol and medication interactions. Alcohol Res. Health 23, 40–54.

Willershausen, B., Callaway, A., Azrak, B., Kloß, C., Schulz-Dobrick, B., 2009. Prolonged in vitro exposure to white wines enhances the erosive damage on human permanent teeth compared with red wines. Nutrit. Res. 29, 558–567.

Williams, A.A., 1975. The development of a vocabulary and profile assessment method for evaluating the flavour contribution of cider and perry aroma constituents. J. Sci. Food Agric. 26, 567–582.

Williams, A.A., 1978. The flavour profile assessment procedure. In: Beech, F.W. (Ed.), Sensory Evaluation—Proceeding of the Fourth Wine Subject Day. Long Ashton Research Station, University of Bristol, Long Ashton, UK, pp. 41–56.

Williams, A.A., 1984. Measuring the competitiveness of wines. In: Beech, F.W. (Ed.), Tartrates and Concentrates—Proceeding of the Eighth Wine Subject Day Symposium. Long Ashton Research Station, University of Bristol, Long Ashton, UK, pp. 3–12.

Williams, A.A., Bains, C.R., Arnold, G.M., 1982. Towards the objective assessment of sensory quality in less expensive red wines. In: Webb, A.D. (Ed.), Grape Wine Cent. Symp. Proc. University of California, Davis, pp. 322–329.

Wilson, D.A., 2003. Rapid, experience-induced enhancement in odorant discrimination by anterior piriform cortex neurons. J. Neurophysiol. 90, 65–72.

Winiarski, W., Winiarski, J., Silacci, M., Painter, B., 1996. The Davis 20-point scale: How does it score today. Wines Vines 77, 50–53.

Winton, W., Ough, C.S., Singleton, V.L., 1975. Relative distinctiveness of varietal wines estimated by the ability of trained panelists to name the grape variety correctly. Am. J. Enol. Vitic .26, 5–11.

Wise, P.M., Olsson, M.J., Cain, W.S., 2000. Quantification of odor quality. Chem. Senses 25, 429–443.

Wollan, D., Pham, D.-T., Wilkinson, K.L., 2016. Changes in wine ethanol content due to evaporation from wine glasses and implications for sensory analysis. J. Agric. Food Chem. 64, 7569–7575.

Wu, T.Z., Lo, Y.R., 2000. Synthetic peptide mimicking of binding sites on olfactory receptor protein for use in 'electronic nose. J. Biotechnol. 80, 63–73.

Yenket, R., Chambers IV, E., Adhikari, K., 2011. A comparison of seven preference mapping techniques using four software programs. J. Sens. Stud. 26, 135–150.

Zarzo, M., 2008. Psychological dimensions in the perception of everyday odors: Pleasantness and edibility. J. Sens. Stud. 23, 354–376.

Zellner, D.A., Stewart, W.F., Rozin, P., Brown, J.M., 1988. Effect of temperature and expectations on liking for beverages. Physiol. Behav. 44, 61–68.

APPENDIX 5.1 AROMA AND BOUQUET SAMPLES

Sample	Amount per 300 mL of base wine
TEMPERATE TREE FRUIT	
Apple	15 mg Hexyl acetate
Cherry	3 mL Cherry brandy essence (Noirot)
Peach	100 mL Juice from canned peaches
Apricot	2 Drops of undecanoic acid y-lactone plus 100 mL juice from canned apricots
TROPICAL TREE FRUIT	
Litchi	100 mL Litchi fruit drink (Leo's)
Banana	10 mg Isoamyl acetate
Guava	100 mL Guava fruit drink (Leo's)
Lemon	0.2 mL Lemon extract (Empress)
VINE FRUIT	
Blackberry	5 mL Blackberry essence (Noirot)
Raspberry	60 mL Raspberry liqueur
Black currant	80 mL Blackcurrant nectar (Ribena)

Sample	Amount per 300 mL of base wine
Passion fruit	10 mL Ethanolic extract of one passion fruit
Melon	100 mL Melon liqueur
FLORAL	
Rose	6 mg Citronellol
Violet	1.5 mgp-Ionone
Orange blossom	20 mg Methyl anthranilate
Iris	0.2 mg Irone
Lily	7 mg Hydroxycitronellal
VEGETAL	
Beet	25 mL Canned beet juice
Bell pepper	5 mL 10% Ethanolic extract from dried bell peppers (2 g)
Green bean	100 mL Canned green bean juice 3 mg 1-Hexen-3-ol
Herbaceous	
SPICE	
Anise/licorice	1.5 mg Anise oil
Peppermint	1 mL Peppermint extract (Empress)
Black pepper	2 g Whole black peppercorns
Cinnamon	15 mg trans-Cinnamaldehyde
NUTS	
Almond	5 Drops bitter almond oil
Hazelnut	3 mL Hazelnut essence (Noirot)
Coconut	1.0 mL Coconut essence (Club House)
WOODY	
Oak	3 g Oak chips (aged 1 month)
Vanilla	24 mg Vanillin
Pine	7.5 mg Pine needle oil (1 drop)
Eucalyptus	9 mg Eucalyptus oil

The recipes are given only as a guide as adjustments will be required based on both individual needs and material availability. They are adequate for most purposes. Where used in research, and purity and consistency of preparations are paramount, details given in Meilgaard et al. (1982) are essential reading. Additional recipes may be found at http://www.nysaes.cornell.edu/fst/faculty/acree/fs430/aromalist/sensorystd.html, or in Lee et al. (2001), Meilgaard (1988), Noble et al. (1987), and Williams (1975). Pure chemicals have the advantage of providing highly reproducible samples, whereas "natural" sources are more complex, but more difficult to standardize. Addition of cyclodextrins may also aid stabilize the compounds and regulate their release. Readers requiring basic information for preparing samples may find Stahl and Einstein (1973), Furia and Bellanca (1975), and Heath (1981) especially useful. Most specific chemicals can be obtained from major chemical suppliers, while sources of fruit, flower, and other essences include wine supply, perfumery, and flavor supply companies. To limit oxidation, about 20 mg of potassium metabisulfite may be added to the samples noted.

Because only 30 mL samples are required at any one time, it may be convenient to disperse the original sample into 30-mL screw-cap test tubes for storage. Parafilm can be stretched over the cap to further prevent oxygen penetration. Samples stored in a refrigerator usually remain good for several months. Alternately, samples may be stored in hermetically sealed vials in a freezer and opened only as required.

1. With whole fruit, the fruit is ground in a blender with 95% alcohol. The solution is left for about a day in the absence of air, filtered through several layers of cheesecloth, and added to the base wine. Several days later, the sample may need to be decanted to remove excess precipitates.

2. All participants should be informed of the constituents of the samples. For example, people allergic to nuts may have adverse reactions even to their smell.

APPENDIX 5.2 BASIC OFF-ODOR SAMPLES[a,b,c,d]

Sample	Amount (per 300 mL neutral-flavored base wine)
CORKED	
2,4,6-TCA	3 ug 2,4,6-Trichloroanisole
Guaiacol	3 mg Guaiacol
Actinomycete	2 mg Geosmin (an ethanolic extract from a *Streptomyces griseus* culture[e])
Penicillium	2 mg 3-Octanol (or an ethanolic extract from a *Hemigera* (*Penicillium*) culture[f])
CHEMICAL	
Fusel	120 mg Isoamyl and 300 mg isobutyl alcohol
Geranium-like	40 mg 2,4-Hexadienol
Buttery	12 mg Diacetyl[g]
Plastic	1.5 mg Styrene
SULFUR	
Sulfur dioxide	200 mg Potassium metabisulfite
Goût de lumière	4 mg Dimethyl sulfide[g] and 0.4 mg ethanethiol
Mercaptan	4 mg Ethanethiol
Hydrogen sulfide	2 mL Solution with 1.5 mg $Na_2S.9H_2O$
MISCELLANEOUS	
Oxidized	120 mg Acetaldehyde
Baked	1.2 g Fructose added and baked 4 weeks at 55°C
Vinegary	3.5 g Acetic acid
Ethyl acetate	100 mg Ethyl acetate
Mousy	Alcoholic extract culture of *Brettanomyces* (or 2 mg 2-acetyltetrahydropyridines)

[a]To limit oxidation, about 20 mg potassium metabisulfite may be added per 300 mL base wine.[b]Because only 30 mL samples are required at any one time, it may be convenient to disperse the original sample into 30-mL screw-cap test tubes for storage. Parafilm can be stretched over the cap to further prevent oxygen penetration. Samples stored in a refrigerator usually remain good for several months.[c]Other off-odor sample preparations are noted in Meilgaard et al. (1982).[d]**Important:** participants should be informed of the chemical to be smelt in the test. For example, some asthmatics are highly sensitive to sulfur dioxide. If so, such individuals should not serve as wine tasters.[e]*Streptomyces griseus* is grown on nutrient agar in 100-cm diameter petri dishes for 1 week or more. The colonies are scraped off and added to the base wine. Filtering after a few days should provide a clear sample.[f]*Penicillium* sp. isolated from wine corks is inoculated on small chunks (1–5 mm) of cork soaked in wine. The inoculated cork is placed in a petri dish and sealed with Parafilm to prevent the cork from drying out. After 1 month, obvious growth of the fungus should be noticeable. Chunks of the overgrown cork are added to the base wine. Within a few days, the sample can be filtered to remove the cork. The final sample should be clear.[g]Because of the likelihood of serious modification of the odor quality of these chemicals by contaminants, Meilgaard et al. (1982) recommend that they be purified prior to use: for diacetyl, use fractional distillation and absorption (in silica gel, aluminum oxide, and activated carbon); for dimethyl sulfide, use absorption.

APPENDIX 5.3 RESPONSE SHEET FOR TASTE/MOUTH-FEEL TEST

	1	2	3

Name: _____ Session: ☐ ☐ ☐

Sample#   **INTENSITY OF SENSATION**

Weak   Medium   Intense

1   _____

Sensations:
Location:

2   _____

Sensations:
Location:

3   _____

Sensations:
Location:

4   _____

Sensations:
Location:

5   _____

Sensations:
Location:

6   _____

Sensations:
Location:

APPENDIX 5.4 OFF-ODORS IN FOUR TYPES OF WINE AT TWO CONCENTRATIONS

Wine	Off-odor	Chemical added	Amount (per 300 mL)
GEWÜRZTRAMINER			
	Oxidized	Acetaldehyde	20, 60 mg
	Sulfur dioxide	Potassium	67, 200 mg
	2,4,6-TCA	2,4,6-Trichloroanisole	2, 10 µg
	Plastic	Styrene	1.5, 4.5 mg

Wine	Off-odor	Chemical added	Amount (per 300 mL)
SAUVIGNON BLANC			
	Vinegary	Acetic acid	0.5, 2 g
	Buttery	Diacetyl	2, 6 mg
	Ethyl acetate	Ethyl acetate	40, 100 mg
	Geranium-like	32, 4 Hexadienol	10, 40 mg
BEAUJOLAIS			
	Geranium-like	2,4-Hexadienol	10, 40 mg
	Buttery	Diacetyl	5, 24 mg
	Ethyl acetate	Ethyl acetate	40, 100 mg
	Oxidized	Acetaldehyde	20, 60 mg
PINOT NOIR			
	Guaiacol	Guaiacol	0.2, 0.6 mg
	Mercaptan	Ethanethiol	5, 24 µg
	2,4,6-TCA	2,4,6-Trichloroanisole	2, 10 µg
	Plastic	Styrene	1.5, 4.5 mg

APPENDIX 5.5 MINIMUM NUMBER OF CORRECT JUDGMENTS TO ESTABLISH SIGNIFICANCE AT VARIOUS PROBABILITY LEVELS FOR THE TRIANGLE TEST (ONE-TAILED, P = 1/3)*

No. of trials (n)	Probability levels						
	0.05	0.04	0.03	0.02	0.01	0.005	0.001
5	4	5	5	5	5	5	
6	5	5	5	5	6	6	
7	5	6	6	6	6	7	7
8	6	6	6	6	7	7	8
9	6	7	7	7	7	8	8
10	7	7	7	7	8	8	9
11	7	7	8	8	8	9	10
12	8	8	8	8	9	9	10
13	8	8	9	9	9	10	11
14	9	9	9	9	10	10	11
15	9	9	10	10	10	11	12
16	9	10	10	10	11	11	12
17	10	10	10	11	11	12	13
18	10	11	11	11	12	12	13
19	11	11	11	12	12	13	14
20	11	11	12	12	13	13	14
21	12	12	12	13	13	14	15

No. of trials (n)	Probability levels						
	0.05	0.04	0.03	0.02	0.01	0.005	0.001
22	12	12	13	13	14	14	15
23	12	13	13	13	14	15	16
24	13	13	13	14	15	15	16
25	13	14	14	14	15	16	17
26	14	14	14	15	15	16	17
27	14	14	15	15	16	17	18
28	15	15	15	16	16	17	18
29	15	15	16	16	17	17	19
30	15	16	16	16	17	18	19
31	16	16	16	17	18	18	20
32	16	16	17	17	18	19	20
33	17	17	17	18	18	19	21
34	17	17	18	18	19	20	21
35	17	18	18	19	19	20	21
36	18	18	18	19	20	20	22
37	18	18	19	19	20	21	22
38	19	19	19	20	21	21	23
39	19	19	20	20	21	22	23
40	19	20	20	21	21	22	24
41	20	20	20	21	22	23	24
42	20	20	21	21	22	23	25
43	20	21	21	22	23	24	25
44	21	21	22	22	23	24	25
45	21	22	22	23	24	24	26
46	22	22	22	23	24	25	27
47	22	22	23	23	24	25	27
48	22	23	23	24	25	26	27
49	23	23	24	24	25	26	28
50	23	24	24	25	26	26	28
60	27	27	28	29	30	31	33
70	31	31	32	33	34	35	37
80	35	35	36	36	38	39	41
90	38	39	40	40	42	43	45
100	42	43	43	44	45	47	49

Source: After tables compiled by Roessler, E.B., Pangborn, R.M., Sidel, J.L., Stone, H., 1978. Expanded statistical tables for estimating significance in prepared-preference, paired-difference, duo-trio and triangle tests. J. Food Sci. 13, 940 943 (Roessler et al., 1978), from Amerine, M.A., Roessler, E.B., 1983. Wines, Their Sensory Evaluation. second ed., Freeman, San Francisco, CA.
*Values (X) not appearing in the table may be derived from $X = (2n + 2.83 z \sqrt{n} + 3/6)$.

APPENDIX 5.6 RANK TOTALS EXCLUDED FOR SIGNIFICANCE DIFFERENCES, 5% LEVEL. ANY RANK TOTAL OUTSIDE THE GIVE RANGE IS SIGNIFICANT

Number of judges	Number of wines										
	2	3	4	5	6	7	8	9	10	11	12
3				4–14	4–17	4–20	4–23	5–25	5–28	5–31	5–34
4		5–11	5–15	6–18	6–22	7–25	7–19	8–32	8–36	8–39	9–43
5		6–14	7–18	8–22	9–26	9–31	10–35	11–39	12–43	12–48	13–52
6	7–11	8–16	9–21	10–26	11–31	12–36	13–41	14–46	15–51	17–55	18–60
7	8–13	10–18	11–24	12–30	14–35	15–41	17–46	18–52	19–58	21–63	22–69
8	9–15	11–21	13–27	15–33	17–39	18–46	20–52	22–58	24–64	25–71	27–77
9	11–16	13–23	15–30	17–37	19–44	22–50	24–57	26–64	28–71	30–78	32–85
10	12–18	14–26	17–33	20–40	22–48	25–55	27–63	30–70	32–78	25–85	37–93
11	13–20	16–28	19–36	22–44	25–52	28–60	31–68	34–76	36–85	39–93	42–101
12	15–21	18–30	21–39	25–47	28–56	31–65	34–74	38–82	41–91	44–100	47–109
13	16–23	20–32	24–41	27–51	31–60	35–69	38–79	42–88	45–98	49–107	52–117
14	17–25	22–34	26–44	30–54	34–64	38–74	42–84	46–94	50–104	54–114	57–125
15	19–26	23–37	28–47	32–58	37–68	41–79	46–89	50–100	54–111	58–122	63–132
16	20–28	25–39	30–50	35–61	40–72	45–83	49–95	54–106	59–117	63–129	68–140
17	22–29	27–41	32–53	38–64	43–76	48–88	53–100	58–112	63–124	68–136	73–148
18	23–31	29–43	34–56	40–68	46–80	52–92	57–105	61–118	68–130	73–143	79–155
19	24–33	30–46	37–58	43–71	49–84	55–97	61–110	67–123	73–136	78–150	84–163
20	26–34	32–48	39–61	45–75	52–88	58–102	65–115	71–129	77–143	83–157	90–170

Source: Adapted by Amerine, M.A., Roessler, E.B., 1983. Wines, Their Sensory Evaluation. second ed., Freeman, San Francisco, CA, from tables compiled by Kahan, G., Cooper, D., Papavasiliou, A., Kramer, A., 1973. Expanded tables for determining significance of differences for ranked data. Food Technol. 27, 64–69 (Kahan et al., 1973).

APPENDIX 5.7 RANK TOTALS EXCLUDED FOR SIGNIFICANCE DIFFERENCES, 1% LEVEL. ANY RANK TOTAL OUTSIDE THE GIVE RANGE IS SIGNIFICANT

Number of judges	Number of wines										
	2	3	4	5	6	7	8	9	10	11	12
3									4–29	4–32	4–35
4				5–19	5–23	5–27	6–30	6–34	6–38	6–42	7–45
5			6–19	7–23	7–28	8–32	6–37	9–41	9–46	10–50	10–55
6		7–17	8–22	9–27	9–33	10–38	11–43	12–48	13–53	13–59	14–64
7		8–20	10–25	11–31	12–37	13–43	14–49	15–55	16–61	17–67	18–73
8	9–15	10–22	11–29	13–35	14–42	16–48	17–55	19–61	20–68	21–75	23–81
9	10–17	12–24	13–32	15–39	17–46	19–53	21–60	22–68	24–75	26–82	27–90
10	11–19	13–27	15–35	18–42	20–50	22–58	24–66	26–74	28–82	30–90	32–98
11	12–21	15–29	17–38	20–46	22–55	25–63	27–72	30–80	32–89	34–98	37–106
12	14–22	17–31	19–41	20–50	25–59	28–68	31–77	33–87	36–96	39–105	42–114
13	15–24	18–34	21–44	25–53	28–63	31–73	34–83	37–93	40–103	43–113	46–123
14	16–26	20–36	24–46	27–57	31–67	34–78	38–88	41–98	45–109	48–120	51–131

Number of judges	Number of wines										
	2	3	4	5	6	7	8	9	10	11	12
15	18–27	22–38	26–49	30–60	34–71	37–83	41–94	45–105	49–116	53–127	56–139
16	19–29	23–41	28–52	32–64	36–76	41–87	45–99	49–111	53–123	57–135	62–146
17	20–31	25–43	30–55	35–67	39–80	44–92	49–104	53–117	58–129	62–142	67–154
18	22–32	27–45	32–58	37–71	42–84	47–97	52–110	57–123	62–136	67–149	72–162
19	23–34	29–47	34–61	40–74	45–88	50–102	56–115	61–129	67–142	72–156	77–170
20	24–36	30–50	36–64	42–78	48–92	54–106	60–120	65–135	71–149	77–163	82–178

Source: Adapted by Amerine, M.A., Roessler, E.B., 1983. Wines, Their Sensory Evaluation. second ed., Freeman, San Francisco, CA, from tables compiled by Kahan, G., Cooper, D., Papavasiliou, A., Kramer, A., 1973. Expanded tables for determining significance of differences for ranked data. Food Technol. 27, 64–69 (Kahan et al., 1973).

APPENDIX 5.8 MULTIPLIERS FOR ESTIMATING SIGNIFICANCE OF DIFFERENCE BY RANGE. TWO-WAY CLASSIFICATION. A, 5% LEVEL; B, 1% LEVEL

Number of judges	Number of wines										
	2	3	4	5	6	7	8	9	10	11	12
A											
2	6.35	2.19	1.52	1.16	0.94	0.79	0.69	0.60	0.54	0.49	0.45
	6.35	1.96	1.39	1.12	0.95	0.84	0.76	0.70	0.65	0.61	0.58
3	1.96	1.14	0.88	0.72	0.61	0.53	0.47	0.42	0.38	0.35	0.32
	2.19	1.14	0.90	0.76	0.67	0.61	0.56	0.52	0.49	0.46	0.44
4	1.43	0.96	0.76	0.63	0.54	0.47	0.42	0.38	0.34	0.31	0.29
	1.54	0.93	0.76	0.65	0.58	0.53	0.49	0.45	0.43	0.40	0.38
5	1.27	0.89	0.71	0.60	0.51	0.45	0.40	0.36	0.33	0.30	0.28
	1.28	0.84	0.69	0.60	0.53	0.49	0.45	0.42	0.40	0.38	0.36
6	1.19	0.87	0.70	0.58	0.50	0.44	0.39	0.36	0.33	0.30	0.28
	1.14	0.78	0.64	0.56	0.50	0.46	0.43	0.40	0.38	0.36	0.34
7	1.16	0.86	0.69	0.58	0.50	0.44	0.40	0.36	0.33	0.30	0.28
	1.06	0.74	0.62	0.54	0.48	0.44	0.41	0.38	0.36	0.34	0.33
8	1.15	0.86	0.69	0.58	0.50	0.44	0.40	0.36	0.33	0.30	0.28
	1.01	0.71	0.59	0.52	0.47	0.43	0.40	0.37	0.35	0.33	0.32
9	1.15	0.86	0.70	0.59	0.51	0.45	0.40	0.36	0.33	0.31	0.29
	0.97	0.69	0.58	0.51	0.46	0.42	0.39	0.36	0.34	0.33	0.31
10	1.15	0.87	0.71	0.60	0.51	0.45	0.41	0.37	0.34	0.31	0.29
	0.93	0.67	0.56	0.50	0.45	0.41	0.38	0.36	0.34	0.32	0.31
11	1.16	0.88	0.71	0.60	0.52	0.46	0.41	0.37	0.34	0.32	0.29
	0.91	0.66	0.55	0.49	0.44	0.40	0.38	0.35	0.33	0.32	0.30
12	1.16	0.89	0.72	0.61	0.53	0.47	0.42	0.38	0.35	0.32	0.30
	0.89	0.65	0.55	0.48	0.43	0.40	0.37	0.35	0.33	0.31	0.30

Number of judges	Number of wines										
	2	3	4	5	6	7	8	9	10	11	12
B											
2	31.83	5.00	2.91	2.00	1.51	1.20	1.00	0.86	0.75	0.66	0.60
	31.83	4.51	2.72	1.99	1.59	1.35	1.19	1.07	0.97	0.90	0.84
3	4.51	1.84	1.31	1.01	0.82	0.70	0.60	0.53	0.48	0.43	0.39
	5.00	1.84	1.35	1.10	0.94	0.83	0.76	0.69	0.65	0.61	0.57
4	2.63	1.40	1.04	0.83	0.69	0.59	0.52	0.46	0.42	0.38	0.35
	2.75	1.35	1.04	0.87	0.76	0.68	0.63	0.58	0.54	0.51	0.48
5	2.11	1.25	0.95	0.77	0.64	0.56	0.49	0.44	0.40	0.36	0.33
	2.05	1.14	0.90	0.77	0.68	0.61	0.56	0.52	0.49	0.46	0.44
6	1.88	1.18	0.91	0.74	0.63	0.54	0.48	0.43	0.39	0.36	0.33
	1.71	1.02	0.82	0.71	0.63	0.57	0.52	0.49	0.46	0.43	0.41
7	1.78	1.15	0.89	0.73	0.62	0.54	0.48	0.43	0.39	0.36	0.33
	1.52	0.95	0.77	0.66	0.59	0.54	0.50	0.46	0.44	0.41	0.39
8	1.72	1.14	0.89	0.73	0.62	0.54	0.48	0.43	0.39	0.36	0.33
	1.40	0.90	0.73	0.63	0.57	0.52	0.48	0.45	0.42	0.40	0.38
9	1.69	1.14	0.89	0.73	0.62	0.54	0.48	0.43	0.39	0.36	0.33
	1.31	0.86	0.71	0.61	0.55	0.50	0.46	0.43	0.41	0.39	0.37
10	1.67	1.14	0.89	0.74	0.63	0.55	0.48	0.44	0.40	0.36	0.34
	1.24	0.83	0.68	0.59	0.53	0.49	0.45	0.42	0.40	0.38	0.36
11	1.67	1.15	0.90	0.74	0.63	0.55	0.49	0.44	0.40	0.37	0.34
	1.19	0.80	0.67	0.58	0.52	0.48	0.44	0.41	0.39	0.37	0.35
12	1.67	1.15	0.91	0.75	0.64	0.56	0.50	0.45	0.41	0.37	0.35
	1.15	0.78	0.65	0.57	0.51	0.47	0.43	0.41	0.38	0.36	0.35

Source: After Kurtz, T.E., Link, T.E., Tukey, R.F., Wallace, D.L., 1965. Short-cut multiple comparisons for balanced single and double classifications. Technometrics 7, 95–165 (Kurtz et al., 1965). Reproduced with permission of Technometrics through the courtesy of the American Statistical Association, modified in Amerine, M.A., Roessler, E.B., 1983. Wines, Their Sensory Evaluation. second ed., Freeman, San Francisco, CA.
The entries in this table are to be multiplied by the sum of the ranges of differences between adjacent wine scores to obtain the difference required for significance for wine totals (use upper entry) and/or judge totals (use lower entry).

APPENDIX 5.9 MINIMUM NUMBERS OF CORRECT JUDGMENTS TO ESTABLISH SIGNIFICANCE AT VARIOUS PROBABILITY LEVELS FOR PAIRED-DIFFERENCE AND DUO-TRIO TESTS (ONE-TAILED, $p = 1/2$)[a]

No. of trials (n)	Probability levels						
	0.05	0.04	0.03	0.02	0.01	0.005	0.001
7	7	7	7	7	7		
8	7	7	8	8	8	8	
9	8	8	8	8	9	9	
10	9	9	9	9	10	10	10
11	9	9	10	10	10	11	11
12	10	10	10	10	11	11	12
13	10	11	11	11	12	12	13
14	11	11	11	12	12	13	13
15	12	12	12	12	13	13	14
16	12	12	13	13	14	14	15
17	13	13	13	14	14	15	16
18	13	14	14	14	15	15	16
19	14	14	15	15	15	16	17
20	15	15	15	16	16	17	18
21	15	15	16	16	17	17	18
22	16	16	16	17	17	18	19
23	16	17	17	17	18	19	20
24	17	17	18	18	19	19	20
25	18	18	18	19	19	20	21
26	18	18	19	19	20	20	22
27	19	19	19	20	20	21	22
28	19	20	20	20	21	22	23
29	20	20	21	21	22	22	24
30	20	21	21	22	22	23	24
31	21	21	22	22	23	24	25
32	22	22	22	23	24	24	26
33	22	23	23	23	24	25	26
34	23	23	23	24	25	25	27
35	23	24	24	25	25	26	27
36	24	24	25	25	26	27	28
37	24	25	25	26	26	27	29
38	25	25	26	26	27	28	29
39	26	26	26	27	28	28	30
40	26	27	27	27	28	29	30
41	27	27	27	28	29	30	31
42	27	28	28	29	29	30	32
43	28	28	29	29	30	31	32

No. of trials (n)	Probability levels						
	0.05	0.04	0.03	0.02	0.01	0.005	0.001
44	28	29	29	30	31	31	33
45	29	29	30	30	31	32	34
46	30	30	30	31	32	33	34
47	30	30	31	31	32	33	35
48	31	31	31	32	33	34	36
49	31	32	32	33	34	34	36
50	32	32	33	33	34	35	37
60	37	38	38	39	40	41	43
70	43	43	44	45	46	47	49
80	48	49	49	50	51	52	55
90	54	54	55	56	57	58	61
100	59	60	60	61	63	64	66

Source: After tables compiled by Roessler, E.B., Pangborn, R.M., Sidel, J.L., Stone, H., 1978. Expanded statistical tables for estimating significance in prepared-preference, paired-difference, duo-trio and triangle tests. J. Food Sci. 43, 940–943 (Roessler et al., 1978), from Amerine, M.A., Roessler, E.B., 1983. Wines, Their Sensory Evaluation. second ed., Freeman, San Francisco, CA.
[a]Values (X) not appearing in table may be derived from: $X = (z \sqrt{n} + n + 1)/2$. See text.

6

Qualitative Wine Assessment

In Chapter 5, tasting was discussed for differentiating and characterizing wine based on a critical analysis of its sensory properties. In this chapter, the primary concern is the expression of a wine's varietal, stylistic, regional, or purely aesthetic qualities, i.e., those features that tend to attract wine lovers, from interested novice to aficionado. This usually involves some form of qualitative ranking. Despite the impression often given by wine critics, detection of the features noted above is often far from easy, and more difficult in some regards than the tasks requested of sensory panels in Chapter 5. Most wines do not express a readily detectable varietal or regional character. Thus, the search for these traits is often fruitless. Only fine examples clearly show the prototypic features attributed to a variety or region. Experience provides the knowledge base for recognizing these attributes, when present, and how their expression can be modified by style and age. Somewhat easier to detect are stylistic characteristics. The method of fermentation or processing usually donates a distinct set of sensory properties, in many ways more pronounced than varietal or regional characteristics. General aesthetic attributes are also comparatively readily recognized, with experience and adequate attention. More difficult is describing these features in meaningful words, and agreeing on how wines should be rated. Because reference standards for complexity, development, balance, body, finish, etc., do not exist (or meet with universal approval), precise definition of such terms is difficult to impossible. Despite these difficulties, and possibly because of them, wine provides much pleasure and intrigue, and an endless source of discussion (and disagreement).

TASTING ROOM

Where possible, any room used for tastings should be bright, of neutral color, quiet, and odor-free. Wine cellars may be "romantic" sites for a tasting, but their permeation with vinous and moldy odors makes any reliable assessment impossible. Pitchers of water (to rinse glasses), and opaque buckets for disposal of samples should always be present.

Because tasting should permit the taster to obtain an unbiased opinion of the wine, direct and indirect physiological and psychological hindrances to impartial judgment should be eliminated or reduced to a minimum. For example, the presence of winemakers or sales staff may be informative and interesting to attendees, but will almost undoubtedly compromise making an honest assessment. "Authority figures" not only can direct opinion in wine tastings (Aqueveque, 2015), but their influence is even more shocking relative to obedience (Milgram, 1963; Caspar et al., 2016). Contextual factors are another significant factor potentially biasing opinion, as discussed in earlier chapters. In contrast, if the intent is to enhance or induce appreciation, discussion is encouraged. Participants of wine societies, for example, often appear to appreciate suggestions and support. Also, where expensive or rare wines are sampled, cold critical analysis is probably counterproductive.

INFORMATION PROVIDED

What information, if anything, is provided about the wines before or during a tasting depends on its objective. For example, if the wines are intended to be ranked on their expression of varietal or regional attributes, then the tasters need this information. It is unreasonable to expect tasters to guess the intent of a tasting as well as perform the task. Admittedly, knowing what is desired can exaggerate a property's presence. Nonetheless, necessary information should also highlight those wines legitimately expressing the expected attributes. However, where the intent is to rank wines on their intrinsic (aesthetic) qualities, regardless origin or price, then is best to withhold identification of the wines to avoid extrinsic factors unduly swaying the evaluation. Such information appears to be less influential with those possessing little wine experience. It is in those with experience that prior knowledge about a wine's origin or price tends to activate wine memories that can prejudice perception. Models of wine characteristics, developed with expensive experience, are particularly potent in influencing perception, potentially generating attributes that may not be present, as color can distort the identification of a flavored solution. Experiments have repeatedly demonstrated that perception of a wine can be affected by suspected origin or price (e.g., Brochet and Morrot, 1999; Plassmann et al., 2008). This influence can be so strong as to see renowned but faulty wines being praised. Whether this is due to peer pressure, or actual suppression of what should be obvious, is unclear. However, where wine promotion is the primary goal, as is the case of in-store or most producer-run tastings, identity, price, and repute are intentionally promoted to achieve the maximal psychological boost.

If the purpose of a tasting is to obtain consumer opinion, leading questions must be avoided. For this, indirect questioning is best. For example, asking which wine the taster would prefer serving a business colleague or close friend is likely to be less prejudicial, and more revealing, than directly asking which sample is preferred (when, in fact, none may be). Indirect questioning has been used with apparent success in detecting subtle differences in beer (Mojet and Köster, 1986) and milk (Wolf-Frandsen et al., 2003).

Wrapping bottles in paper bags is a traditional means of concealing wine identity at informal tastings. Nonetheless, stretch-fabric black bags that fit the bottle tightly, and possess color coding, possess a more professional look. Prepouring or decanting is often impractical and unwarranted. Although bagging neither conceals bottle color, shape, nor neck design, most people are unaware of their potential to provide clues as to provenance.

An alternative technique is to remove the label. In wine-appreciation courses, this technique has the advantage of obtaining labels for reproduction on the tasting sheet, or for distribution after the tasting. Regrettably, with the increasing popularity of heat-activated glues, label removal is not the simple task it used to be. An alternative source of wine labels is now the website of the producer.

Because bottles are usually opened just before or during the tasting session, it may be best for the instructor or coordinator to remove the corks. If this is performed before a class or participants of a wine society, it not only provides an excellent opportunity to demonstrate the proper use of various models of cork screws, but also avoids participants from identifying the wines from inscriptions on the cork. Identification of the wines in advance is appropriate only in the most elementary (or commercial) of tastings.

SAMPLE PREPARATION

Decanting and Breathing

Originally, the primary function of decanting was to separate wine from sediment that had accumulated in the bottle. Disturbance during pouring resulted in later samples being clouded and tasting somewhat gritty. This applied particularly to old red wines, but also occurred with some long-aged white wines. With modern fining techniques, decanting is unnecessary (the major exception being vintage ports that are bottled early and unfined). If decanting precedes the sampling of old wines, it should occur immediately before sampling. The fragrance of most old wines (>40 yr) tends to dissipate rapidly upon opening, due to the depauperated state of the wine's aromatic composition, the wine's aromatics having either degraded in some manner or escaped through the cork.

Possibly because decanting's secondary role is only vaguely remembered, much misinformation surrounds recommending wine aeration. During the 1800s, exposure to air during decanting had the benefit of dispelling or degrading reduced-sulfur off-odors, or the "bottle stink" to which wines of the time were apparently often afflicted. The turbulence generated would have facilitated hydrogen sulfide escaping from the wine, as well as mercaptans. We can be thankful that modern wines are neither so plagued nor do they develop copious amounts of sediment. Thus, few modern wines need or benefit from decanting. Despite this, opening bottles in advance is still frequently recommended, even if this involves just pulling the cork a few minutes before pouring. This act is viewed as allowing the wine to "breathe." However, the surface area of an open, 750-mL bottle of wine is about 3.5 cm^2. If the headspace gases in the bottle neck are replaced with air, it would possess only 5 mg O_2. Because oxygen diffuses into wine at a very slow rate, no detectable oxidation is possible during the time separating cork removal and pouring.

Even when a sample is poured, for example, into a tulip-shaped ISO glass, the wine's surface area increases to only about 28 cm². Swirling the glass more than doubles the wine's surface-to-air contact. If one assumes 30 mL of wine, benefits associated with air contact should occur several hundred times faster in the glass than in a bottle (if based on surface-area-to-volume contact). The increased wine–air contact favors more the release of aromatics from the wine than their oxidation. These changes are generally referred to as the wine's opening (development). Thus, the benefits of simply pulling the cork before serving are illusionary; any perceived benefit probably is the equivalent of the placebo effect in medical studies, or a situation of "my mind's made up, don't confuse me with the facts." Equally, the current popularity of wine aerators is little more than a gimmick. Swirling the wine in the glass is far more effective in increasing wine–air contact and aromatic release. It makes far more sense to increase volatilization of wine constituents in the glass, where they can be detected, than into some container where they dissipate undetected.

Unfortunately, experimental data on sensory changes during tasting are few and far between. In one study, the effect of up to a 30-min exposure to air was studied (Russell et al., 2005). During this period, there were detectable chemical changes in several phenolic compounds, notably gallic acid. However, these were not sensorially discernible. Identifiable oxidative changes in aroma have been detected in sound wine only with periods of air exposure considerably longer than typically experienced during a tasting (Roussis et al., 2005). Eventually, exposure to air results in a loss of perceived quality, but it takes several hours to become noticeable (Ribéreau-Gayon and Peynaud, 1961). Hopefully, application of techniques, such as temporal dominance of sensations (TDS), gas chromatography–olfactometry (GC-O), or retronasal aroma-trapping (Muñoz-González et al., 2014) will add to what little data we have on the sensory changes subsequent to bottle opening and wine pouring.

Although the physicochemistry of the opening of a wine's fragrance is still largely speculation, it probably relates to equilibrium changes between aromatics in the headspace gases above the wine and their dissolved and weakly bound states in the main body of the wine (matrix). Mannoproteins (Lubbers et al., 1994), amino acids (Maier and Hartmann, 1977), sugars (Sorrentino et al., 1986), and reductones (Guillou et al., 1997) are examples of matrix constituents that can weakly bind aromatics, making them temporarily nonvolatile. Upon bottle opening and pouring, aromatics in the headspace above the wine diffuse into the surrounding air. This changes the equilibrium between volatile, dissolved, and bound states of wine aromatics, shifting the balance toward additional volatilization. Because this complex, dynamic equilibrium effect varies with the compound, the relative concentration of a wine's aromatics in the air above the wine will also be in a constant state for flux. Ideally, this is experienced as a dynamically changing, kaleidoscopic, aromatic experience.

As noted, volatilization is accentuated as a consequence of swirling. Ethanol evaporation from the surface of the wine, and the thin film that coats the sides of the glass, modifies the vapor pressure of aromatics (Williams and Rosser, 1981) and changes its surface tension. What are popularly termed wine "breathing," "opening," and "development" are simply named stages of a single phenomenon. Because this is one of the more fascinating features of quality wines, it should not be relegated to transpire undetected in the bottle or a decanter.

Depending on the wine, its aromatic character may dissipate within a few minutes to several hours. Few white wines develop significantly in the glass, except for those produced from cultivars such as Riesling or Chardonnay. Very old wines, regardless of color, do not improve aromatically upon opening, usually losing quickly whatever fragrance they still possess. In contrast, younger red wines that age well often show a fascinating aromatic development in the glass.

For old wines, decanting often involves an elaborate ritual. This may include the use of decanting machines to minimize agitation during the process. It also makes for a good show, heightening the anticipation of the assembled participants. Decanting must of necessity occur slowly and meticulously, to minimize turbulence that would disturb any sediment present as the wine flows out the bottle. To facilitate this, the fulcrum of the decanting machine should be near the neck of the bottle. Otherwise, the mouth of the bottle can pass through more than a 60-degree angle during decanting. This has the disadvantage of requiring a continual readjustment of the carafe to receive the wine as it flows out. Pouring terminates when sediment begins to reach the neck. Although decanting machines are effective, anyone with a steady hand can manually decant a bottle effectively, if performed with caution. Checking for sediment flow toward the neck usually needs to begin only when about 80 percent of the wine has been poured. The presence of a candle, to see the sediment as it moves toward the neck is usually part of the performance, but is required only if decanting occurs under dimly lit conditions.

Prior to decanting, any bottle needing decanting needs to be moved delicately to the tasting site several hours to days in advance and positioned upright. This facilitates any loose sediment settling to the bottom before decanting commences.

Wine Temperature

White wines are generally served at between 8 and 12°C, with sweet versions being presented at the lower end of the range, and dry versions near the higher end. Nonetheless, dry white wines often show well up to 20°C. Red wines are generally sampled at between 18 and 22°C. Only light, fruity, red wines, such as Beaujolais, express well at about 15°C. Rosé wines also express their attributes well around 15°C.

Sparkling wines are generally served at about 4°C. This enhances their toasty fragrance and favors the slow, gentle, steady release of bubbles. Cool temperatures also maximize the prickling sensation of carbon dioxide and generate a refreshing mouth-feel. Furthermore, a cool temperature has an advantage during bottle opening. As carbon dioxide is more soluble at cool temperatures, there is less likelihood of gushing upon opening, or during pouring. The one disadvantage of cool temperatures, when the humidity of the room is high, is the almost immediate fogging of the glass. The result is an inability to assess, and more importantly, appreciate the pirouette of bubbles as they shimmy their way to the surface.

Sherries are generally presented at between 6 and 8°C. This mellows their intense bouquet and diminishes the potential cloying sweetness of some sherries. In contrast, ports are served about 18°C. This diminishes the burning sensation of alcohol and facilitates the release of their complex aromatics.

Some of these preferences are probably associated with tradition and habituation, namely, the house (storage) temperature typical of many European homes prior to central heating. Nonetheless, serving preferences also reflect the influence of heat on volatility and gustatory sensitivity. The increase in volatility, associated with a rise in temperature, varies considerably with the compound. This can diminish or enhance the perception of particular compounds, due to differential masking or synergistic effects of the wine's matrix. This may partially explain why white wines generally appear more pleasant at cooler than warmer temperatures.

Temperature also has pronounced effects on gustatory sensations. Cool temperatures may reduce the sensitivity to sugars, but enhance the perception of acidity. Consequently, dessert wines appear more balanced at cooler temperatures. Coolness can also generate an agreeable freshness in the mouth. In contrast, warmer temperatures tend to reduce perceived bitterness and astringency. This seemingly helps explain why red wines are generally preferred above 18°C.

Where possible, wines should be brought to a desirable temperature before tasting. If necessary, wines may be quickly brought to a selected temperature by immersion in cold or warm water (Fig. 5.17B).

Because wine begins to warm toward ambient after pouring, presenting it at a temperature lower than desired has value in a tasting room situation. It results in the temperature remaining within a desirable range during the assessment period.

Wine temperature seldom is measured directly. Normally, it is assumed to be correct if the bottle has been stored for several hours at the chosen temperature. Alternatively, it can be assessed indirectly with a device that records the surface temperature of the bottle. It uses a colored plastic strip that becomes translucent across a range of temperatures, similar to liquid-crystal temperature strips used to measure forehead temperature. If the wine has had sufficient time to equilibrate, the strip accurately measures the wine's temperature.

Wine Glasses

Wine glasses should ideally be clear, uncolored, and have sufficient capacity and shape to permit swirling. The International Standard Organization (ISO) wine-tasting glass (Fig. 1.2) fully satisfies these requirements. Its narrow mouth helps concentrate aromatics within the bowl, while the broad base and sloping sides expedite vigorous swirling. Terms for the various part of a glass are illustrated in Fig. 6.1.

The problem with ISO glasses can be their fragility (thin crystal). Although donating an elegance at home, where the ring of crystal is pleasing when toasting with friends or family, thicker glass versions may be preferable in wine classes, or under commercial situations. Examples are Libbey's Royal Leerdam #9309RL (Plate 5.13, left), the Durand Viticole Tasting Glass, and the Libbey Citation #8470 (Plate 5.13, right).

For sparkling wines, tall slender flutes (Plate 5.15) permit optimal viewing of the wine's effervescence. This involves careful observation of the chains of bubbles (size, tightness, number, and persistence), the mound of bubbles (*mousse*) in the center of the glass, and the ring of bubbles (*cordon de mousse*) around the edge of the glass. The wide, shallow coupes, once popular for champagnes, are probably the worst possible glass for appreciating the wine's qualities, complicating assessment of the fragrance, accentuating carbon dioxide loss (Liger-Belair et al., 2009), and making any serious observation of its effervescence impossible.

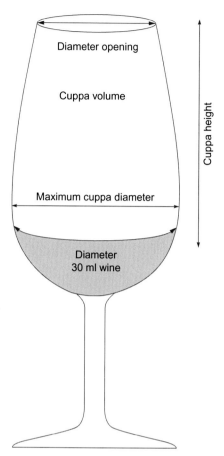

FIGURE 6.1   Wine-glass terminology.

Except for flute-shaped glasses, only tulip-shaped glasses are used for analytic wine tasting. This does not mean that other glasses do not have their value at the dinner table. Changing glass shape for a series of wines can have the same refined appeal as varying the size or design of the chinaware used throughout a meal (e.g., Stewart and Goss, 2013). Whether these changes are considered appropriate depends on personal preference, not objective considerations. However, because glass shape can influence perceived quality (Fischer, 2000; Cliff, 2001), all participants should use identical glasses for any particular wine.

Because glasses can pick up odors from the environment, glassware should be adequately cleaned, rinsed, and stored. Detergents are usually required to remove oils, tannins, and pigments that may adhere to the glass surface. For this, only nonperfumed detergents, water-softeners, or antistatic agents should be used. Equally important is adequate rinsing to remove any residues. Hot tap water is usually adequate. If the glasses are hand dried, odorless tea towels should be used.

Although it may appear to be a fetish, smelling glasses before filling can avoid regrettable instances of off-odor contamination. Stale odors can seriously adulterate a wine's fragrance. It is discouraging to think how often glassware odors can ruin the perception of a wine's quality.

The removal of detergent residues is particularly important when tasting sparkling wines. Formation of continuous chains of bubbles is largely dependent on the presence of nucleation sites along the bottom and sides of the glass. These sites consist primarily of minute dust particles, lint fibers, microscopic tartrate crystals, or miniature rough edges or scratches in the glass (Liger-Belair et al., 2002). Their cavities entrap minute pockets of air during pouring. Diffusion of carbon dioxide into these nucleation sites requires considerably less activation energy than incipient bubble formation within the body of the wine. Detergent residues can leave a molecular thin film over these sites, severely limiting or preventing carbon dioxide diffusion into nucleation sites, and, thereby, bubble enlargement and release. To promote bubble formation, some aficionados etch a star or cross on the bottom of their flutes (Polidori et al., 2009). However, the practice does not appear to enhance the wine's perceived quality (White and Heymann, 2015).

Once glasses have been cleaned and dried, they should be stored upright in a dust- and odor-free environment. The upright position limits odor contamination on the insides of the glass. Hanging glasses upside down is acceptable only for short-term storage in restaurants and wine bars, where it may reduce the likelihood of marking the bowl of the glass with finger prints.

Although not essential for tasting, stemmed wine glasses greatly facilitate wine assessment. They also limit soiling of the bowl during handling, as well as donate an elegance that tumblers just do not have. Cupping the bowl of the glass in one's hands would be conscionable only if the wine were served too cold (something that should not happen). The thinner glass of most stemmed wine glasses also enhances the gracefulness of the tasting experience. This may seem like elitism, but it more a means of elevating an ordinary occasion into something special. Life is all too short not to accentuate those aspects that bring sensory pleasure, and do no harm.

That specific glass shapes are correlated with enhanced perception of particular wines is unestablished, and seems more a ploy to enhance glass sales. Nonetheless, if their use is considered by the user to increase appreciation, is the delusion harmful? In contrast, there appears to be no conceivable rationale of the use of tumblers, currently being promoted for wine use by some glass manufacturers, unless one wishes to look at one's wine through glass with fingerprints. The lily-shaped, cut-crystal glasses of old were fine for water but not wine. Forget any real chance of assessing the wine's character in such glasses.

Sample Number and Volume

The number of samples appropriate for tasting depends on the occasion. In wine courses and tasting societies, six wines are standard, i.e., not too few for the effort involved, nor too many relative to cost, or the potential for over-consumption. In contrast, where sampling and note taking are brief (large trade or commercial tastings), upwards of 30 wines are often sampled, succeeding only in getting a glimpse of their immediate sensory appeal. Regrettably, any potential for assessing the wines' development, duration, or finish are out of the question.

Samples of about 35–50 mL are fully adequate for most sit-down tastings. For quick assessments involving one or two sips, such as at trade tastings, volumes of 15–20 mL are fully adequate, whereas volumes for a full and detailed assessment should be in the range of 50 mL.

Where glasses are being reused for multiple sit-down tastings, etching a line around the outer circumference of the glasses simplifies denoting when an appropriate volume has been poured. This is particularly useful in wine courses or societal tastings, where keeping costs down demands that each participant receive not more than what is minimally necessary for the purpose of assessment. Keeping each sample to about 35 mL makes it possible for a 750-mL bottle to serve up to 20 participants. Limiting sample volume also leaves an unspoken message that tastings are exactly that: tastings, not occasions for drinking.

Cork Removal

Corkscrews come in an incredible range of shapes, sizes, and modes of action. Those with a helical coil are generally the easiest and most efficient to use. The waiter's corkscrew is a classic example. Its major liabilities are the physical effort required to extract the cork and its lack of elegance. Through an adjustment, the same action that forces the screw into the cork also lifts it out of the neck, simplifying extraction. One of most popular of this type is the Screwpull (Plate 5.9). Its Teflon-coated screw makes cork removal particularly easy. Lever (Plate 5.10) and pocket versions are also available.

Regardless of design, removing corks from old bottles is always a conundrum. Over time, the cork loses its resiliency and tends to split and/or crumble upon extraction. The two-prong, U-shaped Ah-so (Fig. 6.2A) can occasionally be invaluable in these situations. If the cork's adherence to the neck has not already become loose, a gentle, back-and-forth application of pressure nudges the prongs down between the cork and the neck. Combined twisting and pulling may be able to remove the fragile stopper without incident. Alternately, one of the pump devices possessing a long, hollow needle can prove effective in removing the cork. The buildup of air pressure inside the bottle can force the cork out of the bottle, assuming the air does not escape between the cork and the bottle neck, or the pressure of inserting the needle does not push the cork into the bottle. If the cork does not budge after 5–10 strokes of the pump, pumping should stop, the needle removed, and another means used to remove the cork.

A final device in the arsenal of most wine aficionados is one of several gadgets that can remove corks inadvertently pushed into the bottle. If so, hopefully the inappropriate direction of cork movement does not occur quickly, otherwise you and your surroundings may get a wine shower. One particularly useful model for removing an inadvertently submerged cork possesses several, flexible, plastic, appendage-like flanges, attached around a hollow core (Fig. 6.2B). The flanges are held together as they are inserted into the neck of the bottle. Past the neck, the flanges

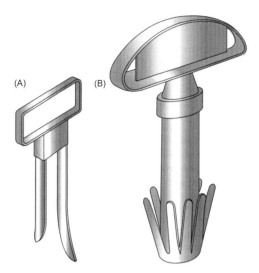

FIGURE 6.2   Illustration of the U-shaped Ah-so (A) and Wine Waiter (B) corkscrews. *Courtesy of H. Casteleyn.*

flare out. As the device is slowly pulled back, the angled barbs on the flanges bite into the cork. Slow, steady pulling extricates the mischievous cork.

More potentially embarrassing are situations where the cork crumbles upon extraction, fragments falling into the wine. If the incident occurs out of sight of your guests, the fragments may be removed by filtering. Coffee filters are usually readily available and fully adequate to the task, but inelegant. Rinsing the bottle permits it to be refilled with the now filtered wine, permitting one to serve the wine from its own bottle, with no one the wiser. Otherwise, one could invent a legend that bits of cork floating in the wine signifies good luck. If nothing else, it will cause a bit of hilarity at the dinner table.

Opening bottles of sparkling wine can be tricky, but presents an excellent opportunity for those giving courses or running wine societies to demonstrate an appropriate technique. Occasionally, people entirely remove the foil covering and wire mesh before beginning to remove the cork. If the wine has been properly chilled, this should not be a problem, but is more risky than needs be. It is safer to hold one hand over the loosened wire mesh, holding the base of the bottle tightly in the other hand. Twisting the bottle at the base, while holding the cork solidly, is physically easier than twisting the cork while holding the bottle still. Once rupture of the seal between the cork and the bottle neck is detected, the cork should start to slowly rise. Continued slow twisting of the bottle permits controlling the assent of the cork. This procedure assures that the cork is always under control. As the cork reaches the mouth of the bottle, a gentle hiss denotes successful and safe removal, without the loss of wine or a ricocheting cork.

The former use of lead-tin capsules occasionally was associated with an accumulation of a crust around the lip of the neck, especially when the wine had been stored in damp cellars. Leakage of wine around the cork can occur due to newly filled and corked bottles being laid on their sides too early (before gas under pressure in the headspace has escaped or become equilibrated within the wine), creases in the cork created during insertion, or inherent structural faults in the cork. Crust formation was typically associated with fungal growth on wine residues or cork dust. This produced a dark deposit on the cork and lip of the neck, as well as promoted capsule corrosion and solubilization of lead. If present on old bottles of wine, it can be wiped away with a damp cloth. Modern aluminum or plastic capsules have a looser fit or possess perforations that minimize moisture buildup, thereby limiting fungal growth. Use of a wax plug, in lieu of a capsule, is even less likely to permit development of a dark growth of fungi on the cork.

Palate Cleansing

Where wines are sampled in rapid succession, as at trade tastings, small cubes of white bread, unsalted crackers, or water are usually provided as palate cleansers. They refresh the mouth for the next sample. Presenting cheese, fruit, and cold cuts, however, gives the tasting more of a social aspect. Their presence tends to convert wine assessment into wine consumption. Where marketing and public relations are the prime (but unarticulated) focus of the presentation (most public tasting), the presence of food may have some justification. Although food and most cheeses interfere with the detection of subtle differences among wines, they can mask the bitter/astringent sensations of red wines (see Chapter 9).

Language

The deficiencies people have in adequately verbalizing sensory experiences were discussed in Chapter 5. Odor memory and recognition are based on experience, not neuronal hardwiring. The specific terms applied to odors are typically environmentally and culturally encrypted. For example, English is considered more active (direct, transitive) than Latin-based languages, which are more indirect (passive, intransitive) (Saussure, 2011). In a similar tack, Depierre (2009) interprets English expressions as being more inductive (based on experience), whereas French is considered more deductive (derived from analysis). How language construct influences perception and cognitive skills is an active field of research (Boroditsky, 2011), and thereby another potential source of olfactory (Berglund et al., 1973) and sensory bias (Bécue-Bertaut and Lê, 2011).

Regrettably, these limitations are often poorly recognized, or purposely ignored, setting the stage for memory illusions (Melcher and Schooler, 1996). Most wine language has the sophistication of a child's drawing, i.e., stick men and ball-and-trunk trees. If this analogy seems harsh, reflect on how many centuries it took humans to feel the need, or develop the skills, to draw lifelike representations of people or physical perspective. For descriptive sensory analysis, the solution has been to develop a specific vocabulary for the wines being studied. Despite this, technical descriptors rarely, if ever, adequately represent the hedonic perceptions wine connoisseurs experience.

Wine language has been extensively discussed by Lehrer (1983, 2009). Amerine and Roessler (1983) and Peynaud (1987) also have reviewed the topic, albeit from markedly different perspectives. Regardless, the primary function of language is to effect clear and precise communication of time, location, appearance, motion, gender, possession, opinion, or emotion. However, in many situations, language has a secondary role, lubricating social interaction. In most wine clubs, descriptions function more as a vehicle for amiable exchange rather than enunciating wine's attributes in an effective manner. At worst, winespeak can veil attempts to purport superior sensory skill or knowledge. Published wine descriptions (magazines, newspapers, websites) usually incorporate metaphoric, figurative, anthropomorphic, and emotional elements, as well as aspects relating to vintage, winemaking conditions, geography, and aging potential (Brochet and Dubourdieu, 2001; Silverstein, 2004; Suárez-Torte, 2007)—much of this being unrelated to what is sensed in the glass. Even seemingly precise terms (sour, bitter) may refer more to disguised holistic impressions than legitimate taste sensations. There is also a preponderance of motion verbs, such as exploding, burst, leap, unfold, and extend (Caballero, 2007). Frequently, odor memory seems divorced from a semantic base (e.g., see Parr et al., 2004). In addition, published descriptions seem more designed to entertain (avoid reader boredom) than inform. Asher (1989) is likely correct in his assessment that the reader "...wants the writer's response to the wine, not his analysis of it." If the florid language on back labels, or expounded in wine columns, inspires consumers to purchase and appreciate wine, then the poetic imagery employed may not be inappropriate (if legitimate). Regrettably, for some consumers, inability to detect the attributes so glowingly described can be disheartening to the novice (Bastian et al., 2005), discouraging a broadening of wine's consumer base. Although world consumption of wine is slowly increasing, in many countries, per capita wine consumption is either static or declining (OIV, 2014). The latter is not the trend desired by the wine industry.

In contrast, trained panels are taught to use specific, nonoverlapping terms (corresponding to physicochemical standards), and may be couched in a common and consistent use of holistic terms, such as complex, balance, finish. The attributes of intensity and duration tend to be expressed simply with quantitative adjectives, such as mild, moderate, marked, and short versus long, respectively, or better along a linear scale from undetectable to intense. Regrettably, this degree of training, and the presence of standards, are rarely available outside the laboratory. Figs. 1.3 and 1.4 list examples of potential standard reference samples used for fragrance and off-odor terms. Depending on need, tasters may also be trained to recognize the fundamental attributes that distinguish the range of varietal, regional, or stylistic wines found in commerce.

Despite the desirability for precision, much of popular wine parlance is anything but, apparent objectivity being cloaked in subjectivity. The terms used may not even permit the originator, let alone anyone else (Lawless, 1984), to recognize the wine described. Different tasters presumably detect similar attributes, but may not describe them similarly. For example, chalky, earthy, and metallic may refer to the same or different impressions. Culture (social conditions) and upbringing (aromatic experiences) also significantly modify what features may be recognized (Chrea et al., 2004), or how they are enunciated (Majid and Burenhult, 2014). Thus, it is not surprising that, although wine writers may use their own expressions in a relatively consistent manner, little consistent precision exists in describing wines (Brochet and Dubourdieu, 2001; Sauvageot et al., 2006). What wine writers possess in common is a tendency to relate description to presumptive types (a paradigm), rather than an enunciation of a wine's actual sensory properties (Brochet and Dubourdieu, 2001). This view is also supported by data from Solomon (1997) and Ballester et al. (2005). For example, when tasters misidentified a Pinot gris wine as Chardonnay, they described it in terms appropriate for a Chardonnay. Sauvageot et al. (2006) suggest that the absence of a common lexicon may reflect wine critics viewing themselves as independent standards, needing no justification to refer to external sources.

Term use tends to reflect the background and interests of the user. For example, wine critics accentuated qualitative terms, such as great, enjoy, amazing, whereas winemakers were inclined to use a predominance of winemaking terms, such as oxidized, yeasty, woody (Brochet and Dubourdieu, 2001). Term use by critics also tended to fall into two categories, relating to how the wine's attributes were liked or not. Categorizing odors by the degree of pleasure or displeasure is a common human trait (Zarzo, 2008; Yeshurun and Sobel, 2010). Few aromatics are innately pleasant, with more inherently unpleasant. Most consumers describe wine complexity in terms of personal and subjective responses, relating to flavor, enjoyment, and appreciation (Parr et al., 2011). Thus, hedonic categorization should be expected, where experience/training has not associated particular odor qualities with descriptors. Wine writers may have a more florid language than novices, but both are similar in kind, i.e., hedonic and conceptual. In addition, not only does the pleasantness of odors vary widely between individuals, but also among cultures, often depending on familiarity, intensity, and emotional association with visual and verbal clues. Descriptive terms imply not only a resemblance to the object or experience noted, but also tend to be imbued with an emotional veneer (especially evident when spoken). Thus, words often express not only the principal meaning, but are also embellished with subtleties. How a person describes a wine often expresses as much about the describer as the described.

English possesses few uniquely odor-related terms, though this may just reflect a cultural trait of most European languages (Majid and Burenhult, 2014). Those English abstract odor terms that exist generally express marked emotional connotations, for example putrid, pungent, and acrid. Usually, though, descriptive terms can be grouped into classes of aromatic objects, such as floral, resinous, burnt, or spicy. The illusion of improved precision is donated by the addition of suffixes to specific objects (e.g., iris-like, beet-like, peachy, mushroomy). If tasters agree on and consistently use these terms, then odor communication has the potential to be effective. Typically, this requires considerable training and discussion before agreement and consistent use is possible. Repeat exposure to unfamiliar odors tends to reduce the number and range of qualitative terms used to denote their qualitative aspects (Mingo and Stevenson, 2007).

One of the regrettable limitations of wine language is that it seldom clearly and precisely enunciates the taster's perceptions. The terms used are usually encumbered with underlying emotional connotations (Lehrer, 2009). For example, terms such as acidic, vegetal, leather, and oaky can have either positive or negative implications, depending on the user. Equally, tart and sharp refer to the same sensation, but may be selectively used to denote the taster's emotional response to the wine's acidity/pH. Furthermore, terms such as weak versus delicate, and austere versus hard, although appearing to be descriptive, may signify more the taster's hedonic response, or current attention to a particular attribute, than the intensity of the sensation. This probably explains why the same odor may generate different responses in terms of pleasantness, depending on whether its identity is known in advance (Bensafi et al., 2007), or is recognized (Degel et al., 2001). These conclusions are supported by the observation that tasters who liked a particular wine described it as fruity, with no mention of any sour or bitter attributes; whereas those who disliked the wine noted its sour or bitter character, without commenting on fruitiness (Lehrer, 1975). Similar findings have been noted by Lesschaeve and Noble (2005).

An example of a lack of clarity is the currently popular term "minerality." What it clearly does not refer to is the wine's mineral content, which is flavorless (Maltman, 2008, 2012). Thus, the soil's mineral content can not contribute to any sensory aspect of *terroir*. Other than some minerals being involved in crystal formation, haze generation, or oxidation reactions, the significance of a wine's mineral content is limited to potentially providing an elemental fingerprint of the wine's provenance (Baxter et al., 1997; Serapinas et al., 2008; Zou et al., 2012).

Several papers have appeared on the use of minerality by professionals. The results indicate a highly variable use, with tasters usually falling into one of several groups. The one consistent feature about the term is its multimodal nature, incorporating both olfactory and gustatory aspects (Ballester et al., 2014; Heymann et al., 2014). In the realm of taste, aspects relating to acidity, and occasionally bitterness and astringency have been cited. Relative to fragrance, the shades of meaning are even more diverse, with some tasters incorporating fruit/floral attributes, others gunflint, lactic, and iodine, and even others noting new oak, smoky, and spicy aspects (Ballester et al., 2014). Only benzyl mercaptan (benzene methane thiol) has been experimentally associated with flinty/mineral/smoky attributes (Tominaga et al., 2003). Any relationship of this compound to the perception of minerality is unknown. In contrast, detection of metallic tastes in wine appears to be a retronasal odor, misinterpreted as coming from the mouth.

Oddly, no one has investigated why wine writers use the term minerality (other than popularity), or what experiences may have led them to consider minerality a relevant term. Except for table salt (the ground form of the mineral halite), only geologists are likely to have experienced other mineral tastes (notably water-soluble minerals such as borax, epsomite, glauberite, melanterite, and sulvite).

One of the problems in investigating how terms are applied to particular attributes is that asking questions about the feature can influence its perception. For example, asking tasters about a particular fault can significantly augment reports of its presence. Is perceived presence partially dependent on having a term for a particular feature or being asked about it? Thus, relative to minerality, does requesting tasters to specify what is meant by the term increase its perceived presence, and how often it is noted? Minerality may just be a currently fashionable expression, used to lump impressions seemingly not adequately covered by existing terms (dumping), or a fad fanned by popularity.

Because a wine writer's hedonic impressions of the wines tasted appears to be what consumers want, the predominance of emotion-packed jargon is understandable. Oro-olfactory sensations pass through the limbic system, associated with emotional responses, before reaching the higher, cognitive centers of the brain. Thus, the metaphoric, emotive, and often poignant illusions supplied have their legitimate place, despite their inherent imprecision. Silverstein (2003) notes that the ritual of tasting and language use appears to be "culturally eucharistic." In the process, tasters second unto themselves many of the symbolic qualities of what they imagine are present in the wine, if only vicariously.

Part of the problem associated with precision in wine description involves the absence of adequately representative terms. Exceptions are the bell pepper and litchi aspects of Cabernet Sauvignon and Gewürztraminer, respectively. More commonly, descriptors only vaguely resemble the odor of the object or experience named. In addition, specific fruits and flowers, used to describe wine aromas, rarely have a single odor quality. For example, apple and rose cultivars have their respective common olfactory elements, but frequently each cultivar possesses a unique and distinguishable aroma. It is the fascinating diversity that gives wine one of its most endearing and captivating, but descriptively irritating, properties.

For several widely grown grape cultivars, descriptors for their aromas have become fairly standard (see Tables 7.2 and 7.3). Experience with their use may explain the improved ability of trained tasters to recognize varietal wines (Hughson and Boakes, 2002), but this is uncertain (Levine et al., 1996; Köster, 2005; Dijksterhuis et al., 2006). Cultivars with only local significance are rarely, or insufficiently known to be, associated with unique aromas (and possess appropriate descriptors in English). Traditional descriptors for particular cultivar wines may also differ from language to language. For example, blackcurrant is traditionally associated with Cabernet Sauvignon wines in English, whereas violet is more common in French. Some terms also seem to suggest regional/cultural/agricultural aspects, such as the use of truffles in referring to some wines from Piedmont. As already noted, descriptor use frequently reflects the personal/cultural/regional heritage of the user.

Most descriptive terms noted in aroma charts (e.g., Fig. 1.3) attempt to delineate aromatic qualities in terms of a series of categories and subcategories of increasing specificity. Regrettably, these arrangements have too often been viewed as a guide to wine flavors. Many wine lovers have been indoctrinated into believing that their use is the passageway to connoisseurship, to the point where books on the subject have been written (Moisseeff, 2006). However, such lists were first developed and intended to be used as memory prompters in sensory analysis. Misappropriation of a research tool in the popular realm can deflect attention from the more useful activity, i.e., encouraging consumers to develop archetypic odor memories for the flavor attributes of distinctive grape cultivars, wine styles, effects of aging, or any regional differences that may exist. This is hard enough, but encourages the purchaser to search for distinguishable features that characterize the paradigms of various wine types or its qualities. They can be used as models against which the quality of other similar wines can be assessed.

Nonetheless, if generating a list of descriptors is how a person enjoys wine, so be it. Personally, it seems the equivalent of basing food appreciation on identifying the ingredients used in the recipe, rather than appreciating the meal itself. The ultimate value of focusing on wine fragrance is that this is where its most complex, informative, and fascinating attributes lie.

For the majority of consumers, at least in wine courses or tasting societies, olfactory descriptors often act as "trial balloons," as the originator searches for conformation. They generate self-esteem when there is agreement, and can be easily disregarded when greeted with skepticism. As noted in de Wijk et al. (1995), terms seem to organize odor perception. As long as the expressions represent honest feelings, they fulfill a psychological and communication need. However, they may also be used to impress or sway, rather than inform. As Samuel Johnson noted:

> This is one of the disadvantages of wine, it makes man mistake words for thoughts.

Terms that are best avoided are those with anthropomorphic connotations, such as feminine, fat, or aggressive. Their meanings possess so many emotional overtones, as well as being imprecise, that their interpretation depends more on the user or reader than the characteristics of the wine. Nonetheless, expressions such as body, legs, nose, and aging have become so entrenched as to be a fixture of winespeak. In addition, terms such as *goût de terroir* (to describe flavors supposedly derived from particular vineyard sites) are figments of the imagination. Thankfully so, as anyone who has put their nose to soil can attest (most of a soil's distinctive odor comes from sesquiterpenes produced by soil-inhabiting fungi imperfecti and actinomycetes).

Normally, descriptions should be kept short, recording only noteworthy attributes. Where time permits, and there is reason, complete notes on all sensory features may be taken. Typically, though, this is unnecessary. Use of one of the many score sheets available (see the following section) encourages the efficient and economic use of time and words.

WINE SCORE SHEETS

Detailed score sheets (e.g., Figs. 5.28–5.32) are appropriate if analysis is intended. However, for most informal tastings, data analysis is neither necessary nor warranted. Under these conditions, a form such as that illustrated in Fig. 1.5 may be adequate. Fig. 6.3 illustrates a variation of the former, used in a wine-appreciation course. The form is normally enlarged and reproduced on 11×17-inch sheets. Labels, reduced in size, are added to aid in remembering the wines tasted. Because the sheet illustrated is intended for the introductory lesson, verbal descriptions of the wines' characteristics and varietal origin are noted to assist the learning process. In addition, the wines' presentation sequence usually corresponds to that of the descriptions. In subsequent lessons, the presentation order is varied from that of the descriptions. This is to encourage students to correlate the wines with the descriptions. In the early stages of a course, little space is provided for student comments, as they usually have difficulty verbally expressing their perceptions, and write little. In subsequent tastings, description of the wines may be eliminated and more space allotted for comments (Fig. 1.5). Descriptive material may also be projected on a screen before, during, or after the tasting, depending on the instructor's intentions. To encourage focusing on the wine's attributes, quality concepts and flavor descriptors are noted to the left of the sheet.

Although most tasting sheets provide space for numerical scoring, ascribing undue significance to the values generated is best avoided. Frequently, repeat sampling generates different values. Equally, ranking wines out of 100 gives an unjustifiable sense of precision. However, even if ranking to such a degree of accuracy were possible, the differences often reported are often minimal (5 units or less). That is equivalent to a difference of 5 percent or less (often considered statistically insignificant). In addition, it is unlikely that most people (aficionado or not) could do so, especially if the ranked wines were not sampled together. Thus, such a ranking system does the consumer injustice, despite their apparent fascination with the method. It may offer the busy purchaser a means of avoiding the tyranny of choice, but to what value if the denoted score is, as probably, unjustifed. Admittedly, entering a well-stocked wine store must be daunting for the novice, faced with hundreds, if not thousands, of choices from the four corners of the globe. No wonder simple numerical ranking is popular. Nonetheless, it abrogates the consumers' right and opportunity (if not duty) to learn and understand their own personal preferences. Genuflecting to any authority is demeaning, especially so when it is unnecessary, and probably unwarranted.

Grading in terms of letters (Fig. 5.32) would do more justice to the relative merits of wine. It would imply broad categories, where fine distinctions, based solely on individual interpretation, are reduced. For additional discussion of ranking scales see Cicchetti and Cicchetti (2010) and Stone et al. (2012).

SENSORY TRAINING EXERCISES

The training and testing of wine-tasting panels were covered in Chapter 5. In this chapter, exercises are provided to help those with little professional experience in recognizing the diversity and complexities of differentiating amongst a wine's sensory aspects. Occasionally, such exercises may also be incorporated into training wine judges.

Several exercises have been proposed by authors such as Marcus (1974) and Baldy (1995). They typically commence with a simple demonstration of the basic taste/touch sensations present in wine (Table 5.4). More detailed exercises on gustatory attributes are found in Appendices 6.1–6.4. This is usually followed by a series of exercises to illustrate how these sensations may interact (Appendix 6.5). Where desired, the intricacies of gustatory interactions can be reinforced by additional series with wines to demonstrate the influence of its matrix on perception.

Because taste sensations are so influenced by other gustatory and olfactory stimuli, in most situations there is little value in having students establish their own sensitivity thresholds. For example, the perception of wine sweetness is a function not only of the wine's glucose and fructose contents, but also the acid, glycerol, alcohol, and phenolic contents, the presence of specific aromatic compounds, carbon dioxide, and the wine's temperature. In addition,

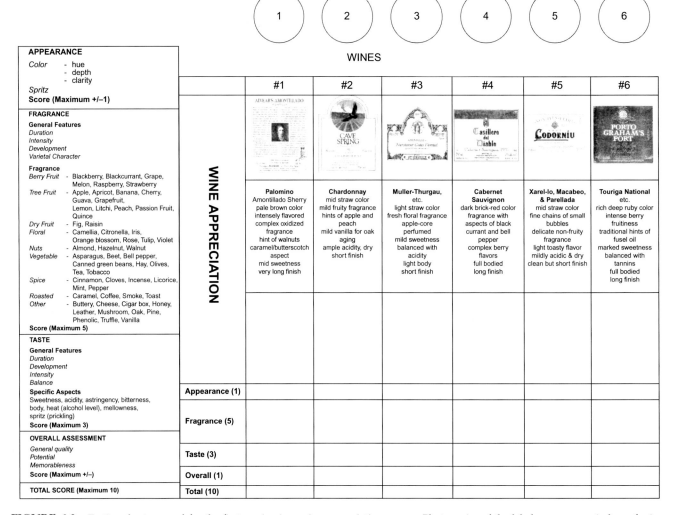

| | 1 | 2 | 3 | 4 | 5 | 6 |

WINES

APPEARANCE
Color - hue
      - depth
      - clarity
Spritz
Score (Maximum +/−1)

FRAGRANCE
General Features
Duration
Intensity
Development
Varietal Character

Fragrance
Berry Fruit - Blackberry, Blackcurrant, Grape, Melon, Raspberry, Strawberry
Tree Fruit - Apple, Apricot, Banana, Cherry, Guava, Grapefruit, Lemon, Litchi, Peach, Passion Fruit, Quince
Dry Fruit - Fig, Raisin
Floral - Camellia, Citronella, Iris, Orange blossom, Rose, Tulip, Violet
Nuts - Almond, Hazelnut, Walnut
Vegetable - Asparagus, Beet, Bell pepper, Canned green beans, Hay, Olives, Tea, Tobacco
Spice - Cinnamon, Cloves, Incense, Licorice, Mint, Pepper
Roasted - Caramel, Coffee, Smoke, Toast
Other - Buttery, Cheese, Cigar box, Honey, Leather, Mushroom, Oak, Pine, Phenolic, Truffle, Vanilla
Score (Maximum 5)

TASTE
General Features
Duration
Development
Intensity
Balance
Specific Aspects
Sweetness, acidity, astringency, bitterness, body, heat (alcohol level), mellowness, spritz (prickling)
Score (Maximum 3)

OVERALL ASSESSMENT
General quality
Potential
Memorableness
Score (Maximum +/−)

TOTAL SCORE (Maximum 10)

WINE APPRECIATION

	#1	#2	#3	#4	#5	#6
	Palomino Amontillado Sherry pale brown color intensely flavored complex oxidized fragrance hint of walnuts caramel/butterscotch aspect mid sweetness very long finish	**Chardonnay** mid straw color mild fruity fragrance hints of apple and peach mild vanilla for oak aging ample acidity, dry short finish	**Muller-Thurgau,** etc. light straw color fresh floral fragrance apple-core perfumed mild sweetness balanced with acidity light body short finish	**Cabernet Sauvignon** dark brick-red color fragrance with aspects of black currant and bell pepper complex berry flavors full bodied long finish	**Xarel-lo, Macabeo, & Parellada** mid straw color fine chains of small bubbles delicate non-fruity fragrance light toasty flavor mildly acidic & dry clean but short finish	**Touriga National** etc. rich deep ruby color intense berry fruitiness traditional hints of fusel oil marked sweetness balanced with tannins full bodied long finish
Appearance (1)						
Fragrance (5)						
Taste (3)						
Overall (1)						
Total (10)						

FIGURE 6.3 Tasting sheet as used for the first session in a wine-appreciation course. Photocopies of the labels serve a reminders of wines tasted. Description of the essential attributes of the wines act as guides. In subsequent tasting, the verbal descriptions are placed at random to force the students to match the descriptions to the wines. Fig. 1.5 is a related form without added wine descriptions but ample space for student comments on the wines' sensory attributes. *From Jackson, R.S., 2000. Wine Science: Principles, Practice, Perception, second ed. Academic Press, San Diego, CA, reproduced by permission.*

individual taste sensations are less important to quality perception than their integration into holistic perceptions, such as balance and body, as well as with visual and olfactory inputs (Sáenz-Navajas et al., 2016). Finally, gustatory perceptions are less complex and interesting than those of fragrance and flavor. Nevertheless, it is vital that tasters realize how interpretation of information from our senses can easily be influenced by context (for example, Fig. 6.4). Contextual influences equally affect olfactory stimuli and quality perception.

Exercises designed to demonstrate the different perceptions of major flavorants in wine, such as esters, lactones, or pyrazines could be prepared, but are usually not particularly instructive or appreciated by students. However, for those pursuing a career in wine they have definite value, if only in training them to identify odors of significance to their future vocation. Aldrich Chemical Co. has kits containing small samples of a variety of aromatic compounds, some of which occur in wine. Aldrich also produces a catalog specifically dealing with flavors and fragrances, from which specific constituents may be obtained. Although more variable and short-lived, samples can be prepared from commercial fruit and floral essences, or from the actual object (e.g., mushrooms, bell peppers, beets, raspberries, cherries), especially for instructors without access to a chemical laboratory. Simpler again, but more limiting, are commercial wine-flavor kits.

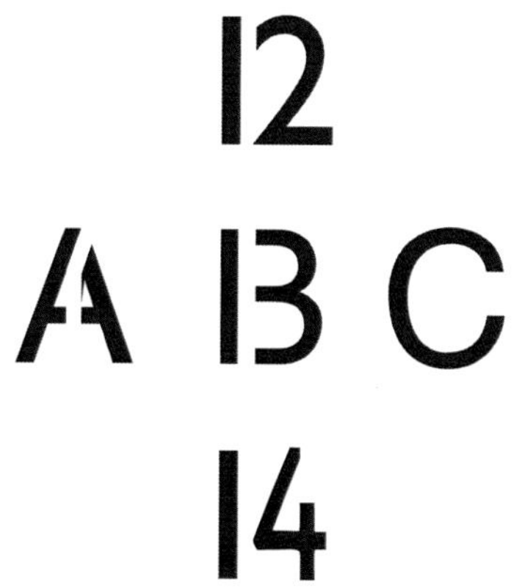

12

A 13 C

14

FIGURE 6.4   An illustration of how context, in this case visual, can influence perception. Reading vertically interprets the center symbol as 13, whereas reading horizontally results in its interpretation as a B.

Staff Training for Medium to Small Wineries

Most small wineries do not have the resources to establish a panel of tasters. Due to demands on a small staff, only the winemaker may be available to make decisions on urgent matters. However, this does run the risk (or advantage) of making the wine on his or her palate. No matter how skilled, everyone has days when their acuity is off. Thus, it is not only useful, but often essential to train other staff members on issues of sensory analysis. An integral part of this training is the development of a consistent and common vocabulary for describing critical sensory attributes. Not only does this permit useful dialog among the staff, but also can facilitate consistent communication with clients.

As part of developing a common terminology, training should include sampling standardized taste and odor samples as described in Chapter 5, but especially those terms most appropriate for those cultivars used. Although it may be traditional in some wineries to sample wines in the cellar, this is not good practice. Cellars are too often contaminated with odors that could jeopardize valid assessments. All training and serious sampling should occur in a brightly lit, quiet, odor-free environment, under comfortable seating conditions.

During term acquisition and training, the staff should sample with black glasses (Plate 6.1). This is especially useful in developing an odor memory independent of visual or other external clues. It is also helpful for those who meet the public in reception/sales rooms. Recognizing their wines immediately by odor alone frees them to concentrate more on the social activities of their job. Not only is this useful in responding to questions, but will give customers the desirable impression that they are dealing with professionals.

Another essential aspect of staff training is learning to recognize faults. This is crucial not only for winemakers, but also for other staff members. If a faulty wine is questioned by a client, the staff should be sufficiently knowledgeable to recognize this, and replace it with a sound sample. Admittedly, everyone has their limits in detecting faults. Thus, it is important that staff members know their limits, and direct questions for which they are uncertain of the correct answer to the person best able to respond.

One of the most difficult aspects of staff training is learning how the attributes of barrel samples blend, be they from free- and press-run samples, different lots from the same or different vineyards, and/or different cultivars. It is equally important to develop the skill to predict how the finished wine will develop after bottling. Currently there are no exercises to teach such skills, other than experience with a knowledgeable cellar master. At this point, art seems to take over from science.

As part of training, it would be instructive for the staff to sample the local competition, as well as some of the more popular or reputed wines from other regions. The more knowledgeable the staff, the more likely they will express legitimate competence, and instill the confidence important to repeat sales.

Another aspect of training could relate to the effects of various temperatures on wine attributes. This is important as consumers often query staff on this issue. If time permits, and the staff sufficiently inclined, practical experience on the influence of context should be included. For example, wines could be sampled with a few cheeses to determine which appears to improve the impression of the wines, white wines could be secretly colored red (with tasteless anthocyanins) to give the staff experience with the biasing influence of color on perception, or

PLATE 6.1   Black ISO wine-tasting glasses. *Photo courtesy of Midnightsun Designs, Norrköping, Sweden.*

wines could be falsely identified (more and less expensive wines interchanged on separate occasions) and their apparent ranking compared. In the latter two examples, it is important to explain, after the fact, the rationale for the exercises. Knowing that these tastings were included only to make staff members aware of how easily people can be influenced by extraneous factors should negate any embarrassment potentially caused. Forewarned (by experience) is forearmed.

Another aspect of training, especially for anyone dealing with customers, is the issue of food and wine combination. The general dictum of red with red and white with white has rationale, based on relatively similar flavor intensities, but is so imprecise as to be of little practical use. If the staff understands the principles underlying food and wine pairings (Chapter 9), they can adjust their responses to the perceived expertise and preferences of the client, without being intrusive or cavalier. Knowledge is a great mitigator, encouraging sensitivity to the needs and feeling of others.

If staff members are part of a tasting panel assessing the relative quality grades of the winery's wines, it is important that at least some simple statistics be applied to the results. The best wine (highest ranked) may in reality be statistically indistinguishable from the second- or third-ranked wines, despite their rank order. One has to be cautious in applying undue significance to ranking results.

Staff may also be trained in the effects of procedures used in wine production. Examples could involve assessing the effects of different additives, the influences of different yeast or lactic acid strains, the level of oaking or microoxygenation on sensory quality, blending formulation, and basic chemical tests. The list of potentially useful exercises is almost endless.

Beyond issues associated with a limited number of tasters is a tendency for the staff to show "parental" pride. This may lead to minimizing problems, possibly due to habituation. For example, frequent exposure to low levels of off-odors can lead to their going undetected. One technique to offset this situation could be the occasional inclusion of samples showing a fault. Another useful exercise could be the unannounced inclusion of samples from other wineries in a standard tasting of the winery's products. It is important that such outliers be changed frequently. Panel members must not become accustomed to and too easily detect an outlier. Although it might be impolitic to occasionally include the owner of the winery (if not the winemaker) in these tastings, it might help curb some of the inflated egos too often shown by owners, and encourage them to support the efforts of their staff in improving their wines.

For another view on sensory analysis in a winery see Robichaud and de la Presa Owens (2002).

TASTING SITUATIONS

Wine Competitions

Wine competitions, properly conducted, can benefit both the consumer and producer. They are a relatively inexpensive way for assessing the commercial potential for a new or modified product. However, the principal rationale for most wine competitions appears to be providing wineries increased market exposure, and collecting medals that hopefully will impress cellar visitors. When successful, enhanced sales and recognition are the major benefits, but if Lockshin et al. (2006) are correct, this is only with less frequent buyers and within the lower- to middle-price brackets.

Government-conducted competitions may be more professionally run, with judges and tasting procedures more regulated, but this has not been confirmed by published research. Langstaff (2010) considers that the United States possesses "no commonly accepted 'tests' to ensure a certain degree of tasting expertise." Nonetheless, the process for qualifying judges at the California State Fair is considered better than most. Qualifying tests for judges are administered several weeks ahead of the competition. In Australia, the AWRI (Australian Wine Research Institute) annually presents a 4-day Advanced Wine Assessment Course. Its purpose is to develop a corps of experienced wine tasters for the local industry. Experience is no sure indicator of expertise (Ashton, 2012). As long as the chances of a wine winning gold "can be statistically explained by chance alone" (Hodgson, 2009), no faith can be placed in a competition's results. The consumer would not, if they knew.

The lack of a full complement of qualified judges may explain the discouraging analysis of most competition results. In one such study, only 10 percent of the judges were found to score consistently (Hodgson, 2008). Individual judges also often scored differently from one year to the next (Fig. 6.5). Equally disheartening is the large number of gold medals awarded (47 percent of 2,440 wines), with multiples of other medals. Amazingly, about 84 percent of wines winning an award in one competition received none in another (Hodgson, 2009). Under such conditions, it is not surprising that analyses have demonstrated that the results are insignificant and demonstrate judge inconsistency (e.g., Scaman et al., 2001; Cicchetti, 2004).

To help rectify this dismal situation, Cao and Stokes (2010) suggest measuring three factors that appear to be at the root of the problem: judge bias, inability to discriminate, and variation. Criteria for accrediting wine judges are discussed in Hodgson and Cao (2014), and examples of assessing judge consistency are noted in Olkin et al. (2015). In summary, options for improving competition veracity should include the following:

- Assessment of judging ability, based on an investigation of their sensory skills in recognizing standards (prototypical examples of the wines to be ranked).
- Insertion of unannounced replicates to ascertain if the wines are given an equivalent score. Consistency in scoring is essential if any faith is to be placed in the results.
- Instruction of those chosen on how they should score wines, with examples of what exemplify the various quality categories.
- If wines are assessed under conditions where the judges see each other, the wines should be presented in randomized order, with the judges knowing this in advance.
- The summed results should be subjected to simple statistical analysis, to ascertain the confidence values of the ranking. Scores showing marked variation could be sampled again and/or the judges required to justify their scoring. If the justification is considered inappropriate, aberrant scores should be removed.

One of the problems with any qualifying test is the specter of disqualification. This could deal a serious blow to the image of a wine writer or critic, especially if peers realize why. A potential solution could be to partially base qualifying sensory tests on suggestions supplied by the candidates, with each person submitting one or more potential qualifying tests. This might reduce some of the apprehension about the tests, as they would feel actively involved in the selection process.

Another problem with qualification tests is the number of judges that may be required. Because individual tasters can be expected to taste only a limited number of wines accurately, within the time frame of the competition, this may lead to the number of judges ranking any category being small, likely too small for appropriate statistical analysis. Ideally there should be six to seven qualified tasters per set of samples (admittedly a goal potentially difficult to achieve in reality). The problem of an inadequate number of judges could be further aggravated if judges were to assess only wines within their range of expertise (a legitimate requirement). Thus, the number of judges needing assessment may be so large that conducting serious qualification tests may be herculean.

Many competitions award medals based simply on achieving a preset minimum score (Hodgson, 2008). Typically, the score range differences between bronze, silver, and gold medals are small. In OIV (International Organization of

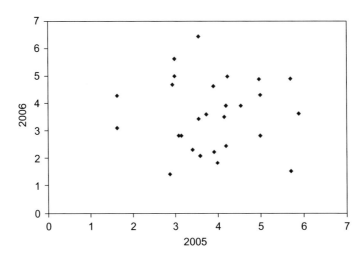

FIGURE 6.5 Scatter diagram of pooled standard deviation in the scores given by 26 judges who participated in both 2005 and 2006 wine competitions, correlation = 0.01. *From Hodgson, R. T., 2008. An examination of judge reliability at a major U.S. wine competition. J. Wine Econ. 3, 105–113, reproduced by permission of Cambridge University Press.*

Vine and Wine) competitions, the score ranges for bronze, silver, and gold are 80–81, 82–84, and 85–91, respectively (Stevenson, 2015). As many people think of awards in the same way they view Olympic medals, it would be better to permit a maximum of three medals in any category. Because numerical scoring often tends to be statistically insignificant, an alternative would be to request judges to select their top three in a category. The wines so chosen would then be sampled a second time and ranked in order of first, second, and third. If there is no immediate consensus, discussion should ensue as to why each judge ordered the top three wines the way they did, to come to an agreement on the arrangement of the category's award-winning wines. However, there should be no compulsion to have any medals if no wines really merit being singled out. In addition, there could be a final tasting of the best-ranked wine out of related categories (e.g., dry red, dry white, dessert white). From these, a few special awards, such as best-in-class, could be awarded. Although this may be considered equivalent to comparing apples with oranges, people do choose between wines of different character. Thus, there is rationale for comparing across categories, especially when assessed on holistic attributes. If the awards warrant mention in the press, they should truly represent wines worthy of note. Having multiple medals in any category makes the competition seem like a charade.

That any ranking by competent judges may not reflect the preferences of most consumers is a legitimate issue, but impossible to surmount. Wine judges, by the very nature of the attributes that qualify them as judges, are precluded from representing the average consumer (Schiefer and Fischer, 2008). However, the consumers most likely to be influenced by published competition results are those searching for wines of superior quality. Even newspaper column recommendations, based only on the writer's opinion, are thought to augment sales of the wines they promote (Horverak, 2009; Priilaid et al., 2009). However, this may only apply to a small, but significant, coterie of impressionable consumers (Chaney, 2000). Regrettably, for many people, the illusion of quality is often more important than actual sensory quality (Gleb and Gelb, 1986; Siegrist and Cousin, 2009), though this hopefully changes with experience (Rahman and Reynolds, 2015).

Despite all their flaws, wine competitions can have benefits, e.g., providing an opportunity for new wineries, or little-known regions, to achieve public awareness. The recognition might enhance a winery's reputation and sales (Lima, 2006) and, by reflection, the region from which the wine comes. An example is the reputational boost to Ontario wines, following awards given its icewines at international competitions. Wine competitions act as one of the few venues by which winemakers and producers can receive acclaim for their efforts. If the results are sufficient to attract major media attention, such as the Paris Wine Tasting of 1976 (between Californian and Bordeaux and Burgundy wines), or when Torres Black Label was rated better than Bordeaux Gran Cru wines in 1979, the influence goes far beyond just an award.

In addition, wines in competitions are at least tasted by category, thereby minimizing significant halo effects. These can occur when wines of markedly different character or quality are tasted in close sequence. Also, when a category contains many samples, it is usually subdivided into subgroups of no more than 10 wines. An alternative solution would be to rapidly sample the wines, eliminating all but above-average quality wines. These could then be subjected to a more thorough assessment.

In comparison with the published studies on competition results, the consistency of ratings reported in wine magazines and related venues have rarely been examined. These sources purport to give expert advice on which wines are the best buys, clearly a dilemma many consumers experience. Gokcekus and Nottebaum (2011) compared the rating of several well-known American expert sources with CellarTracker, an online listing of tasting notes from consumers, relative to Bordeaux wines. Ashton (2013) compared the rankings supplied by several well-known wine critics in France, England, and the United States, also for Bordeaux wines. In the latter instance there was considerable variability between views relative to some vintages but not others, and marked disagreement between some critics, but overall relative agreement among the critics studied. Stuen et al. (2015) investigated the correlation between the rankings noted in three major wine magazines in the United States for Washington and California Merlot, Cabernet Sauvignon, and Chardonnay wines. They also found considerable agreement among the critics. That there is this degree of agreement amongst these various sources might be interpreted that the consumer should have confidence in the rankings provided. However, is the accordance due to training or experience, generating a common percept, peer pressure inducing conformity with a common paradigm, or truly independent views coinciding. Either way, this apparent homogeneity of opinion conflicts with the results from judges in wine competitions noted above. Are the differing opinions in the latter situation a reflection of poor qualification, inconsistency in scoring, or diverging views of what a particular wine type should be? Only further research may be able to ferret out the truth among these and other possible interpretations.

Consumer Preference Tastings

Consumer tastings are much more common in the food industry than with wine. Although consumer tastings may play a major role in preparing new brands for worldwide distribution, their proprietary nature undoubtedly excludes their being published. In addition, understanding the various origins of consumer preferences is still largely in its infancy. Much of the literature related to wine has been limited to extrinsic rather than intrinsic factors, notably country of origin, brand name, price, vintage, grape cultivar, reviews, food pairing, or advertising (Dodd et al., 2005; Hollebeek et al., 2007; de Magistris et al., 2011; Ginon et al., 2014; Thach and Olsen, 2015). Even studies on the influence of bottle-closure type (Marin and Durham, 2007) and bottle weight (Piqueras-Fiszman and Spence, 2012) have been performed. In addition, the relative importance of extrinsic factors on purchase is often strongly influenced by individual factors, such as wine familiarity, past purchase behavior, lifestyle, gender, and cultural and education background (Goodman, 2009; Bruwer et al., 2011; de Magistris et al., 2011). The importance of extrinsic factors was evident in a study where red wines were tasted blind, or with prior knowledge of the wine's price and origin (Priilaid, 2006). When the wines' price and region were known, price explained most of the perceived quality. Correspondingly, most marketing focuses on extrinsic factors, and the self-image the wine can supply the consumer (Schlinder, 1999; Rushkoff, 2000). For example, the unspoken message in the Penfolds' advertisement where bottles of their wine were stacked between brightly lit bars of gold is immediately clear. Sensory aspects, if noted, are relegated to the back label, maybe because producers realize back labels have little consumer significance (Mueller et al., 2010a).

Consumer studies investigating the sensory attributes that may influence repeat sales are comparatively few (Parpinello et al., 2009; King et al., 2010; Bindon et al., 2014). Even less frequent are studies investigating the relationship between extrinsic and sensory factors involved in wine preference and selection (e.g., Lattey et al., 2010; Mueller et al., 2010b). New approaches to identifying the main drivers of consumer appreciation are outlined in Bi and Chung (2011).

As with all consumer studies, preference tests are complex to design well, and may be laden with numerous fallacies and hidden caveats (see Köster, 2003; Rousseau, 2015). Köster (2009) noted several fallacies that researchers need to avoid:

- uniformity: that people's responses vary in degree but not kind
- consistency: that people's behavior is consistent
- conscious choice: that people's choice is guided by rational and conscious motives
- perceptual: that only what is perceived will be remembered
- situational: that consumption conditions can be well and objectively defined

Of all wine assessments, simple preference or acceptance rankings are probably the most sensitive to misinterpretation—most being conducted under laboratory conditions, unrelated to real-life purchase or consumption conditions. What people report in a survey or test situation may bear little resemblance to actual practice. For example, most people acknowledge the benefits of a healthy lifestyle, but only a small portion act upon that realization in a meaningful way. Although consumer wine preference is essentially holistic, wine purchase decisions often appear to be directed by factors unrelated to past sensory experience (due to lack of name recall?), except where distinctly positive or negative. Impulse buying, a frequent phenomenon, is often influenced by herd phenomena (what others are thought to like or want), current mood, confusion based on excessive choice, and anchoring bias (unconscious reliance on the first piece of information provided or detected).

Because consumer testing is partially allied with designing better marketing strategies (and adjusting wine-production procedures), the benefits of situation-oriented studies need to be noted (Köster, 2003). Situation-oriented studies involve arranging tasting conditions to evoke images of the conditions under which wine would normally be consumed. This can include the use of both visual and/or auditory cues. In contrast, most tastings are conducted in wine stores, laboratories, or other conditions possessing none of the usual ambiance of consuming wine. The intent of situation-oriented sampling is to induce the consumer to conjure up realistic situations associated with wine consumption. The importance of situation is often likened to the difference between touching and being touched, or as Mark Twain put it in another context:

The difference between the right word and the almost right word is the difference between lightning and a lightning bug.

The experimenter might request the consumer to rank the appropriateness of a series of wines with a particular situation, or select situations in which a particular wine might be most appropriate (e.g., with guests, in a restaurant, on a picnic, or alone). It is likely too much to expect consumers to relate individual wines with multiple situations. To maximize the value of the procedure, wines should be chosen carefully to include those likely to be purchased by the group being studied. Any questionnaire employed should induce the recall of personally significant images of the conditions under which the wines might be appropriate.

Evoking the appropriate context usually requires considerable forethought by researchers. Simple suggestion is usually inadequate. Experience indicates that imagining appropriate situations demands more effort than most consumers are willing to donate. Thus, supplying appropriate environmental stimuli is usually required to create the virtual reality desired. Projection of photos or videos are often effective, but with situations such as tasting at home, the photos may be too detailed to elicit the apposite image. In this case, relating a hypothetical story in which the consumer is encouraged to be present may be effective. Doing so activates the candidate's imagination, generating the personal details suitable to the situation.

The principal advantage of situation-oriented tasting is that it arouses reactions that relate to veridical situations. It can also be used to extract information on the frequency with which consumers would likely purchase similar wines. The data derived may be further used to suggest descriptive and/or chemical analyses to discover those features important to the participants in the tasting. Finally, the technique avoids many of the potential pitfalls of consumer research—assuming consumer uniformity or consistency, questions are indirect and relate to concrete, realistic settings, and complications associated with imprecision in term use.

Meillon et al. (2010b) provide an example comparing in-home and laboratory tasting conditions, involving a white and red wine, with alcohol contents unadjusted (13.3 and 13.2 percent) and reduced (9.1 and 9 percent) respectively, and with and without information relative to alcohol content. The same order of preference was obtained under both conditions, but significantly higher hedonic scores were achieved in the in-house condition. Expectation also influenced the scores (Fig. 6.6). However, without prior knowledge of the alcohol content, consumer liking seemed little influenced by the alcohol content (Meillon et al., 2010a). For further discussion of the importance of site context relative to preference or acceptance studies see Stone et al. (2012).

Preselection of candidates for testing often involves obtaining data about critical factors that are thought to influence purchase decisions. Cultural background is often significant, followed by education level, yearly income, and gender. The most-mentioned generalities about consumer preferences are based on these determinants. Use of the Wine Neophobia Scale is another option in preselection of participants (Ristic et al., 2016).

Adequate representation of the diversity of consumer groups requires large numbers of individuals (e.g., Hough et al., 2006). It is equally important to choose only representative individuals (McDermott, 1990; Martínez et al., 2006; Bruwer et al., 2011). For example, less-affluent social groups and infrequent wine drinkers tend to consider properties of sweetness and freedom from bitterness and astringency of particular importance (Williams, 1982). In the latter study, consumers also tended to dislike oakiness or spiciness, whereas most wine experts appreciated these attributes (in moderation). However, conflicting data have been reported by Binders et al. (2004) and Hersleth et al. (2003). The importance of color and aroma to acceptability appears to increase with consumer age and consumption frequency. Not surprisingly, experience also influences the words people use to describe wine (Dürr, 1984; Solomon, 1990). Teetotalers may express an opinion, but one that is irrelevant.

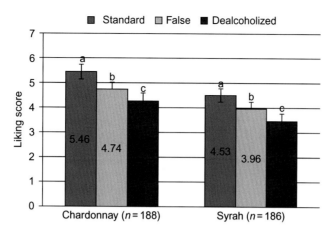

FIGURE 6.6 The effect on alcohol reduction on the liking (means) of wine: Standard (unmodified and noted to be 13.5%), False (dealcoholized by about 4% but noted as being 13.5%), and Dealcoholized (dealcoholized by about 4% and noted as 9.5% alcohol). Confidence intervals ($\alpha = 5\%$). *Reprinted from Meillon, S., Urbano, C., Guillot, G., Schlich, P., 2010. Acceptability of partially dealcoholized wines – Measuring the impact of sensory and information cues on overall liking in real-life settings. Food Qual. Pref. 21, 763–773, with permission from Elsevier.*

Individuals involved in consumer studies should also be used only once, to avoid entrainment. Other than the inducement of sampling wine, consumers deserve a full explanation of the value of their participation, and should be presented with some small token of appreciation. This may involve a monetary reward, or where permissible, a bottle of the wine preferred in the sampling. Procedures should also clearly favor the selection of potential purchasers, not just passers by. Careful scrutiny of the spread and potential data clumping may highlight unsuspected consumer subgroupings (e.g., Tables 5.2 and 5.3). This could be very helpful in generating reliable data to understand the chemical basis of consumer choice.

It is a common finding that women generally do better at aroma recognition than men (Figs. 3.10 and 3.26). In addition, women purchase about 80 percent of wine consumed in the home (Atkin et al., 2007). Thus, it might be appropriate to have more women involved in consumer studies. Some evidence also suggests that men may do more purchasing in specialty stores than women. Women appear to do most of their wine purchasing while shopping for food (Ritchie, 2009). However, in countries such as Canada, where most wine sales are controlled by provincial-run monopolies, and isolated from groceries, wine sales appear to be gender independent (Bruwer et al., 2012).

When collecting sensory data from consumers, a 9-point hedonic ranking scale is often used (Fig. 5.44). Its use is intuitive, permitting it to be used with minimal instruction. Another option is a simple, open-ended questionnaire, where participants are asked to describe several aspects of the wines they liked and disliked (Fig. 6.7). If the number of comments requested on the form is limited (≤ 3), consumers generally do not consider the task overly demanding. The data derived usually highlights those aspects the participants can name and probably found pleasurable or undesirable, generating useful insights that other scoring procedures may not provide, or as easily. The principal limitation in most questionnaires is the inability of less-frequent wine consumers to enunciate their opinions in meaningful words. Although Lawless and Heymann (2010) are probably correct in considering that most consumers respond to products holistically, more knowledgeable consumers are likely to have reasonably defined, less holistic preferences. Good correspondence may be found between these consumers and professionals (see Symoneaux et al., 2012), simplifying and economizing on studies of consumer preferences. Care must always be taken to avoid forced consensus and data aggregation.

Another issue of concern in interpreting consumer results is to acknowledge that consumers may not use or interpret terms in expected ways, even with coaching. Thus, the importance that gustatory sensations appear to play in consumer studies may only denote the inability of consumers to express their impressions in any other way, a phenomenon termed sensory dumping. However, coaching in term use is itself a problem, potentially distorting or changing preferences as a result of training.

Occasionally, differentiation studies are incorporated into preference tests. Although of academic interest, the ability of consumers to differentiate between samples is probably of little practical value. It is uncommon for consumers to comparatively sample wines at the dinner table. Even experienced tasters are not that proficient at remembering the characteristics of several wines that were compared several days before.

WINE QUESTIONNAIRE

Name: _____   Date: _____

Sample # _____

Describe three (3) aspects of the wine you particularly like:

a) _____

b) _____

c) _____

Describe three (3) aspects of the wine you particularly dislike:

a) _____

b) _____

c) _____

FIGURE 6.7   Descriptive wine questionnaire.

Trade Tastings

Trade tastings are normally organized to expose those within the wine industry (e.g., wholesalers, retailers, wine writers, wine producers) to a particular selection of wines. Such tastings are often hosted (funded) by national or regional delegations. Because of the effort and expense, the trade component is usually held in the afternoon, with the evenings set aside for the paying public.

Because of the large number of wines and attendees, extensive organization is required. Serving staff need to be attentive, cordial, and fully cognizant with the wines they are representing. Responses to questions should be quick and accurate. Glasses should be of a type specifically designed for wine tasting. Single glasses are typically provided to each participant upon entering the tasting hall. Pitchers of water for rinsing, and buckets for collecting wine remnants need to be at the ends of each table. The pitchers will need frequent refilling, as well as the buckets emptied on a regular basis.

Most trade tastings are stand-up affairs, where attendees move from table to table, inspecting and sampling wines of their choice. To facilitate note taking, an accurate and complete wine list should be available, organized by table. Space should be available to record impressions. Ideally, this should be available in booklet form, with a sufficiently rigid cover to facilitate writing notes while standing. Although it might be preferable for the wines to be arranged on a varietal or stylistic basis, it is more typical for them to be organized relative to the merchant/producer/appellation supplying the wines.

Occasionally, trade tastings can be sit-down affairs. In this situation, the wines are presented in a particular sequence, with a separate glass for each wine. One or more speakers guide the participants through the wines, to draw attention to the features they wish to emphasize. Typically, only 6–10 wines are presented, usually the wines from a single producer.

Organizers should have no illusions about the complex elements that influence wine purchase (Thach and Olsen, 2015). Sensory subtleties may be important to the serious connoisseur, but rarely to the average consumer. This is not because they are unimportant, but due to lack of sufficient knowledge and experience. For most consumers, convenience, familiarity, and cost are the major factors influencing buying habits. These features usually involve preferences for wines of a particular color (Bruwer et al., 2011), ideological percepts (e.g., organic wines), or cultural/historical identification. For example, Robert Drouin (Burgundy *négociant*) is reported to have said something to the effect that when one samples a Burgundy, one is not just drinking a wine but identifying with its history and culture. Additional motivators can involve peer pressure, advertising, and opinions expressed by wine authorities. Thus, to have their major impact, organizers of trade tastings must have a clear idea of who will attend and what factors are most significant to their purchase behavior.

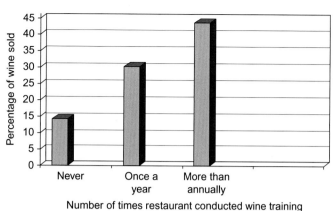

FIGURE 6.8 Amount of wine-service training offered and percentage of wine sold as a percent of total alcohol sales. *Reprinted from Gultek, M. M., Dodd, T. H., Guydosh, R. M., 2006. Attitudes towards wine-service training and its influence on restaurant wine sales. Int. J. Hospitality Manage. 25, 432–446, with permission from Elsevier.*

Another, but much more insular, type of trade tasting involves staff training in restaurants. Clients usually appreciate suggestions from a clearly knowledgeable waiter (or sommelier in more upscale restaurants). Wine sales can have a significant influence on restaurant profitability (Fig. 6.8).

In-Store Tastings

Most in-store tastings are the equivalent of food samplings provided in grocery stores. Although the amounts of wine supplied are generally correct, the plastic cups used are more appropriate for juice than wine sampling. Regrettably, the personnel typically employed are there only to dispense, not inform. Whether such tastings have real value is a moot point.

In contrast, well-administered in-store tastings tend to positively affect sales, as well as supply useful information to the customer. In addition, it can enhance the store's image as run by professionals. Thus, tastings should be conducted by the most knowledgeable staff, willing and able to lucidly describe the wine's characteristics, without being arbitrary or pretentious. The glasses should be adequate to illustrate the qualities of the wine and be impeccably clean. The timing and duration of tastings should correspond to the period when the more serious and affluent customers frequent the store. Ideally, tastings should be held at similar times, so that regular customers come to know when to expect tastings. Because of the intimate nature of the tastings, only one or two wines should be featured at a time, usually newly arrived wines. If the wines are expensive, it is reasonable to charge a nominal fee, applied to any wines purchased. Because promoting sales is as important as education, the identity of the wine should be clearly evident.

Knowledge of a wine's presumed quality and price can significantly influence taster opinion, tending to enhance appreciation, rating, and the endowment of attributes associated with quality. For example, Plassmann et al. (2008) found that both perceived pleasure and activity in the pleasure centers of the brain were correlated with knowledge of the wine's price. In another study, the scoring of champagnes could be predicted from the price, when it was known in advance (Fig. 6.9). However, expensive wines were considered only marginally better when both the identity and price of the wines were withheld, and then, primarily only by those with more wine experience (Goldstein et al., 2008). Furthermore, enjoyment of a wine by wine club members (tasted blind) was often independent or negatively correlated with price (Ashton, 2014). Encouragingly, Almenberg and Dreber (2011) found that knowing in advance that the wines to be tasted were inexpensive did not automatically downgrade their rating.

Further evidence of the complex interplay of extrinsic factors on quality perception is illustrated in Fig. 6.10. Here, the influence of price was dependent on when the information was provided, as well as the taster's gender. The influence of extrinsic information is frequently highly dependent on wine knowledge (Brochet and Morrot, 1999). For example, when the wine's origin was suspected by enology students, based on a nearby empty bottle, the wine's interpreted quality was markedly affected, reflecting whether the sample was thought to be a general Bordeaux blend or from a highly reputed Bordeaux estate.

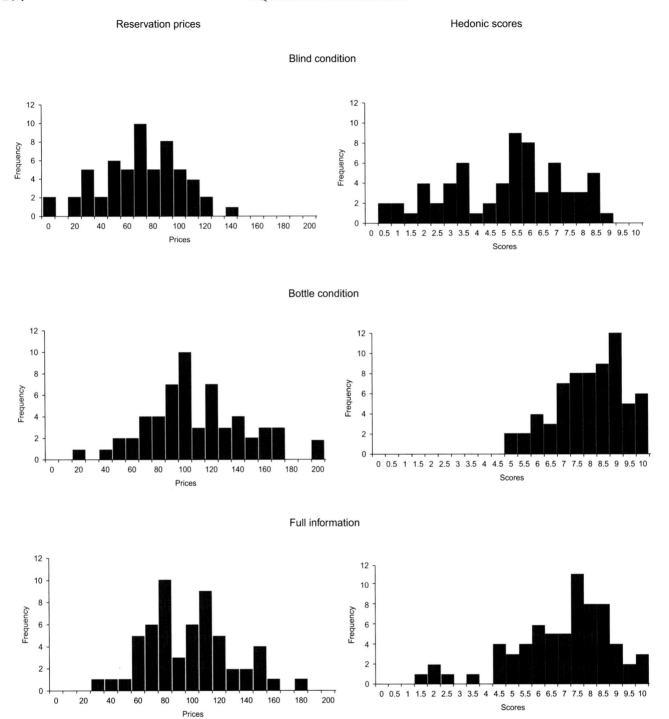

FIGURE 6.9 Comparison of the perception of the same wines as affected by knowledge (*Blind*, knowing only that they were nonvintage brut champagnes; *Bottle*, assessment based only on handling each bottle but without tasting; and *Full*, tasting associated with seeing the actual bottle from which the wine came) on the price they would be willing to pay for the wine (reservation price) and the score given the champagnes (hedonic) by consumers in the Champagne region. *Reprinted from Lange, C., Martin, C., Chabanet, C., Combris, P., Issanchou, S., 2002. Impact of the information provided to consumers on their willingness to pay for Champagne: comparison with hedonic scores. Food Qual. Pref. 13, 597–608, with permission from Elsevier.*

Perceived quality can also be influenced by expressions of quality from acknowledged authorities, be they from magazines, newspapers, or whatever. However, the degree to which this opinion affects perception is largely unknown. A recent study suggests that the influence of expert opinion is most pervasive relative to inexpensive wines, for both knowledgeable and novice consumers, but applies only to novices when it came to expensive wines (Aqueveque, 2015).

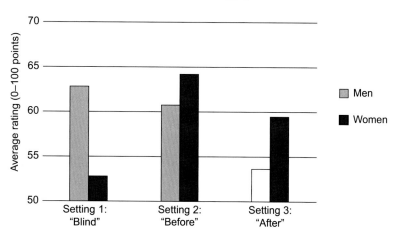

FIGURE 6.10   Average rating of a $40 wine by gender and experimental setting: *Blind*, no price supplied; *Before*, price given before tasting; *After*, price supplied after tasting. *From Almenberg and Dreber, 2011, reproduced by permission of Cambridge University Press.*

The aura surrounding higher prices has probably been used purposely by certain producers to suggest greater quality. By charging an inflated price, it can appeal to those seeking to advertise sophistication to their friends, even if the wine is not sensorially exceptional. For example, when similar wines were tasted with differences in suggested price, consumers indicated that they would be willing to pay more for the apparently higher-priced sample and less for the apparently inexpensive wine (Lewis and Zalan, 2014).

Wine-Appreciation Courses

Training and knowledge are not prerequisites for appreciation, but they can alert people to subtleties of which they were formerly unaware. Recognizing a wine's style, varietal origin, or relative age can enhance intellectual appreciation. Nevertheless, these features are only accessory to actual sensory enjoyment, just as identifying plants or birds is incidental to appreciating walking in a forest. Enjoyment requires only the basic rudiments of sensory acuity. Superior skill improves differentiation, but whether it increases enjoyment is a moot point. Is the critical assessor more appreciative? Like perfect pitch, exceptional acuity can be a two-edged sword: it enhances detection of both the finest and worst attributes of music, potentially downgrading satisfaction with the norm.

Except in those instances where wines are donated, or students are willing to pay high course fees, wines must be selected for their optimal price/quality ratio. They also need to clearly demonstrate the attributes desired. Repeatedly finding prototypic representatives in all categories, within a limited budget, is one of the most daunting tasks the instructor must perform. In choosing samples, the instructor should also attempt to avoid inserting personal biases. It is far too easy to create undue trust, or distrust, in the value of appellation-control laws, vintage charts, regional quality, and other sacred cows. It behooves the instructor to guide students to their own hopefully rational conclusions about the origins and nature of wine quality, as well as discover their own preferences.

Wine-appreciation courses run the gamut from one-session illustrations on how to taste wine, to elaborate multiple-week university courses. Most courses involve from 6 to 10 sessions, dealing with topics such as sensory perception, wine evaluation, wine production, major grape varieties, wine regions, label interpretation, appellation-control laws, wine aging, wine storage, wine and health, and wine and food combination. Combined with the instruction are tastings that, to varying degrees, complement the lecture material. Because many people take these courses as much for enjoyment as education, the instructor should conscientiously avoid verbosity and unnecessary detail. Typically, the first tasting involves a sampling of the major wine styles (white, rosé, red, sparkling, sherry, and port). Subsequent tastings explore various wine styles and cultivars, variations in their regional expressions, and quality differences. They may also include training exercises to illustrate basic taste and/or flavor sensations, and how they interact with one another. Presenting examples of wine faults may be less than appealing, but can provide a unique opportunity to demonstrate what are considered legitimate faults, justifying rejection or return. Too often, consumers are ill informed as to the difference between wine faults and personal dislikes. Thus, presenting examples of wine faults does not only do consumers a service, but also the wine industry. Table 6.1 lists examples of tastings designed for wine-appreciation courses. Additional or alternative tastings may be based on suggestions mentioned in Table 6.2.

TABLE 6.1   Examples of Various Tastings for a Wine-Appreciation Course

Testing	Example
Introduction	Amontillado sherry, Riesling Kabinett, Chardonnay, Cabernet, sparkling, Porto
Taste detection	Sweet, acid, bitter, astringent, astringent-bitter, heat, spritz
Odor detection	Off-odors (e.g., oxidized, sulfur, volatile acidity, fusel, baked, corked, buttery); fragrance (e.g., fruits, flowers, vegetal, spices, smoky, nuts, woody)
Quality detection	Pairs of average and superior quality sweeter (e.g., Riesling) and drier (e.g., Chardonnay) white wines and a red wine (e.g., Cabernet Sauvignon)
White varietals	Pinot Grigio, Riesling, Sauvignon blanc, Chardonnay, Gewürztraminer, Viura
Red varietals	Zinfandel, Sangiovese, Tempranillo, Pinot noir, Merlot, Shiraz
Varietal variation I	Three examples of different expressions of two white varietals
Varietal variation II	Three examples of different expressions of two red varietals
Sparkling wines	Regional expressions (e.g., Champagne, Cava, Sket, Prosecco, Californian, Australian) or production technique (method champenoise, transfer, charmat)
Fortified wines	Fino, amontillado, and oloroso sherries plus ruby, tawny, and vintage ports
Winemaking technique I	Several examples each of the use of procedures such as carbonic maceration (Beaujolais), the *recioto* process (Amarone), or the *governo* process (some Chianti)
Winemaking technique II	Several examples each of the effect of procedures such as oak exposure, *sur lies* maturation, prolonged in-bottle aging

TABLE 6.2   Examples of Various Tastings for Wine Society Tastings

Type	Subtype	Example
Intervarietal comparison	White *vinifera* cultivars	Six samples showing the distinctive varietal aroma of the representative cultivars (e.g., Chardonnay, Riesling, Sauvignon blanc, Parellada, Gewürztraminer, Rhoditis)
	Geographic expression	German Riesling, Silvaner, Müller-Thurgau, Ehrenfelser, Gewürztraminer, Ortega
	Red *vinifera* cultivars	Six samples showing the distinctive varietal aroma of the representative cultivars (e.g., Cabernet Sauvignon, Tempranillo, Sangiovese, Nebbiolo, Pinot noir, Malvasia Nero)
	Red non-*vinifera* cultivars	Chambourcin, Marechal Foch, Baco noir, de Chaunac, Concord, Norton
Intravarietal comparison	International	Six similarly priced representatives (e.g., Australian, Californian, French, Italian, Canadian, Bulgarian)
	Regional	Six similarly priced representatives (e.g., Napa, Sonoma, Mendocino, Santa Barbara, Santa Clara, Central Valley)
	Producers	Six regional wines from the same producer (e.g., Rioja, Chianti, or Bordeaux)
	Vintages	Six different vintages wines from the same producer (e.g., Don Miguel Cabernet Sauvignon, or Wolf Blass Chardonnay)
	Quality	Representatives of the quality scale of German wines (e.g., QbA, Kabinett, Spätlese, Auslese, Beerenauslese, Trockenbeerenauslese) or different grades of wines from the same procedures (e.g., Robert Mondavi Cabernet wine: Reserve, Oakville and Stags Leap Districts, Coastal, Woodbridge, and Opus One)
Winemaking style	White	Several different representatives of one or several of the following different styles: oaked and unoaked Chardonnay; trocken, halbtrocken and traditional auslese Riesling; modern and traditional white Rioja wines; effects of the use of different yeast or malolactic bacterial strains[a]
	Red	Several different representatives of one or several of the following different styles: *recioto* (Amarone vs Valpolicella); *governo* (older vs modern Chianti); carbonic maceration (Beaujolais vs non-Beaujolais Gamay noir), fermentor (rotary vs stationary tank)

[a]*Examples such as these are obtainable only from winemarkers of researchers who are conducting experiments with these. Although difficult for most groups to obtain, the often marked effect of the use of different microbial strains makes the effort in obtaining the samples well worthwhile.*

TABLE 6.3 Some Examples of Descriptors Frequently Associated with the Aroma of Particular Varietal Wines

Descriptor	White cultivar	Descriptor	Red cultivar
Apricot	Chardonnay	Almond	Dolcetto, Schiava
Apple	Chardonnay, Parallada, Torbato	Beet	Pinot noir
Almond	Arinto, Garganega, Greco	Bell pepper	Cab. Sauvignon, Cab. Franc
Banana	Arinto, Viura	Berry	Barbera, Schiava, Zinfandel
Butterscotch	Viura	Blackcurrant	Cab. Sauvignon, Merlot, Chancellor
Buttery	Chardonnay	Cherries	Aglianico, Pinot noir, Sangiovese
Camellia	Chenin blanc	Cigar box	Cab. Sauvignon, de Chaunac
Fig	Sémillon	Citrus	Tempranillo
Guava	Chenin blanc	Clove	Grignolino
Citrus	Parellada, Verdicchio	Daffodil	Corvina
Cinnamon	Parellada	Ginseng	Nebbiolo
Hazelnut	Fiano	Green bean	Cab. Sauvignon, Cab. Franc
Herbaceous	Sauvignon blanc	Ham	Pinot noir
Lemon	Arinto	Incense	Tempranillo
Licorice/anise	Parellada	Leather	Cabernet Sauvignon (aged)
Melon	Chardonnay, Sémillon	Licorice	Sangiovese
Musky	Grillo, Viognier	Mint	Pinot noir, Touriga Nacional
Nuts	Grillo, Viognier	Pepper	Shiraz (Syrah)
Olive	Sauvignon blanc	Quince	Dolcetto
Orange blossom	Concord	Raspberry	Gamay noir, Syrah, Zinfandel
Parmesan cheese	Inzolia	Rose	Nebiolo
Peach	Chardonnay, Rousanne	Spice	Aglianico
Bell pepper	Sauvignon blanc	Tar	de Chaunac, Nebbiolo
Passion fruit	Pinot grigio, Sauvignon blanc	Tea	Baco noir
Pomade	Seyval blanc	Tobacco	Baco noir
Pine	Riesling	Tulip	Corvina, Rougeon
Romano cheese	Pinot blanc	Vanilla	Schiava
Rose	Riesling, Scuppernong	Violet	Cab. Sauvignon, Nebbiolo, Schiava

Descriptors, such as those supplied in Tables 7.2 and 7.3, can help students recognize various varietal aromas. Table 6.3 presents some of the same data, but based on descriptor associations with cultivars, rather than the reverse. However, as noted previously, students should not view these suggestions as adequately representing the *actual* aroma characteristics of the varieties mentioned, nor should anyone come to believe that description of wine fragrance is an end in itself. The terms only allude to aspects of the fragrance that may remind one of odor memories of the named objects or situations. However, for instructors feeling that training in descriptor use is important (or desired by the students), odor samples can be prepared using formulas such as those noted in Appendix 5.1. In addition, colorful charts illustrating wine colors, aromas, and grape varieties are available from Bouchard Aîné et Fils, Beaune, France (http://www.bouchard-aine.fr/en/tours-and-tastings.r-16/the-wine-shop.r-106/our-wine-posters.r-109/?valid_legal=1).

Encouraging students to concentrate on developing odor memories for the major grape cultivars, wine styles, and the holistic attributes that distinguish fine wines is far more beneficial. Excessive doting on finding descriptors diverts attention from the aromatic blueprints that characterize different wines. Coming to recognize these central features is enriching, freeing the consumer from slavishly following the opinions of others.

It may be useful to note that we remember faces or melodies without the need for describe them in words. Experience, not term use, is all that seems necessary for the development of long-term olfactory memories (Engen and Ross, 1973). Avoiding word descriptors may even be beneficial in developing odor memories (Olsson, 1999; Degel et al., 2001). Thus, there may be at least two forms of odor memory, one involving odor–word association and the other with episodic–experience association (even when the odor connection is not conscious) (Köster, 2005; see also Lehrner et al., 1999). Episodic odor memory appears to show neural plasticity and can change with new experience-based associations (Gottfried, 2008). Thus, the ability to name specific descriptors is not a prerequisite for recognizing either distinctive wine styles or cultivars, or appreciating wine, and may be detrimental (Levine et al., 1996; Dijksterhuis et al., 2006). Furthermore, even the most seasoned wine experts have difficulty identifying the varietal or regional origin of wines more often than they would like to admit (e.g., Winton et al., 1975).

It is questionable whether, if a wine too closely resembles its descriptors, the wine is lacking in subtlety and complexity. For example, most Cabernet Sauvignon wines show a bell pepper aspect. If this aspect is marked, it is often viewed as negative attribute. It probably indicates that the fruit was insufficiently ripe when harvested. Fully ripened grapes of this cultivar should yield wines expressing a rich, blackcurrant fragrance, with multiple other subtle favor complexities. There should be no more than just a hint of bell pepper, detected primarily in the aftertaste.

Wine enthusiasts often want to know how professionals seemingly divine wine origin. The popular press propagates the myth that professionals can identify wine origin, occasionally down to the proverbial north- or south-facing slope. With extensive experience, dedication, and a well-developed odor memory, tasters can often detect the basic aspects of style and variety (their gestalt). However, there is no magical Rosetta Stone for identification. Because environmental, viticultural, and enologic conditions can markedly modify the expression of the sensory aspects that can be used to divine wine origin, it is often impossible to ascertain (with certainty) the varietal, regional, or occasionally the stylistic origin of the majority of wines by tasting. Ideal archetypal examples are necessary for learning, in the same sense as a child will never accurately interpret visual input when vision is clouded.

Identification skill depends both on odor memory and its deductive and inductive use, that is, searching both for those features that a particular wine ideally should and should not expresses. It is through experience that one can come to recognize differences that initially were not detected. Thus, wine origin is often divined by both the processes of elimination and inclusion. Thankfully, not every distinctive aspect of certain wines need be present – similar to how one can recognize the Dalmatian dog in Fig. 3.13. Thus, if one were presented with a palish red French wine, possessing a mild fruity but nondistinctive fragrance, it might lead the taster to suspect a less than exemplary example of a Pinot noir (regrettably all too common for Pinot noir). With this supposition, the taster could then scan their bank of Pinot noir memories, in search of other characteristics for confirmation or rejection. The more extensive the taster's archetypal memory, the greater are the chances that a reasonable pattern match might be found. Nonetheless, such deductive and inductive reasoning has its limits. For example, Cabernet Sauvignon and Merlot both show a blackcurrant fragrance. Differentiation might be sought based on the degree of astringency and the presence of a bell pepper fragrance, Merlot often being less marked in both attributes. Another related variety, Cabernet Franc, tends to possess less blackcurrant but a more intense astringency and bell pepper character. Regrettably, these tendencies are just that: tendencies. Incidental information, such as country of origin, vintage, price, etc., helps the process of elimination, but provides little honor to the taster's divining skills. Relative to provenance, some people claim to recognize Australian wines by the presence of a eucalyptus odor. There may be some justification in that claim: 1,4-cineole (a major constituent in eucalyptus oil) is an aromatic marker in some Australian wines (Fig. 6.11). This feature was less obvious in Shiraz than Pinot noir or Cabernet Sauvignon. Furthermore, regional differences occur. Intriguingly, 1,4-cineole can contribute to a hay, dried herbs, and blackcurrant character.

Developing odor models for particular cultivars, regions, and styles must also, eventually, accommodate modifications induced by wine aging and various vintages. Clearly exceptional sensory acuity is required to recognize those properties and to keep these well defined in memory. There are no shortcuts. While training can and does improve potential, it cannot substitute for a lack of inherent sensory acuity, reasoning skill, and superior memory. Motivation and practice help, but alone will not ensure achieving the desired end. While a goal for most connoisseurs, the naming game is just that: a game, enjoyable when one succeeds, but irrelevant to appreciating the pleasures wine can provide at the dinner table.

One of the major difficulties consumers have in learning the archetypical characteristics of particular wines is simply lack of sufficient and frequent exposure to exemplary examples. This is where access to a serious wine society can be of considerable help. Wine societies provide an opportunity to sample many wines. Where no local society exists, forming a small group of like-minded individuals is an option. Alternatively, a dispensing machine (Plates 5.10, 9.4), holding upwards of 16 wines, permits the sampling of a series of wines on a schedule of the owner's choosing.

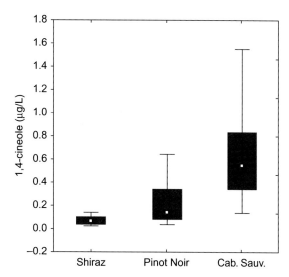

FIGURE 6.11 Distribution of 1,4-cineole concentrations in Australian red wines represented as box plots with the minimum, maximum, median, and quartiles. *Cab. Sauv* ., Cabernet Sauvignon. *Reprinted with permission from Antalick, G., Šuklje, K., Blackman, J. W., Meeks, C., Deloire, A., Schmidke, L. M., 2015. Influence of grape composition on red wine ester profile: Comparison between Cabernet Sauvignon and Shiraz cultivars from Sustralian warm climate. J. Agric. Food Chem. 63, 4664–4672, Copyright 2015, American Chemical Society.*

Surprisingly, conscious effort is not necessary for learning odor associations. Searching and learning for odor patterns is an inherent property of how our brain functions, possibly analogous to how young children learn language. What clearly helps, though, is frequent and repeated exposure, to fix an odor's common elements in a memory profile. With wines, odor memories incorporate multiple sensory elements, often assimilating associated taste, tactile, and visual aspects. Because such memory traces develop as an entity, identification is optimal when all the components are activated simultaneously. Eliminating one (such as masking the color) can seriously deteriorate recognition. Tasting in black glasses may be useful in sensory analysis under laboratory conditions, but is detrimental to developing holistic models of a range of wines.

Examples of cultivars whose wines are easier to identify are Cabernet Sauvignon, Shiraz, Riesling, and Gewürztraminer. Other cultivars produce wines with distinctive fragrances, but are less easily recognized, or vineyard and winemaking conditions more markedly influence their varietal expression. Increased variation complicates learning a wine's flavor gestalt. Examples of varietal wines more challenging to recognize are those from Negro Amaro, Tempranillo, Pinot noir, and Sangiovese (for reds), and Viura, Sémillon, Trebbiano, Chenin blanc, and Pinot gris (for whites).

Stylistic differences are more easily illustrated and recognized, e.g., oaked versus nonoaked, or wines with and without malolactic fermentation. Red wines produced using standard procedures, in contrast to carbonic maceration (*nouveau* wines), or having undergone the *recioto* process (Amarone) can be clearly demonstrated. Some regional styles, such as sherries, ports, and madeiras are so distinctive that anyone can rapidly come to recognize their basic sensory characteristics. Differences due to geographic origin are more problematic to exemplify with ease. For example, Cabernet wines from Bordeaux tend to be more tannic and possess fewer fruit flavors (at least early in their maturation), Californian versions tend to be as tannic but express their fruit flavors earlier, whereas South Australian versions tend to show fewer tannins and possess fruit-forward flavors. However, these differences may reflect as much preferences in production as they do provenance. New Zealand, Californian, and French Sauvignon blanc wines tend to show clear and relatively consistent differences in flavor. Detailed sensory evaluation, conducted under laboratory conditions, usually can identify subtle variations among wines from different locales within a designated appellation (Schlosser et al., 2005; Jouanneau et al., 2012; Lawrence et al., 2013). The problem is that individual wines often vary sufficiently from site to site and year to year to make consistent recognition of regional origin nigh unto impossible.

Because of the marketing importance of a wine's typicity, this is an active field of research; see Fischer et al. (1999); Pardo-Calle et al. (2011), Cadot et al. (2012), Tomasino et al. (2013), Rutan et al. (2014) relative to regional typicity, and Jaffré et al. (2011) and Lawrence et al. (2013) for cultivar typicity. Regrettably, most typicity studies have not been followed up with investigations of their chemical basis. Without this, understanding their physiological/biochemical origins cannot commence. Correspondingly, how meso- and microclimatic conditions influence these factors remains a mystery. If we are to achieve better control over the viticultural origins of wine typicity, and adjust for global warming, this information is essential.

The effects of aging on color and flavor are most easily illustrated if the instructor has collected the same wine over several vintages. For this to be clear, other variables, such as cultivar composition, geographic origin, and production style, need to be similar, if not identical.

With all the stress on sensory perception and developing odor memories, it is easy for the student to forget that most wines poorly express the crucial attributes that permit identification of origin, either varietal or provenance. Thus, it is valuable to note that most wines do not merit the meticulous dissection and search for identifying attributes that are emphasized in class. This should hopefully allay feelings of insufficiency in those experiencing difficulties in achieving the level of competency they might want or expect.

Wine-Tasting Societies

Wine societies come in almost as many variants as there are examples. Regrettably, some are more interested in social contacts and/or gastronomic interests than assessing wine quality. Nevertheless, that such societies benefit wine sales is without question. They also help to spread consumer interest in wine, albeit occasionally not the most appropriate.

From a learning perspective, definitive tasting societies are preferable. They adhere to a protocol, resembling that used in sensory analysis (see Chapter 5), where all wines are tasted blind, poured at the same time, into identical glasses, in equivalent small amounts (about 30 mL), with the wines sampled repeatedly over about 20 minutes, under a veil of relative silence, and notes taken (by some). In addition, food is limited to small cubes of bread (occasionally slices), used theoretically as palate cleansers. Cheese, if present, is restricted until most of the formal tasting session is complete, to avoid unduly influencing the wines' sensory characteristics.

Only the names of the wines are typically noted in advance, but never their order of presentation. To add an element of challenge, participants are encouraged to match the samples with the wines listed. Once everyone has had sufficient time (about 20–30 min) to fully assess the wines' sensory attributes (including development and duration), the wines are ranked according to preference, or some other criteria. Because participants are usually interested in how their ranking correlates with others in the group, this can be noted by a showing of hands for their first and second favorites among the wines. Only then is the identity and price of the wines revealed.

Price is often the most eagerly awaited piece of information. Frequently, people's first and second choices are not the most expensive, in answer to the wine taster's prayer, "May I prefer the inexpensive wines." This awareness can encourage members to broaden their wine purchases, especially within the price range most can afford. It also has the salutary message that quality is expressed in the wine's sensory attributes, not extrinsic factors such as public accolade, exalted price, lauded vintage, provenance, varietal origin, rarity, or venerable age. If pairing with food is a component of the society, this should occur only after formal and detailed assessment of the wines is complete.

Such a system worked well for the author for many years, with the society still active and larger then when he retired to devote his time to writing. To avoid dickering about the wine volume (since members poured their own wine), all members were required to purchase six wine glasses, etched with a ring to indicate the 30-mL level, identical to those of every other member. As a treat, the society did permit one tasting with a meal per year, but the assessment of the wines did precede the meal (Plate 6.2). We were a tasting society!

To increase the educational value of each tasting, every session should be focused around a particular theme. There is an endless array of options, but most fall into one of several categories. A popular theme focuses on differentiating among wine varietal characteristics. Such tastings not only act as a useful refresher for seasoned members, but also function as valuable training for new participants. Other popular topics involve comparing various expressions of a single cultivar or style; probing the nature of quality; or exploring the expression of wines from a single producer, region, or country. Other tastings may examine the effects of winemaking technique or vintage differences. Specific examples could be a comparison of varietal icewines, the various styles of sherry (e.g., Sanlúcar de Barrameda, Jerez de la Frontera, and Montilla finos); variations between *nouveau*, standard, village, and *cru* Beaujolais (storing the nouveau at refrigerator temperature will help to retain its character, permitting it to be compared with other Beaujolais of the same vintage). Table 6.2 lists other potential tastings. The possibilities are limited only by the imagination of the organizers, wine availability, and, of course, money. Variation in availability obviates the value of presenting precise recommendations herein. Occasionally, wine stores will sponsor tastings. However, they may be uneasy about allowing their wines to be tasted blind, devoid of information that may focus opinion in a direction the store owners would prefer.

PLATE 6.2   Brandon Wine Society tasting where the wines were sampled before the meal.

For the organizers, wine societies provide an excellent opportunity to expose members to valuable experiences they would unlikely receive elsewhere, and rarely supplied by the popular press. A perfect example of this educational role can arise when a sample demonstrates an off-odor. Admittedly, it is difficult to inspire most members about wine faults. However, when a fault is detected, displeasure can come with a silver lining: enlightenment. Comparing a faultless with a faulty version is an ideal chance for members to learn from personal experience. The effects of wine age on color and flavor are other easy examples to present, whereby valuable experience can be infused painlessly. Simple examples of how food (e.g., cheese) influences wine perception provides another learning experience that is easily incorporated into a tasting. A clear example is the bitterness reduction of tannic red wines following a sampling of salty cheese (e.g., Parmigiano-Reggiano).

Depending on the level of wine expertise, and the members' willingness to experiment, a tasting relating to the influence of color on perception could be conducted with white wine colored with tasteless anthocyanins. Malvin, the most frequently occurring wine anthocyanin, can be purchased from Sigma-Aldrich Chemical, and undoubtedly other chemical or science supply companies. It may also be possible to obtain anthocyanins in some pharmacies or health food stores. Another informative tasting could involve sampling several wines in both black and clear glasses, without knowing which black and clear samples correspond to each other. This would give members another experience on how color (presence or absence) affects perception. Because black glasses, identical to clear versions, are not readily available (but can be purchased via Amazon.fr or wineware.co.uk), a temporary solution is to spray the exterior of a set of glasses with black paint. Painting them while inverted protects the inner surfaces. It usually takes several days for the paint odor to dissipate. The paint is easily removed following the tasting by soaking the glasses in hot, soapy water. Although somewhat messy, the educational value of this approach may well be worth the effort. Another unique opportunity is for organizers to flavor an unoaked wine in natural, light, medium-plus, and heavy toasted oak chips for several weeks. Submerging the chips in a weak alcohol solution for several hours, before adding them to the wine, expels air in the oak, minimizing any potential oxidative changes to the wine. Amounts in the range of 5–10 g/750-mL bottle are adequate for an initial trial. Employing a white wine permits the oak character to become apparent earlier. Decanting the wines off the oak chips is best conducted shortly before tasting. Tastings of these types will give the society a degree of sophistication not achieved elsewhere in the public sector.

Home Tastings

Tastings at home typically fall into one of two categories: games of identification or comparing wines before or during a meal. Identification games are popular, but usually have more to do with guessing (unless the differences are marked) than wine assessment. However, conducted more seriously, the sessions can be informative. Ideally, the wines should be sampled blind, to avoid extrinsic knowledge from unduly influencing opinion. However, this presupposes the possession of black tasting glasses.

Exercises often involve a series of wines, provided by the host or jointly by the participants. For example, with black glasses, small stickers can be attached under the base of each glass identifying the wine. The bottles are removed and someone, who did not see the wines being poured, rearranges the glasses, affixing stickers with different colors (or numbers) to the top of the base. After the participants have sampled the wines, noting their perceptions, their identity is revealed. To add more intrigue, if four wines are being sampled, six samples can be presented, two of the wines being represented twice, or one wine three times. Alternately, one or two "mystery" wines, not otherwise suspected, could be included. However, in this instance, participants need to be given a chance to sample the principal four wines in advance, but not the mystery wine(s). Thus, not only do they need to identify the wines they have tried, but which glasses contain the mystery wine(s).

Another genre of informative tasting is to sample a series of wines, followed by a repeat tasting with a piece of one (or more) cheeses (e.g., salty, hard, smooth). It can be a wine and cheese party, but with an education goal. If this has appeal, the tasting could be followed by subsequent sessions to assess the significance of particular constituents on wine perception. For example, salt is known to reduce the perception of the tannins. Salt is also recognized for its ability to enhance the flavor of foods, but few have assessed its effect on wine flavor. Thus, a distinctly novel experiment would be to add trace amounts of salt to a series of wine glasses of one or more wines, with controls. After swirling (to dissolve and distribute the salt), it is up to the participants to determine if any aromatic differences can be detected. There is no end to the educational fun one could have with those genuinely interested in wine.

In contrast, sampling two or more wines with a meal can be enjoyable, but because so many influences are at play that can affect perception, it has a limited educational value. Although food flavors can enhance appreciation, primarily by masking certain attributes of the wines, it conflicts with assessing the pure sensory attributes of the wine(s). Social interaction, combined with the effects of alcohol consumption, also make impartial assessment of the wine's qualities complicated. Thus, it is typically better to taste the wines for their characteristics before the meal. Serious tasting tends to distract from the primary purpose of getting together for a meal, i.e., enjoyment and social interaction.

Postscript

The chapter has dealt primarily with tastings relating more to appreciation than sensory analysis. As such, it covered a diversity of forms, from formal competitions to sampling at home. Each has its rightful place in the pantheon of consumer-oriented tastings. Although a critical approach has the benefit of detecting the full spectrum of sensory properties a wine may possess, it also can be a mixed blessing, exposing its weaknesses or faults. There is also evidence that analyzing attributes may not be advantageous, and even identifying descriptors may disrupt episodic memory, valuable in learning how to identify wine type. However, knowing these possibilities may allow tasters to offset some of these effects, as they may negate the effect of wine color by selectively choosing to ignore it.

Acknowledging the problems associated with attempting to describe a wine's attributes in words may free consumers from feeling deficient, leaving them to concentrate on developing more useful odor models for varietal, stylistic and holistic quality characteristics. Since the terms people use to describe wines illuminate more about the taster and their life experiences than the wine, it is probably more useful to focus on the features that distinguish the quality attributes they appreciate the most, as a guide for future purchases. In addition, by focusing on the wine's aesthetic sensory pleasures, it can enhance a person's quality of life and encourage them to view overconsumption as an insult to both the wine and oneself. It is also useful to remember how significant the context of a tasting is to wine's perception and appreciation, especially the dinner table, where conditions are ideally suited to wine enjoyment.

Although it is evident I am not a proponent of focusing primarily on the use of descriptors with consumers, it may have one benefit for some. It can, even falsely, provide a sense of connoisseurship, and this enhances wine sales. Such a result would be good news to grape growers, winemakers, wine merchants, and anyone in the wine business. It puts a new twist on the expression: "No wine, no dine."

Consumers and aficionados, if only for economic reasons, consume ordinary wine as a pleasant meal accompaniment. Savoring special wines, those gifts of the gods, is not for every day, any more than feasting at some famous Michelin restaurant. If special becomes standard, then exceptional wines are not so. To paraphrase Jeffrey Steingarten (*It Must Be Something I Ate*):

"We passionate tasters elevate, ennoble the bestial impulse to drink into a sublime activity, an art, the art of sensory ecstacy."

Suggested Readings

Amerine, M.A., 1980. The words used to describe abnormal appearance, odor, taste, and tactile sensations of wines. In: Charalambous, G. (Ed.), The Analysis and Control of Less Desirable Flavors in Foods and Beverages, Academic Press, New York, pp. 319–351.

Baldy, M.W., 1995. The University Wine Course, second ed. Wine Appreciation Guild, San Francisco.

Broadbent, M., 1998. Winetasting: How to Approach and Appreciate Wine. Mitchell Beazley, London.

Brochet, F., Dubourdieu, D., 2001. Wine descriptive language supports cognitive specificity of chemical senses. Brain Lang. 77, 187–196.

Köster, E.P., 2005. Does Olfactory Memory Depend on Remembering Odors? Chem. Senses 30, i236–i237.

Lehrer, A., 2009. Wine and Conversation, second ed. Oxford University Press, Oxford, UK.

Lesschaeve, I., 2007. Sensory evaluation of wine and commercial realities: Review of current practices and perspectives. Am. J. Enol. Vitic. 58, 252–258.

Peynaud, E., 1996. The Taste of Wine: The Art Science of Wine Appreciation (M. Schuster, Trans.), second ed. Wiley, New York.

Teil, G., 2001. La production du jugement esthétique sur les vins par la critique vinicole. Sociol. Travail 43, 67–89.

References

Almenberg, J., Dreber, A., 2011 When does the price affect the taste? Results from a wine experiment. J. Wine Econ. 6, 111–121.

Amerine, M.A., Roessler, E.B., 1983. Wines, Their Sensory Evaluation, second ed. Freeman, San Francisco, CA.

Antalick, G., Tempère, S., Šuklje, K., Blackman, J.W., Deloire, A., De Revel, G., et al., 2015. Investigation and sensory characterization of 1,4-cineole: A potential aromatic marker of Australian Cabernet Sauvignon wine. J. Agric. Food Chem. 63, 9103–9111.

Aqueveque, C., 2015. The influence of experts' positive word-of-mouth on a wine's perceived quality and value: the moderator role of consumers' expertise. J. Wine Res. 26, 181–191.

Asher, G., 1989. Words and wine. J. Sens. Stud. 3, 297–298.

Ashton, R.H., 2012. Reliability and consensus of experienced wine judges: Expertise within and between? J. Wine Econ. 7, 70–87.

Ashton, R.H., 2013. Is there consensus among wine quality ratings of prominent critics? An emperical analysis of red Bordeaux, 2004–2010. J. Wine Econ. 7, 225–234.

Ashton, R.H., 2014. Wine as an experience good: Price versus enjoyment in blind tastings of expensive and inexpensive wines. J. Wine Econ. 9, 171–182.

Atkin, T., Nowak, L., Garcia, R., 2007. Women wine consumers: information search and retailing implications. Int. J. Wine Bus. Res. 19, 327–339.

Baldy, M.W., 1995. The University Wine Course, second ed. Wine Appreciation Guild, San Francisco.

Ballester, J., Dacremont, C., Le Fur, Y., Patrick Etiévant, P., 2005. The role of olfaction in the elaboration and use of the Chardonnay wine concept. Food Qual. Pref. 16, 351–359.

Ballester, J., Mihnea, M., Peyron, D., Valentin, D., 2014. Perceived minerality in wine: a sensory reality? Wine Vitic. J. 29 (4), 30–33.

Bastian, S., Bruwer, J., Alant, K., Li, E., 2005. Wine consumers and makers: are they speaking the same language? Aust. NZ Grapegrower Winemaker 496, 80–84.

Baxter, M.J., Crews, H.M., Dennis, M.J., Goodall, I., Anderson, D., 1997. The determination of the authenticity of wine from its trace element composition. Food Chem. 60, 443–450.

Bécue-Bertaut, M., Lê, S., 2011. Analysis of multilingual labeled sorting tasks: application to a cross-cultural study in wine industry. J. Sens. Stud. 26, 299–310.

Bensafi, M., Rinck, F., Schaal, B., Rouby, C., 2007. Verbal cues modulate hedonic perception of odors in 5-year-old children as well as in adults. Chem. Senses 32, 855–862.

Berglund, B., Berglund, U., Engen, T., Ekman, G., 1973. Multidimensional analysis of twenty-one odors. Scand. J. Psychol. 14, 131–137.

Bi, J., Chung, J., 2011. Identification of drivers of overal liking – determination of relative importances of regressor variables. J. Sens. Stud. 26, 245–254.

Binders, G., Pintzler, S., Schröder, J., Schmarr, H.G., Fischer, U., 2004. Influence of German oak chips on red wine. Am. J. Enol. Vitic. 55, 323A.

Bindon, K., Holt, H., Williamson, P.O., Varela, C., Herderich, M., Grancis, I.L., 2014. Relationships between harvest time and wine composition in Vitis vinifera L. cv. Cabernet Sauvignon 2. Wine sensory properties and consumer preference. Food Chem. 154, 90–101.

Boroditsky, L., 2011. How language shapes thought. Sci. Am. 304 (2), 63–65.

Brochet, F., Dubourdieu, D., 2001. Wine descriptive language supports cognitive specificity of chemical senses. Brain Lang. 77, 187–196.

Brochet, F., Morrot, G., 1999. Influence du contexte sure la perception du vin. Implications cognitives et méthodologiques. J. Int. Sci. Vigne Vin 33, 187–192.

Bruwer, J., Saliba, A., Miller, B., 2011. Consumer behaviour and sensory preference differences: Implications for wine product marketing. J. Consum. Market. 28, 5–18.

Bruwer, J., Lesschaeve, I., Campbell, B.J., 2012. Consumption dynamics and demographics of Canadian wine consumers: Retailing insights from the tasting room channel. J. Retail. Consum. Serv. 19, 45–58.

Caballero, R., 2007. Manner-of-motion verbs in wine description. J. Pragmatics 39, 2095–2114.

Cao, J., Stokes, L., 2010. Evaluation of wine judge performance through three characteristics: Bias, discrimination, and variation. J. Wine Econ. 5, 132–142.

Caspar, E.A., Christensen, J.F., Cleeremans, A., Haggard, P., 2016. Coercion changes the sense of agency in the human brain. Curr. Biol. 26, 585–592.

Chancy, I.M., 2000. A comparative analysis of wine reviews. Brit. Food J. 102, 470–480.

Chrea, C., Valentin, D., Sulmont-Rossé, C., Mai, H.L., Nguyen, D.H., Abdi, H., 2004. Culture and odor categorization: Agreement between cultures depends upon the odors. Food Qual. Pref. 15, 669–679.

Cicchetti, D., 2004. Who won the (1976) blind tasting of French Bordeaux and US Cabernets? Parametrics to the rescue. J. Wine Res. 15, 211–220.

Cicchetti, D.V., Cicchetti, A.F., 2010. Wine rating scales: Assessing their utility for producers, consumers, and oenologic researchers. Int. J. Wine Res. 1, 73–83.

Cliff, M.A., 2001. Impact of wine glass shape on intensity of perception of color and aroma in wines. J. Wine Res. 12, 39–46.

Degel, J., Piper, D., Köster, E.P., 2001. Implicit learning and implicit memory for odors: The influence of odor identification and retention time. Chem. Senses 26, 267–280.

de Magistris, T., Groot, E., Gracia, A., Albisu, L.M., 2011. Do Millennial generation's wine preferences of the "New World" differ from the "Old World"?: A pilot study. Int. J. Wine Bus. Res. 23, 145–160.

Depierre, A., 2009. English and French taste words used metaphorically. Chem. Percept. 2, 40–52.

de Wijk, R.A., Schab, F.R., Cain, W.S., 1995. Odor identification. In: Schab, F.R., Crowder, R.G. (Eds.), Memory of Odors. Lawrence Erlbaum, Mahwah, New Jersey, pp. 21–37.

Dijksterhuis, A., Bos, M.W., Nordgren, L.F., van Baaren, R.B., 2006. On making the right choice: The deliberation-without-attention effect. Science 311, 1005–1007.

Dodd, T.M., Laverie, D.A., Wilcox, J.F., Duhan, D.F., 2005. Differential effects of experience, subjective knowledge and objective knowledge on sources of information used in consumer wine purchasing. J. Hospit. Tour. Res. 29, 3–19.

Dürr, P., 1984. Sensory analysis as a research tool. In: Proc. Alko Symp. Flavour Res. Alcoholic Beverages, Helsinki 1984. (L. Nykänen, and P. Lehtonen, eds.) Foundation Biotech. Indust. Ferm. 3, 313–322.

Engen, T., Ross, B.M., 1973. Long-term memory of odors with and without verbal descriptions. J. Exp. Psychol. 100, 221–227.

Fischer, U., 2000. Practical applications of sensory research: Effect of glass shape, yeast strain, and terroir on wine flavor. In: Proc. ASEV 50th Anniv. Ann. Meeting. Seattle, WA, June 19–23, 2000. (J. M. Rantz, ed.), pp. 3–8.

Fischer, U., Roth, D., Christmann, M., 1999. The impact of geographic origin, vintage and wine estate on sensory properties of Vitis vinifera cv. Riesling wines. Food Qual. Pref. 10, 281–288.

Ginon, E., Ares, G., Issanchou, S., Laboissière, L.H.E.D., Deliza, R., 2014. Identifying motives underlying wine purchase decisions: Results from an exploratory free listing task with Burgundy wine consumers. Food Res. Int. 62, 860–867.

Gleb, B.D., Gelb, G.M., 1986. New Coke's Fizzle – Lesson for the rest of us. Sloan Manage. Rev. 28 (Fall), 71–76.

Gokcekus, O., and Nottebaum, D., 2011. The buyer's dilemma – Whose rating should a wine drinker pay attention to? AAWE Working Paper 91.

Goldstein, R., Almenberg, J., Dreber, A., Emerson, J.W., Herschkowitsch, A., Katz, J., 2008. Do more expensive wines taste better? Evidence from a large sample of blind tastings. J. Wine Econ. 3, 1–9.

Goodman, S., 2009. An international comparison of retail consumer wine choice. Int. J. Wine Bus. Res. 21, 41–49.

Gottfried, J.A., 2008. Perceptual and neural plasticity of odor quality coding in the human brain. Chem. Percept. 1, 127–135.

Guillou, I., Bertrand, A., De Revel, G., Barbe, J.C., 1997. Occurrence of hydroxypropanedial in certain musts and wines. J. Agric. Food Chem. 45, 3382–3386.

Gultek, M.M., Dodd, T.H., Guydosh, R.M., 2006. Attitudes towards wine-service training and its influence on restaurant wine sales. Int. J. Hospitality Manage. 25, 432–446.

Hersleth, M., Mevik, B.-H., Naes, T., Guinard, J.-X., 2003. Effect of contextual factors on liking for wine – use of robust design methodology. Food Qual. Pref. 14, 615–622.

Heymann, H., Hopfer, H., Bershaw, D., 2014. An exploration of the perception of minerality in white wines by projective mapping and descriptive analysis. J. Sens. Stud. 29, 1–13.

Hodgson, R.T., 2008. An examination of judge reliability at a major U.S. wine competition. J. Wine Econ. 3, 105–113.

Hodgson, R.T., 2009. An analysis of concordance among 13 U. S. wine competitions. J. Wine Econ. 4, 1–9.

Hodgson, R., Cao, J., 2014. Criteria for accrediting expert wine judges. J. Wine Econ. 9, 62–74.

Hollebeek, L.D., Jaeger, S.R., Brodie, R.J., Balemi, A., 2007. The influence of involvement on purchase intention for New World wine. Food Qual. Pref. 18, 1033–1049.

Horverak, Ø., 2009. Wine journalism–marketing or consumers' guide? Market. Sci. 28, 573–579.

Hough, G., Wakeling, I., Mucci, A., Changers IV, E., Gallardo, I.M., Alves, L.R., 2006. Number of consumers necessary for sensory acceptability tests. Food Qual. Pref. 17, 522–526.

Hughson, A.L., Boakes, R.A., 2002. The knowing nose: The role of knowledge in wine expertise. Food Qual. Pref. 13, 463–472.

Jackson, R.S., 2000. Wine Science: Principles, Practice, Perception, second ed. Academic Press, San Diego, CA.

Jaffré, J., Valentin, D., Meunier, J.-M., Siliani, A., Bertuccioli, M., Le Fur, Y., 2011. The Chardonnay wine olfactory concept revisited: A stable core of volatile compounds, and fuzzy boundaries. Food Res. Int. 44, 456–464.

Jouanneau, S., Weaver, R.J., Nicolau, L., Herbst-Johnstone, M., Benkwitz, F., Kilmartin, P.A., 2012. Subregional survey of aroma compounds in Marlborough Sauvignon Blanc wines. Aust. J. Grape Wine Res. 18, 329–343.

King, E.S., Kievit, R.L., Curtin, C., Swiegers, J.H., Pretorius, I.S., Bastian, S.E.P., et al., 2010. The effect of multiple yeasts co-inoculations on Sauvignon Blanc wine aroma composition, sensory properties and consumer preference. Food Chem. 122, 618–626.

Köster, E.P., 2003. The psychology of food choice: Some often encountered fallacies. Food Qual. Pref. 14, 359–373.

Köster, E.P., 2005. Does Olfactory Memory Depend on Remembering Odors? Chem. Senses 30, i236–i237.

Köster, E.P., 2009. Diversity in the determinants of food choice: A psychological perspective. Food Qual. Pref. 20, 70–82.

Lange, C., Martin, C., Chabanet, C., Combris, P., Issanchou, S., 2002. Impact of the information provided to consumers on their willingness to pay for Champagne: Comparison with hedonic scores. Food Qual. Pref. 13, 597–608.

Langstaff, S.A., 2010. Sensory quality control in the wine industry. In: Kilcast, D. (Ed.), Sensory Analysis for Food and Beverage Quality Control. Woodhouse Publ., Oxford, UK, pp. 236–261.

Lattey, K.A., Bramley, B.R., Francis, I.L., 2010. Consumer acceptability, sensory properties and expert quality judgements of Australian Cabernet Sauvignon and Shiraz wines. Aust. J. Grape Wine Res. 16, 189–202.

Lawless, H.T., 1984. Flavor description of white wine by expert and nonexpert wine consumers. J. Food Sci. 49, 120–123.

Lawless, H.T., Heymann, H., 2010. Sensory Evaluation of Food: Principles and Practices, 2nd ed. Springer, New York.

Lawrence, G., Symoneaux, R., Maitre, I., Brossaud, F., Maestrojuan, M., Mehinagic, E., 2013. Using the free comments method fro sensory characterisation of Cabernet Franc wines: Comparison with classical profiling in a professional context. Food Qual. Pref. 30, 145–155.

Lehrer, A., 1975. Talking about wine. Language 51, 901–923.

Lehrer, A., 1983. Wine and Conversation. Indiana University Press, Bloomington, IN.

Lehrer, A., 2009. Wine and Conversation, second ed. Oxford University Press, Oxford, UK.

Lehrner, J.P., Glück, J., Laska, M., 1999. Odor identification, consistency of label use, olfactory threshold and their relationships to odor memory over the human lifespan. Chem. Senses 24, 337–346.

Lesschaeve, I., Noble, A.C., 2005. Polyphenols: Factors influencing their sensory properties and their effects on food and beverage preferences. Am. J. Clin. Nutr. 81, 330S–335S.

Levine, G.M., Halberstadt, J.B., Goldstone, R.L., 1996. Reasoning and the weighting of attributes in attitude judgements. J. Personal. Soc. Psychol. 70, 230–240.

Lewis, G., Zalan, T., 2014. Strategic implications of the relationship between price and willingness to pay: Evidence from a wine-tasting experiment. J. Wine Econ. 9, 115–134.

Liger-Belair, G., Vignes-Adler, M., Voisin, C., Robillard, B., Jeandet, P., 2002. Kinetics of gas discharging in a glass of champagne: The role of nucleation sites. Langmuir 18, 1294–1301.

Liger-Belair, G., Villaume, S., Cilindre, C., Polidori, G., Jeandet, P., 2009. CO_2 volume fluxes outgassing from champagne glasses in tasting conditions: Flute versus coupe. J. Agric. Food Chem. 57, 4939–4947.

Lima, T., 2006. Price and quality in the California wine industry: An empirical investigation. J. Wine Econ. 1, 176–190.

Lockshin, L., Jarvis, W., d'Hauteville, F., Perrouty, J.-P., 2006. Using simulations from discrete choice experiments to measure consumer sensitivity to brand, region, price, and awards in wine choice. Food Qual. Pref. 17, 166–178.

Lubbers, S., Voilley, A., Feuillat, M., Charpontier, C., 1994. Influence of mannoproteins from yeast on the aroma intensity of a model wine. Lebensm.-Wiss. u. Technol. 27, 108–114.

Maier, H.G., Hartmann, R.U., 1977. The adsorption of volatile aroma constituents by foods. VIII. Adsorption of volatile carbonyl compounds by amino acids. Z. Lebensm. Unters. Forsch. 163, 251–254.

Majid, A., Burenhult, N., 2014. Odors are expressible in language, as long as you speak the right language. Cognition 130, 266–270.

Maltman, A., 2008. The role of vineyard geology in wine typicity. J. Wine Res. 19, 1–17.

Maltman, A., 2013. Minerality in wine: A geological perspective. J. Wine Res. 24, 169–181.

Marcus, I.H., 1974. How to Test and Improve your Wine Judging Ability. Wine Publications, Berkeley, CA.

Marin, A.B., Durham, C.A., 2007. Effects of wine bottle closure type on consumer purchase intent and price expectation. Am. J. Enol. Vitic. 58, 192–201.

McDermott, B.J., 1990. Identifying consumers and consumer test subjects. Food Technol. 44, 154–158.

Meillon, S., Dugas, V., Urbano, C., Schlich, P., 2010a. Preference and acceptability of partially dealcoholized white and red wines by consumers and professionals. Am. J. Enol. Vitic. 61, 42–52.

Meillon, S., Urbano, C., Guillot, G., Schlich, P., 2010b. Acceptability of partially dealcoholized wines – Measuring the impact of sensory and information cues on overall liking in real-life settings. Food Qual. Pref. 21, 763–773.

Melcher, J.M., Schooler, J.W., 1996. The misrembrance of wines past: Verbal and perceptual expertise differentially mediate verbal overshadowing of taste memory. J. Memory Lang. 35, 231–245.

Milgram, S., 1963. Behavioral study of obedience. J. Abn. Soc. Psychol. 67, 371–378.

Mingo, S.A., Stevenson, R.J., 2007. Phenomenological differences between familiar and unfamiliar odors. Perception 36, 931–947.

Moisseeff, M., 2006. Arômes du vin. Hachette Practique, Paris.

Mojet, J., Köster, E.P., 1986. Research on the appreciation of three low alcoholic beers.(Confidential resport). Psychological Laboratory, University of Utrecht, The Netherlands, (in Dutch). Reported in Köster (2003).

Mueller, S., Lockshin, L., Saltman, Y., Blanford, J., 2010a. Message on a bottle: The relative influence of wine back label information on wine choice. Food Qual. Pref. 21, 22–32.

Mueller, S., Osidacz, P., Francis, I.L., Lockskin, L., 2010b. Combining discrete choice and informed sensory testing in a two-stage process: Can it predict wine market share? Food Qual. Pref. 21, 741–754.

Muñoz-González, C., Rodríguex-Bencomo, J.J., Moreno-Arribas, M.V., Pozo-Bayón, M.A., 2014. Feasibility and application of a retronasal aroma-trapping device to study in vivo aroma release during the consumption of model wine-derived beverages. Food Sci. Nutr. 2, 361–370.

Noble, A.C., Bursick, G.F., 1984. The contribution of glycerol to perceived viscosity and sweetness of white wine. Am. J. Enol. Vitic. 35, 110–112.

OIV (2014). The wine market: Evolution and trends. www.oiv.int/oiv/info/en_press_conference_may_2014?lang=en.

Olkin, I., Lou, Y., Cao, J., 2015. Analyses of wine-tasting data: A tutorial. J. Wine Econ. 10, 4–30.

Olsson, M.J., 1999. Implicit testing of odor memory: Instances of positive and negative repetition priming. Chem. Senses 24, 347–350.

Pardo-Calle, C., Segovia-Gonzalez, M.M., Paneque-Macias, P., Espino-Gonzalo, C., 2011. An approach to zoning in the wine growing regions of "Jerez-Xérès-Sherry" and "Manzanilla-Sanlúcar de Barrameda" (Cádiz, Spain). Sp. J. Agric. Res. 9, 831–843.

Parpinello, G.P., Versari, A., Chinnici, F., Galassi, S., 2009. Relationship among sensory descriptors, consumer preference and color parameters of Italian Novello red wines. Food Res. Int. 42, 1389–1395.

Parr, W.V., White, K.G., Heatherbell, D., 2004. Exploring the nature of wine expertise: What underlies wine expert's olfactory recognition memory advantage? Food Qual. Pref. 15, 411–420.

Parr, W.V., Mouret, M., Blackmore, S., Pelquest-Hunt, T., Urdapilleta, I., 2011. Representation of complexity in wine: Influence of expertise. Food Qual. Pref. 22, 647–660.

Peynaud, E., 1987. The Taste of Wine. The Art and Science of Wine Appreciation (M. Schuster, Trans.). Macdonald & Co., London.

Piqueras-Fiszman, B., Spence, C., 2012. The weight of the bottle as a possible extrinsic cue with which to estimate the price (and quality) of the wine? Observed correlations. Food Qual. Pref. 26, 41–45.

Plassmann, H., O'Doherty, J., Shiv, B., Rangel, A., 2008. Marketing actions can modulate neural representations of experienced pleasantness. PNAS 105, 1050–1054.

Polidori, G., Beaumont, F., Jeandet, P., Liger-Belair, G., 2009. Ring vortex scenario in engraved champagne glasses. J. Visualiz. 12, 275–282.

Priilaid, D.A., 2006. Wine's placebo effect. How the extrinsic cues of visual assessments mask the intrinsic quality of South African red wine. Int. J. Wine Marketing 18, 17–32.

Priilaid, D., Feinberg, J., Carter, O., Ross, G., 2009. Follow the leader: How expert ratings mediate consumer assessment of hedonic quality. S. Afr. J. Bus. Manage. 40, 15–22.

Rahman, I., Reynolds, D., 2015. Wine: Intrinsic attributes and consumer's drinking frequency, experience, and involvement. Int. J. Hospitality Manage. 44, 1–11.

Ribéreau-Gayon, J., and Peynaud, E., 1961. Traité d'Oenologie. Tome 2. Berenger, Paris.

Ristic, R., Johnson, T.E., Meiselman, H.L., Hoek, A.C., Bastian, S.E.P., 2016. Towards development of a Wine neophobia scale (WNS): Measuring consumer wine neophobia using an adaptation of The Food Neophobia Scale (FNS). Food Qual. Pref. 49, 161–167.

Ritchie, C., 2009. The culture of wine buying in the UK off-trade. Int. J. Wine Bus. Res. 21, 194–211.

Robichaud, J., de la Presa Owens, C., 2002. Application of formal sensory analysis in a commercial winery. In: Blair, R.J., Williams, P.J., Høj, P.B. (Eds.), 11th Aust. Wine Ind. Tech. Conf. Oct. 7–11, 2001, Adelaide, South Australia. Winetitles, Adelaide, Australia, pp. 114–117.

Rousseau, B., 2015. Sensory discrimination testing and consumer relevance. Food Qual. Pref. 43, 122–125.

Roussis, I.G., Lambropoulos, I., Papadopoulou, D., 2005. Inhibition of the decline of volatile esters and terpenols during oxidative storage of Muscat-white and Xinomavro-red wine by caffeic acid and N-acetyl-cysteine. Food Chem. 93, 485–492.

Rushkoff, D., 2000. Advertising. In Coercion: Why We Listen to What "They Say." Riverhead Books, New York, NY pp. 162–192.

Russell, K., Zivanovic, S., Morris, W.C., Penfield, M., Weiss, J., 2005. The effect of glass shape on the concentration of polyphenolic compounds and perception of Merlot wine. J. Food Qual. 28, 377–385.

Rutan, T., Herbst-Johnstone, M., Pineau, B., Kilmartin, P.A., 2014. Characterization of the aroma of Central Otago Pinot noir wines using sensory reconstitution studies. Am. J. Enol. Vitic. 65, 424–434.

Sáenz-Navajas, M.P., Avizcuri, J.M., Echávarri, J.F., Ferreira, V., Fernández-Zurbano, P., Valentin, D., 2016. Understanding quality judgements of red wines by experts: Effect of evaluation condition. Food Qual. Pref. 49, 216–227.

Saussure, F., 2011. Course in General Linguistics (R. Harris, Trans.). Open Court, Chicago, IL.

Sauvageot, F., Urdapilleta, I., Peyron, D., 2006. Within and between variations of texts elicited from nine wine experts. Food Qual. Pref. 17, 429–444.

Scaman, H., Dou, J., Cliff, M.A., Yuksel, D., King, M.C., 2001. Evaluation of wine competition judge performance using principal component similarity analysis. J. Sens. Stud. 16, 287–300.

Schiefer, J., Fischer, C., 2008. The gap between wine expert ratings and consumer preferences: Measures, determinants and marketing implications. Int. J. Wine Bus. Res. 20, 335–351.

Schlosser, J., Reynolds, A.G., King, M., Cliff, M., 2005. Canadian terroir: Sensory characterization of Chardonnay in the Niagara Peninsula. Food Res. Int. 38, 11–18.

Serapinas, P., Venskutonis, P.R., Aninkevičius, V., Ežerinskis, Ž., Galdikas, A., Juzikiené, V., 2008. Step by step approach to multi-element data analysis in testing the provenance of wines. Food Chem. 107, 1652–1660.

Siegrist, M., Cousin, M.-E., 2009. Expectations influence sensory experience in a wine tasting. Appetite 52, 762–765.

Silverstein, M., 2003. Indexical order and the dialectics of sociolinguistic life. Lang. Communic. 23, 193–229.

Silverstein, M., 2004. Cultural concepts and the language-culture nexus. Curr. Anthropol. 45, 621–652.4.

Smith, A.K., June, H., Noble, A.C., 1996. Effects of viscosity on the bitterness and astringency of grape seed tannin. Food Qual. Pref. 7, 161–166.

Solomon, G.E.A., 1990. Psychology of novice and expert wine talk. Am. J. Psychol. 103, 495–517.

Solomon, G.E.A., 1997. Conceptual change and wine expertise. J. Learn. Sci. 6, 41–60.

Sorrentino, F., Voilley, A., Richon, D., 1986. Activity coefficients of aroma compounds in model food systems. AIChE J. 32, 1988–1993.

Stewart, P.C., Goss, R., 2013. Plate shape and colour interact to influence taste and quality judgments. Flavour. J. 2, 27 (1–9).

Stone, H., Bleibaum, R., Thomas, H.A., 2012. Descriptive testing Sensory Evaluation Practices, 4th ed. Academic Press, London, UK pp. 233–289.

Suárez-Torte, E., 2007. Metaphor inside the wine cellar: On the ubiquity of personification schemas in winespeak. www.metaphorik.de 12, 53–63.

Symoneaux, R., Galmarini, M.C., Mehinagic, E., 2012. Comment analysis of consumer's likes and dislikes as an alternative tool to preference mapping. A case study on apples. Food Qual. Pref. 24, 59–66.

Thach, L., Olsen, J., 2015. Profiling the high frequency wine consumer by price segment in the US market. Wine Econ. Pol. 4, 53–59.

Tomasino, E., Harrison, R., Sedcole, R., Frost, A., 2013. Regional differentiation of New Zealand Pinot noir wine by wine professionals using Canonical Variate Analysis. Am. J. Enol. Vitic. 64, 357–363.

Tominaga, T., Guimbertau, G., Dubourdieu, D., 2003. Contribution of benzenemethanethiol to smoky aroma of certain Vitis vinifera L. wines. J. Agric. Food Chem. 51, 1373–1376.

White, M.R.H., Heymann, H., 2015. Assessing the sensory profiles of sparkling wine over time. Am. J. Enol. Vitic. 66, 156–163.

Williams, A.A., 1982. Recent developments in the field of wine flavour research. J. Inst. Brew. 88, 43–53.

Williams, A.A., Rosser, P.R., 1981. Aroma enhancing effects of ethanol. Chem. Senses 6, 149–153.

Winton, W., Ough, C.S., Singleton, V.L., 1975. Relative distinctiveness of varietal wines estimated by the ability of trained panelists to name the grape variety correctly. Am. J. Enol. Vitic. 26, 5–11.

Wolf-Frandsen, L., Dijksterhuis, G., Brockhoff, P., and Martens, M. (2003). Use of descriptive analysis and implicit identification test in investigating subtle differences in milk. Paper presented at the 10th Food Choice Conference, 30 June–3 July, 2002, Wageningen, The Netherlands, http://dx.doi.org/10.1016/S0950-3293(03)00013-2.

Yeshurun, Y., Sobel, N., 2010. An odor is not worth a thousand words: Form multidimensional odors to unidimensional odor objects. Annu. Rev. Psychol. 61, 219–241.

Zarzo, M., 2008. Psychological dimensions in the perception of everyday odors: Pleasantness and edibility. J. Sens. Stud. 23, 354–376.

Zou, F.-F., Peng, Z.-X., Du, H.-J., Duan, C.-Q., Reeves, M.J., Pan, Q.-H., 2012. Elemental patterns of wines, grapes, and vineyard soils from Chinese wine-producing regions and their association. Am. J. Enol. Vitic. 63, 232–240.

APPENDIX 6.1 SWEETNESS IN WINE

All wines possess some sugar. In dry wines, these consist of grape sugars that were, or could not be, fermented by the yeasts during fermentation. Because they occur at concentrations below their detection thresholds, they do not induce sweetness. Wine acidity and bitterness also suppress the perception of sweetness. Conversely, some aromatic compounds may generate the sensation of sweetness, even in the absence of detectable concentrations of sweet-tasting compounds.

The compounds most frequently inducing actual sweetness in wine are glucose, fructose, ethyl alcohol and glycerol. Glucose and fructose come from grapes, ethanol is produced during fermentation, and glycerol may be derived from grapes or be synthesized during fermentation. Other potentially sweet compounds are the grape sugars arabinose and xylose, and by-products of fermentation such as butylene glycol, inositol, and sorbitol. Mannitol and mannose occur in wine in detectable concentrations only as a result of bacterial spoilage. Table 6.4 illustrates the normal concentration range of the major sweet-tasting compounds found in wines.

To demonstrate the relative sweetness of the major sweet-tasting compounds found in wine, prepare aqueous solutions of glucose (20 g/L), fructose (20 g/L), ethanol (32 g/L), and glycerol (20 g/L). Participants arrange the samples (presented at random) in order of ascending relative sweetness. The form given in Table 6.5 can be used to record relative sweetness.

TABLE 6.4 Typical Concentration of Sweet-Tasting Substances Commonly Found in Wine

Group	Compound	Wine type	Concentration (g/L)
Sugars	Glucose	Dry Sweet	up to 0.8 up to >30
	Fructose	Dry Sweet	up to 1 up to >60
	Xylose		up to 0.5
	Arabinose		0.3 to 1
Alcohols	Ethanol		70 to 150
	Glycerol		3 to 15[a]
	Butylene glycol		up to 0.3
	Inositol		0.2 to 0.7
	Sorbitol		0.1

[a]In botrytized grapes.

TABLE 6.5 Assessment Sheet for Sweetness

Sample	(Driest) 1	2	3	(Sweetest) 4	Additional taste or odor sensations
A					
B					
C					
D					

APPENDIX 6.2 SOURNESS

Acids are characteristic of all wines, being important to its taste, stability, and aging potential. Sourness is a complex function of the acid, its dissociation constant, and wine pH. Acids that occur as salts do no effect sourness.

Of the many acids in wine, the most significant are organic acids. The principal examples are natural grape constituents (tartaric, malic, and citric acids), or yeast and bacterial by-products (acetic, lactic and succinic acids). Only acetic acid is sufficiently volatile to affect wine fragrance—usually negatively. The taste perceptions of grape acids are similar, possessing only subtle differences—tartaric acid being viewed as 'hard,' malic acid considered 'green,' and citric acid deemed to possess a 'fresh' taste. Those acids of microbial origin generate more complex responses. Lactic acid is reckoned to possess a light, fresh, sour taste; acetic acid shows an sharp sour taste and a distinctive odor; whereas succinic acid (seldom present at above threshold levels) exhibits salty, bitter side tastes. Gluconic acid typically occurs only in wines made from moldy grapes. By itself, it does not affect either taste or odor. These compounds frequently occur in table wines within the ranges indicated in Table 6.6.

To demonstrate their differences, prepare the following aqueous solutions: tartaric acid (1 g/L), malic acid (1 g/L), citric acid (1 g/L), lactic acid (1 g/L), acetic acid (1 g/L), and succinic acid (1 g/L). Present them in random order for arranging in order of increasing sourness. Their relative sourness and any distinctive attributes should be recorded on a form similar to that provided in Table 6.7.

To roughly assess ability to detect differences in acidity, prepare aqueous solutions containing 0, 0.5, 1.0, 2.0, and 4.0 g tartaric acid per liter. Alternatively, the acids may be dissolved in a 10% ethanol, 0.5% glucose solution to provide a more wine-like medium. Participants arrange the samples in ascending order of acidity. For accuracy, the test should be conducted on several occasions and the results averaged.

TABLE 6.6   Typical Ranges of Several Acids in Wine

Acid	Concentration (g/L)
Tartaric	2–5
Malic	0–5
Citric	0–0.5
Gluconic	0–2
Acetic	0.5–1
Lactic	1–3
Succinic	0.5–1.5

TABLE 6.7   Assessment Sheet for Wine Acidity

Sample	(Least sour) 1	2	3	4	5	(Most sour) 6	Additional taste or odor sensations
A							
B							
C							
D							
E							
F							

APPENDIX 6.3 PHENOLIC COMPONENTS

Bitterness and Astringency

Phenols and their phenyl derivatives are the principal source of bitter–astringent sensations in wine. As discussed previously in Chapter 4, these fall into two major categories: flavonoid and nonflavonoids. Polymers of these compounds generate a complex group of compounds called tannins. The flavonoid tannins are generally more stable (do not breakdown to their monomers in wine) and generally increase in size during aging. If present in sufficient amounts, they can contribute to the formation of sediment in red wine. Nonflavonoid-based tannins are less stable in wine and may breakdown to their individual monomers. As such, they can contribute to the continuing bitterness of some red wines. Tannins comprising both flavonoid and nonflavonoid constituents are generally stable and do not breakdown to their component monomers.

Because of their diverse origins, composition, changing concentration, and chemical makeup, it is not surprising that tannins (and their subunits) differ in sensory quality. Some generalities tend to apply, though, such as complex tannins tend to be the most astringent; tannin monomers tend to be primarily bitter, showing little astringency; and moderately sized polymers tend to be both bitter and astringent. Anthocyanins are generally neither bitter nor astringent.

The perception of tannins depends both on their absolute and relative concentrations. For example, some tannins appear to be more astringent at higher concentrations, but more bitter at lower concentrations. Aging affects their sensory characteristics by modifying their chemical composition and concentration. The smaller monomeric phenolics are less affected by aging.

To obtain practical experience with some of their sensations, prepare aqueous or 10% ethanolic solutions of grape tannins (a flavonoid tannin complex) and tannic acid (a hydrolyzable tannin complex) at 0.01, 0.1, and 0.5 g per liter. These samples demonstrate not only their taste and touch sensations, but also illustrate their olfactory attributes. To represent monomeric flavonoid and nonflavonoid phenolics, prepare aqueous or 10% ethanolic solutions of quercetin (30 and 100 mg/L) and gallic acid (100 and 500 mg/L), respectively.

Oak

Another informative exercise involves preparing oaked wine samples. Add 10 g/L oak chips (presoaked in 10% ethanol) to either unoaked white or red wine (most inexpensive wines are unoaked). After one week, decant the wine, add 30 mg/L potassium metabisulfite (as an antioxidant), and store in sealed glass bottles for at least 3 months. During storage, acetaldehyde produced following inadvertent exposure to air reacts with sulfur dioxide or combines with other wine constituents to neutralize its mild oxidized (acetaldehyde) character.

This exercise can be expanded to include observing the various effects of different oak species (American versus European) or the effects of the degree of toasting (produced during barrel construction). In general, American oak donates more of a coconut character than do European oaks. Toasting progressively reduces the natural woody fragrance supplied by the oak, successively generating vanilla-like flavors, and then smoky, spicy odors. Toasting also reduces the extraction of tannins from the wood during maturation. Small samples often can be obtained from commercial barrel suppliers.

APPENDIX 6.4 ALCOHOLIC WINE CONSTITUENTS

Ethanol

Many alcohols occur in wine, but only a few are present in sufficient concentrations to affect its characteristics. Although fusel alcohols can induce off-odors (see Appendix 5.2), they are unlikely to influence taste perception. Only ethyl alcohol is generally present in sufficient amount to be of sensory significance. As noted previously, ethanol generates a complex of sensory perceptions; it possesses a distinctive odor, activates the perception of sweetness, and stimulates the sensations of heat and weight in the mouth. Ethanol can also mask or modify other wine sensations.

To illustrate these complex effects, and how concentration influences their perception, prepare solutions of 4, 8, 10, 12, 14, and 18% ethanol. These solutions also can be used to detect individual sensitivity to ethanol by asking participants to arrange the samples in ascending order of concentration (Table 6.8).

TABLE 6.8　Assessment Sheet for Ethanol Sensations (Rank in Ascending Order of Alcohol Concentration)

| Sample | (Least alcoholic) | | | (Most alcoholic) | | | Additional taste or odor sensations |
	1	2	3	4	5	6	
A							
B							
C							
D							
E							
F							

TABLE 6.9　Assessment Sheet for Glycerol Sensations

| Sample | None | | | | Maximal | Taste of odor effect |
	1	2	3	4	5	
A						
B						
C						
D						
E						
F						

Glycerol

Glycerol (a polyol containing three alcohol groups) is the next most common alcohol in wine. Because of its low volatility, glycerol has no detectable odor. Glycerol possesses a sweet taste, but it is so mild that it is likely to affect sweetness only in dry wines, if the concentration surpasses 5 g/L. Because glycerol is viscous, it has commonly been thought to noticeably affect wine viscosity. However, glycerol rarely reaches a concentration that perceptibly affects viscosity (~26 g/L) (Noble and Bursick, 1984). At lower concentrations, it may reduce the perception of astringency (Smith et al., 1996) and contribute to the perception of body (Table 6.9).

To assess the effect of glycerol on the sensory characteristics of wine, prepare solutions containing 0, 2, 4, 8, 12, and 24 g of glycerol per liter in a simulated wine solution (3 g glucose, 4 g tartaric acid, and 100 g ethanol). Participants arrange the samples in ascending order of glycerol content, and remark on those features that permit them to make this arrangement.

Glycerol was once believed to be involved in the production of wine tears. Swirling the ethanol and glycerol samples in this exercise quickly demonstrates that it is ethanol, not glycerol, that generates tears.

APPENDIX 6.5 TASTE INTERACTION

The most frequent taste interactions result in mutual suppression. The reduction in perceived bitterness by the presence of sugar (and vice versa) is well known; less recognized is the influence of personal taste acuity on one's subjective response (appreciation versus disappreciation) to particular sensations. This can easily affect a taster's rating of wines.

The following exercises allow participants to recognize their particular reactions to several taste interactions. In addition, they help potential tasters understand their personal sensory biases. This knowledge may encourage tasters to excuse themselves from certain tastings. Sensory idiosyncrasies can make it difficult for tasters to be impartial judges of certain wines.

TABLE 6.10   Assessment Sheet for Sweet-Bitter Interactions

Sample	Bitterness[a]	When first detected (seconds after tasting)										
		0	1	2	3	4	5	6	7	8	9	10
A												
B												
C												
D												

[a]Scale from 0 to 10.

TABLE 6.11   Assessment Sheet for Sweet-Sour Balance

Sample	Intensity[a] of sweetness	Duration of sweet sensation(s)					Intensity[a] of sourness	Duration of sour sensation(s)				
		2	4	6	8	10		2	4	6	8	10
A												
B												
C												
D												
F												

[a]Scale from 0 to 10.

Sweet–Bitter Interactions

Provide participants with a series of aqueous bitter/astringent solutions containing 1 g/L grape tannins. The samples also contain 0, 20, 40, and 80 g/L sucrose. Tasting the samples in random order—at least 2 min should separate each tasting to allow the mouth to reestablish its baseline sensitivity—have the participants rank the samples both on the speed of bitterness detection and maximum perceived intensity (0–10) on a form as provided in Table 6.10. A crude time–intensity graph may be prepared for each sample to better illustrate the dynamics of sugar's influence on perceived bitterness (see Fig. 4.14). The time–response curve of the loss of bitterness can also be determined after expectoration. Quinine (0.1 g/L) can be substituted for grape tannins if only a bitter sensation is desired (grape tannins induce both bitter and astringent sensations). The data generated from all participants can be combined to illustrate the degree of variation that exists within the group.

This exercise can be reversed to observe the effect of tannic bitterness on the perception of sweetness. For this, the amount of sugar is held constant while adjusting tannin content. An example of such a test could use 40 g sucrose combined with either 0, 0.5, 1, 2, or 4 g grape tannins.

Because ethanol has a sweet attribute, an alternative exercise could involve substitution of alcohol for sugar. For example, dissolve 4 g grape tannins in a series of solutions containing 0, 4, 8, 10, 12, and 14% ethanol. These solutions will clearly demonstrate the effect of alcohol content on both the bitter and astringent characteristics of tannins.

Sweet–Sour Balance

Balance between the various sapid compounds is an important quality attribute in wines. This interaction can be demonstrated in the same manner as sweet–bitter interactions noted previously. However, it may be more interesting if the procedure is changed.

In this exercise, provide participants with at least six aqueous solutions (Table 6.11). One pair contains sucrose (20 and 40 g/L), a second pair contains tartaric acid (0.7 and 1.4 g/L), and the third pair combines different sugar and acid combinations, but at the same ratio (20 g/L sucrose + 0.7 g/L tartaric acid, and 40 g/L sucrose + 1.4 g/L tartaric acid).

Participants taste the samples at random, noting the relative intensity (0–10) of the sweet and/or sour sensations on a form as provided in Table 6.11. The duration(s) of each sensation may also be recorded for comparison. When the exercise is complete, suggest combining the lower sugar concentration with the higher acid solution and vice versa. This will give a further indication of the complexities of sweet–sour balance.

7

Styles and Types of Wine

Organization is a human trait. This is particularly obvious with language, where objects and ideas are codified in words. However, producing a coherent, logical classification of wine based on evolutionary principles is impossible. Wine is an eclectic mix of beverages, having evolved in various places, at different times, under diverse environmental and cultural conditions, produced from many distinctly different cultivars, and in styles that may bear little developmental commonality, or are similar but have developed independently. Modern technical advances and understanding have made great strides in improving wine quality, while a resurgence of interest in older techniques is contributing to new and interesting variations in style. The result is a diversity of fermented grape beverages of multifarious origins and characteristics. The solution has been to group wines based on their most obvious differences, i.e., color, carbon dioxide content, and fortification. Although simplistic, long use has given the system a feeling of familiarity and accreditation. It does provide the consumer with a crude idea of the wine's sensory attributes. For example, white wines typically have milder flavors, are more acidic, and often possess a fruity/floral aspect. In contrast, red wines tend to be more flavorful and astringent, with more jammy aspects. Red wines are rarely sweet or even semisweet, whereas whites may be both. The association between anything sparkling and celebration is longstanding and one of the prime reasons why sparkling wines are grouped together in a special category. Separating fortified wines as a distinct category also has a rationale. They are consumed in smaller quantities, either before or after a meal. These categorizations have also been used by some governments in levying taxes, sparkling and fortified being taxed at a higher rate than table wines.

In spite of issues relating to an evolutionary tree of wine, a crude timeline of technological advancements is possible. As usual, the further back one goes the more imprecise the timeline becomes. That the first wines would have been red is most likely. Wild grapevines are deep blue-purple colored, with white- and rose-colored cultivars being mutated derivatives, having lost their functional anthocyanin genes. It is also highly likely that the first wines, produced some 8000 years ago, were accidental consequences of grapes being left unattended in containers for several days to weeks. Grape autofermentation would have generated up to about 2 percent alcohol, resulting in a weakening of the cell walls, resulting in juice release, and its subsequent fermentation by the epiphytic yeast flora. The result would have been a frothy, yeasty, aromatic wine with a short "shelf life," crudely resembling a Beaujolais nouveau. Eventually, someone would have had the bright idea that it would be easier to crush the grapes (presumably underfoot) to release the juice so that fermentation could commence earlier. Seeing no reason to remove the skins and seeds, the ferment would proceed with any whole grapes that remained. This would have generated a wine resembling a semicarbonic maceration wine, due to an autofermentation of any whole grapes that had not been crushed. This situation continued to varying degrees up into the late 1800s, when the first fully efficient mechanical crushers became available.

When pressing after fermentation began is unknown. Nonetheless, it presumably was grueling, if illustrations on the walls of ancient Egyptian tombs are accurate. Fermented grapes are shown being twisting in large elongated sacks by poles at both ends or other arduous methods (Darby et al., 1977). Other tomb illustrations have been interpreted as suggesting that the wines were cloudy, requiring tubes to suck wine out of amphoras. Long-term storage of wine was presumably out of the question, if only because stoppering amphoras at the time was extremely primitive; stuffed straw and clay are too porous to effectively keep oxygen out. Covering the wine with a layer of oil has been suggested but clear evidence of the practice is missing. In addition, early amphoras required the application of an inner lining of pitch, or possibly wax, to prevent wine from seeping out through the container. Storing wine in a sheepskin might have permitted wine to be transported out into the field for laborers, but does not seem an appealing idea to the modern connoisseur.

Conditions improved significantly during the Roman period, when a technique was developed that could form a vitreous lining on the insides of amphoras. Ancient Greece and Rome were familiar with cork, and cut slabs about 2.5 cm thick to seal amphoras. Thus, the stage was set for the discovery of the advantages of aging, at least wines from the best vintages and sites. Wine almost a century old was extolled for its quality, but was apparently so concentrated that it needed (or at least benefitted from) dilution before drinking. Although a shocking idea today, diluting wine was a common practice in Roman times, a practice still used in many places until recently.

Like so much else in the story of wine, when white wines first began to be produced is just speculation. Nonetheless, their occurrence first required the isolation and selective cultivation of white mutants. Thus, the production of white wines probably has occurred repeatedly, and in different places, over the millennia. The first evidence of white wine production is noted in the inscription on a wine amphora found in Tutankhamun's tomb (1325 B.C.) (Guasch-Jané et al., 2006). The ability to produce white wines from deeply colored grapes (e.g., the original still wine from Champagne) had to await the invention of an efficient means of pressing intact grapes (early 1700s). Without efficient presses, it is likely that most white grapes were fermented similarly to red grapes, i.e., in the presence of the seeds and skins. The practice continued up into the middle of the 20th century in some regions.

Whether early wines were dry, as is the current fashion, is unknown but somewhat dubious. Until *Saccharomyces cerevisiae* became associated with wine production, it is unlikely that the indigenous yeast epiphytic flora would have been able to metabolize grape sugars to dryness. Epiphytic yeasts tend to cease growth and metabolism early, due to the toxicity of the accumulating alcohol content. In contrast, *S. cerevisiae* (and its progenitor *S. paradoxus*) are much less sensitive to ethanol, and can ferment to dryness. Strains of both wild species can produce wine, but their sensory qualities seem less than appealing to modern taste (Fig. 7.1).

By a fortuitous set of circumstances, the natural habitat of *S. paradoxus* (and wild strains of *S. cerevisiae*) is the sap of oak and related genera (Fay and Benavides, 2005; Naumov et al., 1998). Oaks grow sympatrically with wild

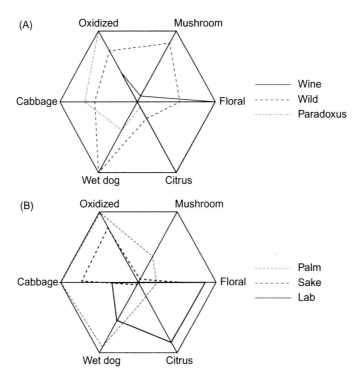

FIGURE 7.1   Illustration of the sensory differences attributable to wine and nonwine strains of *Saccharomyces cerevisiae*. (A) Class means for wine strains, wild strains, and *Saccharomyces paradoxus* strains, and (B) means for the palm, sake, and laboratory strains relative to six wine attributes that distinguish wine strains from nonwine strains. Means were scaled from 0 (center) to 1 (spokes), where 0 represents the lowest mean score, and 1 represents the highest mean score for any class. *Reprinted from Hyma, K. E., Saerens, S. M., Verstrepen, K. J., and Fay, J. C., 2011. Divergence in wine characteristics produced by wild and domesticated strains of Saccharomyces cerevisiae. FEMS Yeast Res. 11, 540–551, by permission of Oxford University Press.*

grapevines in the region where it is suspected that grapevines were first brought into cultivation and the first wines produced. Thus, the proximity of oak trees, supporting wild grapevines and bearing both *Saccharomyces* spp., probably permitted the frequent "contamination" of grapes and fermenting must with both yeast species. Neither are typical grape epiphytes. Early farmers also collected acorns, not only for human consumption but also to feed livestock becoming domesticated around the same time.

The hot climate of many regions thought to be involved in early wine production makes the complete fermentation of grape sugars, even with *S. cerevisiae*, unpredictable. That sweet wines were being produced in ancient times is clear from inscriptions on Egyptian wine amphoras (Lesko, 1977), and written accounts by ancient Greek and Roman authors. A problem for centuries in southern regions (before refrigeration) was the premature termination of fermentation, leaving the wine semisweet and, thereby, especially susceptible to early spoilage. Thus, desired or not, much early wine was probably at least semisweet and of short shelf life. A preference for dry wines may have developed later, notably in Central Europe, where grape sugar contents at harvest were lower, fermentation occurred more slowly, and went to completion under cool autumn conditions.

With the decline of the Roman Empire, amphora use went into decline. North of the Alps, oak barrels took over from amphoras as the storage vessel. Equally, oak tanks became the preferred fermentation vessel (in lieu of clay *dolia*, which are essentially very large, bulbous-shaped, handleless amphoras). Nonetheless, the use of *dolia* as fermentation vessels has continued in some parts of Spain, the Aegean, and Georgia up to the present day.

Without a central government to maintain transport systems, wine largely returned to a beverage produced and consumed locally. Under subsistence-farming conditions, the demand for better-quality wine was largely limited to the small ruling class and monasteries. Even here, without containers to limit oxidation, or effective means of controlling microbial spoilage, the benefits of long-term wine storage largely faded from memory. Nonetheless, improvements in pressing grapes, either before or after fermentation, during Roman times were retained and further improvements in press design occurred periodically during the next millennium.

Near the end of the medieval period, alchemists began to experiment with, and improve, distillation practices. One offshoot was the production of *eau de vie*. When aged in oak barrels, the harsh-tasting distillate became smoother, and brandy was born. The product became popular, notably in the more northerly parts of Europe. For economy of shipping, the Dutch were the first to encourage distillation onsite, i.e., Armagnac and Cognac.

With the beginning of the Renaissance, and the slow beginnings of the Industrial Age, a middle class developed who had disposable income. This had a beneficial effect on all aspects of trade, including wine, and profit for the grape grower/winemaker. For the abundant wines produced in southern regions, there was a serious problem, i.e., they tended to spoil rapidly, especially during transport (the jostling of barrels undoubtedly lead to increased oxygen uptake). The principal solution was to add wine spirits. In southern Spain, this slowly led to the elimination of red wine production, and the transformation of their white wines into what we know as sherry. Other forms of white fortified wines arose in Malaga (Spain) and Marsala (Italy). In Portugal, the principal exported (Duoro) red evolved into porto, whereas on the island of Madeira, madeira evolved. Except for sherry, most of these fortified wines were distinctly sweet. In mainland Italy, a fortified medicinal wine appears to have been the progenitor of vermouth.

Deforestation in England during the late 1500s and early 1600s, associated with the production of charcoal, was approaching crisis conditions for the British navy's ship building. The situation induced James I to restrict the use of wood for charcoal production, used as fuel in the burgeoning glass industry. Conveniently, England was well supplied with a ready substitute, coal. For glass production, coal provided a more steady, hotter, heat source, especially combined with the use of a bellows. The conditions were now set for the production of strong glass, and the adoption of glass bottles as the preferred transport and storage vessel for wine. With a sufficiently strong neck, cork could again be effectively (and safely) used as a bottle closure. The impermeable properties of glass, combined with a cork closure, reintroduced the minimum conditions required for wine aging, and thereby the opportunity for its benefits to be rediscovered. Over a period (c. 1650–1850), the original, bulbous-shaped bottle evolved into the mallet form, and finally into a cylindrical shape, the precursor of modern bottle shapes (Dumbrell, 1983). With a shape that could be easily reclined, the cork stayed in contact with the wine, retaining its moisture content, elasticity, and comparative gas impermeability.

The use of coal also supplied sulfur dioxide that, during glass tempering, reacted with sodium oxide. This generated a layer of sodium sulfate on the insides of the bottle, reducing weak zones in the glass. Combined with thicker glass, bottles of sufficient strength became available that could withstand the internal gas pressures generated during sparkling wine production, a precondition for its effective commercialization. Nonetheless, it was only with developments in chemistry that it was possible to predict and regulate the production of carbon dioxide. Once the role of sugar in the second, in-bottle fermentation was known, the amount of sugar required to achieve the desired carbon dioxide content could be calculated precisely. This led to a dramatic reduction of the heretofore significant losses due to bottle explosion (due to excessive CO_2 production). Another important step was discovering a means of eliminating the yeast sediment produced during the second fermentation (riddling and disgorgement). These labor-intensive processes finally gave way to mechanization during the latter part of the 20th century.

As noted, retention of unfermented sugars markedly increases the tendency of the wine to spoil. However, accidental events resulted in the production of a style of sweet wine that did not suffer this fate (combined with bottling and cork closure). Under cool fall conditions, where foggy evenings alternated with dry, sunny days, infection with *Botrytis cinerea* induced a chemical transformation and partial dehydration of the grapes. When pressed slowly, the concentrated juice fermented into a glorious wine with beatific properties. This discovery has occurred at least three times, first in the Tokaj region in Hungary (c. 1560), about two centuries later in the Rheingau, Germany (c. 1775), and somewhat later in Sauternes, France. The stability of the wine was also assisted by another discovery, i.e., the beneficial effect of burning sulfur-containing wicks in barrels before adding wine. Existing documentation suggests that the practice began in Germany, at least by about 1500 (Anonymous, 1986). Although the microbial cause of most wine spoilage was to remain a mystery for another 350 years, the benefit in maintaining wine in a drinkable state became obvious. Nonetheless, burning sulfur wicks in empty barrels, prior to filling (and eventually adding the active ingredient, sulfur dioxide, to wine) was surprisingly slow, not becoming standard practice until the 20th century.

The use of sulfur dioxide, and the development of refrigeration, have been two of the most significant developments in producing microbially stable wines. Sulfur dioxide not only is the safest and most effective antimicrobial agent known, but it is also a very effective antioxidant, valuable in limiting the premature browning of white wines. Refrigeration has also given the cellar master control over the temperature at which fermentation occurs, thereby influencing the production and survival of fruit-smelling esters. Refrigeration has eliminated one of the most common causes of premature termination of fermentation, i.e., overheating and yeast inactivation. This has been particularly valuable in regions where temperatures during harvest are warm to hot. Not only could the temperature and rate of the ferment now be regulated, but the grapes could be cooled if necessary on reaching the winery.

Some of the current advances, allowing new styles of wine to be developed, have depended on the production of new efficient forms of crushers, presses, and fermentors. Modern crushers are so effective that no grapes are left unbroken to undergo autofermentation. Combined with stemmers, grape stems no longer remain with the juice, leaching rough tannins into the wine. In addition, modern presses no longer need stems to form channels for the juice/wine to escape. The presses also apply less force on the crushed grapes or fermented must, extracting fewer solids that need to be removed to avoid unfavorably influencing fermentation or wine maturation. Furthermore, modern fermentors are far easier to cleanse after fermentation, can effectively regulate the fermentation temperature, can be made of inert material (avoiding flavorant extraction from the cooperage), and can function as storage (maturation) vessels.

Where carbonic maceration is desired, as in the production of *nouveau*-style wines, this can now be precisely regulated in sealed containers. They can be put under carbon dioxide pressure and heated to the desired temperature. Thus, an ancient process that produced readily drinkable red wines shortly after fermentation was perfected. Other older techniques are also seeing renewed and expanded interest. Examples are *sur lies* maturation (notably but not exclusively used for white wines), cold prefermentation maceration (mostly for red wines), cofermentation of several cultivars, and *appassimento* (the equivalent of late harvest, but under controlled conditions after harvest). In its original form, at least in Veneto, Italy, *appassimento* was frequently associated with a slow, but atypical, reactivation of nascent *Botrytis* infection in some of the grapes during the cool storage. It is this infection that gives Amarone its distinctive (and for me, its superlative) character.

Originally, most wines were designated based on the region from which they came. This was fully adequate at the time, as most wines were consumed locally, obviating any need to mention the grape variety or varieties used. Typically, only one or a few cultivars were grown throughout a particular region. Noting the grape varietal name on the label did occur in Germany and a few other German-speaking regions, where many regions grew and produced several varietal wines. The habit of noting the cultivar(s) used in wine production subsequently was adopted in the New World, where individual vineyards often grew a range of cultivars, and simultaneously produced several single and blended cultivar wines. Earlier, cultivar designation was less important as the wines were given European regional names, ostensibly (but too often falsely) implying similar characteristics to the named region. In contrast, most European wines have flavor profiles identified by their regional and varietal origins and see no need for varietal designation. It is only comparatively recently that some European wines are now varietally designated, presumably to aid export sales.

Although designation by varietal name has much in its favor, its use may be limited by the various regional synonyms under which many cultivars are known. For example, Syrah in France is Shiraz in Australia, Cannonau in Sardinia is Garnacha in Spain, and Grenache, Alicante, or Carignane in France, Pinot noir in France is Blauer Burgunder in Germany, and Pinot nero in Italy, and Zinfandel in California, is Primitivo in southern Italy, and Crljenak kastelanski in Croatia. The list could go on almost ad infinitum (Alleweldt, 1989; Robinson et al., 2012). Adding to potential confusion is the application of the same name, or variants thereof, to genetically unrelated cultivars, for example Riesling has been appended to a host of varietal names.

Regardless of varietal or regional designation, these are of little value to consumers without prior knowledge of the characteristic flavor profile of the designated region or cultivar. Because buyers often lack this knowledge, repute or familiarity with the country of origin or producer often gives a degree of confidence (even if it may be unjustified). Color is another indicator considered "safe." Regrettably, color gives only the roughest indication of a wine's sensory attributes. Although many European wines carry geographic appellations that have become familiar, consistency in flavor is often more an illusion than reality. In many regions, the wine can be a blend from several (but delimited) cultivars. In addition, the proportional mix can change from producer to producer, as well as vary from year to year, depending on the vintage conditions and the views of the current winemaker. In addition, with the spread and adoption of techniques from around the globe, regional distinction has blurred even further.

Geographic origin (appellation) is often subliminally assumed to guarantee quality. In reality, appellation is little more than a guarantee of provenance and, depending on the region, the cultivars and procedures used. Even as a geographic indicator, appellation designation can be a major hurdle to many consumers. Typically, the names of smaller regions bear no logical relationship to one another. For example, Pommard and Pauillac do not obviously suggest that the first is from Burgundy and the second from Bordeaux. Learning these appellations is often considered a rite of passage to connoisseurship. However, wine shops simplify matters for the majority of purchasers by physically organizing their wines by country and subregion.

In the New World, geographic designation is simply that. Only occasionally is it associated with a readily distinguishable style, e.g., Marlborough Sauvignon blanc from New Zealand, or icewines from Canada. Most New World regions (and even wineries) produce a variety of wine styles from a diversity of grape cultivars. Although not perfect, varietal origin is usually a better indicator of probable flavor characteristics than regional origin.

The former tendency of naming New World wines after European regions has almost disappeared. This pleases not only producers in the named regions, but also removes a source of confusion. In addition, it no longer promotes recognition of foreign appellations. For example, the prolonged generic use of champagne for sparkling wines encouraged consumers to consider champagne *the* standard, against which all other sparkling wines should be compared. This denigrates other producers and benefits only those in Champagne.

With the increasing importance of New World wines, geographically based wine classifications were losing some of their former appeal. Regrettably, the *terroir* mentality has spread and infected wine producers worldwide. This has led to a flood of research attempting to establish if sensory distinctions exist between regionally produced wines. The results demand much faith, faith that seems largely unjustified. Even if it were possible to consistently detect minor differences between regions, does this have practical relevance (except to marketing)? Why do consumers have to be duped so often?

Even more prevalent is the illusion of *terroir* (estate) distinctiveness. It undoubtedly has marked marketing advantage, implying a uniqueness that is irreproducible elsewhere. Despite this, universalization of cultivar plantings is a worldwide phenomenon (Fig. 7.2). For the thoughtful consumer, the question must often arise, what is the most important quality factor: cultivar, region, vintage, producer, or age?

For those desirous of a diversity of sensory experiences, the ubiquitous spread of a few well-known cultivars is a regrettable development. If wine critics would cease their relentless promotion of a few cultivars and regions, the dispersion of other, equally worthy cultivars and regions might have a chance, to the benefit of tasters eager for not just another variation on a theme, wonderful though well-known cultivars and appellations can be. Concentration on a few cultivars is the equivalent to considering that only a select few composers or artists are worthy of note, to the detriment of equally talented individuals. Conservatism diverts consumers from savoring a much wider palette of flavors and sensations. *Vive la différence!*

Possibly a hopeful sign is the increasing success of blends, with no indication of cultivar origin. They are made to be fruit-forward, avoid aggressive astringency or acidity, and be ready to drink (with enjoyment) once purchased. Aficionados may be willing to wait 10, 20, or more years for a wine to develop its finest attributes, but not most consumers (including me at my age).

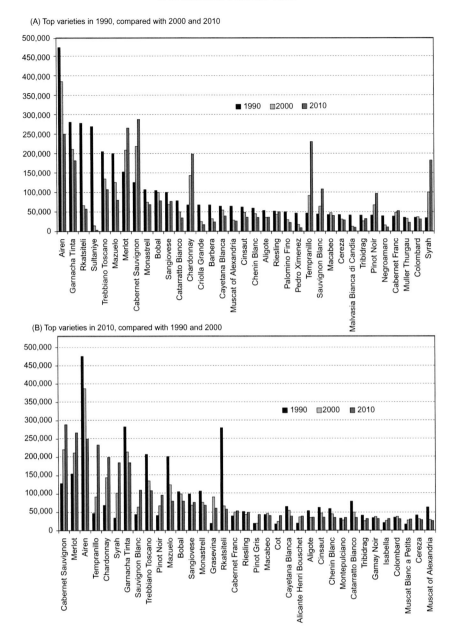

FIGURE 7.2   World's top 35 grape varieties in 1990, 2000, 2010 (ha). *From Anderson, K., 2014. Changing varietal distinctiveness of the world's wine regions: Evidence from a new global database, J. Wine Econom. 9, 249–272 (Anderson, 2014), reproduced by permission Cambridge University Press.*

STILL TABLE WINES

Still table wines constitute the largest category of wines, and correspondingly require the most extensive number of subcategories (Table 7.1). Following tradition, this grouping is initially divided based on color, reflecting major differences in use, flavor, and production technique.

Wines possessing a distinctive varietal aroma generally increase in flavor complexity after bottling, at least initially. Correspondingly, they are highly esteemed and command a higher price. Depending on their geographic origin, the label may or may not indicate varietal origin. Although most New World wines are commonly varietally designated, a new trend is for a return to naming wines without varietal designation. Although particularly common with economically priced wines, the phenomenon is also occurring with higher priced wines (e.g., Opus One, Dominus, Conundrum, Grange Hermitage, Tignanello, San Giorgio, Marzieno). Such designation may reflect the

TABLE 7.1   Categorization of Still Wines Based on Stylistic Differences

A. White[a]

Better when consumed young (seldom matured in oak cooperage)		Potential for aging well (often matured in oak cooperage)	
Varietal aroma atypical	*Varietal aroma typical*	*Varietal aroma atypical*	*Varietal aroma typical*
Trebbiano	*Müller-Thurgau*	Sauternes	*Riesling*
Muscadet	*Kerner*	*Vernaccia* di San Gimignano	*Chardonnay*
Folle blanche	*Pinot grigio*	Vin Santo	*Sauvignon blanc*
Chasselas	*Chenin blanc*		*Parellada*
Aligoté	*Seyval blanc*		*Sémillon*

B. Red[a]

Better when consumed young		Potential for aging well	
Varietal aroma atypical	*Varietal aroma typical*	*Maturation in-tank traditional*	*Maturation in-barrel*
Gamay	*Dolcetto*	*Tempranillo*	*Cabernet Sauvignon*
Grenache	*Grignolino*	*Sangiovese*	*Pinot noir*
Carignan	*Baco noir*	*Nebbiolo*	*Syrah*
Barbera	*Lambrusco*	Garrafeira	*Zinfandel*

C. Rosé[a]

Dry	**Sweet**
Tavel	Mateus
Cabernet rosé	Pink Chablis
White *Zinfandel*	Rosato
Some blush wines	Some blush wines

[a]*Representative examples in italics refer to the names of grape cultivars used in the wine's production.*

producer's desire to indicate their distinctiveness, and not *just* another varietal wine. It also liberates the producer from the legal restrictions associated with varietal or appellation designation.

As noted in Chapter 8, the most significant factor in wine production is the winemaker. How the wine develops depends on their decisions. However, unless one knows the winemaker, and his or her stylistic preferences, this knowledge is of limited value in purchase decisions. Beyond the winemaker, the grape variety (or varieties) set outer limits on the sensory attributes the wine may possess. Thus, the grape variety is to wool, what the winemaker is to the weaver, and the fermentor to the loom. Correspondingly, wine classification should ideally be based on the cultivar(s) used. Unfortunately, many grape cultivars do not, or are not widely known to, possess a readily distinguishable varietal aroma. This may only reflect an unfortunate coincidence, leading to their origin and cultivation being outside France or Germany. The cultivars first taken to, and forming the basis of, New World viticulture and wine production were primarily those grown in France or Germany, imported either by French or German immigrants, or lovers of their wines (the English).

The following section briefly details these cultivars as well as others that could greatly expand the flavor profile of wines, to the enhanced enjoyment of leagues of deeply interested wine lovers. Those described are no more than a personal selection of well- and lesser-known vinous gems, the latter still to be brought to broader awareness. Other cultivars, almost assuredly of equal worth, coming from Portugal and Eastern Europe, are still insufficiently known to the author to be included. Varietal aromas are noted, but should be taken with a more than just a grain of salt.

White Cultivars

Airen was, until recently, the most widely grown cultivar in the world, but its name is largely unknown to connoisseurs, let alone the average consumer. This conundrum relates to its predominant cultivation in Spain, and use either in blends (to lighten their character) or in brandy production. It is well adapted to growing in poor soils, under hot, dry conditions, and is a good producer. Regrettably, it has a relatively neutral character, limiting its value as a varietal wine.

Alvarinho (*Albariño*) appears to have originated in northeastern Portugal or adjacent Spanish Galicia. It is considered to be characterized by fragrances resembling various tree flowers (e.g., basswood and orange) and a diverse assortment of citrus and other fruits. While still primarily cultivated in northern Portugal and Spain, it is also grown in small amounts in the Pacific West of the United States and in New Zealand. Australian *Albariño* is apparently a mislabeled clone of *Traminer*.

Arinto is extensively grown in Bucelas, as well as elsewhere in Portugal. When aged in oak, it can develop a banana-like flavor. It is typically characterized as expressing hints of lemon and almond.

Aurore is a Seibel French-American hybrid. It has been used to provide the northeastern United States with one of its own distinctive wines, being not just another *vinifera* wine. Its fragrance is mild, appealing, and somewhat citrus-like. It is produced in both dry and off-dry styles, and has been used in producing sparkling wines.

Cayuga White is one of the more successful Geneva, New York hybrid cultivars. Although grown predominantly in New York State, it is also cultivated in neighboring states. It produces an excellent dry wine with hints resembling some of the milder flavors of muscat and occasionally Riesling (none of which occur in its ancestry). It has also been used to successfully produce a sparkling wine.

Chardonnay probably produces the most ubiquitous white wine and is seemingly destined to become the most cultivated white grape, being second currently only to *Airen*. Chardonnay not only generates wines with an appealing fruity fragrance, but also tends to do well under diverse climatic conditions (far away from its ancestral Burgundian home). Both characteristics have favored its now global cultivation. In addition to producing fine table wines, it is an important component in the blending of one of the most well-known sparkling wines, champagne. Under optimal conditions, the wine develops aspects reminiscent of various fruits, frequently said to resemble apple, peach, or melon.

Fiano is one of the better ancient southern Italian cultivars, doing well not only in Italy but now in Australia. It is appreciated for its acceptance of hot, dry conditions. Although not considered to have a pronounced varietal aroma, evidence from Australia suggests otherwise. There, it is considered to have lemony and pineapple aspects.

Furmint is the most famous Hungarian variety, due to its use in producing the first definitively known botrytized wine. Its susceptibility to late infection by *Botrytis cinerea*, high acidity, thick skin, and the cool, moist nights and warm, sunny days during autumn in the Tokaj region favor the development of noble rot. The resulting sugar concentration, balanced with ample acidity, produces stably sweet *aszú* wines. These have been famous for some 500 years, with *eszencia* being legendary. Relatively similar botrytized wines are also produced from this cultivar in Rust, Austria, and the adjoining region in Slovakia. Elsewhere, Furmint produces flavorful dry wines described as possessing lime, pear, and smoky notes.

Garganega is the primary cultivar used in the production of Soave wines from Veneto, Italy, occasionally blended with some Trebbiano and Chardonnay. Garganega produces a delicately to distinctly aromatic wine, considered by some to possess hints of almond and citrus.

Inzolia is a locally important cultivar used in the production of newer-style Sicilian wines, such a white Corvo. Inzolia is also an important cultivar used in the production of Marsala. It is considered to be characterized by a mild, fruity aspect, combined with hints of Parmesan cheese.

Müller-Thurgau is possibly the most widely grown of modern *V. vinifera* cultivars, constituting about 30 percent of German hectarage. It is the progeny of a cross between *Riesling* and *Madeleine Royale*. Its mild acidity, subtle fruity–floral fragrance, as well as early maturity make it ideal for producing light wines in cooler climates, those sneered at by critics. Despite this, it is widely cultivated throughout cooler areas of France and northern Italy, as well as New Zealand and Japan.

Muscat blanc is the major member and a parent of many other Muscat cultivars. They are generally grouped together due to their distinctive aroma, described in terms of the epithet, muscaty. Nonetheless, unrelated cultivars may also produce the terpenes that generate a muscaty aroma (e.g., Chardonnay muscaté and Viognier). That their wines have a marked floral character is not surprising, since terpenes are major floral aromatics. Because of the intense fragrance, slight bitterness, and tendency to oxidize, most Muscat cultivars are usually used to produce early-drinking, sweet to semisweet wines. *Muscat of Alexandria* is another major cultivar. *Moscato bianco* is the primary cultivar used in the flourishing sparkling wine industry located around Asti, Italy. Muscat cultivars are grown extensively in almost all southern regions of Europe, as well as many New World countries. The reduced bitterness and lower oxidation susceptibility of *Symphony*, a new Muscat cultivar developed at Davis, CA, permits the production of dryer wines with better aging potential.

Palomino is a variety primarily cultivated in southern Spain and used predominantly in the production of sherry. It develops only a mild varietal aroma. This is one of its advantages, as a stronger varietal character could conflict with the aged aroma so desired in sherries.

Parellada is a variety originating and extensively grown in Aragon, the region just west of Catalonia, Spain. It is one of the three cultivars typically used in producing cava sparkling wines. It produces a dry aromatic wine with apple- to citrus-like fragrances, occasionally showing hints of licorice or cinnamon.

Pinot gris (*grigio*) and *Pinot blanc* (*bianco*) are color mutants of Pinot noir. Both are cultivated throughout the cooler climatic regions of Europe for the production of dry, botrytized, as well as sparkling wines. Both cultivars are also popular in much of the New World. Amazingly, being color mutants of Pinot noir, they bear little aromatic resemblance to their parent. Anthocyanins, which give red cultivars their color, do not possess flavor. Pinot gris tends to yield a richer, more fragrant wine, with aspects of passion fruit, whereas Pinot blanc is more subtly fruity, occasionally suggested to resemble hard cheese. Imagination is wonderful.

Rhoditis (*Roditis*) is a pink cultivar particularly popular in Greece. Despite its color, it is typically used in the production of white wines. Several more or less distinct clones tend to occur in vineyards together. As was once common, vines are thought to often be virally infected. This not only affects coloration but may also contribute to aromatic complexity. Its wines are considered to possess a diversity of aromatic attributes, from apple and melon to citrus.

Riesling is without doubt Germany's most highly regarded grape variety. Other than probably originating in Germany, possibly the Rheingau, its parentage is still under speculation. Riesling produces fresh, aromatic, well-aged wines, which can vary from dry to sweet (notably when botrytized). Its complex floral aroma, commonly reminiscent of roses and pine, has made it acclaimed throughout central Europe and much of the world. Its character is lovely when young, but improves markedly with several years in-bottle. Botrytized versions age exceptionally well, but tend to express more the flavors donated by *Botrytis cinerea* than Riesling. Of any botrytized wines, those based on Riesling are the lowest in alcohol content. Outside Germany, the largest plantings of Riesling appear to be in California and Australia.

Sauvignon blanc is the major white cultivar in the upper Loire Valley, and the most important white variety in Bordeaux (along with *Sémillon*). It has become particularly popular in California and New Zealand. The cultivar seems particular susceptible to environmental influences, showing regional distinctiveness. Unique Marlborough, New Zealand, California, and Loire styles are recognized. The aromatic character of the wines can vary from complexly floral to markedly herbaceous, with elements of green peppers, especially in cooler climates. Several thiols donate much of the typical aroma of this (and some other cultivars), being described as resembling passion fruit and cat urine. Maybe the thiols possess these odors as pure chemicals, but these qualities do not seem to character-ize the cultivar's fragrance (personally). Other characteristic attributes of the cultivar are derived from pyrazines, donating a herbaceous, bell pepper aspect.

Sémillon is one of the two varieties cultivated, along with Sauvignon blanc in Sauternes. Both cultivars are apparently closely related. However, Sémillon has never achieved the popularity of its supposed relative. Sémillon is often the principal cultivar involved in producing one of the most well-known botrytized wines, sauternes. In flavor, Sémillon does not express the herbaceousness of Sauvignon blanc, but is more lemony and lanolin-like.

Torrontés is a name applied to a diverse set of cultivars with little if any relationship. Nonetheless, under the name *Torrontés Riojano*, the cultivar is one of the prominent white grapes producing wines in Argentina. Its parentage appears to be Muscat of Alexandria and Listán Prieto. Correspondingly, the wine possesses a muscat character.

Traminer (*Savagnin blanc*) is a distinctively aromatic cultivar grown throughout the cooler regions of Europe and much of the world. Its origin is still in doubt, but is considered to be ancient. Although white clones exist, most develop a rosy blush in the skin. Nonetheless, it is treated as a white cultivar in wine production. It wine is pro-duced in both dry and sweet styles, depending on regional preferences. Intensely fragrant, slightly rosé-colored (*Gewürztraminer*) clones, generally possess an aroma resembling the fragrance and aromatic chemistry of litchi fruit. Despite the prefix *gewürz-*, implying spiciness, it only indicates the intensity of its fragrance and acidity. Gewürztraminer is likely the most widely cultivated clonal subtype of the variety, occurring in many European regions as well as the New World. Its popularity is somewhat surprising as most critics describe cultivars with pronounced aromas as second grade, if not worse. If it is acceptable, why not Concord?

Traminette is the offspring of a complex American hybrid, derived from a crossing between Gewürztraminer and a Joannes Seyve hybrid. It was produced in Illinois, but selected and propagated in Geneva, New York. As its parentage would suggest, it has the aromatic heritage of Gewürztraminer, but less of its fragrant intensity. It is one of the more popular modern American hybrids, being grown not only on the East Coast of the United States but also in the central and Midwestern states.

Trebbiano is a name applied to several unrelated Italian cultivars, but the most widely cultivated is *Trebbiano Toscano*. It appears to be an offspring of Garganega, another venerable Italian cultivar. By itself, it has a relatively neutral fragrance and acidic character, producing light, quaffable, occasionally surprisingly pleasant (relative to critics' disparagement) white wines. In Cognac and Armagnac, it is extensively grown as the source of their brandies, but under the name *Ugni blanc*. Another synonym under which its wines may be sold is St Emilion.

Viognier is a minor French cultivar that has become popular in the United States and Australia, after languishing in the Rhône Valley since the phylloxera epidemic of the late 1800s. The wine matures quickly and is characterized by the development of a fragrant, muscat-like, peach to apricot aroma.

Viura is the main white variety cultivated in Rioja. Under the synonym *Macabeo*, it is grown in Catalonia. It is related to *Xarel·lo*, one of the three cultivars used in producing cava sparkling wines. In cooler regions, Viura produces a fresh wine possessing a subtle floral aroma with aspects of citron. After extended aging in a large wood cooperage, it develops a golden color and rich butterscotch to banana fragrance that epitomizes the traditional white wines of Rioja. It has a special place in my heart as it was the first wine for which "the heavens opened." (A Marques de Murrieta Ygay blanc, 1973, sampled in our apartment in Ithaca, NY, Wednesday, October 24, 1979, with chicken popovers, 6:32 PM, thereabouts—a veritable supernaculum.)

Red Cultivars

Aglianico is an important cultivar in southern Italy, its presumptive site of origin (due to similarity to other local cultivars). Under ideal conditions, it produces dark red, tannic wines that benefit from considerable aging. It is said to yield wines with plum and chocolate flavors. Because of Aglianico's preexisting adaptation to hot climates, it is beginning to be grown in equivalent regions in Australia and the American Pacific West.

Baco noir is one of the most successful of French-American hybrids, producing excellent wines in North America, notably Ontario and New York. It produces a rich, berry-flavored wine that could become a signature wine to these regions, were its qualities not obscured by minds closed to anything other than *vinifera* cultivars.

Baga is a Portuguese cultivar thought to have originated in the Dão region, where it is still cultivated. However, it is also the major cultivar grown in the neighboring region of Bairrada. Although demanding to cultivate, it produces some of the finest Portuguese reds. For the consumer, though, it often requires patience. When young it is often "closed in," opening up only after a decade in-bottle. The wine possesses fragrances envisioned to resemble plums in youth that evolve into tobacco and leather with age.

Barbera is an Italian cultivar of unknown origin. It is most well known as a source of wine in Piedmont, but is also grown extensively throughout much of Italy. Outside of Italy, it is cultivated primarily in California and Argentina. Its acidity helps the wine to retain a bright red color into age. Some consider it to have an aroma with a likeness to cherries, an attribute ascribed almost indiscriminately to many red wines. Ascribing a cherry attribute seems curious as cherries have one of the mildest, least distinctive fragrances of almost any fruit. Maybe that is why it is so easy to imagine its presence in a wine with a similar deep red color.

Cabernet Sauvignon is the most well-known member of the Carmenet family of grape cultivars. Its renown comes from its involvement in most Bordeaux wines (and equivalents elsewhere). Other members of the family include *Merlot* and *Malbec*. Their inclusion in Bordeaux blends moderates the tannin content donated by Cabernet Sauvignon. The tendency of Merlot to mature more quickly has made it a popular substitute. Under optimal conditions, Cabernet Sauvignon yields a fragrant wine resembling blackcurrants (or violets). Under less-favorable conditions, it may possess a pronounced bell pepper odor. The cultivar appears to be the offspring of an accidental crossing between grapes related to, if not identical with, Cabernet Franc and Sauvignon blanc. As noted in Fig. 7.2, Cabernet Sauvignon has become the most widely planted grape cultivar. Cabernet Sauvignon is a relatively forgiving cultivar, which allows it to be grown with success in a range of climates. Given enough sun and full maturity, its grapes produce a wine possessing its distinctive varietal character, and in the hands of a capable winemaker, it produces superb wines. Nonetheless, such wines may take several to many years to show their mettle.

Chambourcin is one of the most successful of French-American hybrids, possessing a deep red color and luscious flavor, somewhat resembling Shiraz. It is even permitted to be grown in France. Nonetheless, it is most popular in Australia, with limited amounts being grown in the eastern United States and Canada.

Corvina is an old Veronese cultivar, being one of the varieties used in producing Valpolicella. Corvina is also a parent to one of the other cultivars (*Rondinella*) used in producing Valpolicella. In addition, Corvina is crucial in producing the *recioto* version of Valpolicella, Amarone. This relates to it being more susceptible to the reinitiation of nascent *Botrytis cinerea* infections during *appassimento* (Usseglio-Tomasset et al., 1980). While the infection may degrade the grape's varietal (cherry-almond) character, it generates Amarone's distinctive, oxidized phenol odor.

Gamay noir is the primary, white-juiced, Gamay cultivar (other clonal variants possess pink juice). Its reputation shot to fame in association with the temporary popularity of Beaujolais wines, notably the *nouveau* version. Crushed and fermented by standard procedures, Gamay produces a light red wine with few redeeming qualities. However, when processed by carbonic maceration, it yields a distinctively fruity wine. Most of these features come from the grape-berry fermentation phase of carbonic maceration.

Graciano is one of the important cultivars contributing to the perfumed fragrance of Rioja wines, as well as those of neighboring Navarra. Its principal advantages are its fragrance and resistance to drought. It is also popular in Sardinia, but under the synonyms *Bovale Sardo* and *Cagnulari*.

Grignolino is a minor but pleasing cultivar largely restricted to cultivation in parts of Piedmont, Italy. It produces a relatively pale-colored wine, reportedly possessing hints of clove, herbs, and alpine flowers.

Lambrusco refers to a range of distinct but apparently ancient cultivars that were selected in northeastern Italy, and now primarily grown in Emilia-Romagna. Lambrusco is often denigrated by critics due to its predominant use in a frothy, fruity, sweet, red wine. Nonetheless, its dark coloration and pleasing fragrance can be a most enjoyable accompaniment to a meal. Is not wine, in essence, a beverage to be normally savored without fanfare?

Lemberger (*Blaufränkisch*) is an old cultivar that may have arisen in Austria, but is now grown in significant amounts in Germany, Hungary, and adjacent regions. It produces a fragrant, dark red wine with fruity aspects resembling those of carbonic maceration wines. Some versions are sufficiently aromatic to be considered spicy. Limited amounts are grown in British Columbia and neighboring Washington State.

Nebbiolo (*Spanna*) (*Chiavennasca*) is generally acknowledged as producing the most highly regarded red wine in northwestern Italy. With traditional vinification, it produces a wine high in acidity and tannin content. Correspondingly, it demands much patience for the wine to mellow. It appears to be of ancient, but uncertain, lineage, and to have been a parent of several Piedmontese and Valtellina cultivars. Similar to Pinot noir, it is a finicky cultivar, demanding just the right conditions for its wines to express its potential. Possibly because of this feature, it has seen little cultivation outside of northwestern Italy (a feature that has not limited the cultivation of Pinot noir, however). The color has a tendency to oxidize rapidly. Common aroma descriptors include tar, violets, faded roses, and truffles.

Negroamaro is one of the more desirable cultivars in southern Italy, producing rich flavorful wines that age well, but can also be made to mature early. It is not uncommonly blended with one of its progeny, Malvasia Nera. It has only recently been tentatively planted in Australia.

Pinot noir (*Blauer Burgunder*) is the famous and ever-so-punctilious red grape of Burgundy. It is particularly environmentally sensitive, producing its typical fragrance (beets and cherries) only under ideal conditions. Possibly due to the absence of acylated anthocyanins, Pinot noir wines are some of the palest red wines seen in commerce. Were its occasional sensory splendor not already well known, it is unlikely it would be given the special treatment it is given, or grown as extensively. Even then, it seldom lives up to its reputation, regardless of price. Indicative of the cultivar's antiquity, there are many visually and sensorially distinctive clones. The more prostrate, lower-yielding clones produce the more flavorful wines. The upright, higher-yielding clones are more suited to the production of rosé and sparkling wines. The South African cultivar *Pinotage* is a cross between Pinot noir and Cinsaut.

Sangiovese is another ancient cultivar, also low in acylated anthocyanins, and correspondingly not particularly dark colored when produced as a single varietal wine. Typical of ancient cultivars, many distinctive clones are known, many of which are grown throughout central Italy. It appears to a cross between a Tuscan cultivar, Ciliegiolo, and a now rare Calabrian cultivar, Calabrese Di Monenuovo. Sangiovese is most well known for the light- to full-bodied wines from Chianti, but also produces many fine red wines elsewhere in Italy. Under optimal conditions, it yields a wine possessing an aroma considered reminiscent of cherries, violets, and licorice. *Sangiovese* is also labeled under local synonyms, such as *Brunello* and *Prugnolo*, used in producing Brunello di Montalcino and Vino Nobile di Montepulciano wines, respectively. Although primarily grown in Italy, it is also being grown with success in California and other Pacific Western states.

Sousão (*Vinhão*) is one of the more important Portuguese cultivars, being used extensively in the Duoro and Minho, where it is thought to have originated. In Minho, it is the principal cultivar used in the region's red wines. In the Duoro, it is appreciated for its intense coloration and acidity. Its cultivation outside of Portugal is limited, with small amounts used in California, primarily for making port. Limited amounts are also grown in South Africa and Australia. It is supposed to possess a wild fruit–cherry-like fragrance.

Syrah (*Shiraz* in Australia) is a classic northern Rhône cultivar that has achieved fame for its deep red, flavorful wine produced in Australia. With this realization, Syrah has begun to regain the prominence it once had in France. Syrah appears to be the progeny of two minor cultivars in the Rhône-Alps (Mondeuse blanche and Dureza). Its wines are distinguished by pepperiness (derived from rotundone), with aspects supposedly reminiscent of violets, raspberries, and currants.

Tempranillo is probably the finest Spanish cultivar, and certainly the most widely planted red grape in Spain. Under favorable conditions, it yields a delicate, subtle wine that ages well. It is the most important cultivar in producing the red wines of Rioja, either alone or with several others cultivars. It is thought to have originated in the neighboring region of Ribera del Douro. In the Douro, Portugal, it goes under the name of *Tinta Roriz*, whereas in California it usually goes under the name *Valdepeñas*. Outside Spain, it is primarily grown in Argentina. Tempranillo generates an aroma distinguished by a complex berry-jam fragrance, with nuances of citrus and incense.

Touriga National is thought to have originated in Dão, Portugal. It is one of the preferred reds used in producing porto, but increasingly is used as a varietal in producing dry Portuguese red wines. It produces rich, flavorful wines said to possess resemblances to rosemary, roses, and violets. Its tannic content demands patience on the part of the consumer to develop its full potential.

Zinfandel is extensively grown in California, from whence comes its current fame. Its origin is unknown, but is currently thought to be Croatian, due to it being a parent of several Croatian cultivars. This variety is known under the name of *Primitivo* in Italy and *Crljenak kastelanski* in Croatia. Zinfandel is used to produce a wide range of wines, from ports to light blush wines. In rosé versions, it shows a raspberry fragrance, whereas full-bodied red wines possess rich berry flavors.

Production Procedures

Cultivar characteristics set limits on the attributes its wine may possess. However, these properties are often modified by the growing conditions, as well as grape health and maturity at harvest. The cultivar's properties are even more dramatically modified and transformed by the procedures chosen by winemaker during fermentation and maturation.

Because production procedures have often arisen independently, in various locations, and at different times, they seem more logically discussed relative to their application before, during, and after alcoholic fermentation. Many procedures are designed primarily to compensate for deficits in grape or wine attributes, e.g., clarification, enzyme addition, hyperoxidation, acidity adjustments, and chaptalization. Since these inherently do not result in stylistic changes, they are not discussed here. What are noted below are those procedures that affect the wine's stylistic characteristics.

Prior to Fermentation

Before the development of efficient stemmer-crushers in the late 1800s, some grapes remained whole throughout much, if not all of, the fermentation process. Before breaking open and releasing their juice, they underwent grape-berry fermentation. Although the alcohol content in the grapes rose to only about 2 percent, it activated the production of distinctive aromatic compounds. If grape clusters are collected and piled together without crushing, most of the grapes undergo a grape-cell fermentation. Flushing the air surrounding the grapes with carbon dioxide and heating the volume to 30°C or above facilitates rapid grape-cell autofermentation. The process is termed carbonic maceration (Fig. 7.3). A simpler, more ancestral, version is used in much of Beaujolais to produce its wines. Either way, the wine develops a distinctive fruity fragrance, often likened to that of strawberries, raspberries and occasionally pears, and typically unrelated to any varietal aroma the used cultivar(s) may possess. The *nouveau* version is the lightest, producing an easily drinkable wine shortly after completion of both alcoholic and malolactic fermentations. However, the process does come with a distinct handicap. It typically loses its pleasantness within 6 months to a year. Although occasionally used to produce white wines, carbonic maceration is primarily employed in the production of red wines.

In some regions, harvested grapes are stored whole for a period of postmaturation and partial dehydration. The procedure has recently become popularized under the name *appassimento*. Except for the longer, cooler, more stable storage temperature, and avoidance of the risks of inclement weather in the vineyard, it differs little from leaving the grapes on the vine to overmature for several weeks to months. In Europe, postharvest drying has primarily been

Grapes harvested in the heat of the day and whole clusters
dumped into wide shallow fermentor

Top of vat sealed and air flushed out with CO_2 (optional), allowed to grapes
to self-ferment at temperatures reaching 30°C or above

Some berries break under the weight of those above them, the juice released
begins yeast-induced fermentation

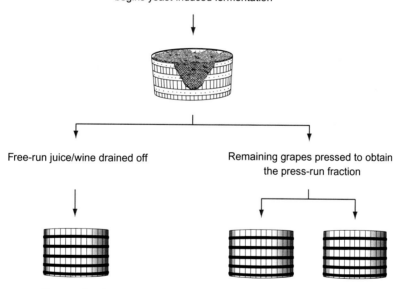

Free-run juice/wine drained off   Remaining grapes pressed to obtain
the press-run fraction

Ferment fractions separately at cooler temperatures of cellar and
blend as desired for style

FIGURE 7.3   Flow diagram of carbonic maceration wine production.

used for the production of certain white wines. In some instances, the grapes are left to partially dry in the sun, markedly increasing their relative sugar content, potentially generating higher alcohol contents in the finished wine. In some locations, this led to the production of naturally fortified wines, e.g., the sherry-like wines from Montilla, Spain. Other classic wines made from partially dehydrated white grapes are *vin santo* (Italy) and *vin de paille* (France). Partial dehydration has also been used in the production of red wines. It was partially used in the production of old-style Chianti wines. The portion of the harvest so treated was crushed and added to wine produced from the majority of the crop, inducing a second fermentation. For reasons that are still unclear, this technique produced a lighter, earlier-drinking wine. The procedure, termed *governo*, produced much of the jug Chianti that for decades was synonymous with easy-drinking wine.

A distinctive wine style, based on partially dried grape clusters, or their "wings" (the smaller side-branch of a cluster) is also used in Veneto and Lombardy (Fig. 7.4, Plate 7.1). The treatment appears to activate several stress-related genes (Zamboni et al., 2008). In addition, nascent *Botrytis* infections may reactivate (Plate 7.2, Usseglio-Tomasset et al., 1980). This initiates chemical changes that resemble those that occur during the noble-rotting of white grapes. These include marked increases in glycerol, gluconic acid, and relative sugar content. Surprisingly, the anthocyanin content is not as oxidized as might be expected. *Botrytis cinerea* has the potential to produce a powerful polyphenol oxidase, laccase. Despite this, the color may be only more brickish than usual for a red wine of equivalent age, and not brown as would occur with Botrytis bunch rot. Nonetheless, the sharp tulip and daffodil

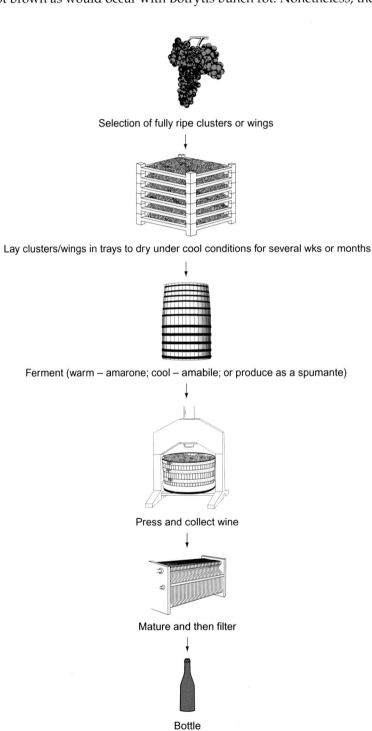

Selection of fully ripe clusters or wings

Lay clusters/wings in trays to dry under cool conditions for several wks or months

Ferment (warm – amarone; cool – amabile; or produce as a spumante)

Press and collect wine

Mature and then filter

Bottle

FIGURE 7.4   Flow diagram of *recioto* wine production.

PLATE 7.1   Storage location for the progressive slow grape drying for the production of *recioto* wines. *Photo courtesy of Masi Agricola S.p.a., Italy.*

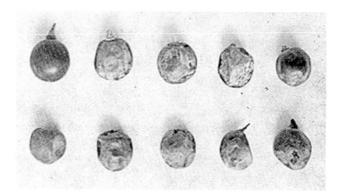

PLATE 7.2   Grapes exposed to progressive drying in the *recioto* process: *upper row*, healthy grapes; *lower row*, grapes infected with *Botrytis cinerea*. *Photo courtesy Dr. Usseglio-Tomasset, Instituto Sperimentale per l'Enologia, Asti, Italy.*

odor that characterizes traditional *recioto* wines, such as Amarone, probably originates from the phenols that are oxidized by laccase. The sharp odor does not apply to most *appassimento*, or even many modern Amarone wines, the latter resembling late-harvested red wines more than Amarones, except for their high alcohol contents.

Currently, it is typical for harvested grapes to be destemmed and crushed shortly after harvest. With some modern harvesters, both procedures can occur immediately after harvesting with white grapes, only the juice being delivered to the winery. More typically, the released juice from white grapes remains in contact with the seeds and skins for a short period at cool temperatures, before pressing and the commencement of fermentation. In contrast, the released juice from red grapes is allowed to ferment in the presence of the seeds and skins for several days to weeks, depending on the desires of the winemaker. Occasionally, white grapes are pressed whole (without prior destemming and crushing), notably in the production of sparkling and botrytized wines. However, there is renewed interest in whole-grape pressing from the producers of dry white wines. Its primary benefit appears to be reducing the extraction of tannins or other undesirable constituents from the skins. Depending on the character of the desired wine and the cultivar, it can also limit the extraction of flavorants.

Before pressing crushed grapes (or the fermenting pomace of red grapes), a fraction is normally allowed to escape under the action of gravity; this is called the free-run. Additional juice (or wine) is subsequently extracted by pressing the remaining wet solids one or more times; these are the press-run fractions. The equipment used affects their chemical composition, and thereby the number of successive pressings deemed acceptable. Thus, the press, how it is used, and the proportion of free- to press-run juice (wine) can be used to significantly affect the style and characteristics of the wine.

Crushed white grapes may be left in contact with the seeds and skins for several hours after crushing. The process, termed maceration, helps extract grape flavorants, most of which are located in the skin. To limit the growth of microbes during this interval, maceration is usually conducted at cool temperatures. The same procedure may also be used in the production of rosé wines, in this case to limit color extraction. Cool maceration is also used with some red wines, notably Pinot noir. The procedure has been correlated with improved wine coloration and flavor development. In this instance, prefermentative maceration can last several days. Why coloration is improved is unclear, but it may be associated with the extraction of other phenolics. Because Pinot noir wines tend to be relatively poorly colored, extracted copigments appear to stabilize the anthocyanins from oxidative color loss.

During Fermentation

Fermentation conditions tend to influence the stylistic characteristics of red wines more than white wines. This undoubtedly results from the variable contact between the seeds and skins (pomace) and the fermenting juice. For standard red wines, this can vary from 2 to 5 days. Shorter periods usually involve more active mixing of the fermenting juice and pomace. Without periodic mixing, most of the pomace would float to the surface, forming a cap. The temperature and alcohol content in the cap can become considerably different from that of the main ferment. Mixing limits this stratification, as well as facilitates the extraction of anthocyanins and flavorants from the skins. In the past, when mixing of the juice and pomace was periodic, and less efficient, it was common for wineries to leave the pomace in contact with the fermenting juice until fermentation came to completion. Depending on the size of the fermentor and the temperature of both the winery and the grapes entering the facility, fermentation might take from a few days to several weeks. When contact with the pomace was prolonged, sufficient tannins were extracted that extensive fining and long aging were required before the wine lost its extreme astringency. Few wineries currently use prolonged skin contact, preferring shorter, more efficient, gentle mixing to extract intense color and rich flavor, without the high tannin contents of old. By the judicious selection of skin-contact time, mixing, and blending (of free- and press-run fractions), the winemaker can significantly alter the style of the wine, from light and fruity to heavy and jammy.

Fermentation temperature is another means by which winemakers adjust wine style. It can be used in the production of both red and white wines, but most commonly white wines. With few exceptions, red wines begin their fermentation at about 20°C, rising into the mid- to high-20s. Cooling is employed if fermentation is too rapid and the temperature approaches that which might kill the yeasts. The higher fermentation temperatures used for red wines favors color extraction. For white wines, winemakers frequently select a temperature favoring the style they wish to produce. For fruitier, lighter wines, temperatures may be held at or below 15°C. This is frequently the case for wines produced from grapes that are fairly neutral in aroma, and intended for early consumption. For white wines meant to be aged for several years, higher temperatures are preferred, but still below 20°C. These wines generally are more full bodied and favor the expression of varietal aromas. Cooling can also be used to induce a premature termination of fermentation, leaving the wine with sufficient residual sugar to generate a sweet finish.

Spontaneous versus induced fermentation, and the choice of yeast strain(s), are other techniques by which stylistic properties can be directed. This is especially the case with neutral-flavored grapes, where most of the wine's aromatic character comes from yeast byproducts. Thus, wines can be given a fruity aspect they would not otherwise possess. Inoculation with a particular yeast strain can also avoid the production of undesirable concentrations of hydrogen sulfide or acetic acid, which can come from the metabolism of indigenous yeasts. Nonetheless, spontaneous fermentations allow a greater diversity of yeast species to influence the wine's flavor profile. Whether this is worth the potential risks depends on the views of the winemaker, and luck.

Malolactic fermentation can also be used to subtly (or occasionally not so subtly) modify wine character. Originally, malolactic fermentation was favored to reduce excess wine acidity and the perceived astringency of red wines. Subsequently, malolactic fermentation has become viewed more as a tool in structuring the wine's sensory character. This is most well known due to its donation of a buttery attribute. Nonetheless, various strains of *Oenococcus oeni*, the principal bacterium inducing malolactic fermentation, can differ markedly in the aromatic compounds they contribute to the wine. Examples are the production of acetic and succinic acids, as well as a variety of higher alcohols, acetaldehyde, and ethyl acetate. Through changing the wine's chemistry, malolactic fermentation can modify perceived bitterness, fruitiness, and overall quality. Because the sensory consequences of malolactic fermentation are often unpredictable, bacterial strains of known characteristics are often chosen to favor the chances of achieving the characteristics desired.

After Fermentation

Wines designed for early consumption are usually aged in stainless steel. This is especially so for white and rosé wines, where maturation in oak cooperage might supply flavors that could mask the wine's delicate fragrance. However, this tendency is largely dictated by habituation and/or tradition. For example, wines from Riesling and Sauvignon blanc are seldom aged in oak, whereas those from Chardonnay and Pinot grigio commonly are. The character transferred from oak varies, depending on how the wood has been seasoned, toasted during coopering, cooperage size, contact time, and reuse frequency. The species of oak used in barrel production, and the conditions during tree growth, can also influence the flavors donated, but are often diminished significantly as a result of blending wine from hundreds of different barrels. Depending on selection and use, oak can provide the winemaker with a veritable palate of subtle flavors from which to choose, varying from hints of coconut, vanilla, caramel to smoke. In addition, oak flavors change as much during in-bottle aging as do the sensory attributes of the wine itself. There is no right or wrong, just personal preference.

Another procedure that may be used to adjust wine flavor is *sur lies* maturation. This old procedure is currently receiving much attention, in the hopes of adding a *je ne sais quoi* attribute to a producer's wines. It typically involves delaying racking (removing wine from the yeast lees that settle during maturation). Small oak barrels (~225 L) are the fermentation/storage cooperage of preference. To limit the production of reduced-sulfur off-odors forming in the lees, the wine (and lees) are periodically stirred to permit the absorption of trace amounts of oxygen. *Sur lies* maturation is viewed as a means of enhancing flavor complexity.

White Wine Styles

White wines, in some ways, come in a wider diversity of styles than reds. The major production options leading to these styles are outlined in Fig. 7.5.

The most common styles are dry, with a clean refreshing taste and fruity bouquet. This is derogatively termed by some critics the "international style." Cool fermentation temperatures favor their generation, by enhancing both the production and retention of fruit esters (ethyl esters of low-molecular-weight fatty acids). These fermentation byproducts are synthesized in amounts greater then their equilibrium constants permit. As they slowly hydrolyze back to their alcohol and acid moieties, a wine whose fragrance is based on fruit esters loses much of its initial aromatic character. Storage under cool conditions markedly slows the hydrolytic process, helping to retain the fragrance donated by esters. Because fruit esters are produced primarily by yeast metabolism, grape variety has limited effect on their production. Consequently, the yeast strain chosen to induce fermentation is particularly important when producing white wine from cultivars lacking a marked or distinctive varietal aroma.

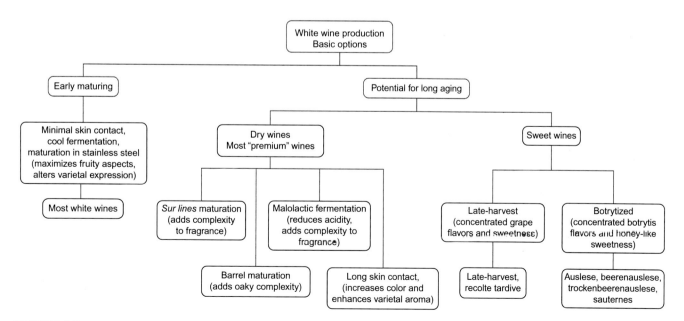

FIGURE 7.5   Classification of white wines based on basic production options.

Only a comparatively few white grape varieties are noted for their propensity to age well, notably Riesling, Chardonnay, and Sauvignon blanc. As their varietal aroma fades, it is often replaced by a pleasing aged bouquet. The chemical nature of this transformation still remains essentially a mystery.

Most white wines are dry, as befits their primary use as a food beverage. Their fresh crisp acidity achieves balance in combination with food, enhancing food flavors and reducing the fishy character of some seafoods. The lower flavor intensity of most white wines also suits their combination with relatively mild-flavored foods. Premium-quality white sweet wines also have bracing acidity. It provides the balance the wines otherwise would lack. The lighter of these semisweet wines are often taken cold, and by themselves as "sipping" wines. In contrast, the sweeter versions typically replace dessert. Consequently, the expression "dessert wine" refers more to their substitution for, rather than compatibility with dessert.

Although the presence of a varietal aroma is important to most premium white wines, a few are characterized by the loss of their varietal aroma. A classic example involves noble-rotted grapes (Fig. 7.6, Plate 7.3). Under a unique set of autumn conditions, infection by *Botrytis cinerea* leads to juice concentration and degradation of most varietal aromatics. This loss is replaced with a distinctive botrytized fragrance, characterized by a rich, luscious, apricoty, honey-like aroma. The principal examples of botrytized wines are the *ausleses, beerenausleses*, and *trockenbeerenausleses* of Germany and Austria, the tokaji aszus of Hungary, and the sauternes of France. In the New World, they may be variously called botrytized or selected-later-harvest wines. Additional styles that age well, without the benefit of a marked varietal character, are *vin santo* and icewines (*eisweins*). Fig. 7.7 illustrates characteristics that can distinguish different icewines, whereas Plates 7.4–7.6 illustrate some of the issues associated with their production. Fig. 7.8 outlines the procedures leading to the major sweet white wine styles.

As noted previously, recognizing the varietal character of a wine is often difficult. Factors such as vintage, fermentation, and maturation conditions can reduce or modify a wine's varietal character. The most readily identifiable varieties are those depending primarily on monoterpene alcohols (e.g., Muscats, Viognier, Torrontés, Gewürztraminer) or several labrusca cultivars (e.g., Niagara, Glenora, Concord). Table 7.2 provides a compilation of descriptors often mentioned in popular literature as characterizing particular white grape cultivars.

Red Wine Styles

Most red wines are dry. The absence of detectable sweetness is consistent with their use as a food beverage (as generally viewed from a European perspective). The bitter and astringent compounds characteristic of most red wines bind with food proteins, producing a balance that otherwise would not develop for years. In contrast, the attributes of well-aged red wines are more easily detected, and therefore appreciated, assessed on their own. Their diminished tannin content does not require food to develop smoothness, and their subtly complex bouquet is not compromised by extraneous food flavors.

Most red wines that age well are matured in oak (Fig. 7.9). Storage in small oak cooperage (~225 L) usually speeds wine maturation, while supplying complementary oaky, vanilla, caramel, or spicy/smoky flavors. Following maturation, the wines generally receive additional in-bottle aging, either at the winery or by the consumer. Where less oak character is desired, used barrels or large (1000- to 10,000-L) oak tanks are used. Barrel aging appears to accentuate the varietal character of several major red cultivars, such as Cabernet Sauvignon, whereas the traditional longer maturation in large oak cooperage is often viewed as mellowing varietal distinctiveness. Alternatively, the wine may be matured in inert tanks to avoid oxidation or the uptake of accessory flavors.

The procedures used in red wine production often depend on the consumer group for whom the wine is intended. Wines expected to be consumed early tend to be made accentuating fresh fruit flavors, whereas those designed for extended aging aim for more jammy flavors. These develop as their earlier harsh tannic and bitter attributes mellow. Beaujolais nouveau is a perfect example of a wine destined for early consumption. Its production by carbonic maceration, and the inclusion of little press-wine, gives it a distinctive fresh, fruity flavor. However, as noted, early drinkability comes at the cost of short shelf-life. Nouveau wines seldom retain their typical attributes for more than 12 months, often fading noticeably within 6 months. In contrast, premium Cabernet Sauvignon and Nebbiolo wines illustrate the other extreme. They may require one to several decades before they develop a smooth mouth-feel and refined bouquet. Some of the basic options in red wine production are illustrated in Fig. 7.10.

The reasons for these differences in aging potential are poorly understood. Features that favor fruit ripening, such as adequate temperature and sun exposure, moisture and nutrient conditions, and optimal fruit/leaf ratio are clearly important. Vinification at moderate temperatures, in the presence of the seeds and skins, followed by skillful maturation is also essential. Nevertheless, these techniques alone cannot explain why most red cultivars do not produce wines that age well, even following best-practice procedures. Part of the answer undoubtedly relates to the

Select infected cluster or portions, infected berries, or dried infected berries (depending on style)

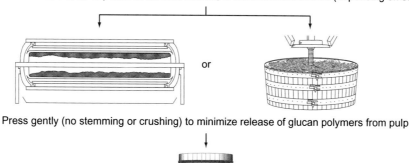

or

Press gently (no stemming or crushing) to minimize release of glucan polymers from pulp

Add nutrients and ferment at 20–22°C or 10–15°C
(Depending on style)

Mature

Filter

Bottle

FIGURE 7.6   Flow diagram of botrytized wine production.

amounts, types, and ratios of particular anthocyanins, tannins, and copigments. Red cultivars vary markedly in their anthocyanin and tannic compositions (Van Buren et al., 1970; and Bourzeix et al., 1983). Retention of sufficient acidity and the judicious uptake of oxygen during maturation favor color retention. Distinctive aromatic constituents, such as methoxypyrazines, eventually oxidize, isomerize, hydrolyze, or polymerize to less-aromatic compounds, while (ideally) new aromatic compounds slowly form. These may be the origin of the cigar-box, leather, mushroomy scents that distinguish the aged bouquets of premium red wines.

Although most red wines owe much of their character to the cultivar(s) used in their production, some owe most of the character to the production techniques used, e.g., carbonic maceration. Although it has the advantages of

PLATE 7.3   Cluster of botrytized grapes showing berries in different states of noble rot. *Photo courtesy of D. Lorenz, Staatliche Lehr-und Forschungsanstalt für Landwirtschaft, Weinbau und Gartenbau, Neustadt, Germany.*

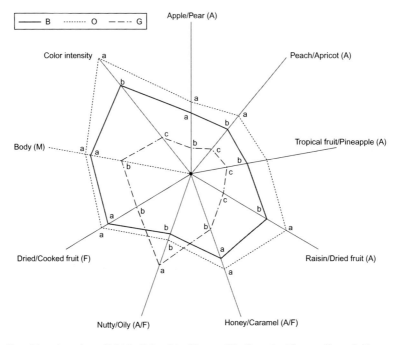

FIGURE 7.7   Sensory profile of icewines from British Columbia (B, n = 13), Ontario (O, n = 9), and Germany (G, n = 3) for appearance (Color Intensity), aroma (A), flavor (F), and mouth-feel (M) attributes. For each attribute, means that are significantly different are identified with different lowercase letters. *From Cliff et al., 2002, reproduced with permission, 12002 American Society for Enology and Viticulture. AJEV 53:46–53.*

PLATE 7.4   Grapes protected with netting prior to harvesting for icewine. *Photo courtesy of CCOVI, Brock University, St Catharines, Ontario, Canada.*

PLATE 7.5   Harvesting grapes for the production of icewine. *Photo courtesy of E. Brian Grant, CCOVI, Brock University, St Catharines, Ontario, Canada.*

rapid capital return, its disadvantages include extensive demands on fermentor volume, special fermentation conditions, manual grape harvesting, and the wine's relatively low price. *Recioto* wine production is another wine style largely dependent on production technique (Fig. 7.4). It concentrates as well as modifies the wine's flavor, ideally generating a distinctive, oxidized, phenolic odor.

The common, popular emphasis placed on a wine's aging potential generates questions regarding when wines are likely to reach their sensory peak. This, however, reflects a misconception that wines are best at some particular stage in maturation. In reality, this relates more to personal choice, as the wine progresses through a spectrum of aromatic changes, many of which can be equally, but distinctly pleasurable. The aging question also relates to how the wine is to be consumed (with a meal or by itself), and whether the consumer prefers the fresh fruity aroma of a young wine or the richer, more subtle, complex attributes of mature wines. It is more appropriate to refer to a wine's plateau than to a peak (Fig. 7.11). One of the major features that distinguish superior wines is the duration

PLATE 7.6 Dumping frozen grapes into a basket press in preparation for juice extraction. *Photo courtesy of E. Brian Grant, CCOVI, Brock University, St Catharines, Ontario, Canada.*

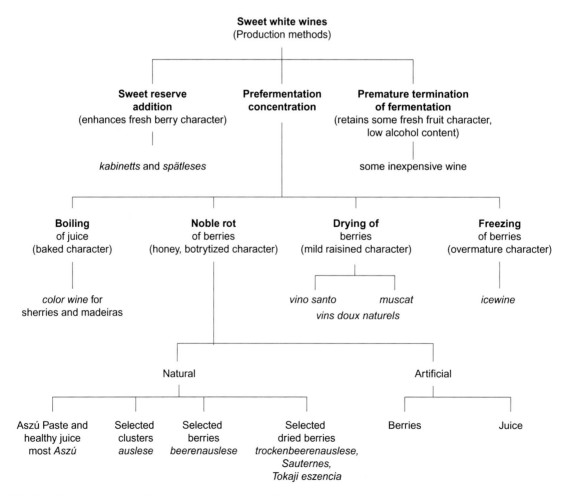

FIGURE 7.8 Classification of sweet white wines based on production method.

TABLE 7.2 Aroma Descriptive Terms for some Varietal White Wines[a]

Grape variety	Country of origin	Aroma descriptors
Chardonnay	France	Apple, melon, peach, almond
Chenin blanc	France	Camellia, guava, waxy
Garganega	Spain	Fruity, almond
Gewürztraminer	Italy	Litchi, citronella, spicy
Muscat	Greece	Muscaty
Parellada	Spain	Citrus, green apple, licorice
Pinot gris	France	Fruity, Romano cheese
Riesling	Germany	Rose, pine, fruity
Rousanne	France	Peach
Sauvignon blanc	France	Bell pepper, floral, herbaceous
Sémillon	France	Fig, melon
Torbato	Italy	Green apple
Viognier	France	Peach, apricot
Viura	Spain	Vanilla, butterscotch, banana

[a]*Note that the varietal aroma frequently has but a faint resemblance to the fragrance of the descriptor. Descriptors often act as anchor terms representing the memory for the varietal aroma. In addition, depending on a range of factors, varietal attributes may or may not be expressed. Finally, the descriptors noted may reflect more the culture and personal experiences of the originators than the wine.*

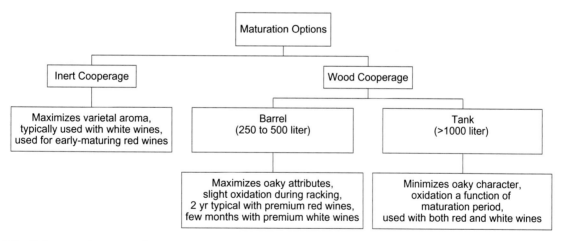

FIGURE 7.9 Oak maturation options.

of the plateau. Wines with little aging potential have comparatively short plateaus, whereas premium wines should have plateaus that can span decades.

Because most grape varieties are still primarily cultivated close to where they evolved, limited information is available on their winemaking potential elsewhere. New World experience has been largely restricted to a few cultivars from France and Germany, reflecting the biases of those who started vineyards in former colonial regions. Consequently, the qualities of the extensive number of cultivars that characterize Italy, Spain, and Portugal, let alone those from Eastern Europe, remain largely unrealized. How many varietal diamonds await discovery only time can tell. This does not negate the qualities of the cultivars that provide the majority of the world's varietally designated wines. It does signal, though, the limited aroma base available for producing most wines. Sadly, the self-perpetuating cycle of consumer, critic, and producer conservatism limits the more extensive investigation of other varieties. Nonetheless, more adventuresome winemakers are investigating not only largely forgotten indigenous cultivars,

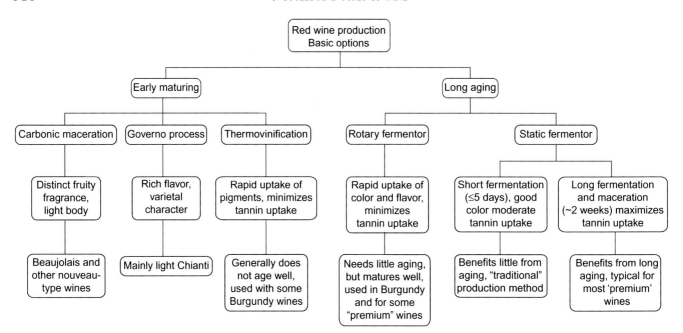

FIGURE 7.10   Classification of red wines based on basic production options.

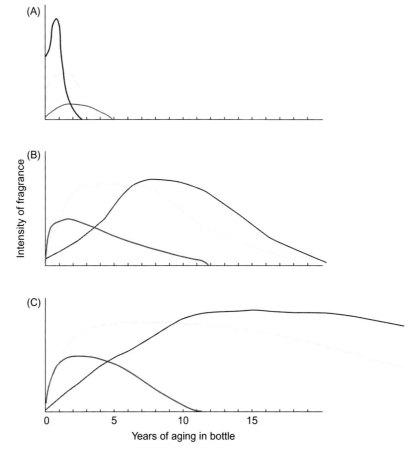

FIGURE 7.11   Diagrammatic representation of the changes in aromatic character associated with aging in (A) nouveau-style wines, (B) standard quality red wines, and (C) premium quality red wines (*darkest line*, fermentation bouquet; *lightest line*, aroma; *medium line*, aged bouquet).

TABLE 7.3 Aroma Descriptive Terms for some Varietal Red Wines[a]

Grape variety	Country of origin	Aroma descriptors
Aleatico	Italy	Cherries, violets, spice
Barbera	Italy	Berry jam
Cabernet Franc	France	Bell pepper
Cabernet Sauvignon	France	Blackcurrant, bell pepper
Corvina	Italy	Berry (tulip/daffodil in Amarone)
Dolcetto	Italy	Quince, almond
Gamay noir	France	Kirsch, raspberry (after carbonic maceration)
Grignolino	Italy	Clove
Merlot	France	Blackcurrant
Nebbiolo	Italy	Violet, rose, truffle, tar
Nerello Mascalea	Italy	Violet
Pinot noir	France	Cherry, raspberry, beet, mint
Sangiovese	Italy	Cherry, violet, licorice
Syrah/Shiraz	France	Currant, violet, berry jam, pepper
Tempranillo	Spain	Citrus, incense, berry jam, truffle
Touriga Nacional	Portugal	Cherry, mint
Zinfandel	Italy	Raspberry, berry jam, pepper

[a]Note that the varietal aroma frequently has but a faint resemblance to the fragrance of the descriptor. Descriptors often act as anchor terms verbally representing the memory for the varietal aroma. In addition, depending on a range of factors varietal attributes may or may not be expressed. Finally, the descriptors noted may reflect more the culture and personal experiences of the originators than the wine.

but also newly developed varieties. Thus, there is increasing hope that consumers may soon not have to depend primarily on subtle differences alone, but sample new and pleasurable vinous delights.

Because aroma expression is often indistinct, clear differences in varietal expression depends on optimal grape-growing, winemaking, and storage conditions. Under these conditions, the aroma characteristics of several red cultivars may approximate those noted in Table 7.3.

Rosé Wine Styles

Except for sparkling rosés, which are made by blending a small amount of red wine into a base white wine, most rosés are derived from red grapes given short maceration. This limits anthocyanin uptake, generating only a slight pinkish coloration. Frequently, this involves gentle crushing, followed by about 12–24h maceration under cool conditions (to delay the onset of fermentation). After the juice has run off, it is fermented as if it were a white wine, typically at cool temperatures. Alternatively, the grapes may be pressed whole to limit color extraction. Where fruit coloration is low, the grapes may be crushed and fermented with the skins until sufficient pigment has been extracted. Alternatively, pectinase enzymes may be added to the crush to enhance color extraction, as well as encourage flavor liberation from the skins. Subsequently, the juice is fermented without further skin contact. Occasionally, rosé production can be a byproduct of red wine production. This typically occurs in poorer years, when there is insufficient color to generate a typical red wine. The technique, called saignée, involves drawing off a portion of the fermenting juice, which is used to produce a rosé. The remaining juice continues its fermentation with the seed and skins, achieving a more intense color than would have been possible otherwise.

Although short maceration limits anthocyanin uptake (achieving the desired color), it even more markedly limits tannin extraction. Thus, rosés tend to show poor color stability. Most of the color is based on unstable, free antho-cyanins, not the more stable tannin–anthocyanin polymers typical of red wines. Despite the relatively low phenolic extraction, they still act as important antioxidants (Murat et al., 2003). For example, they limit the oxidation of 3-mercaptohexan-1-ol (and the rapid loss of a constituent that donates significantly to the fruity fragrance of many rosé wines). Phenethyl acetate and isoamyl acetate are additional typical contributors to the flavor of rosé wines

(Murat, 2005). Other compounds that have been associated with the fragrance of rosé wines are β-damascenone, 3-methylbutyl acetate, and ethyl hexanoate (Wang et al., 2016). These constituents supply characteristics that can range from spicy, savory, green, citrus, tropical fruit to floral.

As a class of wines, rosés have long suffered under the stigma of lacking aging potential. Correspondingly, they have never been taken seriously by wine cognoscenti. Furthermore, rosés have often been sneered at for being considered to possess the negatives attributes of a red wine, but none of the positives of a white wine. Although this may be occasionally true, they escape the strictures and formality of reds, but are more fun than whites. In the past, to mask the wine's mild phenolic bitterness, many rosés were processed semisweet, and mildly carbonated to increase their appeal as a cool refreshing beverage. Both features augmented their denigration by critics. To counter the unfavorable image frequently associated with the name *rosé*, some ostensibly rosé wines were redesignated "blush" wines. Despite these negative images, there has been a recent and marked increase in the popularity of dry rosé wines. Occasionally, consumers wisely ignore the advice of critics, or do not know they are supposedly uncouth in their choice.

Most red grape varieties can be used in producing rosé wines. However, except for *saignée* versions, it is not economically sound to use premium cultivars to produce rosés. Of all cultivars, Grenache is probably the favorite. In California, Zinfandel is also popular for producing rosé (blush) wines. Some Pinot noir wines are so pale as to be *de facto* rosés, notably many Blau burgunder wines from Germany.

SPARKLING WINES

Sparkling wines derive much of their distinctive character from their high carbon dioxide content (~600 kPa, or six times atmospheric pressure). The prickling, tactile sensation of bubbles bursting on the tongue is central to the style. The cold serving temperature further accentuates the tingling pain sensation generated by the dissolved gas turning into carbonic acid. Flat sparkling wine (like decarbonated soft drinks or soda water) loses most of its appeal. For most sparkling wines, base wines are chosen to be nearly colorless, relatively low in alcohol content (about 9 percent, to avoid the final alcohol content exceeding 12 percent), minimal in varietal aroma (favoring expression of a *sur lies*–derived, toasty aspect), and relatively acidic (to heighten the flavor, as lemon juice for fish).

Typically, sparkling wines are classified by production technique (Fig. 7.12) and flavor characteristics (Table 7.4). Most derive their effervescence from retaining carbon dioxide generated during a second, yeast fermentation. Although definitive, this classification system seldom denotes clear sensory differences. For example, both standard (*champenoise*) and transfer methods have typically been used to produce dry to semidry versions. Differences arise primarily from the duration of maturation in contact with the lees following the second fermentation, being shorter for the transfer method. Yeast contact releases compounds that are crucial to the development of the distinctive fragrance, described as toasty. Prolonged contact (typically 1 year, but occasionally 3 or more years) also favors the release of mannoproteins during yeast autolysis. These yeast cell-wall degradation products not only entrap flavorants (releasing them when the wine is in the glass), but also favor the production of a fine, abundant, enduring effervescence. This involves the production of multiple chains of minute bubbles, as well as the *mousse* that forms in the center of the glass and around the meniscus (Plate 1.1). Additional differences in character arise from the cultivars and conditions used in preparing the base wine. This is particularly marked in the production of some Charmat produced wines, for e.g., Asti Spumante. The use of Muscato d'Asti donates the characteristic, marked, muscat aroma of Asti Spumante, while premature termination of fermentation supplies the sweet finish. The Charmat process is also extensively used in the production of many other Italian wines, many of which may be produced in dry, sweet (*amabile*), or slightly sparkling (*frizzante*) versions. The most well-known Italian sparkling wines, other than Asti Spumante, are Prosecco (white) and Lambrusco (red). Occasionally, effervescent wines are produced by adding grape juice (rather than a sugar solution as usual) to a base wine to activate the second, CO_2-generating fermentation. With red versions, the second fermentation is often incomplete, producing a pétillant wine with mild sweetness. Many sparkling wines in Germany (*Sekt*) are (or were) produced using the Charmat method.

The predominant use of the traditional (*champenoise*) method largely reflects consumer image. All three procedures noted on the right side of Fig. 7.12 employ a second, yeast fermentation to produce the entrapped carbon dioxide. Assuming similar-quality base wines, the resultant wines should be relatively similar. The traditional and transfer processes are identical up to disgorging (Fig. 7.12). In the traditional process, the wine remains in the bottle in which the second fermentation occurs. In the transfer process, the wine is discharged into a pressurized tank, just prior to filtering (to remove the lees), and transferred to new bottles. Other than avoiding the expense of

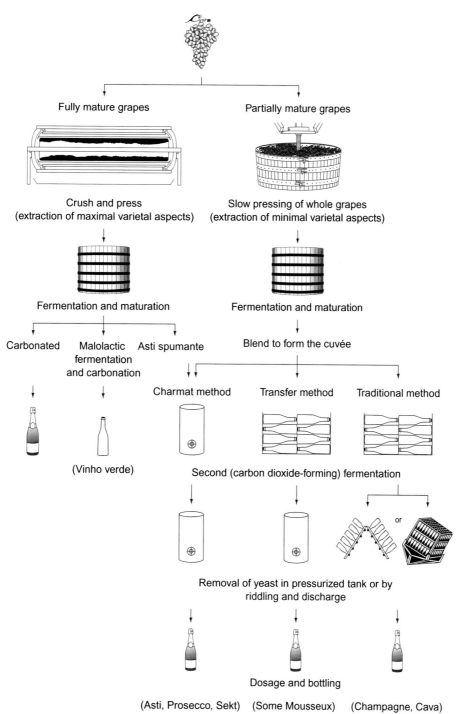

FIGURE 7.12   Flow diagram of the three principal methods of sparkling wine production.

riddling, minimizing bottle-to-bottle variation, and possibly foreshortened lees contact, there are no published data indicating any sensory differences accruing from the choice of procedure. The basic disadvantage of the transfer process is the expense involved in purchasing the pressurized tanks. Automation of the formerly manual riddling process, required in the traditional method, has largely eliminated the financial savings that once favored the transfer method. Riddling transfers the lees to the bottle neck, prior to ejection during disgorging. The Charmat process also uses specially designed tanks, one used for the second fermentation and the other to receive the filtered wine, before it is bottled. Its expense involves both pressurized tanks and the system needed to avoid development of

TABLE 7.4 Categorization of Effervescent Wines

With added flavors	Without added flavors		
Fruit flavored	High aromatic	Subtle flavor	
Carbonated	Muscat based, sweet	Crackling/carbonated	Traditional style, dry/sweet
Coolers	Asti	Perlwein	Champagne
		Lambrusco	Cava
		Vinho Verde	Spumante
			Sekt

TABLE 7.5 Desired Attributes for Sparkling Wines of the Champagne-type

Appearance

Brilliantly clear.
Pale straw-yellow to bright gold.
Slow prolonged release of carbon dioxide that produces many, long, continuous chains of minute bubbles.
Bubbles mound on the surface in the center of the glass (*mousse*) and collect around the edges of the glass (*cordon de mousse*).

Fragrance

Presence of a complex subtle bouquet that shows a hint of toastiness.
Subdued varietal character (to avoid masking the subtle bouquet).
Possess a long finish.
Absence of atypical or grape-like aspects.

Taste

Bubbles that explode in the mouth, producing a tingling, prickling sensation on the tongue.
Possess a zestful acidity without tartness.
Presence of a clean, lingering, aftertaste.
Well balanced.
Absence of a noticeable astringency or bitterness.

a highly reduced environment (and off-odor production) in the accumulating yeast sediment in the fermentation vessel. Properties considered appropriate for dry sparkling wines are outlined in Table 7.5.

Carbonation (incorporating gaseous carbon dioxide under pressure) is the least expensive method of producing sparkling wines. Its initial and still primary application has been in the soft-drink industry. Its subsequent use in the production of flavored pétillant wines has given the process a negative connotation. Pétillant wines, being of lower carbon dioxide content, are often taxed at a lower rate, and consequently possess a price edge over sparkling wines. Although carbonation obviates development of a toasty aspect, it can accentuate a varietal aroma. Absence of yeast contact also minimizes the presence of mannoproteins. This tends to limit the production of the fine effervescence so desired in sparkling wines. Although this aspect is vital to aficionados, most consumers probably would not notice the difference; attention is usually directed toward celebration, not the attributes of the wine. In addition, commercially available mannoproteins could be added to permit the formation of finer bubbles (Pérez-Magariño et al., 2015).

Sparkling wines are predominantly white, due to difficulties posed by the presence of phenolics in red wine. Phenolics can suppress yeast action, especially under the cool temperatures and the progressive buildup of pressure during the second fermentation. Not only can phenolics diminish CO_2 production, but they favor gushing upon cork removal. To partially circumvent these problems, most rosé sparkling wines are produced by blending a small amount of red wine in the assemblage prior to the second fermentation. Sparkling red wines are usually carbonated, to circumvent these issues. They also tend to possess reduced carbon dioxide contents (to reduce the likelihood of gushing).

Blending is used extensively in the production of sparkling wines. Combining wines from many vineyards, as well as several vintages and cultivars, tends to reduce sensory deficiencies in any of the base wines and accentuate their separate qualities. Only in superior years do all base wines come from a single vintage. Knowledge of the benefits of blending has a venerable and ancient lineage, having been described by Theophrastus (371–287 B.C.), and more recently confirmed by Singleton and Ough (1962). Advocating blending is also the only contribution to champagne production that can, with comparative certainty, be ascribed to Dom Pérignon. Examples of cultivars used are Parellada, Xarel-lo, and Macabeo in the production of cava (Catalonia); and Chardonnay, Pinot noir, and Pinot Meunier (Champagne). In other regions, single varieties may be used, e.g., Riesling (Germany), Chenin blanc (Loire), Muscat (Asti) and Glera (Prosecco).

Carbonated wines show a wide diversity of styles. These include dry white wines, such as vinho verde; most crackling rosés; and fruit-flavored coolers. In the case of vinho verde, the wine originally obtained its slight sparkle from a late-onset malolactic fermentation in barrel-stored wine. Carbon dioxide, generated by lactic acid bacterial metabolism during the spring, was trapped in the barrel. As the wine was locally served directly out of the barrels, the wine retained its fizz. However, when the wine began to be bottled for export, filtering (to produce a crystal clear wine) removed most of the carbon dioxide. To reestablish the sparkle, the wine was carbonated before bottling. Because the pressure is relatively low, there is no need to use the strengthened bottles required for most sparkling wines.

Regardless of the carbon dioxide source, the nature and content of mannoproteins appear to largely influence bubble characteristics. When the bottle is opened, the diminished pressure above the wine destabilizes the dissolved carbon dioxide. Without agitation, there is insufficient free energy to activate a rapid gas escape. Instead, the dissolved carbon dioxide would slowly dissipate by diffusion over several hours. However, when the wine is poured, the free energy generated during the pouring permits the well-known, rapid, but short-lived, frothing in the glass. Once this ends, formation of the characteristic chains of bubbles depends on the presence of minute particles on the glass surface, or those suspended in the wine (termed floaters). Lint particles are the most well-studied of these nucleation sites (Liger-Belair et al., 2002). When wine is poured, microscopic air bubbles are trapped within crevices in these particles. Carbon dioxide diffuses easily (requiring little energy) into these nucleation sites. As the volume of gas increases, bubbles bud off and initiate their ascent, generating the effervescent chains so typical of sparkling wine. As they rise, continuing carbon dioxide diffusion enlarges each bubble, and results in the distance between each bubble increasing. On reaching the surface, most bubbles burst, ejecting microdroplets of wine into the air (Plate 1.2). This facilitates the release of some aromatics into the air (Fig. 7.13), but can also suppress the perception of others (presumably by triggering trigeminal receptors in the nose) (Kobal and Hummel, 1988). However, some bubbles remain intact and begin to accumulate (forming what is termed *mousse*) (Plate 7.7). Some form a mound in the center of the glass, whereas other bubbles collect around the edge of the glass (*cordon de mousse*). Because the wine's protein content is limited, bubbles quickly coalesce, forming larger but increasingly unstable bubbles. Thus, sparkling wine does not produce the "head" associated with beer's higher protein content.

For pouring, the method traditionally used for beer is recommended, i.e., slow pouring down the sides of the glass. It reduces the initial frothing and undesirable loss of carbon dioxide, thereby lengthening the duration and

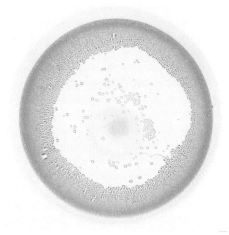

PLATE 7.7   Mound of bubbles (*mousse*) in the center, and *cordon de mousse* around the edge, of a glass of champagne. *Photo courtesy Collection CIVC 1 Alain Cornu.*

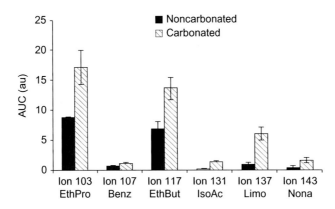

FIGURE 7.13   Comparison of AUC (area under the curve) values of noncarbonated and carbonated model systems, aromatized with a mixture of volatile compounds. *Benz*, benzaldehyde; *EthBut*, ethyl butyrate; *EthPro*, ethyl propionate; *IsoAc*, isoamyl acetate; *Limo*, limonene; *Nona*, 2-nonanone. *From Pozo-Bayón, M.Á., Santos, M., Martín-Álvarez, P.J., Reineccius, G., 2009. Influence of carbonation on aroma release from liquid systems using an artificial throat and a proton transfer reaction-mass spectrometric technique (PRTR-MS). Flav. Frag. J. 24, 226–233, Flavour Frag. J., reproduced by permission of John Wiley and Sons, copyright.*

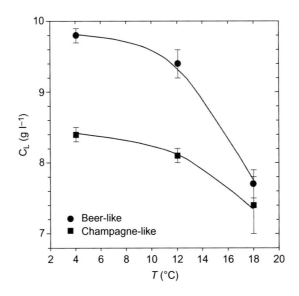

FIGURE 7.14   Comparison of the carbon dioxide remaining in champagne after pouring at three different temperatures and by two different methods: along the side of the glass (beer-like) and vertically down from the top (champagne-like). *Reprinted from Liger-Belair, G., 2016. Champagne and sparkling wines: production and effervescence. In: Cabellero, B., Finglas, P.M., Toldrá, F. (Eds.), Encyclopedia of Food and Health, pp. 526–533, with permission from Elsevier.*

liveliness of the effervescence (Fig. 7.14). Having the wine chilled to ≤4°C also limits the tendency to gush, and slows the rate of carbon dioxide liberation. As one pays enough for those bubbles, one might as well savor them as long as possible.

Where objective assessment of the effervescence is desired, a manual/automatic dispenser (FIZZeye-Robot) can be used. It can be set to dispense identical samples, take images every half-second, as well as analyze the images (Fuentes et al., 2014).

The use of flutes is always recommended for enjoyment. It is difficult to fully appreciate the beauty and elegance of the effervescence otherwise. However, where assessing the wine's fragrance is paramount, standard ISO wine-tasting glasses are superior. Because flutes are typically filled to near the top of the glass (to maximize visualization of the effervescence), vigorous swirling of the wine is not possible (or the wisest idea). As with all wines, allowing it to sit for a few minutes before sampling allows for the accumulation of aromatics in the headspace above the wine (White and Heymann, 2015).

Although some connoisseurs etch as mark on the bottom of their flutes, to produce additional sites for bubble nucleation, this seems gratuitous (White and Heymann, 2015). There are usually sufficient lint particles left on the insides of the glass after drying to provide ample nucleation sites for bubble initiation.

Glass cleaning is of particular importance with flutes, as even traces of detergent can hinder bubble generation. This results from detergent covering nucleation sites with a thin layer of hydrophobic molecules. Consequently, diffusion of carbon dioxide into nucleation sites is retarded or prevented, limiting the formation of continuous chains of bubbles. This can be demonstrated by washing wine glasses with detergent and improperly rinsing or drying them. Using soft drinks, in lieu of sparkling wine, is equally demonstrative, and avoids the loss of good wine.

In Champagne, leaving a partially consumed bottle (properly sealed and chilled) for a day or two is known to have a beneficial effect on the wine's fragrance (Pierre-Jules Peyrat, personal communication). This probably involves the changed dynamic equilibrium between aromatics dissolved and weakly bound to the wine's matrix, and those free in the headspace. Oxidation is not a problem as, during pouring, escaping carbon dioxide expels any air that might enter the bottle. In addition, because the bottle contains sufficient carbon dioxide to fill six bottles at atmospheric pressure, there is ample carbon dioxide left in the wine to generate full effervescence once the remaining wine is poured. Several sparkling wine closure systems are available, but probably the simplest is the ZORK closure. It is used to seal several brands of sparkling wines and can be reused on most other bottles of sparkling wine.

Because champagne is still considered the epitome of what a sparkling wine should be, it would be pleasing if one could describe its attributes clearly. Other than expressing minimal varietal character, and a note of toast, it is very difficult. Expressions noted in Vannier et al. (1999), such as dust, rubbery, moss, animal, floral, apple, buttery, caramel, bitter exotic fruit, fruit, and mold do not inspire enthusiasm, or seem particularly appropriate. Sparkling wine aroma wheels, such as that produced by Noble and Howe (1990, unpublished), are no better, leaving the consumer bemused, if not befuddled. A better, but still unnecessarily complex version can be found in Duteurtre (2010).

FORTIFIED WINES (DESSERT AND APPETIZER WINES)

All terms applied to this category are somewhat misleading. For example, the sherry-like wines from Montilla, Spain, achieve their elevated alcohol contents without fortification (the sugar content of the grapes is sufficient to generate ethanol values within the fortified range). The alternative designation of "aperitif" and "dessert wines" are also somewhat erroneous. Sparkling wines are often viewed as the ultimate aperitifs, and botrytized wines are often considered the preeminent dessert wine, neither being fortified. Table 7.6 provides a categorization of some of the more commonly available, typically fortified, wines.

Regardless of designation, fortified wines are usually taken in small amounts. Thus, individual bottles are rarely consumed in their entirety upon opening. Their high alcohol content limits microbial spoilage, and their marked flavor and oxidation resistance permit them to remain stable for weeks upon opening. Exceptions are fino sherries and vintage ports. Fino sherries lose their distinctive properties within a few months of bottling, whereas the unique character of vintage ports dissipates almost as quickly upon opening as do other aged red table wines.

Fortified wines come in a diversity of styles. Dry and/or bitter versions function as aperitifs before meals (e.g., fino sherries and vermouths). By activating the release of gastric acid secretion, digestion can be promoted (Teyssen et al., 1999). Typically, though, fortified wines are sweet. Major examples are oloroso sherries, ports, madeiras, and marsalas. These wines are traditionally consumed after meals, usually in lieu of, or after, dessert.

Unlike table wines, which have been produced for millennia, fortified wines are of comparatively recent origin. The oldest may to be fino-type sherries. Sherry-like wines may have been made as far back as late Roman times, although there is no documentary evidence of this. Under hot dry conditions, production of wine with alcohol contents above 15 percent is possible without fortification. The extremely low humidity in bodegas (aboveground wine cellars), selectively favors water evaporation from the surfaces of storage barrels. This results in a further increase in the wine's relative alcohol content. The alcohol both suppresses bacterial spoilage and, within a limited alcohol range, favors the development of a yeast pellicle on the wine surface (limiting oxidation). The addition of distilled spirits has the same effect, but more rapidly and consistently. The other major types of fortified wines (ports, madeira, marsala, and vermouth) had to await the perfection of wine distillation.

Distillation, as a means of concentration, is an ancient technique, having been practiced by the Egyptians at least 2500 years ago. However, its use to concentrate alcohol developed much later. In about the 10th century A.D., the Arabs developed efficient stills for alcohol purification. Alcohol distillation is reported to have been practiced in Spain as early as the mid-1200s (Léauté, 1990), and by 1100 in Dalerno, Italy. Nonetheless, wine distillation only began in earnest during the 1500s.

TABLE 7.6   Categorization of Fortified Wines

With added flavors	Without added flavors
Vermouth	Sherry-like
Byrrh	Jerez-Xerès-sherry
Marsala (some)	Malaga (some)
Dubonnet	Montilla
	Marsala
	Château-Chalon
	New World solera & submerged sherries
	Port-like
	Porto
	New World ports
	Madeira-like
	Madeira
	Baked New World sherries & ports
	Muscatel
	Muscat-based wines
	Setúbal
	Samos (some of)
	Muscat de Beaumes de Venise
	Communion wine

FIGURE 7.15   Illustration of an onion-shaped wine bottle, circa 1725.

Fortification with distilled spirits was apparently first used in preparing a herb-flavored medicinal wine called *treacle*. This may be the origin of modern vermouths. Subsequently, distilled spirits, derived from wine, were added to table wines as a preservative, leading to the evolution of most fortified wines. Liqueurs were developed from distilled spirits, to which various flowers, herbs, roots, and other botanicals were added. Modern liqueurs are almost always heavily sweetened. Modern brandies began their evolution when the benefits of long-aging of distilled wine spirits in oak barrels was discovered. Although not likely to have contributed to the discovery of distillation in the West, credit for the first production of distilled alcohol may go to India (~150–350 B.C.) (Allchin, 1979), and was possibly even earlier in China.

The use of wine spirits in sherry production was occasionally practiced by the mid-1600s. Its use in port stabilization began about 1720. By 1750, the practice shifted from fortifying the finished wine from the Duoro to the fermenting must. The resultant premature termination of fermentation retained up to 50 percent of the original grape sugars. Nonetheless, extensive treading (grape crushing and mixing under foot in shallow stone fermentors termed *lagars*) throughout the short fermentation period achieved sufficient pigment extraction to produce a dark red wine. Modern procedures have obviated the need for this old, laborious system of crushing. The tannin, sugar, and alcohol contents of the wine helped supply it with long-aging potential. Combined with the use of cork as a bottle closure, evolution of the original onion-shaped bottle (Fig. 7.15) into its modern cylindrical shape (that can be lain horizontally, keeping the cork moist), was crucial to the rediscovery of the benefits of wine aging. These advantages were clearly recognized by the early 1800s.

Sherry

Sherry is produced in two basic styles: *fino* and *oloroso* (Fig. 7.16). Each comes in a number of subcategories. In *fino* production, the alcohol content of the base wine is raised to between 15 and 15.5 percent before maturation begins, whereas with *oloroso*, the alcohol content is increased to 18 percent. At about 15 percent ethanol, changes in the yeast cell wall enhances buoyancy, resulting in the cells floating to the surface and the formation of a pellicle (*flor*) over the wine. At 18 percent ethanol, all yeast metabolic activity ceases.

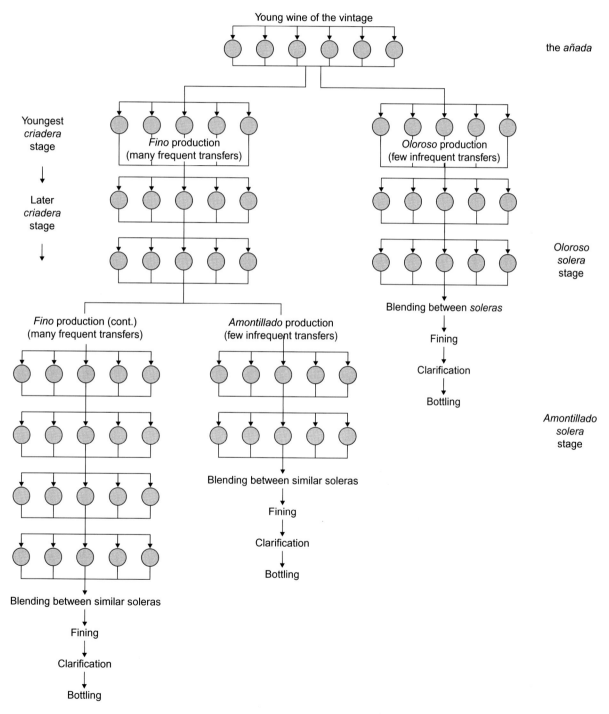

FIGURE 7.16 Flow diagram of sherry production. Shading indicates whether the butts are kept full or partially (~20%) empty. *Reprinted from Wine Science: Principles and Application fourth ed. R. S. Jackson (2014), with permission from Elsevier.*

The principal cultivar grown for sherry production is Palomino fino. Some Pedro Ximénez is also grown to produce a sweetening agent added to most sherries for export. In contrast, most sherries available in Spain are left dry. When both coloring and sweetening are desired, *arrope* is added. It is derived by boiling juice from the second pressing of the grapes down to one-fifth of its original volume.

In fino sherry production, the *flor* pellicle protects the wine from significant oxidation. The wine is stored in partially filled barrels (termed *butts*). While oxidation does occur, it is slow and involves the synthesis of a wide range of aldehydes and acetals. There is also production of important flavorants, such as soloton. It is considered to resemble the fragrance of curry and nuts.

Aging involves factional blending, where fractions of wine from older series of barrels (*criaderas*) are removed and replaced by wine from younger *criaderas* (Fig. 7.16). After about 5 years, the wine has reached an average alcohol content of 16–17 percent. Unlike most other categories, *fino* sherries are typically left unsweetened, even for export. They are pale to very light gold in color, and characterized by a mild walnut-like bouquet. According to Zea et al. (2001), they are also distinguished by floral/fruity, cheesy/rancid, and pungent/apple notes. Manzanilla is the palest and lightest of all *fino* sherries, possessing less of the aldehydic flavor that characterizes most *finos*. Once bottled, *fino* sherries do not age well, and should be consumed shortly after purchase. This situation may change with the adoption of roll-on screw caps. They are more effective at limiting oxygen uptake than the traditional, chamfered, T-corks. *Amontillado* sherries begin their development as a *fino*, but partway through their development, the alcohol content is raised. They subsequently complete their maturation in the manner of an *oloroso* (Fig. 7.16). This style is darker in color and often 1–3 percent stronger in alcohol content than a *fino*. Most *amontillados* are slightly sweetened for export, and are more full-flavored than *finos*, with a clean, nutty bouquet.

Oloroso sherries are the most oxidized of sherry styles. They receive no *flor* protection and undergo minimal fractional blending (Fig. 7.16). Thus, they possess a more pungent, smoky, ethereal, aldehydic, nutty bouquet, which can give a false impression of sweetness in dry versions. *Amoroso* sherries are heavily sweetened versions, usually with a golden-amber to brown color. The dark color comes from melanoid pigments produced during the heat concentration of the sweetening juice. Cream sherries are *amaroso* sherries initially developed for the English market. *Palo cortado* and *raya* sherries are special *oloroso* sherries. They are more subtle and rougher versions, respectively.

Sherry-like wines are produced elsewhere, but seldom approach the diversity and finesse that epitomize the Jerez-Xérès-Sherries from Jerez de la Frontera and vicinity. Some European semiequivalents are noted in Table 7.6. There are few equivalents made outside Europe, in spite of the adopted name. Nonetheless, those bottles that display the name *solera* are likely to be similar. Most non-European sherries used to be baked, producing a caramelized character typical of inexpensive madeiras. They were essentially always sweet.

Port

Porto is typically made from red, but occasionally white grapes in the Duoro region of Portugal. Fermentation of red grapes is stopped prematurely by the addition of nonrectified wine spirits (*aguardente*), retaining about half the original grapes' sugar content. This produces a sweet wine with an initial alcohol content of about 18 percent. Adding the distillate also contributes to the wine's fragrance, by donating fusel alcohols, esters, benzaldehyde, and some terpenes (Rogerson and de Freitas, 2002). These adjuncts supplement the wine's distinctive and complex fragrance. Subsequent maturation defines the various port styles (Fig. 7.17). The blending of wines from various cultivars and locations produces the consistency required for the production of the proprietary brands that characterize most ports. Although there are distinctive port styles, named blends tend to typify port as much as they do sherries and sparkling wines.

Vintage-style ports are blended from wines produced during a single vintage, and aged in inert or oak cooperage. Vintage Port is the most prestigious example. It is produced only in exceptional years, when the grape quality is deemed sufficiently high. Only rarely are the grapes derived from a single vineyard (*quinta*). After maturing for about 2 years, the wine is bottled unfined. Consequently, the wine "throws" considerable sediment. It takes from 10 to 20 years before its famously complex aroma and bouquet are considered to have reached a long-lived plateau. Vintage ports age well for an almost indefinite period. Vintage ports are also the wine with probably the longest finish, deserving intense scrutiny, as the various veils dissipate, revealing previously hidden sensory delights. Late-bottled Vintage Port (L.B.V.P.) is treated similarly, but receives about 5 years' maturation before bottling. By this time, most of the sediment has settled, eliminating the need for the decanting required with vintage port. L.B.V.P. is often called the poor man's vintage port (or those with insufficient patience). L.B.V.P. benefits little from after-bottling aging.

Wood ports are derived from the blending of wines from several vintages. Aging occurs predominantly in oak casks (termed *pipes*). Maturation is not specifically intended to give the wine an oak flavor, but to permit slow oxidation. Consequently the casks are used repeatedly (to minimize the uptake of an oaky character), and not filled completely. The most common wood-aged style is ruby port. The wine receives 2–3 years' maturation before bottling.

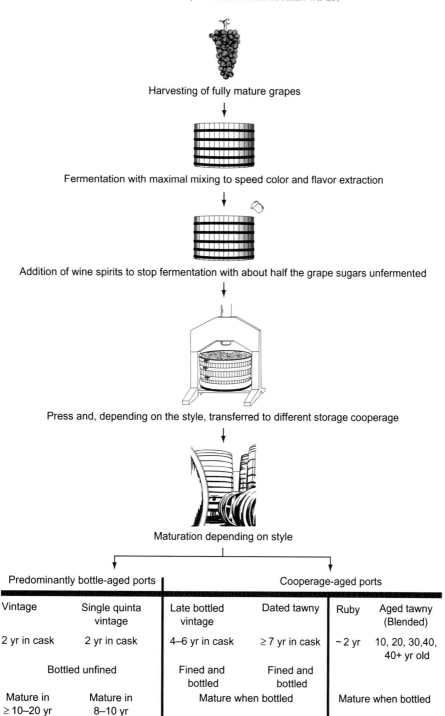

Harvesting of fully mature grapes

Fermentation with maximal mixing to speed color and flavor extraction

Addition of wine spirits to stop fermentation with about half the grape sugars unfermented

Press and, depending on the style, transferred to different storage cooperage

Maturation depending on style

Predominantly bottle-aged ports		Cooperage-aged ports			
Vintage	Single quinta vintage	Late bottled vintage	Dated tawny	Ruby	Aged tawny (Blended)
2 yr in cask	2 yr in cask	4–6 yr in cask	≥ 7 yr in cask	~ 2 yr	10, 20, 30,40, 40+ yr old
	Bottled unfined	Fined and bottled	Fined and bottled		
Mature in ≥ 10–20 yr	Mature in 8–10 yr	Mature when bottled		Mature when bottled	

FIGURE 7.17 Flow diagram of port wine production.

Tawny port is a blend of long-aged ruby ports, which have lost most of their red coloration, having turned a palish brown. The finest tawnies are sold with an indication of their average age, typically more than 10 years. Long-aged tawny ports are, by some, considered to be the pinnacle of ports. As applies to all wine (and essentially everything else), different strokes for different folks. Inexpensive tawny ports are a blend of ruby and white ports. White ports are matured similarly to ruby ports. They may come in dry, semisweet, or sweet styles. They often superficially resemble *amaroso* sherries. A comparatively recent innovation is the production of rosé ports.

Port-like wines are produced in several countries outside Portugal, notably Australia and South Africa. Only seldom are similar cultivars and aging procedures used. Fortification is usually with highly rectified (flavorless)

alcoholic spirits. Thus, they lack the distinctive flavor of Portuguese *porto*. Most ports produced in North America are baked. Thus, they typically possess a madeira-like (caramelized) odor.

Madeira

Fortification was initially used to stabilize the wine for shipment. However, in the case of wine from Madeira, being shipped to the colonial Americas, the often-protracted voyages exposed the wine to prolonged heating as the ships traversed the equator, often several times. When it became clear that colonists preferred the "baked" version, producers began to heat the wine before bottling and shipment (to more than 40°C for up to 3 months), a process called *estufagem*. Maturation occurs in wooden cooperage for several years. To avoid giving the wine an oaked character, the barrels are extensively reused.

Madeira is produced in diverse styles, ranging from dry to very sweet. They may involve the use of a single cultivar and be vintage-dated, or extensively blended and carry only a brand name. Despite these variations, the predominant factor that distinguishes madeira is its pronounced baked (caramelized) flavor. With prolonged aging, a complex, highly distinctive bouquet develops that many connoisseurs adore. In contrast, inexpensive versions are used primarily in cooking, notably in producing the classic Madeira sauce.

Most madeiras are produced from white grapes, the preferred varieties being Malvasia, Sercial, Verdelho, and Bual. Spontaneously induced fermentations are preferred, with its duration depending on the style desired. Malmseys are fortified shortly after the onset of fermentation, retaining a high residual sugar content (~120 g/L). They possess a dark brownish color, with a rich coffee/caramel aspect. Medium-sweet buals (boals) are fortified when about half the grape sugars have been fermented (~95 g/L residual sugar). They also possess a dark color and express raisiny flavors. The fortification of verdelhos is timed to retain about 70 g/L residual sugar. They tend to be distinguished by their high acidity and smoky flavors. Dry sercial styles are fortified near the end of fermentation (retaining 25–50 g/L residual sugars). Fermentation to dryness, when desired, can take upward of 4 weeks under the naturally cool winery conditions. Sercials have a profile characterized by lighter colors, refreshing acidity, and almond flavors. Some of the distinguishing sensory features in these wines are reflected in their chemical differences, noted in Fig. 7.18. Several red cultivars may also be used, notably Tinta Negra Mole. It is commonly used in the production of inexpensive madeiras.

Vermouth

Since ancient times, wine has been used as a carrier (solvent) for medicinal herbs and spices. This presumably is the origin of what eventually evolved into vermouth. Fortification facilitated both the wine's preservation and extraction of flavorants from the many botanicals used in vermouth production.

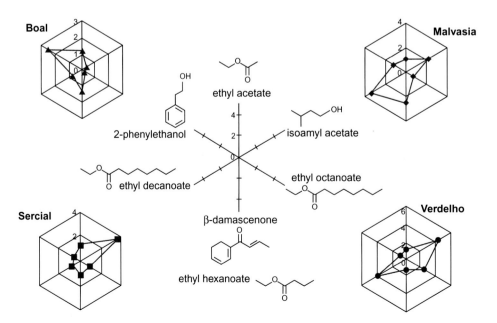

FIGURE 7.18 Polar plot illustration of some of the principal aromatic compounds distinguishing the flavor of young madeira wines. *Reprinted from Perestrelo, R., Albuqerque, F., Rocha, S. M., and Câmara, J. S. (2011) Distinctive characteristics of madeira wine regarding its traditional winemaking and modern analytical methodologies. Adv. Food Nutr. Res. 63, 207–250, with permission from Elsevier.*

Modern vermouths are subdivided into Italian and French styles. Italian versions are fortified to between 16 and 18 percent alcohol, and contain from 4 to 16 percent sugar (for dry and sweet versions, respectively). French vermouths contain about 18 percent alcohol and 4 percent sugar. Sugar helps to partially mask the bitterness of some of the flavorants.

The base wine is often a neutral-flavored white wine, although the best Italian vermouths are produced from Muscato bianco. Upward of 50 herbs and spices may be used in flavoring. The types and quantities employed are proprietary secrets. Nonetheless, typical flavoring agents often include allspice, angelica, anise, bitter almond, cinnamon, cinchona, clove, coriander, juniper, marjoram, nutmeg, orange peel, rhubarb, summer savory, and wormwood. A discussion of aromatic plants and how they are used in alcoholic beverages can be found in Tonutti and Liddle (2010). Extraction may occur in highly rectified alcohol, subsequently used to fortify the base wine (as with most Italian vermouths), or extraction can occur directly in the base wine (typical of French vermouths). After flavorant extraction, the wine is aged for 4–6 months. During this period, the components are considered to meld together. Before bottling, the wine may be sterile filtered or pasteurized.

BRANDY

By definition, brandy is distilled wine. Brandies are produced in almost every wine producing region, but the most well-recognized regional appellations come from southwestern France, i.e., Armagnac and Cognac. They are differentiated by provenance, cultivar use, and the type of still used (Armagnac column and alembic, respectively) (Jackson, 2014). All brandies are aged in oak barrels, with aging brandies from different lots being combined to formulate the proprietary blends that typify the industry. The major brandy designations are based on a combination of the minimum age of the youngest brandy used, and the minimum average age of the blend. Respectively, these are Three Stars (2/2 years); VO, VSOP (4/5years); XO, Extra, Napoleon, Vieille Réserve, Hors d'Age (5/6 years). In addition to these basic designations are select blends that contain much older reserve blends. Examples of color differences that develop with age are illustrated in Plate 7.8. Fig. 7.19 illustrates sensory differences of three cognac blends. Plate 7.9 shows an example of a tasting room in Cognac.

Related distilled grape products, derived primarily from pomace-based wines, include *marc* (France), *grappa* (Italy), *aguardiente* or *orujo* (Spain), *bagaceira* (Portugal), and *tsipouro* (Greece). Depending on the degree of rectification during distillation, they can be more grapy or neutral flavored than most brandies. They tend to be rougher in character, due to their shorter maturation in oak.

By tradition, brandies possess moderate levels of higher alcohols, generally in the range of 65–100 mg/L (donating pungency), aldehydes and acetals (supplying sharpness), oak lactones (furnishing a coconut fragrance), phenolic aldehyde derivatives from lignin degradation (providing vanilla and sweet fragrances), and ethyl esters of C_8 to C_{12} fatty acids (appending fruity/floral notes). The oxidation and transformation of fatty acids into ketones and heat-derived furans, and pyrazines during distillation generate caramel and roasted notes. Excessive amounts

PLATE 7.8 Comparison of cognacs of various ages (2, 12, 22, and 67 years), Remy Martin, Cognac, France. Notice the preferred size and shape of glasses for critically assessing cognac. *Photo courtesy R. S. Jackson.*

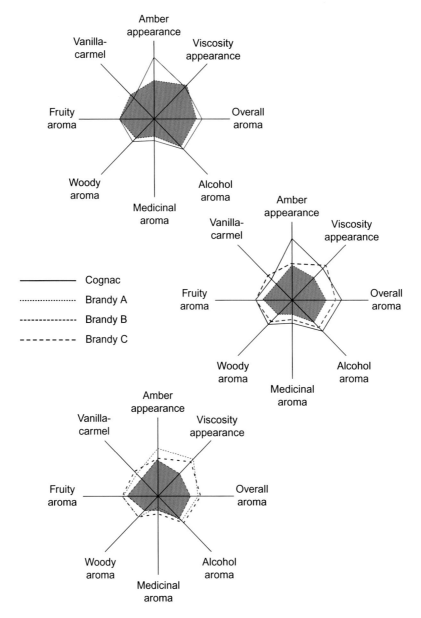

FIGURE 7.19   Sensory comparison of three different cognacs with up to three separate brandies. *Reprinted from Stone, H., Bleibaum, R., and Thomas, H. A. (2012) Sensory Evaluation Practices. 4th ed., Academic Press, London, UK, with permission from Elsevier.*

of low-volatile constituents, such as ethyl lactate and 2-phenylethanol, tend to produce an atypical heavy flavor, whereas highly volatile constituents create sharp, irritating notes. Terpenes typically add their particular character only when Muscat cultivars are used as the base wine.

Of the volatile fraction in cognacs, the most significant flavorants appear to be a ketone (β-damascenone), an aldehyde (methylpropanal), three esters (ethyl (S) 2-methylbutanoate, ethyl methylpropanate, and ethyl 3-methyl butanoate), and ethanol. Nonetheless, additional odor-active compounds were required to reconstruct the basic cognac aroma (Uselmann and Schieberle, 2015).

The sensory analysis of brandies presents conditions distinct from those associated with wines, even fortified wines. Because brandies tend to be sniffed over an extended period, rather than tasted, assessing the fragrance is of particular significance. Consequently, the samples are primarily "nosed," thereby avoiding any alcohol consumption associated issues, even with expectoration. Also, the samples are normally assessed at room temperature, obviating concern about maintaining the samples at a temperature below ambient. ISO glasses are appropriate, but smaller (150 mL) versions are often used (Plate 7.8). Balloon glasses, frequently viewed by consumers as ideal, are considered unnecessary for quality brandies.

PLATE 7.9 Tasting room of Remy Martin, Cognac, France. *Photo courtesy R. S. Jackson.*

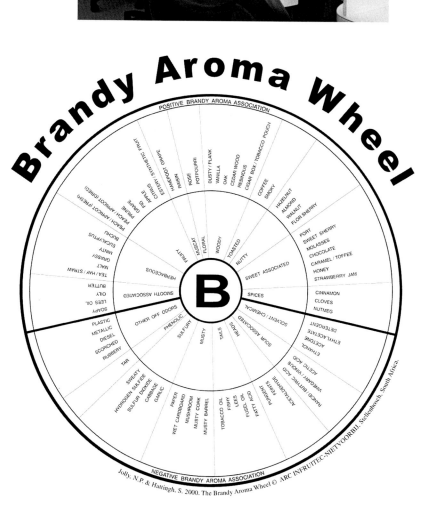

FIGURE 7.20 The brandy wine aroma wheel. *From Jolly, N.P., Hattingh, S., 2001. A Brandy aroma wheel of South African brandy. S. Afr. J. Enol. Vitic., 22, 16–21, reproduced with permission.*

Where sensory analysis is required, use any of the procedures discussed in Chapter 5 is appropriate. Specialized tasting sheets have also been designed for brandies (see Bertrand, 2003). In addition, an aroma wheel for brandies has been developed (Fig. 7.20). Alternative wheels, designed more for marketing, have been developed by the Bureau National Interprofessionel du Cognac (Plate 7.10) and de l'Armagnac (Plate 7.11), respectively. For further discussion of techniques used in assessing other distilled spirits, see Piggott and Macleod (2010).

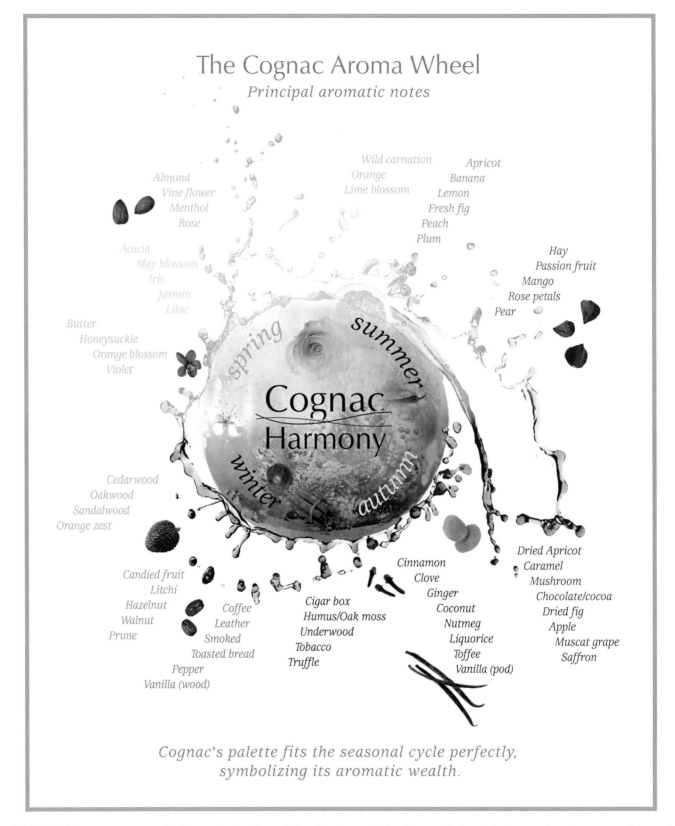

The Cognac Aroma Wheel
Principal aromatic notes

Wild carnation
Orange
Lime blossom

Apricot
Banana
Lemon
Fresh fig
Peach
Plum

Almond
Vine flower
Menthol
Rose

Acacia
May blossom
Iris
Jasmin
Lilac

Hay
Passion fruit
Mango
Rose petals
Pear

Butter
Honeysuckle
Orange blossom
Violet

spring *summer*

Cognac

Harmony

winter *autumn*

Cedarwood
Oakwood
Sandalwood
Orange zest

Dried Apricot
Caramel
Mushroom
Chocolate/cocoa
Dried fig
Apple
Muscat grape
Saffron

Cinnamon
Clove
Ginger
Coconut
Nutmeg
Liquorice
Toffee
Vanilla (pod)

Candied fruit
Litchi
Hazelnut
Walnut
Prune

Coffee
Leather
Smoked
Toasted bread
Pepper
Vanilla (wood)

Cigar box
Humus/Oak moss
Underwood
Tobacco
Truffle

*Cognac's palette fits the seasonal cycle perfectly,
symbolizing its aromatic wealth.*

PLATE 7.10   Cognac aroma wheel. *Reproduced with permission of the Bureau National Interprofessionnel du Cognac – http://www.cognac.fr/cognac/_en/2_cognac/index.aspx?page=aromes.*

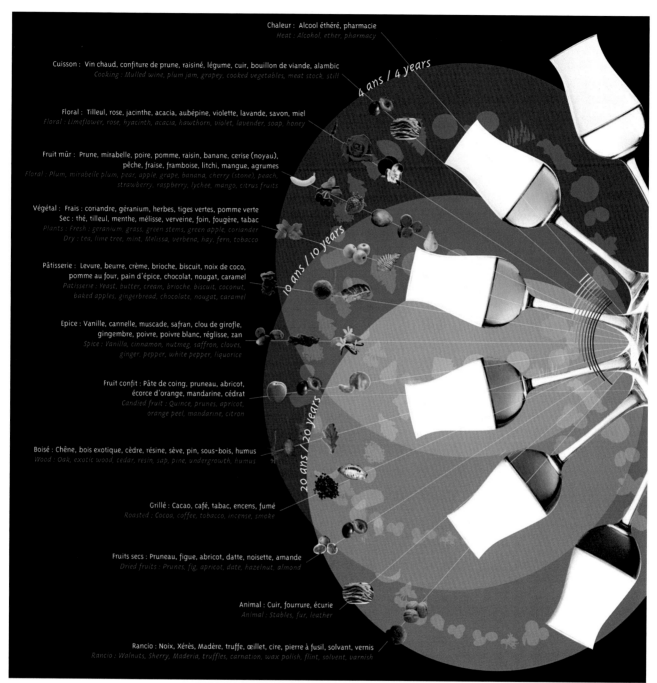

Chaleur : Alcool éthéré, pharmacie
Heat : Alcohol, ether, pharmacy

Cuisson : Vin chaud, confiture de prune, raisiné, légume, cuir, bouillon de viande, alambic
Cooking : Mulled wine, plum jam, grapey, cooked vegetables, meat stock, still

Floral : Tilleul, rose, jacinthe, acacia, aubépine, violette, lavande, savon, miel
Floral : Limeflower, rose, hyacinth, acacia, hawthorn, violet, lavender, soap, honey

Fruit mûr : Prune, mirabelle, poire, pomme, raisin, banane, cerise (noyau),
pêche, fraise, framboise, litchi, mangue, agrumes
Floral : Plum, mirabelle plum, pear, apple, grape, banana, cherry (stone), peach,
strawberry, raspberry, lychee, mango, citrus fruits

Végétal : Frais : coriandre, géranium, herbes, tiges vertes, pomme verte
Sec : thé, tilleul, menthe, mélisse, verveine, foin, fougère, tabac
Plants : Fresh : geranium, grass, green stems, green apple, coriander
Dry : tea, lime tree, mint, Melissa, verbena, hay, fern, tobacco

Pâtisserie : Levure, beurre, crème, brioche, biscuit, noix de coco,
pomme au four, pain d'épice, chocolat, nougat, caramel
Pâtisserie : Yeast, butter, cream, brioche, biscuit, coconut,
baked apples, gingerbread, chocolate, nougat, caramel

Epice : Vanille, cannelle, muscade, safran, clou de girofle,
gingembre, poivre, poivre blanc, réglisse, zan
Spice : Vanilla, cinnamon, nutmeg, saffron, cloves,
ginger, pepper, white pepper, liquorice

Fruit confit : Pâte de coing, pruneau, abricot,
écorce d'orange, mandarine, cédrat
Candied fruit : Quince, prunes, apricot,
orange peel, mandarine, citron

Boisé : Chêne, bois exotique, cèdre, résine, sève, pin, sous-bois, humus
Wood : Oak, exotic wood, cedar, resin, sap, pine, undergrowth, humus

Grillé : Cacao, café, tabac, encens, fumé
Roasted : Cocoa, coffee, tobacco, incense, smoke

Fruits secs : Pruneau, figue, abricot, datte, noisette, amande
Dried fruits : Prunes, fig, apricot, date, hazelnut, almond

Animal : Cuir, fourrure, écurie
Animal : Stables, fur, leather

Rancio : Noix, Xérès, Madère, truffe, œillet, cire, pierre à fusil, solvant, vernis
Rancio : Walnuts, Sherry, Maderia, truffles, carnation, wax polish, flint, solvent, varnish

4 ans / 4 years

10 ans / 10 years

20 ans / 20 years

PLATE 7.11 Armagnac aroma wheel. *Reproduced with permission of the Bureau National Interprofessionnel de L'Armagnac–www.armagnac.fr/ of-men-and-aromas.*

The major tasks of any *maître de chai* is to oversee the progressive reduction in the brandy's alcohol content, from ~70 percent (during maturation) to ~40 percent (for bottling), and the combination of different lots to design the blend formula for the finished, branded products. For this, the cellar master has a stock of reference samples of different qualities, age, and price for comparison. The duo-trio and triangle tests are often used to compare formula blends with standards for each brand.

Postscript

When Linnaeus came to provide humans with a Latin name, he chose *Homo sapiens*: thinking man. At that stage in the development of science, codifying life, and essentially everything else, was the order of the day. Considered overly optimistic (or egotistical) today, based on the unspeakable harm we can do to ourselves and the environment, it does denote one of our strong tendencies, i.e., organizing objects and ideas in a hopefully logical fashion. Preferentially, these are couched in terms of underlying laws or principles. In physics, this is enshrined in mathematical laws, in chemistry in terms of orbital energies, in biology in terms of evolution, in economics in terms of conceptual principles, and in morality in terms of philosophy and religion. It is when we come to categorizing wines that difficulties arise; there seem to be no fundamental principles. What has been used is pure practicality, i.e., how different wines are used. Thus, wines are roughly categorized by color, alcohol, and carbon dioxide content, subsequently being arranged by varietal and geographic origin, producer, and vintage; certainly nothing elegant to celebrate our supposed intellectual prowess, but functional.

Suggested Readings

Amerine, M.A., Singleton, V.L., 1977. Wine—An Introduction. University of California Press, Berkeley, CA.
Anderson, K. (assisted by N. R. Aayal) (2013) Which Winegrape Varieties are Grown Where? A Global empirical picture. University of Adelaide Press, Adelaide, Australia. (www.adelaide.edu.au/wine-econ/databases/winegrapes-front-1213.pdf)
Beckett, F., 2006. Wine by Style. Mitchell Beazley, London.
Halliday, J., Johnson, H., 2007. The Art and Science of Wine. Firefly Books, Buffalo, N.Y.
Jackson, R.S., 2014. Wine Science: Principles and Application, 4th ed. Academic Press, San Diego, CA.
Johnson, H., Robinson, J., 2013. The World Atlas of Wine, 7th ed. Mitchell Beazley, London.
Robinson, J., Harding, J., Vouillamoz, J., 2012. Wine Grapes. HarperCollins, New York, NY.

References

Allchin, F.R., 1979. India: The ancient home of distillation? Royal Anthropol. Inst. Gr. Brit. Irel. New Series 14, 55–63.
Alleweldt, G., 1989. The Genetic Resources of Vitis: Genetic and Geographic Origin of Grape Cultivars, Their Prime Names and Synonyms. Federal Research Centre for Grape Breeding. Geilweilerhof, Germany.
Anderson, K., 2014. Changing varietal distinctiveness of the world's wine regions: Evidence from a new global database. J. Wine Econom. 9, 249–272.
Anonymous, 1986. The history of wine: Sulfurous acid–used in wineries for 500 years. German Wine Rev. 2, 16–18.
Bertrand, A., 2003. Armagnac and wine-spirits. In: Lea, A.G.H., Piggott, J.R. (Eds.), Fermented Beverage Production, 2nd ed. Kluwer Academic/Plenum Publishers, New York, NY, pp. 213–238.
Bourzeix, M., Heredia, N., Kovac, V., 1983. Richesse de différents cépages en composés phénoliques totaux et en anthocyanes. Prog. Agric. Vitic. 100, 421–428.
Darby, W.J., Ghalioungui, P., Grivetti, L., 1977. Food: The Gift of Osiris, Vol. 2 (Figs. 14.3 & 14.4). Academic Press, London.
Dumbrell, R., 1983. Understanding Antique Wine Bottles. Antique Collector's Club, Woodbridge, England, UK.
Duteurtre, B., 2010. Le Champagne: de la tradition à l science. Lavoisier Tec & Doc, Paris, France.
Fay, J.C., Benavides, J.A., 2005. Evidence for domesticated and wild populations of Saccharomyces cerevisiae. PLoS Genet. 1, 66–71.
Fuentes, S., Condé, B., Caron, M., Needs, S., Tesseiaar, B., Hollingworth, H., et al., 2014. What a robot can tell you about the quality of your sparkling wine: Share a glass with FIZZeye-Robot. Wine Vitic. J. 29 (4), 25–29.
Guasch-Jané, M.R., Andrés-Lacueva, C., Jáuregui, O., Lamuela-Raventós, R.M., 2006. First evidence of white wine in ancient Egypt from Tutankhamun's tomb. J. Archaeol. Sci. 33, 1075–1080.
Hyma, K.E., Saerens, S.M., Verstrepen, K.J., Fay, J.C., 2011. Divergence in wine characteristics produced by wild and domesticated strains of Saccharomyces cerevisiae. FEMS Yeast Res. 11, 540–551.
Jackson, R.S., 2014. Wine Science: Principles and Application, 4th ed. Academic Press, San Diego, CA.
Jolly, N.P., Hattingh, S., 2001. A Brandy aroma wheel of South African brandy. S. Afr. J. Enol. Vitic. 22, 16–21.
Kobal, G., Hummel, C., 1988. Cerebral chemosensory evoked potentials elicited by chemical stimulation of the human olfactory and respiratory nasal mucosa. Electroencephalo. Clin. Neurophysiol. 71, 241–250.
Léauté, R., 1990. Distillation in alambic. Am. J. Enol. Vitic. 41, 90–103.
Lesko, L.H., 1977. King Tut's Wine Cellar. Albany Press, Albany.
Liger-Belair, G., 2016. Champagne and sparkling wines: Production and effervescence. In: Cabellero, B., Finglas, P.M., Toldrá, F. (Eds.), Encyclopedia of Food and Health. Elsevier Press, London, UK, pp. 526–533.
Liger-Belair, G., Marchal, R., Jeandet, P., 2002. Close-up on bubble nucleation in a glass of champagne. Am. J. Enol. Vitic. 53, 151–153.
Liu, X., 1988. Ancient India and Ancient China. Trade and Religious Exchanges. A.D. Oxford Univ. Press, Bombay, India.1–600
Murat, M.-L., 2005. Recent findings on rosé wine aromas. Part I: Identifying aromas studying the aromatic potential of grapes and juice. Aust. NZ Grapegrower Winemaker 497a, 64–65. 69, 71, 73–74, 76.
Murat, M.L., Tominaga, T., Saucier, C., Glories, Y., Dubourdieu, D., 2003. Effect of anthocyanins on stability of a key-odorous compound, 3-mercaptohexan-1-ol, in Bordeaux rosé wines. Am. J. Enol. Vitic. 54, 135–138.
Naumov, G.I., Naumova, E.S., Sniegowski, P.D., 1998. Saccharomyces paradoxus and Saccharomyces cerevisiae are associated with exudates of North American oaks. Can. J. Microbiol. 44, 1045–1050.
Perestrelo, R., Albuqerque, F., Rocha, S.M., Câmara, J.S., 2011. Distinctive characteristics of madeira wine regarding its traditional winemaking and modern analytical methodologies. Adv. Food Nutr. Res. 63, 207–250.

Pérez-Magariño, S., Martínez-Lapuente, L., Bueno-Herrera, M., Ortego-Heras, M., Guadalupe, Z., Ayestarán, B., 2015. Use of commercial dry yeast products rich in mannoproteins for white and rosé sparkling wine elaboration. J. Agric. Food Chem. 63, 5670–5681.

Piggott, J.R., Macleod, S., 2010. Sensory quality control of distilled beverages. In: Kilcast, D. (Ed.), Sensory Analysis for Food and Beverage Quality Control. A Practical Guide. Woodhead Publ. Ltd, Cambridge, U.K, pp. 75–96, 262–275.

Pozo-Bayón, M.Á., Santos, M., Martín-Álvarez, P.J., Reineccius, G., 2009. Influence of carbonation on aroma release from liquid systems using an artificial throat and a proton transfer reaction-mass spectrometric technique (PRTR-MS). Flav. Frag. J. 24, 226–233.

Rogerson, E.S.S., de Freitas, V.A.P., 2002. Fortification spirit, a contributor to the aroma complexity of port. J. Food Sci. 67, 1564–1569.

Singleton, V.L., Ough, C.S., 1962. Complexity of flavor and blending of wines. J. Food Sci. 12, 189–196.

Stone, H., Bleibaum, R., Thomas, H.A., 2012. Sensory Evaluation Practices, 4th ed. Academic Press, London, UK.

Teyssen, S., González-Calero, G., Schimiczek, M., Singer, M.V., 1999. Maleic acid and succinic acid in fermented alcoholic beverages are the stimulants of gastric acid secretion. J. Clin. Invest. 103, 707–713.

Theophrastus (371–287 B.C.). Concerning Odours (ΠΕΡΙ ΟΣΜΩΝ 11.2). In Enquiry into Plants. Vol II. Loeb Classical Library (Trans. Sir A. Hort, 1916). William Heinemann, London, UK, p. 373.

Tonutti, I., Liddle, P., 2010. Aromatic plants in alcoholic beverages. A review. Flavour Fragr. J. 25, 341–350.

Uselmann, V., Schieberle, P., 2015. Decoding the combinatorial aroma code of a commercial cognac by application of the sensomics concept and first insights into differences from a German brandy. J. Agric. Food Chem. 63, 1948–1956.

Usseglio-Tomasset, L., Bosia, P.D., Delfini, C., Ciolfi, G., 1980. I vini Recioto e Amarone della Valpolicella. Vini d'Ítalia 22, 85–97.

Van Buren, J.P., Bertino, J.J., Einset, J., Remaily, G.W., Robinson, W.B., 1970. A Comparative study of the anthocyanin pigment composition in wines derived from hybrid grapes. Am. J. Enol. Vitic. 21, 117–130.

Vannier, A., Run, O., Feinberg, M.H., 1999. Application of sensory analysis to champagne wine characterisation and discrimination. Food Qual. Pref. 10, 101–107.

Wang, J., Capone, D., Wilkinson, K.L., Jeffery, D.W., 2016. Chemical and sensory profiles of rosé wines from Australia. Food Chem. 196, 682–693.

White, M.R.H., Heymann, H., 2015. Assessing the sensory profiles of sparkling wine over time. Am. J. Enol. Vitic. 66, 156–163.

Zea, L., Moyano, L., Moreno, J., Cortes, B., Medina, M., 2001. Discrimination of the aroma fraction of Sherry wines obtained by oxidative and biological ageing. Food Chem. 75, 79–84.

8

Nature and Origins of Wine Quality

Inherent in all wine tastings is an exploration of quality. Quality is a concept to which everyone genuflects. However, there is no agreement on what it means, or how it should be assessed or achieved. Reeves and Bednar (1994) consider that quality entails several factors: excellence, conformity with historical prototypes, exceeding expectation, and relative monetary value. These attributes are certainly components of what most experts would agree are essential elements of wine quality, but they themselves are nebulous percepts. Appellation control laws were enacted ostensibly to enshrine traditional views of regional quality, but in reality only act to provide some confidence in the authenticity of providence, and occasionally production features. High price is too often viewed as a predictor of quality. All high price guarantees is an element of exclusivity for the purchaser. Evidence suggests that, without knowing the price (and presumably origin), expensive wines are frequently not associated with enhanced sensory pleasure (Goldstein et al., 2008).

Quality is often viewed in intrinsic (physicochemical, sensory) or extrinsic (price, prestige, context) terms. In an ideal world, only intrinsic factors would have legitimacy. Nonetheless, extrinsic factors typically play a significant, if not preeminent, role in peoples' concept of quality (Hersleth et al., 2003; Verdú Jover et al., 2004; Priilaid, 2006; Siegrist and Cousin, 2009), and in their purchase decisions (see Jaeger, 2006). Although those with extensive wine experience and its assessment tend to associate appreciation with sensory quality (on a comparatively objective basis), this is not necessarily the situation even with consumers who consume wine regularly (Hopfer and Heymann, 2014).

Nonetheless, the most distinctive feature usually ascribed to quality can be encompassed within the expressions "good" versus "bad," or in less moralistic/emotional terms, "desirable" versus "undesirable." Or, in another way, as noted by Prescott (2012) relative to food:

> A food's hedonic value is based in the person, not the food itself.

Clearly this is not particularly helpful when attempting to understand the underpinnings of wine quality, but is an honest acceptance of reality for most people, wine critics included.

Despite the inherent subjectivity of the concept, one of the principal goals of sensory analysis is to understand the physiological, psychological, and chemical bases of quality. However, without a clear definition of what wine quality means, designing experiments to investigate this concept is fraught with difficulties, and success has been incremental at best. Nonetheless, attempts may not necessarily be fruitless. For example, even the concept of facial beauty has been analyzed sufficiently to develop a universal model of this seemingly nebulous attribute (Bronstad et al., 2008). Thus, beauty is not "just in the eye of the beholder." There is a conceptual model of physical beauty that seems hardwired in the human brain. This likely does not apply to wine.

Other than a few innate tendencies, such as a dislike of bitter substances and a liking for sweetness, human sensory responses seem to be primarily based on experience, not reflex. As a consequence, flavor preferences are potentially malleable and primarily culturally based, and there may be few absolutes relative to wine quality. Acquaintance and experience appear to be the predominant factors. For example, familiarity abets odor discrimination, and often enhances perceived intensity and pleasantness (Distel and Hudson, 2001; Mingo and Stevenson, 2007). Furthermore, repeat exposure often modifies preferences, promoting acceptance (Köster et al., 2003). Thus, we have the oft-expressed view that experience induces people to appreciate more complex (better quality?) wines. Although potentially valid, it clearly is not a consistent consequence. People are molded both by "nature" and by "nurture."

The various and changeable nature of what constitutes wine quality, or at least expressed in popularity, makes defining the enigma of quality doubly difficult. For example, it is common to read wine critics complaining about overoaked Chardonnays. In addition, Hersleth et al. (2003) found that the Californian consumers they assayed preferred unoaked Chardonnay. Nonetheless, what people express on a questionnaire, or under an experimental tasting condition, does not necessarily reflect what they practice. For example, two of the most popular brands of Chardonnay in the United States and Canada (Lindeman's Bin 65 and Casella's Yellow Tail) are oaked. There may be as many ideas of wine quality as there are people. A stable and consistent viewpoint of wine quality has usually been considered one of the principal advantages of trained versus consumer panels. Nonetheless, consistency does not validity make, any more than strong correlation assures causation.

For many consumers, quality is a reflection of satisfaction, be that identification with the culture or history of the wine's provenance, the wine's quality/price ratio, or some other attribute. Because quality is a cerebral construct, it should not be unexpected that it has been detected on a neuronal level (Plassmann et al., 2008). The more marked the perceived quality/price differential, the greater was the perceived pleasure (at least under test conditions). The element of surprise also appears to be a major contributor to enhanced pleasure (Berns et al., 2001), potentially elevating enjoyment to stellar heights.

Effective marketing not only influences actions, such as purchasing habits, but also how the brain responds to stimuli, independent of clearly detectable sensory differences (McClure et al., 2004). Cultural influences and apparent sophistication can also override inherent sensory aspects, affecting both how a person views quality and its appropriateness. For many aficionados, quality appears to be a reflection of their desires, such as virtual experience of a wine's romantic heritage (e.g., a geographic region), a sophisticated lifestyle (e.g., refined food and wine), a sense of cultural identity (e.g., French, Italian), connection with individuality and uniqueness (e.g., *terroir*, estate bottling), a sense of physical and/or social warmth (e.g., Provence, Tuscany), feeling of relaxed elegance (e.g., Rheingau auslese), celebration (e.g., champagne), a social statement (e.g., identification as an oenophile, possession of wealth, or of some august vintage), or sensory exultation (e.g., broad flavor palette). For individuals where exclusivity and pride of ownership are paramount, wine may cease to be just a beverage, but a means of fulfilling a symbolic and/or psychological need (see Bhat and Reddy, 1998). That the wine may not live up to its expectation upon opening is often of less significance than the feeling derived during and after purchase, with purchase itself providing a vicarious experience of imagined places or experiences. These extraneous associations, donating a sense of empathy or community with the wine's origin, are not necessarily bad. They often amplify the appreciation and joy derived from whatever is purchased, be it wine, clothing, a car, or a computer tablet (Atkin, 2005). Although worshiping wine can provide its own pleasure, thankfully, most consumers view wine more rationally, as a savory and salubrious beverage.

Appreciation is frequently colored by context. For example, wine in the morning rarely seems appropriate, with the possible exception of champagne and orange juice with a stately breakfast. Although the thought is appealing, it does seem to be a waste of a good champagne (the orange juice masking the refined delicacy of the champagne). Expectation also significantly influences how quality is perceived, as well as how people view wine's combination with food (Wansink et al., 2007). At its most venerable, wine is an aesthetic experience (Charters and Pettigrew, 2005). As such, it is both sensual and ennobling.

Typically, objects considered to possess high quality are considered to be difficult to produce, obtain, or find. This often means that they are rare and expensive (possessing the attribute of exclusivity). For rarity to possess desirability, it must give the purchaser pride of ownership, being a recognized icon or an original (e.g., a painting). Why else would wines from reputed estates command prices frequently out of all proportion to their sensory quality? Hopefully, for those who can afford such wines, knowledge of the price is enough of an incentive to detect exquisite quality, if actual sensory quality is insufficient. The desire for ownership can also develop into a compulsion: a passion for collecting, occasionally masquerading as economic investment (Burton and Jacobsen, 2001).

Objects of any sort considered to be of high quality are viewed as possessing artistic attributes. These features are usually endowed with properties such as complexity, harmony, dynamism, development, duration, elegance, uniqueness, memorability, and pleasure.

For some of these hedonistic expressions, potential physiologic/psychologic explanations are possible. Complexity is almost assuredly related to the development of multiple olfactory patterns in the brain (odor/flavor memories). These may or may not be sufficient marked to trigger connections with objects or experiences, generating appropriate or illusory perceptions. The more a person has developed veridical olfactory models for different wines, the greater the likelihood the taster will perceive complexity. As the concentration of aromatics released from a wine fluctuate, different qualities materialize and fade, the olfactory equivalent of a kaleidoscope. When this dynamic complexity continues throughout the course of a tasting, it is referred to as development. Duration is used to express how long a distinctive aroma/bouquet remains detectable. Regrettably, most wines fail when it comes to significant development and duration. A wine's fragrance should linger, not dissipate. Harmony is harder to define or explain. It often involves how aromatic, gustatory, and trigeminal sensations are integrated in the orbitofrontal cortex, leading to the subjective perception of balance, with no one attribute dominating (to excess). Elegance is another subjective term relating to harmony, but considered of a higher hedonic order, involving activation of the amygdala (a major center for emotional expression). Uniqueness indicates that the sensory pleasures derived are sufficiently marked and distinct to be clearly noticeable. Memorability refers to unanticipated sensory qualities, branding themselves into an unforgettable experience—an apotheosis.

A wine's quality is also intimately linked with flavors that may donate distinctive stylistic, varietal, and possibly regional attributes. Aging potential is another prominent quality concept in the view of most aficionados, but not a feature that can be accurately predicted in young wines. Experience may be a guide, but no more. Aging potential is made manifest only long after purchase. Finally, sensory quality is (or should ideally be) independent of the conditions of sampling. Wine quality is at its most appealing when nothing is known about the wine. Even sampling in black wine glasses has its charm. It forces the taster to trust totally in the sensory qualities the wine itself expresses, the only legitimate indicator of greatness.

Although wine is often associated in peoples' minds with refined living (Lindman and Lang, 1986; Klein and Pittman, 1990), this can have a negative influence on some people. The common affiliation of wine with haute cuisine and musical events, so espoused by many wineries, is anti-chic to the grunge mores of the "X" generation, and countercultural to many left-wing thinkers. For them, beer provides a statement that is more in accord with their social self-image.

In the popular press, most wine critics appear to agree on what constitutes wine quality. This has been interpreted as support for its veracity (Goldwyn and Lawless, 1991). Alternatively, the appearance of a consensus may reflect no more than training, habituation, and acquiescence to accepted norms. In contrast, Brochet and Dubourdieu (2001) found no evidence for a consistent view of wine quality, or at least how qualitative attributes were described. Too often fealty is paid to the opinions of self-proclaimed arbiters of good taste, and far too little to personal sensory perception. In what other field would the ranking of products, developed more than a century and a half ago (the Grands Crus Classés de Bordeaux), be considered of any relevance today? The real damage of sheepishly following pied pipers is that it hinders little-known but superior wines, winemakers, and cultivars from receiving the respect and just financial return they deserve, and depriving consumers of experiencing the full range of sensory pleasures wine can provide.

SOURCES OF QUALITY

Although seldom acknowledged, the most critical factor in the development of a wine's quality is the winemaker (Fig. 8.1). Without the winemaker there would be no wine. It is ultimately on their decisions that a wine will evolve, and the attributes it will eventually possess; sculptors shape stone, winemakers mold grapes. The lack of winemaker credit partially reflects insufficient human sensory acuity, not humility. Human olfactory skills are rarely up to the task of recognizing the subtle features brought to bear on a wine by individual winemakers. In addition, winemaker identity is not yet a marketable commodity, though owner/producer name can be, if sufficiently well known and prominently displayed on the label. Thus, if the enologic equivalents of a Michelangelo or Mozart exist, their finesse largely goes unsung and unnoticed, except to some estate owners, believing that chiasmic "flying" winemakers can raise the quality of their wines into the stratosphere. In addition, maturation and aging modify, and eventually erase, the subtle influences that might distinguish the wines brought to fruition by particular winemakers. Even with considerable training, few individuals recognize the more marked effects of varietal origin. The subtle effects of regional factors are even more difficult to detect consistently, that is, under blind tasting conditions.

Grape cultivar(s) and production style donate more easily detectable differences than regional characteristics. Nonetheless, the vast majority of grape cultivars are not known to produce wine with a distinctive aroma, at least consistently. Even some famous cultivars are notorious for their elusive varietal aroma, Pinot noir being the most well-known example. It is not without reason that Pinot noir has been dubbed the "heartbreak grape." Production style more consistently stamps a distinctive flavor profile on a wine. For example, use of production procedures could convert the same red grapes into a red, rosé, or white wine, that could be dry, sweet, sparkling, or fortified, each potentially appearing in an incredible diversity of substyles.

Aside from varietal attributes, grape quality (maturity, health, flavor content) sets limits on potential wine quality. It is at this level that macro- and microclimate vineyard characteristics have their major impacts on wine excellence.

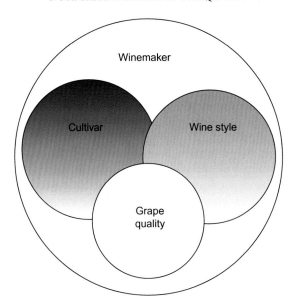

FIGURE 8.1　Diagrammatic representation of the major sources of wine quality.

However, aging eventually diminishes most of a wine's individuality, though, for premium wines, there is a temporary enhancement in character, or at least so interpreted by connoisseurs of aged wines.

What follows is a discussion of those features that influence wine quality. Although trends can be noted, quality involves the complex interaction of innumerable factors, where any one is likely to be influenced by others. For example, wine produced from older vines of Cabernet Sauvignon is considered to be more berrylike, and possess less of a vegetal character (Heymann and Noble, 1987). However, this might simply be a consequence of the vines' lower vigor, smaller berry size, and improved fruit–light exposure, generating more flavor per berry. Also, because of their anticipated greater quality, grapes from older vines may be given preferential treatment during fermentation and maturation. Correlation, as usual, is not necessarily causally related.

In addition, caution must be taken not to overextend the relevance of existing scientific research. Most studies are based on comparatively small samples and a few cultivars, without the control of additional factors typically considered essential in rigorous experimentation. For example, the effect of vine age noted in Heymann and Noble (1987) was based on an analysis of 21 wines. The results were consistent, though, with commonly held beliefs in the wine community, if that is of any significance.

VINEYARD INFLUENCES

Before the microbial nature of fermentation was discovered, wine quality was attributed to the soil and grape-production procedures. Subsequently, the significance of winery technology dominated explanations of wine quality, at least in the New World. More recently, winemakers have returned to ascribing quality to what occurs in the vineyard, possibly as a response to growing public distrust of technology, and a desire to distinguish their wines from the competition. In reality, both viticultural and enologic practices are of importance. In most instances, one would be hard pressed to say either was more important than the other. Although the adage "one cannot make a silk purse out of a sow's ear" applies to wine, without skilled guidance, the finest grapes can produce plonk. In a recent study of Pinot noir wines, winemaking conditions, barrel maturation and vintage appeared more significant in influencing the wines' characteristics than vineyard site or clone (Schueuermann et al., 2016).

Macroclimate

Macroclimate refers to those influences that can be ascribed to major regional or geographic features, such as latitude, proximity to large bodies of water, ocean currents, mountain ranges, or the size of the associated landmass. Clearly, these features have dramatic effects, not only on the styles of wine most easily produced, but also on whether viticulture is economically viable or even possible. For example, early cold winters are as essential to the efficient production of icewines, as is a dry hot climate conducive to sherry production. Equally, cool fall conditions favor acid retention during ripening and slow fermentation; both are conducive to producing stable, dry table wines. However, modern viticultural and enologic procedures can mollify, if not offset, many of the climatic influences that once limited regional wine production. Thus, no region or country can legitimately claim that it is preeminent in producing quality wine, although some may be inherently blessed in producing certain styles or growing particular grape cultivars. Although consumers may have favorites, justifiable quality is more dependent on the judicious and skillful application of technological know-how than geographic origin.

Meso- and Microclimate

Both terms refer to features that are, more or less, under the potential influence of the grower. Mesoclimate is typically defined as those climatic conditions on the scale of hundreds of meters. In contrast, microclimate refers to conditions immediately affecting the region enveloping individual or small groups of vines. In both terms, features such as local soil conditions and topography are involved, as well as the influences produced by vine training and the vine's growth habit. On the scale of individual vines, one is dealing with its immediate soil-atmosphere microclimate (SAM). In the popular wine vernacular, the mesomicroclimate is often termed *terroir*. Unfortunately, the term has been equally used to include local grape-growing and winemaking conditions, and occasionally imbued with elitist intimations verging on the mystic. The acronym SAM is a more precise, and devoid of any underlying supercilious connotations. To bring *terroir* into perspective, Hugh Johnson (Johnson, 1994) noted that "both one's front and back yards have distinct *terroirs*."

That a vineyard's mesoclimate and a vine's microclimate influence grapes, and thereby, wine chemistry, is beyond question. What is in question is whether the few, detectable, minute changes in grape chemistry, effected by the vineyard site, such as the concentration of trace elements (e.g., barium, lithium, and strontium), or the relative proportions of carbon, oxygen, and hydrogen isotopes generate humanly detectable differences. The answer is probably not, even though such chemical differences are the most easily identifiable effects of a vine's meso- and microclimate.

Soil conditions influence grape growth primarily through their effects on heat retention, water-holding capacity, and nutritional status. For example, soil color and textural composition can affect fruit ripening by influencing heat absorption and reradiation. Clayey soils, due to their huge surface area to volume ratio (2 to $5 \times 10^6 \, cm^2/cm^3$), have an incredible water-holding capacity. This means that the soil warms slowly in the spring (retarding vine activation), but provides extra warmth to the vine during the autumn (reducing the likelihood of the localized damaging frosts). However, the small average pore size of clayey soils can induce poor drainage. The result can be waterlogging during rainy spells, and the associated potential for berry splitting and subsequent rotting. Vineyard practices that augment humus content can increase soil drainage and aeration by promoting the development of a fine soil aggregate. Humus is also a major reserve of mineral nutrients. These are held loosely in a form readily accessible to vine roots. This encourages optimal vine growth and fruit ripening.

Only rarely is the geologic origin of the soil of significance. Typically, centuries of weathering have fundamentally transformed the chemistry and structural character of the parental rock material. Thus, famous wine regions are as likely to be situated on geologically uniform (e.g., Champagne, Jerez, Mosel) as on geologically heterogeneous soils (e.g., Bordeaux, Rheingau), or on soils derived from any of the various igneous, sedimentary, or metamorphic types of rock. Homogeneity within a vineyard, however, is of significance. Soil nonuniformity is one of the prime sources of uneven berry ripening throughout a site, generally viewed as likely to lower wine quality.

Limiting vineyard variation to enhance grape uniformity is the principal aim of precision viticulture (PV). To achieve this, selective vineyard modification is applied, such as localized adjustment in soil nutrient and water-retention conditions. In addition, selectively harvesting particular parcels of a vineyard can further improve grape uniformity at the cellar door. Although PV can achieve a more uniform base material, does this necessarily equate to enhanced wine quality? Quality is often associated with aromatic complexity. However, it has yet to be established that grape uniformity at harvest equates to enhanced aromatic wine complexity. Aromatic complexity may in fact be better achieved with some grape nonuniformity (variability in aromatic composition).

Viticultural practices, such as adjusting vine density at planting, or training system modification affecting light penetration and wind flow within and among vines, can significantly affect fruit ripening, disease susceptibility, and flavor development. Each adjusts the potential of the fruit to generate high-quality wine.

Topographic influences, such as vineyard slope and solar orientation, affect the growing environment, and thereby, the potential of the vine to fully ripen its fruit. Sloped sites become increasingly significant the higher the vineyard latitude (providing enhanced solar exposure). Fig. 8.2 illustrates that the slope's major benefit, relative to

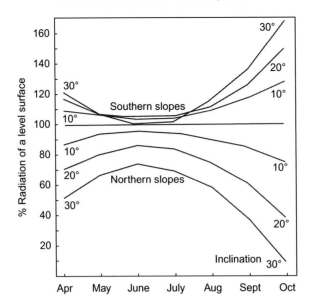

FIGURE 8.2   Reception of direct sunlight during the growing season in relation to position and inclination of slope in the northern hemisphere. This example is in the upper Rhine Valley, Germany: 48°15′N. *From Becker, N., 1985. Site selection for viticulture in cooler climates using local climatic information. In Proc. Int. Symp. Cool Climate Vitic. Enol. In: Heatherbell, D.A., et al. (Eds.), Agriculture Experimental Station Technical Publication No. 7628, Oregon State University, Corvallis, pp. 20–34, reproduced by permission.*

grape maturation, occurs late in the season, when extra solar radiation is most valuable. Slope also increases sunlight exposure reflected off nearby water bodies – a beneficial feature long realized in several famous river valleys. At low sun angles (<10 degrees), reflected radiation off water can amount to almost half the solar energy falling on vines on steep, sun-facing slopes (Büttner and Sutter, 1935). Slopes also facilitate water drainage and can direct cold (frost-inducing) air away from the vines.

Nearby bodies of water also can generate significant climatic influences. These may be beneficial, by reducing both summer and winter temperature fluctuations and their extremes, or detrimental, by shortening the growing season in cool maritime climates. Fog development can also nullify the potential for increased solar exposure associated with a sun-facing slope, as well as increase disease prevalence.

In a few instances, studies have demonstrated detectable sensory differences among wines produced in adjacent regional appellations (Douglas et al., 2001; Kontkanen et al., 2005; Tomasino et al., 2013; Fig. 8.3). Although interesting, whether these effects are consistently detectable (from year to year) has not been established. Features such as the vintage conditions, production procedures, and vine age have the potential to mask regional subtleties (Ribéreau-Gayon, 1978; Noble and Ohkubo, 1989; Schueuermann et al., 2016). In addition, sensory differences, based on averaged regional data, are just that: averages. Not all wines in a region will equally express similar characteristics. Nor can it be assumed that even skilled tasters will be able to distinguish among wines from adjacent regions, based on their sensory expression (Tomasino, 2011). In addition, producers may not wish to use regional designations (due to their restrictions), preferring more general appellation designations to achieve the brand consistency and reputation more important to consumer success. Broader geographic designations are often easier for consumers to grasp than the more ethereal sensory differences supposedly associated with vineyard or regional appellations. Nonetheless, the perception of uniqueness, real or otherwise, can be a powerful force in promoting sales of expensive wines, to the detriment of less well-known regional wines. Supporting local production is a significant factor in enhancing the profitability and survival of small producers. Winery door sales can give the consumer a sense of knowing the producer, and access to unique wine otherwise unattainable, as well as a means of avoiding the tyranny of choice (Schwartz, 2004) in wine outlets. Winery sales can also encourage consumers to sample a diversity of varietal flavors that they might otherwise not experience.

Species, Variety, and Clone

Conventional wisdom implies that only cultivars of *Vitis vinifera* produce wines of quality. Wines produced from other species and interspecies hybrids, even possessing *vinifera* heritage, are often considered unworthy of serious consideration. This prejudice was one of the reasons that provoked laws restricting the cultivation of interspecific hybrids

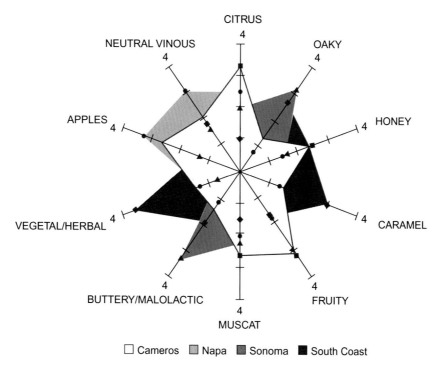

FIGURE 8.3 Polar plot derived from 10 aroma attributes of 1991 Chardonnay wines, showing regional variation within four viticultural areas of California. *Reprinted with permission from Arrhenius, S.P., McCloskey, L.P., Sylvan, M., 1996. Chemical markers for aroma of Vitis vinifera var. Chardonnay regional wines. J. Agric. Food. Chem. 44, 1085–1090. Copyright (1996) American Chemical Society.*

in France and then Europe. This decision partially arose from unfamiliarity with the "foreign" flavors occasionally associated with some interspecies hybrids, potentially threatening the reputations of established appellations. Additional concerns arose due to the higher yield of some hybrids. This became an issue due to excess grape production in Europe. The shortsighted response of governments legislating against planting interspecies hybrids is clear from the quality of the Chambourcin wines from Australia, or the fine Baco noir, Maréchal Foch, Vidal blanc, Cayuga White, and Traminette wines produced in North America. They are examples of the heights attainable with hybrid cultivars.

Even more maligned are non-*vinifera* wines. Part of this rejection may be derived from their association with the syrupy sweet wines produced from them in decades past. Many people came to believe that these were the only styles they could produce. Their non-*vinifera* aromas also conflicted with the sensibilities of those habituated to *vinifera* wines, desirous to emulate European wines, or achieve acclaim from wine critics. The aroma intensity of cultivars such as Concord, Catawba, and Niagara has also been claimed to be a negative attribute. However, if this were so, then *vinifera* cultivars such as Gewürztraminer and Muscat should also be abhorred. They are not. In the southern United States, considerable interest is shown in producing *V. rotundifolia* and *V. aestivalis* wines. Only prolonged experimentation and work by dedicated winemakers, willing to defy eurocentric naysayers, will reveal the full potential of non-*vinifera* wines.

The preponderance of Western European cultivars in world viticulture partially originated from favorable climatic and socioeconomic factors in these regions. The moderate climate of Western Europe provided conditions that permitted the production of wine that could age well (combined with the perquisite technological advances) (Jackson, 2016b). These conditions allowed better cultivars to be recognized as such. Coincidentally, the development of the Industrial Revolution in proximity to these wine-producing regions generated a burgeoning middle class, with the free capital to support the production of finer wines. With colonization and empire building, the views of wine-conscious Europeans spread worldwide. Correspondingly, their biases have significantly influenced the varieties chosen for planting in New World vineyards. Regrettably, Southern Europe (and its grape cultivars) did not enjoy the same benefits, either climatically or socioeconomically. Thus, their premium-cultivar equivalents have remained largely unknown, except locally. Their qualities still remain inappropriately appreciated. In addition, how many distinctive variations on a theme can be produced from the few, so-called premium cultivars that dominate commerce? Other cultivars might rekindle interest, and stimulate the relatively stagnant worldwide wine sales situation. It seems that most wine writers, possibly unintentionally, counteract such a possibility. However, increased varietal diversity (where named) could be counterproductive at the lower end of the market, leading to enhanced consumer confusion. The current bewildering array of essentially identical wines is already bad enough, leading to a depression in wine sales (Drummond and Rule, 2005). The mental paralysis many consumers experience in wine stores is an example of what has been termed Gruen transfer, named after Victor Gruen, the architect who created the concept of the shopping center. For the average consumer, blended wines without varietal designation, and skillful naming, are probably optimal. In such wines, the use of 'new' cultivars would not be noted (avoiding consumer confusion), but add novel and refreshing flavors.

Most cultivars exist as a collection of clones, forms that are genetically identical in all but a few mutations or epigenetic modifications. Occasionally, these differences significantly influence winemaking characteristics, by directly or indirectly affecting fruit flavor. For example, certain clones of Chardonnay possess a muscaty character, whereas particular clones of Pinot noir are better for champagne than red wine production. Clones can also differ significantly in yield. Until recently, growers typically planted a single clone. This is beginning to change, as winemakers search for new ways to increase aromatic complexity. This could generate a feature that might distinguish a producer's wine from those of neighboring wineries.

As part of the ongoing process of eliminating systemic pathogens from grape varieties, there is also selection for clones with both enhanced yield and improved grape quality. Fig. 8.4 shows that yield increase, by almost a factor of 4, has not depressed average fruit quality (as measured by sugar content) since the mid-1920s, except for an unexplained slight decline between the 1950s and 1980s. An aspect that has not been investigated (as almost heresy)

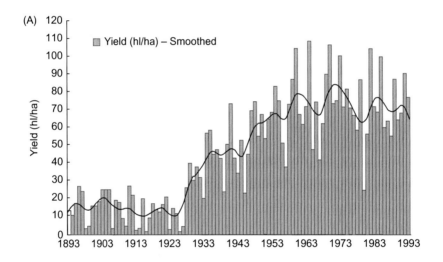

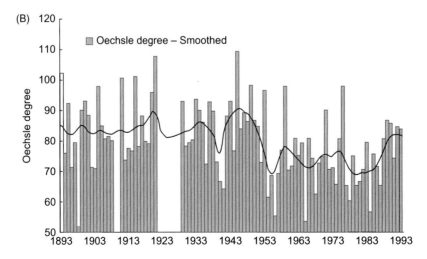

FIGURE 8.4   Grape yield (A) and must quality (B) of Riesling at Johannisberg (Rheingau, Germany) from 1893 to 1993. *From Hoppmann, D., Hüster, H., 1993. Trends in the development in must quality of 'White Riesling' as dependent on climatic conditions. Wein Wiss. 48, 76–80, reproduced by permission.*

in the possibility that eliminating viral infection from vines may have reduced aromatic complexity. For example, the source of the more flavorful clones of Gewürztraminer has been attributed to viral infection (Bourke, 2004), and many vines in Burgundy were once clearly infected with viruses.

Rootstock

Rootstocks produce the root system that supports the majority of the world's grapevines. Although providing the resistance or tolerance to soil-based problems, which permit commercial grape cultivation in most wine regions, rootstocks seldom receive the public acknowledgment they deserve, a case of "out of sight, out of mind." Besides providing the resistance factors noted above, rootstocks may also be used to regulate vine vigor and capacity (and thereby, the vine's potential to ripen its fruit), as well as influence vine nutritional and hormonal balance. Rootstock choice can also affect potential wine quality by improving vine health (donating resistance or tolerance to various aboveground pests, diseases, and unfavorable environmental conditions).

Grafting began in the late 1800s as the only effective means of combating the ravages being caused by phylloxera (*Daktulosphaira vitifoliae*). At the time, the root louse was decimating European vineyards. However, early rootstock selections were not well suited to the alkaline soils of most European vineyards. This may be the origin of the impression that wine quality suffered as a consequence of grafting. The only advantage to own-rooted vines (in the few places where this remains possible) is the economy of escaping the cost of grafting (assuming it is not needed to counter other viticultural limitations).

Yield

The relationship between vine yield and grape (wine) quality is complex. Increased yield has often been correlated with delayed sugar accumulation during ripening, a rough indicator of fruit flavor development. However, as Figs. 8.4 and 8.5 illustrate, the relationship between grape sugar accumulation or potential alcohol content and yield can be both inconsistent and highly variable. Although enhanced flavor is seemingly beneficial to wine quality, the advantages of increased aroma may eventually be offset by an augmentation in aggressive taste (Fig. 8.6).

What is commonly missing in most discussions of yield–quality relationships is acknowledgment of the importance of vineyard and climatic conditions. With vines growing on relatively dry or nutrient-poor hillside sites (common in several European viticultural regions), severe pruning tends to induce early cessation of vegetative growth, resulting in full ripening of the limited fruit crop. These observations presumably led to the generalization that small yields were inherently correlated with quality. However, the same procedure, applied to healthy vines, adequately supplied with nutrients and water (as in most New World vineyards), has the effect of promoting shoot growth, to the detriment of fruit maturation and quality. The error of assuming that a particular European generalization was universal became obvious when new training systems improved light exposure to large vines on moist, fertile soils in the New World. These new systems helped direct the increased growth potential of the vines (vigor) into improved fruit maturity (capacity), rather than enhanced shoot growth.

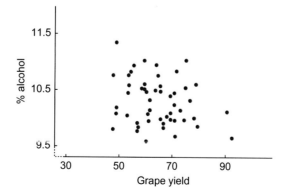

FIGURE 8.5   Relation between grape yield and alcohol content of the wines produced from 51 vineyards in the south of France. Data from each vineyard is averaged over 22 years. *From Plan, C., Anizan, C., Galzy, P., Nigond, J., 1976. Observations on the relation between alcoholic degree and yield of grapevines. Vitis 15, 236–242, reproduced by permission.*

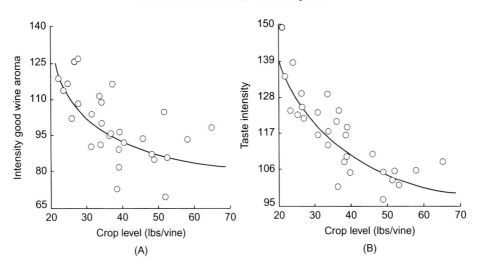

FIGURE 8.6   Relation between crop level of Zinfandel vines and (A) good wine aroma and (B) taste intensity. *From Sinton, T.H., Ough, C.S., Kissler, J.J., Kasimatis, A.N., 1978. Grape juice indicators for prediction of potential wine quality, I. Relationship between crop level, juice and wine composition, and wine sensory ratings and scores. Am. J. Enol. Vitic. 29, 267–271, reproduced by permission.*

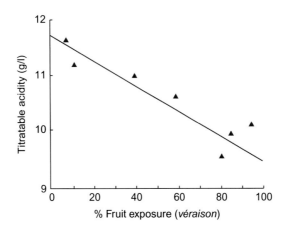

FIGURE 8.7   Relation between fruit exposure at *véraison* and titratable acidity at harvest of Sauvignon blanc. *From Smith, S., Codrington, I.C., Robertson, M., Smart, R., 1988. Viticultural and oenological implications of leaf removal for New Zealand vineyards. In Proc. 2nd Int. Symp. Cool Climate Vitic. Oenol. In: Smart, R.E., et al., (Eds.), New Zealand Society of Viticulture and Oenology, Auckland, New Zealand, pp. 127–133, reproduced by permission.*

Central to most new training systems has been the division of large vine canopies, possible on rich soils, into several, smaller, separate canopies (termed canopy management). The resultant increase in water demand helped to restrict postmidseason shoot growth. Judicious use of shoot topping and devigorating rootstocks further restricted mid- to late-season vegetative growth. The division of the canopy into several thinner canopies also opened the fruit to increased sun and wind exposure, tending to favor early and complete fruit ripening, as well as limit disease development. Fig. 8.7 illustrates how improved sun exposure can reduce excessive fruit acidity. The combined effects of canopy management meant that increased fruit yield was not associated with reduced fruit quality. Although improved light exposure is often associated with enhanced fruit coloration, this feature is cultivar dependent, and may not necessarily be reflected in more intensely colored wine. Fruit shading, however, is often associated with reduced flavor potential.

High-density planting (common in Europe) is an older alternative, tending to achieve the same results as modern, canopy-management training systems. Table 8.1 illustrates the value of high-density planting on color density. Regrettably, these benefits come at considerably enhanced vineyard development and maintenance costs (and clearly reflected in wine price). Canopy management is a more economic means of producing high-quality grapes on fertile soils with an adequate water supply.

TABLE 8.1 Effect of plant spacing on the yield of 3-year-old Pinot noir vines[a]

Plant spacing (m)	Vine density (vine/ha)	Leaf area (m²/vine)	Leaf area (cm²/g grape)	Yield (kg/vine)	Yield (kg/ha)	Wine color (520 nm)
1.0 × 0.5	20,000	1.3	22.03	0.58	11.64	0.875
1.0 × 1.0	10,000	2.7	26.27	1.03	10.33	0.677
2.0 × 1.0	5000	4.0	28.25	1.43	7.15	0.555
2.0 × 2.0	2000	4.0	15.41	2.60	6.54	0.472
3.0 × 1.5	2222	4.5	18.01	2.50	5.51	0.419
3.0 × 3.0	1111	6.3	15.36	4.12	4.57	0.438

From Jackson, R.S., 2014. Wine Science: Principles and Applications, fourth ed. Academic Press, San Diego, CA, data from Rapp, A., Güntert, M., 1986. Changes in aroma substances during the storage of white wines in bottles. In: Charalambous, G. (Ed.), The Shelf Life of Foods and Beverages. Elsevier, Amsterdam, pp. 141–167, reproduced by permission.
[a]*Data from Archer, 1987; and Archer et al., 1988.*

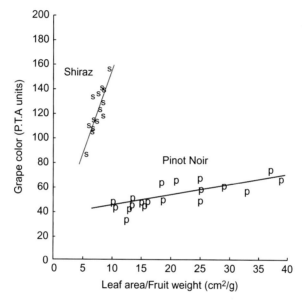

FIGURE 8.8 Relation between grape color and leaf area/fruit (LA/F) ratio for Shiraz and Pinot noir vines; *PTA*, potential total anthocyanin. *From Iland, P.G., Marquis, N., 1993. Pinot noir – Viticultural directions for improving fruit quality. In Proc. 8th Aust. Wine Ind. Tech. Conf. Adelaide, 13–17 August, 1992. In: Williams, P.J., Davidson, D.M., Lee, T.H. (Eds.). Winetitles, Adelaide, Australia, pp. 98–100, reproduced by permission.*

Other means of directing vine vigor into increased capacity involve procedures such as minimal pruning and partial rootzone drying. Minimal pruning allows the vine to grow and self-adjust its size. This usually takes several years, with most cultivars establishing a canopy structure that permits excellent fruit exposure (for optimal maturation), combined with high yield and limited pruning need. It seems particularly suited to comparatively dry climates. Another technique, particularly applicable under arid to semiarid conditions, and where irrigation is often obligatory, is partial rootzone drying. By alternately supplying water to only one side of the vine, the roots send hormonal signals to the shoots that suppress mid- to late-season shoot growth, despite an adequate water supply. The consequence is the increased likelihood of the production of an abundant fruit crop that ripens fully.

An apparently useful correlation, between fruit yield and quality, involves the ratio between the active leaf area of the vine and the mass of fruit produced (LA/F ratio). It focuses attention on a fundamental relationship between energy supply (photosynthesis) and demand (fruit ripening). For many cultivars, an appropriate value falls in the range of 10 cm²/g. However, this value can be influenced by several factors, such as the cultivar (Fig. 8.8), training system, soil nutrient and water supply, and climatic conditions. The ultimate objective of any viticultural procedure is to establish long-term optimal canopy size and fruit placement to promote full fruit ripening.

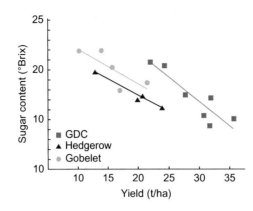

FIGURE 8.9   Relation between yield and soluble solids (5-year average) using three distinct training systems. *From Intrieri, C., Poni, S., 1995. Integrated evolution of trellis training systems and machines to improve grape quality and vintage quality of mechanized Italian vineyards. Am. J. Enol. Vitic. 46, 116–127, reproduced by permission.*

Training System

Training systems refer to techniques designed to position the fruit-bearing shoots to optimize both fruit yield and quality, consistent with long-term vine health. Hundreds of training systems exist, but few have been studied sufficiently to establish their efficacy. In contrast, several modern training systems, such as the Scott–Henry, Lyre, Smart–Dyson, and Geneva Double Curtain (GDC), have been shown to possess clear advantages in improving both fruit yield and quality (Smart and Robinson, 1991). Fig. 8.9 illustrates the influence of divided-canopy systems (GDC) versus older systems (Goblet and Hedgerow) relative to grape sugar accumulation. In addition to increasing vine capacity, improved fruit health (reduced incidence of infection) and cluster location can facilitate mechanical harvesting. These features enhance fruit quality and decrease production costs.

As previously noted, the vine's inherent vigor often needs to be restrained. On relatively nutrient-poor, dry soils, this has traditionally been achieved by dense vine planting (about 4000 vines/ha) and severe pruning (removal of upwards of 90–95 percent of the year's shoot growth). However, on rich, moist, loamy soil, less pruning is preferable, combined with wide vine spacing (about 1500 vines/ha). Under these conditions, it is prudent to redirect increased growth potential into greater fruit production, not prune it away. When an appropriate LA/F ratio is established, increased yield and quality can coexist. It is on rich soils that the newer training systems are most appropriate.

Nutrition and Irrigation

In the popular literature, stressing the vine is often viewed as promoting fine wine production. This view probably arose from the reduced-vigor, improved-grape-quality association views espoused by many renowned European vineyards noted previously. However, balancing vegetative and fruit-bearing functions, not stress, should be the goal. Exposing the vine to physiological stress, due to water or nutrient deficit, especially at particular growth stages, is detrimental. However, limited deficits, early or late in the season, can enhance wine color (Matthews et al., 1990). Flavor influences are more subtle and cultivar specific. Conversely, supplying nutrients and water in excess is detrimental, as well as wasteful.

In practice, regulating nutrient supply to improve grape quality is difficult. Because the yearly nutrient demands of grapevines are surprisingly minor (partially due to the nutrient reserves of the vine's woody parts), deficiency symptoms may express themselves clearly only months or years after deficiency starts. In addition, establishing nutrient availability to vine roots (versus nutrient presence in the soil) is still an inexact science.

Irrigation, as noted earlier (partial rootzone drying), can be used to regulate vine growth and promote optimal fruit ripening. Irrigation water can also supply nutrients and disease-control chemicals directly to the roots, in precisely regulated amounts, and only at times of need. These possibilities are most applicable in arid and semiarid regions, where most of the water comes from irrigation. Thus, climatic conditions that may initially seem unfavorable may be turned to advantage to produce some of the finest wines.

Disease

It may be unfashionable to contemplate disease abatement as a component of wine quality, but disease control is certainly necessary. However, exactly how most vine diseases depress fruit quality is poorly understood. One exception is powdery mildew. When it infects grapes, the fungus incites the development of a bitterish attribute and other flavor perturbations (Fig. 8.10). These effects are enhanced with increased maceration before or during fermentation. Some of these changes may result from the conversion of several ketones to 3-octanone and (Z)-5-octen-3-one (Darriet et al., 2002). The viscous/oiliness of the wine produced from diseased grapes has been unexpectedly correlated with the wine's phenolic content.

In only one instance can a grape infection occasionally be considered a quality feature. This occurs under the conditions of cyclical alternating sunny days with humid/foggy evenings in the autumn. Under these conditions, *Botrytis cinerea*, a normally destructive pathogen, can result in the concentration of grape constituents and synthesize its own special, and appreciated, aromatics. This is termed noble rot. Grapes so affected produce some of the most expensive, luscious, white wines available. Examples are German ausleses, beerenausleses, and trockenbeerenausleses, and French sauternes (Fig. 7.6). Infection with *B. cinerea* is only associated with an improved flavor of red wines when disease development occurs after visually healthy grape clusters are harvested and stored under cool dry conditions for several weeks to months (Fig. 7.4, Plate 7.2).

Although not a direct consequence of pathogenesis, the application of protective chemicals may indirectly affect wine quality. For example, the copper used in Bordeaux mixture can compromise the quality of Sauvignon blanc wine, by reducing the concentration of important varietal thiol aroma compounds, notably 4-mercapto-4-methyl-pentan-2-one. This effect can be reduced by prolonged skin contact (Hatzidimitriou et al., 1996), or more directly

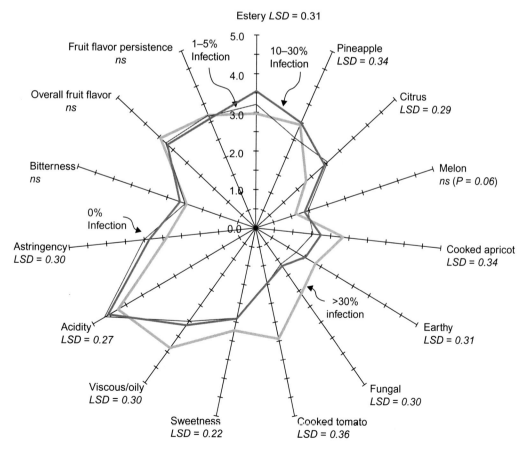

FIGURE 8.10 Mean ratings for sensory attributes of Chardonnay wines made from fruit with varied powdery mildew severity. Each value is the mean score from duplicate fermentation wines that were presented to 16 assessors in two replicate sessions. *LSD*, least-significant difference (*p* = 0.05); *ns*, not significant. *From Stummer, B.E., Francis, I.L., Zanker, T., Lattey, K.A., Scott, E.S., 2005. Effects of powdery mildew on the sensory properties and composition of Chardonnay juice and wine when grape sugar ripeness is standardised. Aust. J. Grape Wine Res. 11, 66–76, reproduced by permission of John Wiley and Sons, copyright.*

by limiting the application of Bordeaux mixture (Darriet et al., 2001). Even the chemicals used in organic viticulture are not devoid of potential detrimental effects. Various soaps and oils (used as organic pesticides) can contaminate wine with off-tastes or off-odors, and sulfur can seriously damage leaves and increase the incidence of some insect pests. Any detrimental effect of chemical pest control, organic or not, can be minimized or avoided if applied only when, and in the amounts, needed.

Maturity

Most vineyard activities are designed to promote optimal fruit maturity. Once maturity is achieved, the grapes are usually harvested and processed into wine. Measuring optimum maturity is, however, far from simple.

Maturity is often estimated by grape sugar and acid contents, their ratio, color intensity, and/or flavor characteristics. Depending on legal constraints, the sugar and acid contents of the juice may be adjusted after harvesting, to account for slight deficiencies or imbalances. However, color and flavor cannot be directly augmented. The only accepted methods of their adjustment relate to procedural or enzymatic techniques that enhance pigment extraction or the volatility of existing flavorants. Thus, there would be considerable interest if color or flavor intensity could be measured with sufficient accuracy to be used as indicators of optimal grape maturity.

For red wines, near infrared spectroscopy (NIRS) measurements correlate well with grape anthocyanin content (Kennedy, 2002). It also provides a good predictor of glycosyl-bound flavorants in Chardonnay grapes. However, fermentation and maturation conditions so affect color development and stability that no direct relationship exists between fruit coloration, and that of the wine. For flavor, monoterpene content is an indicator of potential wine flavor in Muscat cultivars, and those possessing a muscat character. For varieties with aromas not dependent on terpenes, other potential flavor indicators are required. One such indicator is provided my measuring the juice glycosyl–glucose content. Many grape flavorants are loosely bound to one or more sugars. Thus, determining the glycosyl–glucose (G–G)content has been investigated as a gauge of potential wine flavor (Gishen et al., 2002). Under some conditions, the accumulation of glycosyl flavor conjugates correlates well with the accumulation of sugars during ripening. In these situations, measuring sugar content is a simpler means of assessing potential flavor content. However, in cool climatic regions, there may be poor correlation between sugar content and juice flavor potential. Thus, the more laborious measures of free terpene or G–G contents may be required, if these indicators are determined to be valuable measures of optimal fruit maturity.

Once the decisions on maturity and harvest date have been made, the next issue involves the harvest method. In the past, hand-picking was the only option. Even today, for some wine styles, and grape varieties, manual harvesting is the only choice. For example, wine made by carbonic maceration (such as Beaujolais) involves a grape-cell fermentation that must occur before grape crushing. Grapes harvested for champagne production also involve manual harvesting, to permit the pressing of whole clusters (to minimize phenolic and color extraction). Where a significant portion of the crop is diseased, manual harvest is also essential to ensure maximal removal of infected grapes. In most cases, though, the choice between manual and mechanical harvesting has more to do with economics than wine quality. Premium wines can justify the expense of manual harvesting, but increasingly, mechanical harvesters are used for all quality categories of wines. Comparative studies have demonstrated no or negligible differences in the sensory characteristics of similar grapes harvested manually or mechanically (see Clary et al., 1990).

WINERY

Winemaker

Wine is primarily the vinous expression of a winemaker's practical and aesthetic skills. As such, no two winemakers produce identical wines. Every individual brings to the process the culmination of their experience and concept of quality. How well these are transformed into wine defines the difference between the skilled technician and the creative artisan.

Increasingly, the winemaker is in frequent communication with the grape grower. The interaction helps supply the raw materials the winemaker needs, as far as nature permits. Depending on the characteristics of the grapes reaching the winery, the winemaker must make decisions on how best to transform them into wine. Fig. 8.11 illustrates the basic sequence of events involved in this metamorphosis. None of the stages are without choices, the selection of which is likely to affect the wine's sensory attributes, for better or worse. While most decisions affect style, others primarily influence quality. Some of these decisions, and their quality implications, are briefly outlined below.

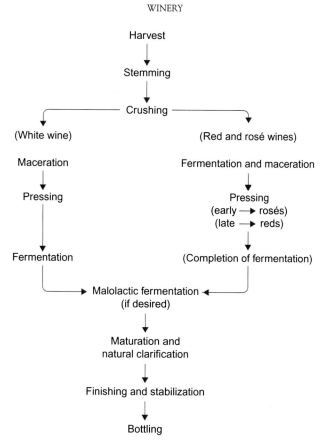

Harvest

↓

Stemming

↓

Crushing

(White wine)   (Red and rosé wines)

Maceration   Fermentation and maceration

↓   ↓

Pressing   Pressing
                                          (early → rosés)
                                          (late → reds)

Fermentation   (Completion of fermentation)

Malolactic fermentation
(if desired)

↓

Maturation and
natural clarification

↓

Finishing and stabilization

↓

Bottling

FIGURE 8.11 Flow diagram of winemaking. *From Jackson, R.S. (2014). Wine Science: Principles and Applications, fourth ed. Academic Press, San Diego, CA, reproduced by permission.*

Prefermentation Processes

Typically, grapes are crushed immediately upon reaching the winery. This disrupts the integrity of grape cells, allowing the release of nutrients, flavorants, and the escape of the juice. Crushing also liberates hydrolytic and oxidative enzymes that begin to react with grape constituents, further aiding nutrient and flavorant release. Until recently, exposure to air during, or after, crushing was considered detrimental (based on the belief that it made the subsequent wine more susceptible to oxidation). Although counterintuitive, early juice aeration limits subsequent oxidation by activating the expeditious oxidation and precipitation of readily oxidized phenolics. Thus, most wine-makers now allow air access during crushing and/or actively aerate the juice after crushing. This enhances the shelf-life of white wines, as well as encourages complete fermentation (supplying a slight amount of oxygen that favors the near-complete metabolism of fermentable grape sugars). Except for some white and dessert wines, sweetness is best donated of sterile grape juice. It is added to dry wine shortly before or during bottling.

Depending on the intent of the winemaker, the juice from freshly crushed grapes may be left in contact with the seeds and skins (pomace) for up to several hours (white wines), or days (red wines). The duration of skin contact depends on the intensity of flavor to be extracted. Up to a point, flavor intensity and aging potential increase with prolonged skin contact. This period, called maceration, occurs before fermentation with white wines, but simultaneously with fermentation for red wines. The difference relates to the much longer period required for anthocyanin and tannin extraction, the primary chemicals that distinguish red from white wines. An example of the close correlation between wine pigmentation and quality is illustrated in Fig. 8.12. Skin contact also favors a quick onset and completion to fermentation, but also alters the synthesis of yeast-produced aromatics. Thus, the fundamental character of a wine is partially determined by the timing and duration of maceration, as well as the temperature at which maceration occurs (Fig. 8.13).

The next major process affecting wine quality is pressing (separation of the juice from the seeds and skins). Ideally this should occur with minimal incorporation of particulate matter (cellular debris and macromolecular complexes). This is achieved by applying pressure over as large a surface area as practical. Most modern presses are elongated,

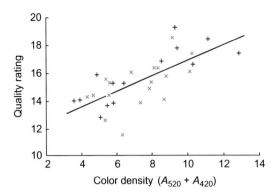

FIGURE 8.12 Relation between quality rating and wine color density in 1972 Southern Vales wines: (+) Cabernet Sauvignon; (×) Shiraz. *From Somers, T. C., Evans, M. E., 1974. Wine quality: Correlations with colour density and anthocyanin equilibria in a group of young red wine. J. Sci. Food. Agric.* **25**, *1369-1379. Copyright Society of Chemical Industry. Reproduced with permission. Permission is granted by John Wiley & Sons Ltd on behalf of the SCI.*

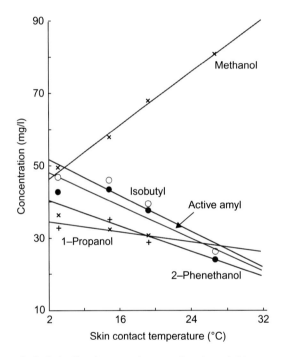

FIGURE 8.13 Concentration of various alcohols in Chardonnay wine as a function of skin-contact temperature. *From Ramey, D., Bertrand, A., Ough, C.S., Singleton, V.L., Sanders, E., 1986. Effects of skin contact temperature on Chardonnay must and wine composition. Am. J. Enol. Vitic. 37, 99–106, reproduced by permission.*

horizontally positioned, cylindrical chambers. Pressure is applied either via air pumped into a membranous bladder against the must (juice, seeds, and skins), or by plates that move in from one or both ends. The older, basket-type presses were positioned vertically and had pressure applied by a plate from the top. The gentler pressure applied by new press designs liberates juice that is less bitter and astringent, but rich in any potential varietal flavors. In contrast, the older presses produced rougher tasting wines with fewer fruit flavors. This partially resulted from the frequent use of grape stalks to produce drainage channels to ease juice drainage.

Regardless of the means of pressing, juice from white grapes usually needs some clarification before the start of fermentation. For this, winemakers have many means at their disposal. The selection typically has more to do with economics and speed than with quality concerns. Most clarification procedures have little effect on wine quality, if not used to excess. Because pressing in red wine production occurs at or near the end of fermentation, rapid clarification is rarely a critical issue. Correspondingly, clarification of red wines is initially by gravity-induced sedimentation.

If the sugar and acid composition of the juice is unavoidably inferior, the winemaker usually attempts to make adjustments at this stage. The addition of sugar and/or the addition of acidity (or its neutralization) can improve the basic attributes of the wine. It cannot compensate for a lack of color or flavor, though. As noted earlier, these desirable qualities frequently develop concomitantly with desirable grape acid and sugar levels.

Fermentation

Fermentor

The first fermentative decision facing the winemaker is the type of fermentor. Typically this will be a closed tank, as large as conveniently possible (economies of scale). However, small producers may choose fermentation in small (~250 L) oak barrels, especially for lots of high-quality juice. Those who favor this option justify the expense by the "cleaner" (less fruity) expression of the wine's varietal aroma. In-barrel fermentation modifies yeast metabolism, and thereby the aromatics they produce, as well as donates compounds extracted from the oak. These differences alone can generate features that can define the differences between wines from adjacent properties.

For the majority of white wines, fermentation occurs in comparatively simple tanks, except for the cooling required to regulate the fermentation temperature and rate. The tanks are usually made from inert materials, typically stainless steel. This permits transformation of the juice into wine without compromising the grape's natural flavors. Red wines are also frequently fermented in inert tanks. However, red wine fermentors vary considerably more in design than white wine fermentors. The difference is imposed by the need to extract color and flavorants from the skins during fermentation. As fermentation progresses, the carbon dioxide generated carries the seeds and skins to the surface of the juice, forming a cap. Various means have been developed to periodically or continuously submerge the cap into the fermenting juice. One of the more recent and effective designs is a rotary fermentor. It commonly possesses several blades attached to a central cylinder. These slowly rotate (or can be set to simply slosh) the fermenting must back and forth to a desired degree, gently mixing the seeds, skins, and juice. This process promotes rapid pigment and flavor extraction, while apparently favoring the extraction of soft tannins, while limiting the liberation of hard tannins. The result is a full-flavored wine, of intense color, but smooth enough to be enjoyed without requiring prolonged aging. Most other fermentors, giving intensely flavored and colored red wines, require several to many years to soften.

Yeasts

The next serious decision facing the winemaker is whether to permit spontaneous fermentation (by yeasts on the grape and winery equipment), or to add one or more commercial yeast strains (induced fermentation). There are advocates on both sides, equally claiming superior results. Spontaneous fermentations may yield more complex wines, but at the risk of wine spoilage. Most of the added complexity seems to come from acetic acid and diacetyl, but this may only reflect our lack of knowledge. At threshold levels, acetic acid and diacetyl can add an element of "sophistication," but at slightly higher concentrations, they generate off-odors. Preference likely depends on personal thresholds and/or habituation. Even where induced fermentation is the choice, the winemaker must decide on which species, strains, or combination thereof to use. As yet, there are no clear guidelines other than experience, and winemaker preference. These effects can be more than just subtle (Figs. 8.14 and 8.15).

If deciding which species or strain to use were not enough, yeast properties can vary with the fermentation conditions. The chemical composition of grapes (which varies from year to year and cultivar to cultivar), as well as the physical conditions of fermentation (e.g., temperature and pH), alter yeast metabolic activity.

Lactic Acid Bacteria

Most wines undergo two fermentations. The first, yeast-induced fermentation generates the alcohol and vinous bouquet that characterize wines. The second, bacteria-induced fermentation converts malic into lactic acid, reducing wine acidity. This alone can modify the wine's perception. However, byproducts released by bacterial metabolism further modify the wine's flavor (Fig. 8.16). In many instances, winemakers encourage malolactic fermentation more for its sensory contribution than its acidity reduction.

Malolactic fermentation is encouraged in most red wines, notably in moderate-to-cool climatic regions. It makes the wine more drinkable by mollifying a potentially overly sour, rough taste of the wine. In contrast, winemakers attempt to limit malolactic activity in warm-to-hot climatic regions, where the grapes (and wine) have a tendency to be too low in acidity. The action of malolactic fermentation could give the wine a flat taste under the latter conditions. In addition, malolactic fermentation has a propensity to generate off-odors in wines of low acidity.

Because malolactic fermentation is often sporadic, especially in wines low in pH, winemakers frequently inoculate their wines with one or more desirable strains (typically *Oenococcus oeni*). As with yeast strains in alcoholic fermentation, bacterial strains differ considerably in their aromatic impact. Fig. 8.17 illustrates the sensory effects

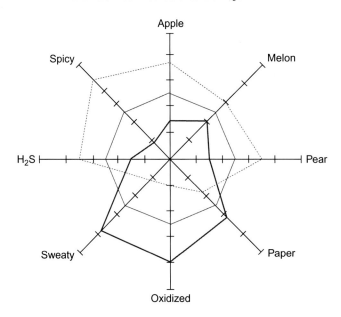

FIGURE 8.14   The effect of spontaneous (---) and induced (■■) fermentation on the sensory characteristics of Riesling wine; (——) mean score: H₂S, hydrogen sulfide. *From Henick-Kling, T., Edinger, W., Daniel, P., Monk, P., 1998. J. Appl. Microbiol. 84, 865–876; reproduced by permission.*

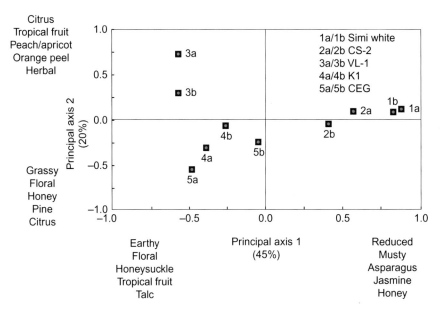

FIGURE 8.15   Profile of aroma of a Riesling wine (after 20 months) fermented with different yeast strain. From Dumont, A., Dulau, L., 1996. The role of yeasts in the formation of wine flavors. In "Proc. 4th Int. Symp. Cool Climate Vitic. Enol. In: Henick-Kling, T., et al. (Eds.), New York State Agricultural Experimental Station, Geneva, NY, pp. VI–24–28, reproduced by permission.

inducible by different strains of *Oenococcus oeni*. From the divergence of opinion demonstrated in Fig. 8.17, it should be no surprise that there are strong and diverse opinions concerning the relative merits of malolactic fermentation.

Postfermentation Influences

Adjustments

Ideally, a wine should require only minimal clarification, such as spontaneous settling and gentle fining before bottling. However, if the wine is bottled early, is imbalanced, or possesses some fault, additional treatment(s) may

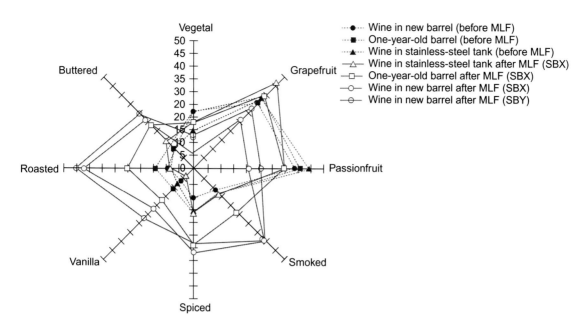

FIGURE 8.16 Differentiation of wine as affected by malolactic fermentation (MLF) in stainless steel or oak cooperage by two strains of lactic acid bacteria (SBX and SBY). *From de Revel, G., Martin, N., Pripis-Nicolau, L., Lonvaud-Funel, A., Bertrand, A., 1999. J. Agric. Food Chem. 47, 4003–4008. Copyright(1999) American Chemical Society).*

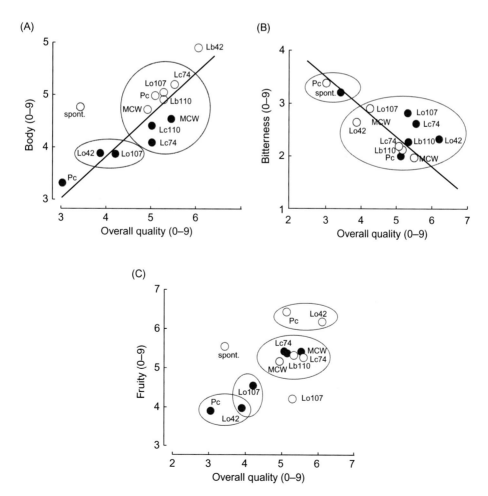

FIGURE 8.17 Relation of body (A), bitterness (B), and fruitiness (C) to overall quality of Cabernet Sauvignon wine fermented with various lactic acid bacteria. The relation was assessed by two different panels, one composed of winemakers (●) and the other a wine research group (○). *From Henick-Kling, T., Acree, T., Gavitt, B.K., Kreiger, S.A., Laurent, M.H., 1993. Sensory aspects of malolactic fermentation. In Proc. 8th Aust. Wine Ind. Tech. Conf. In: Stockley, C.S., et al. (Eds.), Winetitles, Adelaide, Australia, pp. 148–152, reproduced by permission.*

be required. Nevertheless, such treatments need to be kept to a strict minimum. Most forms have the potential to remove or neutralize the subtle distinctiveness that ideally should grace every wine.

Blending

Blending is one of the most misunderstood aspects of wine production, especially for some supposed wine connoisseurs. In one or more forms, blending is involved in the production of every wine. It can vary from the simple mixing of wine from different fermentors, to the complexities of combining wines produced from different cultivars, vineyards, or vintages. For several wines, notably sparkling, sherries, and ports, complex blending is central to their quality and brand distinctiveness. In other regions, blending wines made from several grape varieties supplies their traditional character (e.g., Bordeaux and Chianti). Blending tends to enhance the best qualities of each component wine, while diminishing their individual defects. Fig. 8.18 illustrates the general benefits of blending. It shows that blends between wines of roughly similar character and quality were considered as good or better than their component wines.

The negative connotation often attributed to blending arises primarily from those with a vested interest in appellation control, and smaller wineries promoting their wines' uniqueness (estate bottled). Authenticity of provenance and vintage derivation are marketed as essential ingredients in quality. Whether this is valid depends more on the skills of the producer and grape maturity than delimited geographic origin. Blending wines from different vineyards and regions is no more detrimental to quality than blending between wines produced from different grape varieties. Geographic identity does, however, give the consumer a readily recognizable identifier, and seemingly justify having confidence in the wine's sensory attributes. Much effort is being currently spent on studies to identify unifying sensory features that may characterize particular appellations. A review of techniques investigating wine typicity is found in Maitre et al. (2010). Examples of results are illustrated in Figs. 5.46 and 5.47. Although laudable, one has to wonder if much of the impetus behind these studies is the potential marketing advantage that may accrue to the appellation involved rather than esoteric academe.

Processing

An old processing technique receiving renewed attention is *sur lies* maturation. The process involves leaving white wine in contact with the lees (dead and dying yeast cells) for an extended period. The contact period typically occurs in the same container as did fermentation, traditionally barrels. *Sur lies* maturation can enhance wine stability and increase flavor complexity. This benefit, however, runs the risk of contamination with hydrogen sulfide and other odoriferous reduced-sulfur compounds (released from the lees). To reduce this likelihood, the wine is periodically stirred to incorporate small amounts of oxygen. Unfortunately, this, in turn, can activate dormant acetic acid bacteria that produce acetic acid and ethyl acetate off-odors. Thus, the wine in each barrel must be sampled frequently to assess the wine's development.

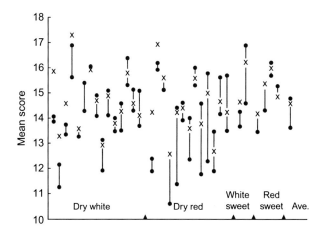

FIGURE 8.18   Mean quality scores of 34 pairs of wines compared (●) with their 50:50 blend (×), indicating the value of a complex flavor derived from blending. *From Singleton, V.L. (1990). An overview of the integration of grape, fermentation, and aging flavours in wines. In Proc. 7th Aust. Wine Ind. Tech. Conf. In: Williams, P.J., et al. (Eds.), Winetitles, Adelaide, Australia, pp. 96–106 [based on data from Singleton and Ough, 1962], reproduced by permission.*

Sparkling wine production also involves a form of *sur lies* maturation. In this instance, the process occurs in the bottle in which the second fermentation takes place. Because the amount of lees involved is small, the generation of reduced-sulfur odors is not a concern. On the contrary, yeast autolysis generates a toasty scent, if the wine remains on the lees sufficiently long, which is typically the case with champagnes. In addition, yeast autolysis liberates colloidal mannoproteins that favor the generation of continuous chains of small, semi-durable bubbles (Feuillat et al., 1988; Maujean et al., 1990). The helical structure of these proteinaceous polymers probably entraps carbon dioxide as it can volatile compounds.

Processing is also crucial to the flavor of most fortified wines. For example, fractional (*solera*) blending provides the consistency of character expected of sherries, and promotes the growth of *flor* yeasts required for *fino* production. Equally, baking (*estufagem*) is essential to development of a typical madeira flavor.

Oak

Wines with sufficient flavor and distinctiveness may be matured in oak cooperage. This can add a desirable element of complexity, as well as occasionally improve varietal expression (Sefton et al., 1993; Fig. 8.19). Fig. 8.20 further illustrates how the presence of one component (oak lactones) correlates with several wine attributes.

Oak can donate a spectrum of flavors, depending on a whole host of factors. These include the oak species (Fig. 8.21), the conditions under which the trees grew (Chatonnet, 1991), the method of wood seasoning (Chatonnet et al., 1994), the degree of "toasting" (heat applied during barrel coopering), and the number of times the barrel has been used (Fig. 8.22). Each aspect modifies the attributes contributed by the oak. For example, light toasting retains most of the basic oak flavors intact (e.g., oaky, coconut attributes); medium toasting partially degrades these, generating some pyrolytic byproducts (notably vanilla and caramel flavorants); and heavy toasting/charring degrades most natural oak flavorants and phenolic and furanilic aldehydes (initially produced on the inner surfaces

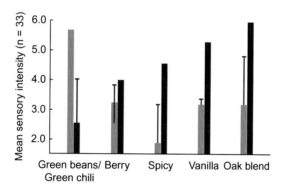

FIGURE 8.19 Mean intensity rating of aroma terms for Cabernet Sauvignon wines aged 338 days in glass (control) and in French oak barrels (*light bars*, control; *dark bars*, oak-aged; 11 judges, 3 replications). *From Aiken, J.W., Noble, A.C., 1984. Comparison of the aromas of oak- and glass-aged wines. Am. J. Enol. Vitic.35, 196–199, reproduced by permission.*

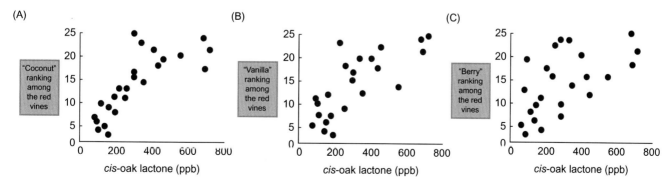

FIGURE 8.20 Correlation of the perception of several flavor characteristics with the presence of *cis*-oak lactones in red wines matured in oak barrels: (A) coconut, (B) vanilla, and (C), berry. The rank correlation was significant in all three cases ($p < .001$). *From Spillman, P.J., Pocock, K.F., Gawel, R., Sefton, M.A., 1996. The influences of oak, coopering heat and microbial activity on oak-derived wine aroma. In: Stockley, C.S., et al. (Eds.), Proc. 9th Aust. Wine Ind. Tech. Conf, Winetitles, Adelaide, Australia, pp. 66–71, reproduced by permission.*

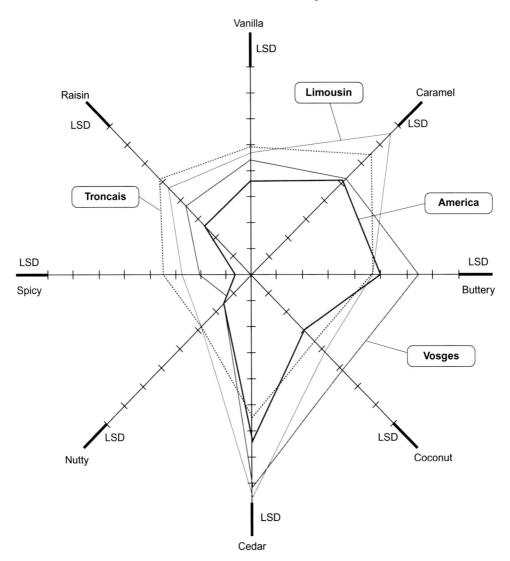

FIGURE 8.21   Polar coordinate graph of mean intensity ratings and least-significant differences (LSD) for descriptors by oak origin (n = 14 judges × 3 reps × 6 samples). *From Francis, I.L., Sefton, M.A., Williams, J., 1992. A study by sensory descriptive analysis of the effects of oak origin, seasoning, and heating on the aromas of oak model wine extracts. Am. J. Enol. Vitic. 43, 23–30, reproduced by permission.*

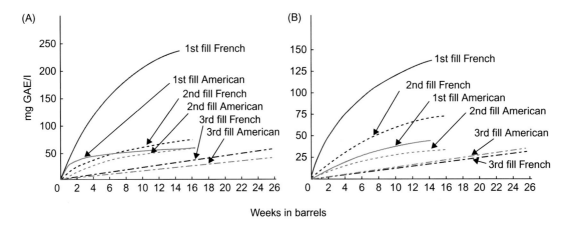

FIGURE 8.22   Changes in (A) total phenolics and (B) nonflavonoids over time for French and American oak barrels. *From Rous, C., Alderson, B., 1983. Phenolic extraction curves for white wine aged in French and American oak barrels: GAE, gallic acid equivalents. Am. J. Enol. Vitic. 34, 211–215, reproduced by permission.*

of the staves), producing volatile phenols with a smoky, spicy aspect (e.g., eugenol, quaiacol, 2-methoxyphenol). The former compounds still occur, but are now located deeper in the wood, requiring longer for their extraction.

The intensity of these respective aspects can be regulated by the duration of wine contact (varying from several weeks to years) and cooperage capacity. Whether the added flavors enhance or detract from the central character of the wine depends more on personal preference than chemistry. In some appellations, law dictates the type and duration of oak maturation. In such cases, a certain oakiness is considered an obligatory attribute.

Where economics demands alternatives to oak barrels, thin slats of oak may be submersed in tanks while the wine is maturing in the cooperage. Even less expensive options can involve adding oak chips or sawdust (in permeable sacks to facilitate removal), or oak extract to maturing wine.

Bottle Closure

For several centuries, cork has been the exclusive closure for wine matured in barrel or in bottle. In this role, cork has several distinctly desirable properties. These include compressibility (with little lateral expansion for ease of insertion), elasticity (rapid return to its original shape after compression), resilience (prolonged exertion of pressure against the bottle neck after insertion), chemical inertness (notably to wine acids and alcohol), relative impermeability to most liquids and gases, and a high coefficient of friction (adheres well to surfaces). Thus, it normally provides a long-lasting, tight seal that limits contamination and loss of aromatics, and severely retards any gas entrance or escape. Most cork stoppers of normal length possess a permeation barrier of about 300–500 cells/cm along their length. Because of the direction in which stoppers are punched out from the cork-oak bark, the principal porous regions of the cork (lenticels and crevices) are positioned at right angles to the stopper's length (Fig. 8.23).

Nevertheless, the dominance of natural cork is being challenged by alternatives. This has arisen primarily due to contamination of the cork with off-odors, either in the forest, during bark storage (seasoning), stopper manufacture, or during transport and storage to wineries. The most well known of these is 2,4,6-trichloroanisole (TCA). In addition to being a potential source of off-odors, cork can scalp (absorb) aromatics from the wine. Finally, variations in the internal structure and morphology of cork stoppers (difficult to assess visually) can affect their protective function. Unexplained examples of apparently random premature wine oxidation have often been attributed to faults in cork anatomy, permitting undue oxygen ingress, but maybe not (Lagorce-Tachon et al 2016). Common alternatives to natural cork for bottle closure include agglomerate cork, synthetic cork, and roll-on (screw) caps.

Premium quality corks possess sealing qualities equivalent to screw caps, but progressively lose their elasticity over time (~20 years). Thus, their sealing properties eventually fail. The deterioration progresses outward, from the end in contact with the wine. Thus, cork length affects how long a cork is likely to effectively seal the bottle, a relationship that has not gone unnoted by some producers, desirous of subtlety denoting their wine's greater quality (or aging potential).

A factor little recognized for its importance to sealing quality is the rate at which the cork tissue grew. Bark derived from cork oak (*Quercus suber*) that has grown slowly contains a higher proportion of resilient, spring-produced, cork cells (and correspondingly shows more annual growth rings in the stopper (Fig. 8.24). Thus, cork derived from trees grown in drier, mountainous regions has better sealing properties than cork derived from trees grown in moister, lowland regions.

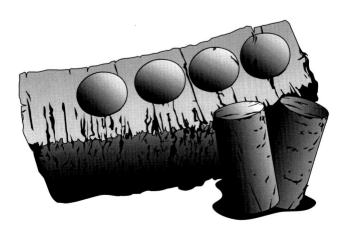

FIGURE 8.23 Slab of cork bark showing the direction of stopper extraction. *Diagram courtesy of H. Casteleyn.*

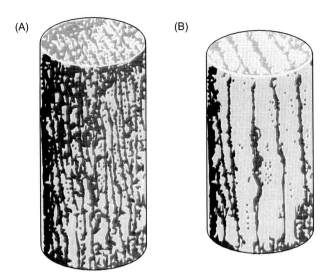

FIGURE 8.24   Representative diagrams of (A) high-quality corks, derived from slow-growing cork oak, and (B) lower-quality cork, derived from rapid-growth cork oak. Note the large number of growth rings (≥9) in the cork on the left compared with the one on the right (≤7) *Diagram courtesy of H. Casteleyn.*

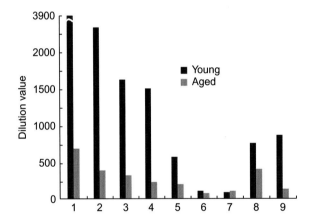

FIGURE 8.25   Loss of aromatic compounds during the aging of a Vidal wine (1, β-damascenone; 2, 2-phenylethanol; 3, fruity; 4, floral; 5, spicy; 6, vanilla/woody; 7, vegetative; 8, caramel/oatmeal; 9, negative aromas). *From Chisholm, M.G., Guiher, L.A., Zaczkiewicz, S.M., 1995. Aroma characteristics of aged Vidal blanc wine. Am. J. Enol. Vitic. 46, 56–62, reproduced by permission.*

Aging

The tendency of wine to improve, or at least change, during in-bottle aging is one of its most intriguing properties. Unlike most commercial food and beverage products, which have a "best before" date, no such precise designation applies to wine (with the exception of nouveau wines) (Jackson, 2016a). Unfortunately, most wines improve only for a few several years, before beginning a show, progressive, and irreversible loss of aromatic character (Fig. 8.25). Only rarely does this involve microbial spoilage, leading to the generation of off-odors or a vinegary character.

During the initial stages (maturation prior to bottling), loss of yeasty odors, excess dissolved carbon dioxide, and the precipitation of particulate material lead to sensory improvements. Additional improvements in character may commence during maturation and continue during aging (postbottling). Examples involve acid-induced liberation of terpenes and other aromatics from nonvolatile glycosidic complexes. Several hundred glycosides have been isolated from varieties such as Riesling, Chardonnay, Sauvignon blanc, and Shiraz.

One of the more obvious age-related changes is a color shift toward brown. Red wines may initially deepen in color after fermentation, but subsequently become lighter and take on ruby and then brickish hues. An initial decrease in color intensity can result from the dissociation of anthocyanin complexes found in grapes. This is usually

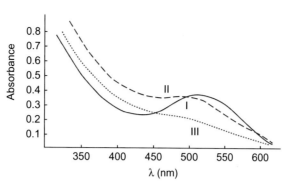

FIGURE 8.26   Absorption spectra of three red wines of different ages: I, 1-year old; II, 10-year old; III, 50-year old. *From Ribéreau-Gayon, P., 1986. Shelf-life of wine. In: Charalambous, G. (Ed.), Handbook of Food and Beverage Stability: Chemical, Biochemical, Microbiological and Nutritional Aspects. Academic Press, Orlando, FL, pp. 745–772, reproduced by permission.*

followed by polymerization between anthocyanins and tannins. These stabilize the color, but initiate a slow loss in color density and shift from red-purple to brickish, and eventually brownish red. These changes are detectable with a spectrophotometer as a drop in optical density, and a shift in the absorption spectrum (Fig. 8.26). Without polymerization, anthocyanins oxidize and the wine would permanently lose its vibrant color. The reasons why different cultivars vary in the rate at which their color changes during aging probably relates to the amounts and types of anthocyanins, and the copigments they possess. In addition, small amounts of acetaldehyde, produced following limited aeration (coincidental consequences associated with racking and other maturation procedures), are known to enhance anthocyanin–tannin polymerization. The storage temperature and pH of the wine are also significant factors affecting the rate of color change.

Age-related color changes in red wines are often assessed by measuring changes in optical density at two wavelengths (520 and 420 nm) (Somers and Evans, 1977). High 520/420 nm values indicate a bright-red color, whereas low values denote a shift to brickish tones. In contrast, white wines darken in color, developing yellow, gold, and eventually brownish shades. The origins of this latter color shift are poorly understood. Nonetheless, it probably involves a combination of phenolic oxidation, metal ion–induced structural changes in galacturonic acid, Maillard reactions between sugars and amino acids, and sugar caramelization.

The slow hydrolytic rupture of glycosidal linkages during aging not only liberates aromatic compounds, but also phenolics such as quercetin and resveratrol. The sensory significance of the release of these phenolics is unclear. However, their reduced solubility as free phenolics may accentuate crystallization and haze generation.

During aging, wines lose their original fresh fruity character. This is especially noticeable when the fragrance depends on fruit esters, whose concentration progressively declines (Figs. 8.27 and 8.28). Although the concentration of volatile carboxylic acid esters (e.g., diethyl succinate) increases, they are poorly volatile. In addition, the fruity/floral aspects of terpenes slowly degrade or oxidize to less aromatic or flavorless compounds. Oxidation of 3-mercaptohexan-1-ol is one of the principal reasons for the comparatively short shelf-life of most rosé wines (and the loss of its fruity fragrance) (Murat, 2005).

Wines that age well tend to show a varietal character that initially becomes more evident (e.g., Riesling), being finally replaced by a subtle, complex, aged bouquet. In white wines, their sensory attributes (if they develop) are often couched in terms such as hay or honey. For red wines, when the mature jammy character begins to fade, it is supplanted by an aged bouquet often described in terms of leather, cigar box, smoky, and truffle. Other than differences correlated with wine color, most aged bouquets seem similar and relatively independent of cultivar origin (see the descriptions of aged wines in Broadbent, 1980).

The chemical nature of most aged bouquets is poorly understood. However, it partially involves degradation byproducts of norisoprenoids and related diterpenes, carbohydrate derivatives, and the synthesis of thiols, and volatile and oxidized phenolics.

Of isoprenoid degradation products, 1,1,6-trimethyl-1,2-dihydronaphthalene (TDN) appears to contribute to the aged bouquet of Riesling wines. Table 8.2 illustrates a few other age-related chemical changes that can occur in Riesling wines. The table also demonstrates the importance of temperature to the aging process, as does Fig. 8.28. Another isoprenoid, (E)-1-(2,3,6-trimethylphenyl)buta-1,3-diene, often accumulates during aging, and can donate a green or cut-grass aroma to several white wines. Although this compound has been isolated from red grapes, it has not been found in red wines. This seeming anomaly may to be due to its reaction with tannins (Cox et al., 2005).

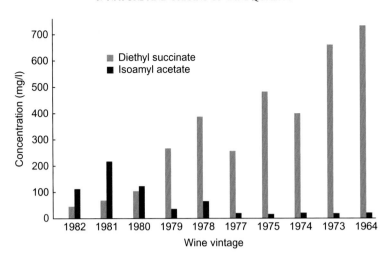

FIGURE 8.27   Examples of the influence of wine age on the concentration of esters, notably acetate esters (isoamyl acetate) and ethanol esters (diethyl succinate). *From Jackson, R.S., 2014. Wine Science: Principles and Applications, fourth ed. Academic Press, San Diego, CA (Jackson, 2014), data from Rapp, A., Güntert, M., 1986. Changes in aroma substances during the storage of white wines in bottles. In: Charalambous, G. (Ed.), The Shelf Life of Foods and Beverages. Elsevier, Amsterdam, pp. 141–167, reproduced by permission.*

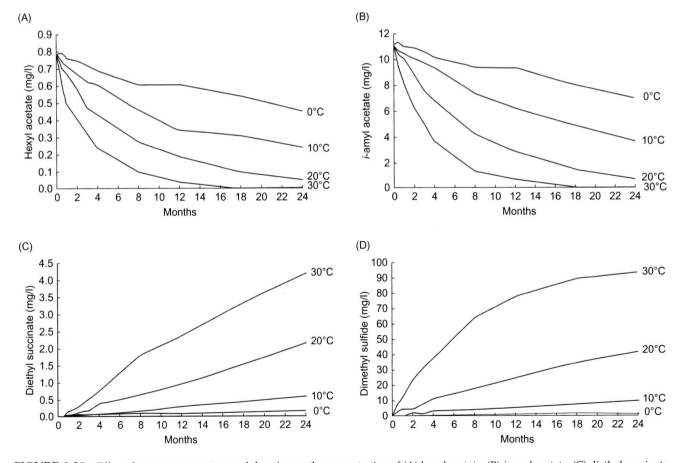

FIGURE 8.28   Effect of storage temperature and duration on the concentration of (A) hexyl acetate, (B) *i*-amyl acetate, (C) diethyl succinate, and (D) dimethyl sulfide in a Colombard wine. *From Marais, J., 1986. Effect of storage time and temperature of the volatile composition and quality of South African Vitis vinifera L. cv. Colombar wines. In: Charalambous, G. (Ed.), Shelf Life of Foods and Beverages. Elsevier, Amsterdam, pp. 169–185, reproduced by permission.*

TABLE 8.2 Changes in bouquet composition from carbohydrate decomposition during aging of a Riesling wine[a,b]

Substance from carbohydrate degradation	Year					
	1982	1978	1973	1964	1976 (frozen)	1976 (cellar stored)
2-Furfural	4.1	13.9	39.1	44.6	2.2	27.1
2-Acetylfuran	—	—	0.5	0.6	0.1	0.5
Furan-2-carbonic acid ethyl ester	0.4	0.6	2.4	2.8	0.7	2.0
2-Formylpyrrole	—	2.4	7.5	5.2	0.4	1.9
5-Hydroxymethylfurfural (HMF)	—	—	1.0	2.2	—	0.5

From Jackson, R.S., 2014. Wine Science: Principles and Applications, fourth ed. Academic Press, San Diego, CA.
[a]Data from Rapp and Güntert (1986).
[b]Relative peak height on gas chromatogram (mm).

Other isoprenoid degradation products, such as vitispirane, theaspirane, ionene, and damascenone appear little involved in the development of an aged bouquet.

Most changes in terpene content are negative, their oxidation not only changing their sensory quality but increasing their thresholds (Rapp and Mandery, 1986). However, accumulation of a monoterpene ketone, piperitone, contributes to the minty character of aged red Bordeaux wines (Picard et al., 2016).

Carbohydrate degradation products, notably Maillard products, develop slowly at ambient temperatures. These contribute to the brownish gold coloration of some white wines, as well as donating various flavors. The most familiar is a caramel fragrance. Another Maillard product, the ethyl ester, 2-(ethoxymethyl)furan, may contribute to a fruity, slightly pungent character. It has been found in aged Sangiovese wines (Bertuccioli and Viani, 1976).

The concentration and nature of reduced-sulfur compounds often change during aging. For example, the accumulation of dimethyl sulfide has been correlated with development of a desirable aged aspect in aged Colombard wines (Fig. 8.28D). Its addition (20 mg/L) to several wines has also been correlated with an increase in the wine's flavor score (Spedding and Raut, 1982), but higher concentrations (≥ 40 mg/L) were considered detrimental. By itself, dimethyl sulfide has a shrimp-like odor. Occasionally, the production of dimethyl sulfide is so marked at warm temperatures that it can mask a wine's varietal character within several months (Rapp and Marais, 1993). Other thiols contributing to aged bouquets, in this case red Bordeaux, are 3-sulfanylhexanol and 2-furanmethanethiol (Picard et al., 2015). 2-Furanmethanethiol forms in oak-aged wine and reportedly possesses toasty, cooked meat, and roasted coffee notes (Tominaga et al., 2000).

In red wines, one of the best-understood aspects of aging relates to the polymerization of bitter, astringent tannin subunits into large complexes. Because increased polymer size usually correlates with enhanced astringency, at least up until precipitation occurs, polymerization has normally been assumed to increase perceived astringency. However, acetaldehyde-induced polymerization may reduce solubility (Matsuo and Itoo, 1982). Thus, the uptake of oxygen during barrel maturation and aging may partially explain the oft-noted progressive mellowing of wine astringency. The reverse process, proanthocyanidin breakdown (Vidal et al., 2002) may also play a role in age-related reduction in astringency. A summary of potential chemical modifications during wine aging is illustrated in Fig. 8.29.

Other than the initial benefits of aging noted earlier, older wines do not necessarily show greater quality than their younger versions. Younger versions demonstrate more fruitiness, and typically express a truer manifestation of any varietal character. As the wine ages, the fragrance tends to become less fruity; red wine fragrances taking on a more jammy character, becoming progressively less varietal, more subtle, and potentially developing an aged aspect. In addition, a well-aged wine often possesses the qualities of flavor development and an exquisite finish. It depends on personal preference whether the fresh young aroma of youth, the fuller, more complex bouquet of a mature wine, or the fully developed aged bouquet is more esteemed. They are certainly different, but equally enjoyable in their separate ways, but, given the choice, the author will always go for the finesse and elegance of the aging monarch (also, there is no shortage of excellent young wines). However, anyone aging wine must be cautious not to go overboard; what fragrance and flavor an old wine retains, if any, may rapidly dissipate upon opening. The wine may be rare, the tasting a unique and historic experience, but the sensory consequences may be bereft of any delight and a bathos. Regrettably, few consumers are likely to have a chance to sample well-aged wines, unless they have aged wine themselves. Finding wines in the range of 10–15 years old on store shelves is now exceptionally

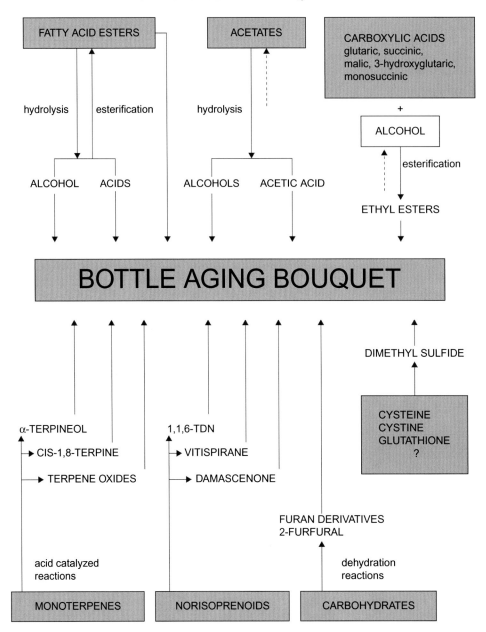

FIGURE 8.29   Illustration of the types of potential chemical changes in wine during aging. *Reprinted from Shelf Life Studies of Foods and Beverages. Chemical, Biological, Physical and Nutritional Aspects. (G. Charakanbous, ed.), Aldave, L., Almy, J., Cabezudo, M. D., Cáceres, I., González-Raurich, M., Salvador, M. D., 1993. The shelf-life of young white wine. pp. 923–943, with permission from Elsevier.*

rare, and, if the wines are not known to be from good producer and have been stored under ideal conditions, they may not be worth buying.

The effects of aging have usually been interpreted as changes in chemical equilibria (e.g., esters to their alcohol and acid constituents), hydrolysis of aromatic glycosides (e.g., terpenes, thiols, and norisoprenoids), oxidation reactions (e.g., formation terpene oxides), reduction reactions (e.g., formation of thiols), and breakdown products (e.g., hydrocarbons). However, other factors to consider are absorption (scalping) of aromatics by, and diffusion and loss of aromatic compounds via, the closure. The rate of loss of any compound is a function of how it is absorbed by, and diffuses through, or around the closure. These properties are affected by the absorptive and diffusion properties of the closure, the ambient temperature, the solubility and volatility of the aromatics (a function of their hydrophobic/hydrophilic properties, degree of hydration, association in ethanol micelles, and weak bonding with other matrix constituents), and the compound's molecular weight. Alone, the molecular weight of the compound suggests that a wine is likely to lose lighter-weight aromatics at a faster rate than heavier compounds. Thus, with age, a wine theoretically should become progressively more characterized by heavier aromatic compounds, i.e., those that tend to possess more ligands, and therefore react with a wider range of olfactory receptors. The result is that an aged bouquet becomes not only quantitatively simpler (being less varied chemically and present at lower concentrations), but also sensorially more complex (activating a wider range of olfactory receptors). Advances in headspace sampling are currently available to confirm and extend these views (Bicchi et al., 2012; Stashenko and Martínez, 2012).

Aging Potential

In most instances, the best guide to a wine's aging potential is experience. However, the problem is that few consumers have the opportunity to gain the requisite experience. Thus, they must depend on advice, which may be of questionable quality. Recommendations often show a distinct cultural bias, even from so-called experts. Relative to Bordeaux wines, earlier consumption is recommended by French experts more than by their British counterparts, with American authorities being somewhere in between. The personal value of such advice can be established only by experimentation. In the case of wines of repute, this would be as expensive as it would be long.

The situation is not helped by a lack of detailed information on the origin of a wine's aging potential. Usually it is considered to depend primarily on the wine's alcohol and phenolic contents, as well as sugar content with sweet wines. Increases in any of these constituents are thought to augment aging potential, up to a point. Varietal origin is also clearly important. For example, varieties depending predominantly on yeast-generated ethyl and acetate esters or most terpenes have truncated aging potentials. Riesling is a clear exception to this generality. In contrast, most red wines have longer aging potentials (often considered to be derived from the antioxidant action of their phenolic constituents).

Although variety and chemical constituents are important to aging potential, how the wines have been made and stored are equally important. Cool storage temperatures dramatically slow the effects of aging (Fig. 8.28), markedly enhancing aging potential. Aging is often considered optimal at about 10°C, but this may simply reflect habituation, many underground cellars possessing temperatures in this range. Having heard that temperature is important, people often asked at what temperature they should age their wines. I respond, somewhat facetiously, by asking how long they expect to live. If the response is long, it is safe to age wine slowly at cool temperatures, otherwise, store at near room temperature. Whether aging can be effectively accelerated at much about this point is a moot point. Heating promotes some aging reactions, but it also activates others that are generally viewed as detrimental (Singleton, 1962). Several commercial products supposedly produce the effects of aging within minutes. Except for magnets, none of these appear to have been subjected to scientific scrutiny. It the case of magnets, the results did not verify the claims of the producer (Rubin et al., 2005). Patience and time are the only known effective procedures.

Vibration has occasionally been reported as detrimental to wine aging. Only minor physicochemical changes have been observed, with marked and prolonged vibration (Chung et al., 2008), and they appear to have had no sensory significance. Another common view is that movement, whether it is from the store to home, or by overseas transport, is detrimental. The only evidence of any veracity to these views relates to potential exposure to temperature extremes during transport, not movement. The effect of temperature fluctuations may also be the origin of the idea that vibration is a detriment.

Unfortunately, consumers rarely have the opportunity to investigate these factors personally. For them, they must depend on extrinsic factors to guestimate aging potential. These typically involve the wine's price, as well as the repute of the vintage, winery, region, and producer. For additional, dubious sources of "precision," there are the plethora of books, magazines, and newspaper articles extolling the virtues of wines and their aging potential. Another indicator, suggesting the views (or hopes) of the winemaker, is the length and quality of the cork used to seal the wine.

CHEMISTRY

Although wine quality is typically framed in terms of age, vintage, style, provenance, varietal origin, prestige, or other attributes, its legitimate quality lies in its sensory characteristics derived from its chemistry. With more than 800 organic constituents known to be potentially present in wine, that chemistry is obviously complex. It also changes over time, as compounds volatilize, degrade, oxidize, reduce, polymerize, depolymerize, and undergo other transformations. Nonetheless, the vast majority of these compounds occur at concentrations below their individual detection thresholds. Even acknowledging synergistic interactions that may enhance detection, the number of sensorially important compounds may be less than 50 in any particular wine. Of these, only a few groups, notably sugars, alcohols, carboxylic acids, esters, and phenolics affect the sensory attributes of essentially all wines. They donate much of the basic vinous character of a wine.

TABLE 8.3   Visual, aroma and in-mouth (taste and mouth-feel) terms linked to high- and low-quality perception. Terms cited by less than 15% of experts have been omitted for clarity. Numbers in parentheses are the frequency of citation for a term expressed in %

	High quality	Low quality
Visual terms	Limpidity/clarity (81), high depth-intensity (71), red–purple color (43)	Oxidzed-brown color (81), turbidity (67), low color intensity (57)
Aroma terms	Fruit (71), integrated wood (71), intense aroma (43), complex aroma (29), varietal aroma (24)	Oxidation (57), reduction (52), dirt 48), low intensity (48), *Brett* (43), excessive old wood (33), fault (33), green/vegetal (24), mold (19)
Taste and mouth-feel terms	Balance (67), volume/body (48), round/smooth tannins (43), persistency (24), fatty mouth-feel (19)	Excessive astringency (67), excessive sourness (52), unbalance (48), light/short (33), green (29), bitterness (29), coarse tannins (19)

Reprinted from Sáenz-Navajas, M.-P., Avizcuri, J. M., Echávarri, J. F., Ferreira, V., Fernández-Zurbano, P., Valentin, D., 2015. Understanding quality judgements of red wine by experts: Effect of evaluation condition. Food Qual. Pref. 48, 216–227, with permission from Elsevier.

The features that distinguish fine from ordinary wine have usually been ascribed to the myriad of minor constituents that may occur in wine. Ritchey and Waterhouse (1999) conducted a fascinating analysis of the chemical differences between high-volume and ultrapremium Californian Cabernet Sauvignon wines. The most marked differences detected were in the wines' phenolics. Ultrapremium versions showed about three times the concentration of flavonols, with cinnamates and gallates being about 60–70 percent higher. Ultrapremium wines were also more alcoholic (14.1 versus 12.3 percent), but lower in residual sugar and malic acid contents. Another comparison of wines, categorized by price (low-standard, high-standard, and premium), looked at term-use frequency during assessment (Sáenz-Navajas et al., 2012). A complex pattern was observed, with several terms, such as dried fruit being noted as positive in premium wines, but negative in the other categories. The sensory pair woody/animal was the most significant in relegating quality status. Table 8.3 shows how terms were used to express the high- and low-quality red Rioja wines. How the wines were assessed (by visual, olfactory, taste, mouth-feel alone, or together) significantly affected how quality was perceived.

Studies on the effect of the molecular size (Zarzo, 2011) and complexity (Kermen et al., 2011) of aromatic compounds have yielded some intriguing findings. There appears to be a correlation between a compound's molecular size and its hedonic perception, with more oxygen atoms correlated with pleasantness, whereas for carboxylic acids and sulfur compounds, increased molecular size increased aversiveness. Structural simplicity was associated with unpleasant odors, whereas complexity was correlated with more numerous olfactory notes and pleasantness (reflecting likely activation of higher numbers of olfactory receptors). Some similar findings have been noted in functional magnetic resonance imaging (fMRI) (Sezille et al., 2015). Does this indicate that molecular complexity, not just diversity of a wine's aromatic constituents, may be a central tenet in wine quality?

Holistic expressions, such as balance, probably arise out of the interaction of sugars, acids, alcohols, and phenolics. Because balance can occur equally in dry, sweet, white, red, sparkling, and fortified wines, it is clear that this interaction is perplexing. For example, the high sugar content of botrytized wines is partially balanced by their acidity, alcohol content, or both. Balance is also influenced by fragrance. In full-bodied red wines, balance may develop as various phenolics polymerize, and depolymerize, during aging, losing many of their former attributes that generate bitterness and astringency. The alcohol content and moderate acidity of red wines are also likely contributors to balance. Balance in light red wines seems to be achieved at a lower alcohol content and higher acidity than full-bodied dark red wines. Phenomena such as duration and development are likely to arise from the action of polysaccharides, mannoproteins and phenolics that loosely fix aromatics, slowly releasing them, once the wine is poured into a glass, as free, volatile compounds (Lubbers et al., 1994).

Progressive sensory adaptation may also play a role in the expression of minor aromatic constituents. Nevertheless, our ability to explain sensory perceptions such as complexity, finesse, and power in precise chemical terms still lies in the future. It is undoubtedly a function of the interaction of multiple aromatic compounds, but at the moment this remains just conjecture. It may be decades before the chemical origins of wine quality yield their secrets.

Postscript

Wine quality can be viewed from many angles, but fundamentally, quality is dependent on the wine's physicochemistry, and how it is detected and processed sensorially. The wine chemistry is based initially on grape biochemistry and physiology, partially transformed by yeast and bacterial metabolism, and subsequent modified by physical and organic chemical changes during maturation, aging, and volatile release after pouring. All of this presupposes that the wine possesses relative chemical and microbial stability, nutritional quality, and safety. However, from a human perspective, wine quality is perceived in terms of sensory and psychological pleasure. Although our senses respond to the visual, taste, mouth-feel, and olfactory stimuli demonstrated by the wine, conscious perception is based on how their impulses are processed and collated in various parts of the brain, where they are integrated, analyzed, and interpreted in relation to experiences, including pertinent social pressures and emotional desires. In addition, for many consumers, extrinsic factors (exclusivity, provenance, price, renown, age, and rarity) are potentially more important than the wine's intrinsic (sensory) quality. Finally, there is the influence of the context in which the wine is tasted, and one's state of relative hunger or satiety.

In analyzing a wine's intrinsic quality, psychological influences should be reduced to the absolute minimum. Nonetheless, psychological factors are a major promoter of sales, especially of higher-end wines. Hopefully, the wine also possesses, or will obtain with aging, a refined character. Its absence probably explains why some expensive wines are euphemistically said to exhibit *subtlety*. Wine quality, like beauty, is quixotic, existing only for a short time in the glass and, as it must, in the eye of the beholder. Also it is sobering to realize that quality does not necessarily correlate with consumer appreciation (Hopfer and Heymann, 2014).

Suggested Readings

Ashton, R.H., 2013. Is there consensus among wine quality ratings of prominent critics? An empirical analysis of red Bordeaux, 2004–2010. J. Wine Econ. 7, 225–234.

Cardello, A.V., 1995. Food quality: Relativity, context and consumer expectations. Food Qual. Pref. 6, 163–1680.

Charters, S., Pettigrew, S., 2007. The dimensions of wine quality. Food Qual. Pref. 18, 997–1007.

Geiger, R., Aron, R.H., Todhunter, P., 2003. The Climate Near the Ground, sixth ed. Rowman & Littlefield, Lanham, MD.

Jackson, R.S., 2014. Wine Science: Principles and Applications, fourth ed. Academic Press, San Diego, CA.

Jaeger, S.R., 2006. Non-sensory factors in sensory science research. Food Qual. Pref. 17, 132–144.

Lattey, K.A., Bramley, B.R., Francis, I.L., Herderich, M.J., Pretorium, S., 2007. Wine quality and consumer preferences: Understanding consumer needs. Wine Indust. J. 22, 31–39.

Matthews, M.A., 2015. Terroir and Other Myths of Winegrowing. University of California Press, Oakland, CA.

Skeenkamp, J.-B.E.M., 1990. Conceptual model of the quality perception process. J. Busin. Res. 21, 309–333.

References

Aiken, J.W., Noble, A.C., 1984. Comparison of the aromas of oak- and glass-aged wines. Am. J. Enol. Vitic. 35, 196–199.

Aldave, L., Almy, J., Cabezudo, M.D., Cáceres, I., González-Raurich, M., Salvador, M.D., 1993. The shelf-life of young white wine. In: Charakanbous, G. (Ed.), Shelf Life Studies of Food and Beverages. Chemical, Biological, Physical and Nutritional Aspects. Elsevier, Amsterdam, pp. 923–943.

Archer, E., 1987. Effect of plant spacing on root distribution and some qualitative parameters of vines. In: Lee, T. (Ed.), Proc. 6th Aust. Wine Ind. Conf. Australian Industrial Publishers, Adelaide, pp. 55–58.

Archer, E., Strauss, H.C., 1985. Effect of plant density on root distribution of three-year-old grafted 99 Richter grapevines. S. Afr. J. Enol. Vitic. 6, 25–30.

Archer, E., Swanepoel, J.J., Strauss, H.C., 1988. Effect of plant spacing and trellising systems on grapevine root distribution, In: The Grapevine Root and its Environment (J. L. van Zyl, comp.), Technical Communication No. 215. Department of Agricultural Water Supply, Pretoria, South Africa. pp. 74–87.

Arrhenius, S.P., McCloskey, L.P., Sylvan, M., 1996. Chemical markers for aroma of *Vitis vinifera* var. Chardonnay regional wines. J. Agric. Food. Chem. 44, 1085–1090.

Atkin, D., 2005. The Culting of Brands: Turn Your Customers into True Believers. Penguin Books Ltd, London, UK.

Becker, N., 1985. Site selection for viticulture in cooler climates using local climatic information. In: Heatherbell, D.A. (Ed.), Proc. Int. Symp. Cool Climate Vitic. Enol. Agriculture Experimental Station Technical Publication No. 7628, Oregon State University, Corvallis, pp. 20–34.

Berns, G.S., McClure, S.M., Pagnoni, G., Montague, P.R., 2001. Predictability modulates human brain response to reward. J. Neurosci. 21, 2793–2798.

Bertuccioli, M., Viani, R., 1976. Red wine aroma: Identification of headspace constituents. J. Sci. Food. Agric. 27, 1035–1038.

Bhat, S., Reddy, S.K., 1998. Symbolic and functional positioning of brands. J. Consumer Market. 66 (July), 1–17.

Bicchi, C., Cordero, C., Liberto, E., Sgorbini, B., Rubiolo, P., 2012. Headspace sampling in flavor and fragrance field, In: Comprehensive Sampling and Sample Preparation, Vol. 4. Extraction Techniques and Applications: Food and Beverage. Academic Press, New York, NY. pp. 1–25.

Bourke, C., 2004. Is Traminer Gewurz, or is it Roter or Rose, and if Bianco, what about Albarino? Goodness only knows! Aust N. Z. Grapegrower Winemaker 488, 19–22, 24.

Broadbent, M., 1980. The Great Vintage Wine Book. Alfred Knopf, New York, NY.

Brochet, F., Dubourdieu, D., 2001. Wine descriptive language supports cognitive specificity of chemical senses. Brain Lang. 77, 187–196.

Bronstad, P.M., Langlois, J.H., Russell, R., 2008. Computational models of facial attractiveness judgments. Perception 37, 126–142.

Burton, B.J., Jacobsen, J.P., 2001. The rate of return on investment in wine. Econ. Inquiry 39, 337–350.

Büttner, K., Sutter, E., 1935. Die Abkühlungsgröße in den Dünen etc. Strahlentherapie 54, 156–173.

Charters, S., Pettigrew, S., 2005. Is wine consumption an aesthetic experience? J. Wine Res. 16, 121–136.

Chatonnet, P. (1991). *Incidences de bois de chêne sur la composition chimique et les qualités organoleptiques des vins. Applications technologiques.* Thesis. Univ. Bordeaux II, Talence, France.

Chatonnet, P., Boidron, J.-N., Dubourdieu, D., Pons, M., 1994. Évolution des composés polyphénoliques de bois de chêne au cours de son séchage. Premieres résultats. J. Int. Sci. Vigne Vin. 28, 337–357.

Chisholm, M.G., Guiher, L.A., Zaczkiewicz, S.M., 1995. Aroma characteristics of aged Vidal blanc wine. Am. J. Enol. Vitic. 46, 56–62.

Chung, H.-J., Son, J.-H., Park, E.-Y., Kim, E.-J., Lim, S.-T., 2008. Effect of vibration and storage on some physico-chemical properties of commercial red wine. J. Food Comp. Anal. 21, 655–659.

Clary, C.D., Steinhauer, R.E., Frisinger, J.E., Peffer, T.E., 1990. Evaluation of machine- *vs.* hand-harvested Chardonnay. Am. J. Enol. Vitic. 41, 176–181.

Cox, A., Skouroumounis, G.K., Elsey, G.M., Perkins, M.V., Sefton, M.A., 2005. Generation of (*E*)-1-(2,3,6-trimethylphenyl)buta-1,3-diene from C13-norisoprenoid precursors. J. Agric. Food Chem. 53, 3584–3591.

Darriet, P., Bouchilloux, P., Poupot, C., Bugaret, Y., Clerjeau, M., Sauris, P., et al., 2001. Effects of copper fungicide spraying on volatile thiols of the varietal aroma of Sauvignon blanc, Cabernet Sauvignon and Merlot wines. Vitis. 40, 93–99.

Darriet, P., Pons, M., Henry, R., Dumont, O., Findeling, V., Cartolaro, P., et al., 2002. Impact odorants contributing to the fungus type aroma from grape berries contaminated by powdery mildew (*Uncinula necator*); incidence of enzymatic activities of the yeast *Saccharomyces cerevisiae*. J. Agric. Food Chem. 50, 3277–3282.

de Revel, G., Martin, N., Pripis-Nicolau, L., Lonvaud-Funel, A., Bertrand, A., 1999. Contribution to the knowledge of malolactic fermentation influence on wine aroma. J. Agric. Food Chem. 47, 4003–4008.

Distel, H., Hudson, R., 2001. Judgement of odor intensity is influenced by subjects' knowledge of the odor source. Chem. Senses 26, 247–251.

Douglas, D., Cliff, M.A., Reynolds, A.G., 2001. Canadian: Sensory characterization of Riesling in the Niagara Peninsula. Food Res. Int. 34, 559–563.

Drummond, G., Rule, G., 2005. Consumer confusion in the UK wine industry. J. Wine Res. 16, 55–64.

Dumont, A., Dulau, L., 1996. The role of yeasts in the formation of wine flavors. In: Henick-Kling, T. (Ed.), Proc. 4th Int. Symp. Cool Climate Vitic. Enol. New York State Agricultural Experiment Station, Geneva, NY. pp. VI–24–28.

Feuillat, M., Charpentier, C., Picca, G., Bernard, P., 1988. Production de colloïdes par les levures dans les vins mousseux élaborés selon la méthode champenoise. Revue Franç. Oenol. 111, 36–45.

Francis, I.L., Sefton, M.A., Williams, J., 1992. A study by sensory descriptive analysis of the effects of oak origin, seasoning, and heating on the aromas of oak model wine extracts. Am. J. Enol. Vitic. 43, 23–30.

Gishen, M., Iland, P.G., Dambergs, R.G., Esler, M.B., Francis, I.L., Kambouris, A., et al., 2002. Objective measures of grape and wine quality. In: Blair, R.J., Williams, P.J., Høj, P.B. (Eds.), 11th Aust. Wine Ind. Tech. Conf. Oct. 7–11, 2001, Adelaide, South Australia. Winetitles, Adelaide, Australia, pp. 188–194.

Goldstein, R., Almenberg, J., Dreber, A., Emerson, J.W., Herschkowitsch, A., Katz, J., 2008. Do more expensive wines taste better? Evidence from a large sample of blind tastings. J. Wine Econ. 3, 1–9.

Goldwyn, C., Lawless, H.T., 1991. How to taste wine. ASTM Standardization News 19, 32–37.

Henick-Kling, T., Acree, T., Gavitt, B.K., Kreiger, S.A., Laurent, M.H., 1993. Sensory aspects of malolactic fermentation. In: Stockley, C.S. (Ed.), Proc. 8th Aust. Wine Ind. Tech. Conf. Winetitles, Adelaide, Australia, pp. 148–152.

Henick-Kling, T., Edinger, W., Daniel, P., Monk, P., 1998. Selective effects of sulfur dioxide and yeast starter culture addition on indigenous yeast populations and sensory characteristics of wine. J. Appl. Microbiol. 84, 865–876.

Hersleth, M., Mevik, B.-H., Naes, T., Guinard, J.-X., 2003. Effect of contextual factors on liking for wine – use of robust design methodology. Food Qual. Pref. 14, 615–622.

Heymann, H., Noble, A.C., 1987. Descriptive analysis of Pinot noir wines from Carneros, Napa and Sonoma. Am. J. Enol. Vitic. 38, 41–44.

Hopfer, H., Heymann, H., 2014. Judging wine quality: Do we need experts, consumers or trained panelists? Food Qual. Pref. 32, 221–233.

Hoppmann, D., Hüster, H., 1993. Trends in the development in must quality of 'White Riesling' as dependent on climatic conditions. Wein Wiss. 48, 76–80.

Iland, P.G., Marquis, N., 1993. Pinot noir – Viticultural directions for improving fruit quality. In: Williams, P.J., Davidson, D.M., Lee, T.H. (Eds.), Proc. 8th Aust. Wine Ind. Tech. Conf. Winetitles, Adelaide, Australia, pp. 98–100. Adelaide, 13–17 August, 1992.

Intrieri, C., Poni, S., 1995. Integrated evolution of trellis training systems and machines to improve grape quality and vintage quality of mechanized Italian vineyards. Am. J. Enol. Vitic. 46, 116–127.

Jackson, R.S., 2014. Wine Science: Principles and Applications, fourth ed. Academic Press, San Diego, CA.

Jackson, R.S., 2016a. Shelf life of wine. In: Subramaniam, P., Wareing, P. (Eds.), The Stability and Shelf Life of Food, second ed. Woodhead Publishing, Cambridge, UK. pp. 311–346.

Jackson, R.S., 2016b. Innovations in Winemaking. In: Kosseva, M.R., Joshi, V.K., Panesar, P.S. (Eds.), Science and Technology of Fruit Wine Production. Academic Press, San Diego, CA. pp. xx–xx.

Jaeger, S.R., 2006. Non-sensory factors in sensory science research. Food Qual. Pref. 17, 132–144.

Kennedy, A.M., 2002. An Australian case study: Introduction of new quality measures and technologies in the viticultural industry. In: Blair, R.J., Williams, P.J., Høj, P.B. (Eds.), 11th Aust. Wine Ind. Tech. Conf. Oct. 7–11, 2001, Adelaide, South Australia. Winetitles, Adelaide, Australia, pp. 199–205.

Kermen, F., Chakirian, A., Sezille, C., Joussain, P., Le Goff, J., Ziessel, A., et al., 2011. Molecular complexity determines the number of olfactory notes and the pleasantness of smells. Sci. Reports 1, 206 (1–5).

Klein, H., Pittman, D., 1990. Drinker prototypes in American society. J. Substance Abuse 2, 299–316.

Kontkanen, D., Reynolds, A.G., Cliff, M.A., King, M., 2005. Canadian terroir: Sensory characterization of Bordeaux-style red wine varieties in the Niagara Peninsula. Food Res. Int. 38, 417–425.

Köster, E.P., Couronne, T., Léon, F., Lévy, C., Marcelino, A.S., 2003. Repeatability in hedonic sensory measurement: A conceptual exploration. Food Qual. Pref. 14, 165–176.

Lagorce-Tachon, A., Karbowiak, T., Paulin, C., Simon, M-M., Gougeon, R. D., and Bellat, J.-P., 2016. About the role of the bottleneck/cork interface on oxygen transfer. J. Agric. Food Chem. 64, 6672–6675.

Lindman, R., Lang, A.R., 1986. Anticipated effects of alcohol consumption as a function of beverage type: A cross-cultural replication. Int. J. Psychol. 21, 671–678.

Lubbers, S., Voilley, A., Feuillat, M., Charpontier, C., 1994. Influence of mannoproteins from yeast on the aroma intensity of a model wine. Lebensm.–Wiss. u. Technol. 27, 108–114.

Marais, J., 1986. Effect of storage time and temperature of the volatile composition and quality of South African *Vitis vinifera* L. cv. Colombar wines. In: Charalambous, G. (Ed.), Shelf Life of Foods and Beverages. Elsevier, Amsterdam, pp. 169–185.

Matsuo, T., Itoo, S., 1982. A model experiment for de-astringency of persimmon fruit with high carbon dioxide treatment: *In vitro* gelatin of kaki-tannin by reacting with acetaldehyde. Agric. Biol. Chem. 46, 683–689.

Matthews, M.A., Ishii, R., Anderson, M.M., O'Mahony, M., 1990. Dependence of wine sensory attributes on vine water status. J. Sci. Food Agric. 51, 321–335.

Maujean, A., Poinsaut, P., Dantan, H., Brissonnet, F., Cossiez, E., 1990. Étude de la tenue et de la qualité de mousse des vins effervescents. II. Mise au point d'une technique de mesure de la moussabilité, de la tenue et de la stabilité de la mousse des vins effervescents. Bull. O.I.V. 63, 405–427.

McClure, S.M., Li, J., Tomlin, D., Cypert, K.S., Montague, L.M., Montague, P.R., 2004. Neural correlates of behavioral preference for culturally familiar drinks. Neuron. 44, 379–387.

Mingo, S.A., Stevenson, R.J., 2007. Phenomenological differences between familiar and unfamiliar odors. Perception 36, 931–947.

Murat, M.-L., 2005. Recent findings on rosé wine aromas. Part I: Identifying aromas studying the aromatic potential of grapes and juice. Aust. NZ Grapegrower Winemaker 497a, 64–65. 69, 71, 73–74, 76.

Noble, A.C., and Ohkubo, T. (1989). Evaluation of flavor of California Chardonnay wines. In: First International Symposium: Le Sostanze Aromatiche dell'Uva e del Vino, S. Michele all'Adige, 25–27 July, 1989. pp. 361–370.

Picard, M., Thibon, C., Redon, P., Darriet, P., de Revel, G., Marchand, S., 2015. Involvement of dimethyl sulfide and several polyfunctional thiols in the aromatic expression of the aging bouquet of red Bordeaux wines. J. Agric. Food Chem. 63, 8879–8889.

Picard, M., Lytra, G., Tempere, S., Barbe, J.-C., de Revel, G., Marchand, S., 2016. Identification of piperitone as an aroma compound contributing to the positive mint nuances perceived in aged red bordeaux wines. J. Agric. Food Chem. 64, 451–460.

Plan, C., Anizan, C., Galzy, P., Nigond, J., 1976. Observations on the relation between alcoholic degree and yield of grapevines. Vitis. 15, 236–242.

Plassmann, H., O'Doherty, J., Shiv, B., Rangel, A., 2008. Marketing actions can modulate neural representations of experienced pleasantness. PNAS 105, 1050–1054.

Prescott, J., 2012. Taste Matters: Why We Like the Food We Do. Reaktion Books, London, UK.

Priilaid, D.A., 2006. Wine's placebo effect. How the extrinsic cues of visual assessments mask the intrinsic quality of South African red wine. Int. J. Wine Marketing 18, 17–32.

Ramey, D., Bertrand, A., Ough, C.S., Singleton, V.L., Sanders, E., 1986. Effects of skin contact temperature on Chardonnay must and wine composition. Am. J. Enol. Vitic. 37, 99–106.

Rapp, A., Güntert, M., 1986. Changes in aroma substances during the storage of white wines in bottles. In: Charalambous, G. (Ed.), The Shelf Life of Foods and Beverages. Elsevier, Amsterdam, pp. 141–167.

Rapp, A., Mandery, H., 1986. Wine aroma. Experientia 42, 873–880.

Rapp, A., Marais, J., 1993. The shelf life of wine: Changes in aroma substances during storage and ageing of white wines. In: Charalambous, G. (Ed.), Shelf Life Studies of Foods and Beverages. Chemical, Biological, Physical and Nutritional Aspects. Elsevier, Amsterdam, pp. 891–921.

Reeves, C.A., Bednar, D.A., 1994. Defining quality: Alternatives and implications. Acad. Manage. Rev. 19, 419–445.

Ribéreau-Gayon, P., 1986. Shelf-life of wine. In: Charalambous, G. (Ed.), Handbook of Food and Beverage Stability: Chemical, Biochemical, Microbiological and Nutritional Aspects. Academic Press, Orlando, FL, pp. 745–772.

Ritchey, J.G., Waterhouse, A.L., 1999. A standard red wine: Monomeric phenolic analysis of commercial Cabernet Sauvignon wines. Am. J. Enol. Vitic. 50, 91–100.

Rous, C., Alderson, B., 1983. Phenolic extraction curves for white wine aged in French and American oak barrels. Am. J. Enol. Vitic. 34, 211–215.

Rubin, G.J., Hahn, G., Allberry, E., Innes, R., Wessely, S., 2005. Drawn to Drink: A double-blind randomised cross-over trial of the effects of magnets on the taste of cheap red wine. J. Wine Res. 16, 65–69.

Sáenz-Navajas, M.-P., González-Hernández, M., Campo, E., Fernández-Zurbano, P., Ferreira, V., 2012. Orthonasal aroma characteristics of Spanish red wines from different price categories and their relationship to expert quality judgements. Aust. J. Grape Wine Res. 18, 268–279.

Sáenz-Navajas, M.-P., Avizcuri, J.M., Echávarri, J.F., Ferreira, V., Fernández-Zurbano, P., Valentin, D., 2015. Understanding quality judgements of red wine by experts: Effect of evaluation condition. Food Qual. Pref. 48, 216–227.

Schueuermann, C., Khakimov, B., Engelsen, S.B., Bremer, P., Silcock, P., 2016. GC-MS metabolite profiling of extreme southern Pinot noir wines: Effects of vintage, barrel maturation, and fermentation dominate over vineyard site and clone selection. J. Agric. Food Chem. 64, 2342–2351.

Schwartz, B., 2004. The tyranny of choice. Sci. Amer. 290 (4), 71–75.

Sefton, M.A., Francis, I.L., Williams, P.J., 1993. The volatile composition of Chardonnay juices: A study by flavor precursor analysis. Am. J. Enol. Vitic. 44, 359–370.

Sezille, C., Ferdenzi, C., Chakirian, A., Fournel, A., Thevenet, M., Gerber, J., et al., 2015. Dissociated neural representations induced by complex and simple odorant molecules, Neuroscience 287, 23–31.

Siegrist, M., Cousin, M.-E., 2009. Expectations influence sensory experience in a wine tasting. Appetite 52, 762–765.

Singleton, V.L., 1962. Aging of wines and other spirituous products, acceleration by physical treatments. Hilgardia 32, 319–373.

Singleton, V.L., 1990. An overview of the integration of grape, fermentation, and aging flavours in wines. In: Williams, P.J. (Ed.), Proc. 7th Aust. Wine Ind. Tech. Conf. Winetitles, Adelaide, Australia, pp. 96–106.

Sinton, T.H., Ough, C.S., Kissler, J.J., Kasimatis, A.N., 1978. Grape juice indicators for prediction of potential wine quality, I. Relationship between crop level, juice and wine composition, and wine sensory ratings and scores. Am. J. Enol. Vitic. 29, 267–271.

Smart, R.E., Robinson, M., 1991. Sunlight into Wine. A Handbook for Winegrape Canopy Management. Winetitles, Adelaide, Australia.

Smith, S., Codrington, I.C., Robertson, M., Smart, R., 1988. Viticultural and oenological implications of leaf removal for New Zealand vineyards. In: Smart, R.E. (Ed.), Proc. 2nd Int. Symp. Cool Climate Vitic. Oenol. New Zealand Society of Viticulture and Oenology, Auckland, New Zealand, pp. 127–133.

Somers, T.C., Evans, M.E., 1974. Wine quality: Correlations with colour density and anthocyanin equilibria in a group of young red wine. J. Sci. Food Agric. 25, 1369–1379.

Somers, T.C., Evans, M.E., 1977. Spectral evaluation of young red wines: Anthocyanin equilibria, total phenolics, free and molecular SO$_2$ "chemical age" J. Sci. Food. Agric. 28, 279–287.

Spedding, D.J., Raut, P., 1982. The influence of dimethyl sulphide and carbon disulphide in the bouquet of wines. Vitis. 21, 240–246.

Spillman, P.J., Pocock, K.F., Gawel, R., Sefton, M.A., 1996. The influences of oak, coopering heat and microbial activity on oak-derived wine aroma. In: Stockley, C.S. (Ed.), Proc. 9th Aust. Wine Ind. Tech. Conf. Winetitles, Adelaide, Australia, pp. 66–71.

Stashenko, E.E., Martínez, J.R., 2012. *In vivo* Sampling of flavor, In: Comprehensive Sampling and Sample Preparation, Vol. **4**. *Extraction Techniques and Applications: Food and Beverage*. Academic Press, New York, NY. pp. 1–25.

Stummer, B.E., Francis, I.L., Zanker, T., Lattey, K.A., Scott, E.S., 2005. Effects of powdery mildew on the sensory properties and composition of Chardonnay juice and wine when grape sugar ripeness is standardised. Aust. J. Grape Wine Res. 11, 66–76.

Tomasino, E., 2011. Characterization of regional examples of New Zealand Pinot Noir by means of sensory and chemical analysis. PhD Thesis. Lincoln University, New Zealand.

Tomasino, E., Harrison, R., Sedcole, R., Frost, A., 2013. Regional differentiation of New Zealand Pinot noir wine by wine professionals using Canonical Variate Analysis. Am. J. Enol. Vitic. 64, 357–363.

Tominaga, T., Blanchard, L., Darriet, Ph, Dubourdieu, D.A., 2000. A powerful aromatic volatile thiol, 2-furanmethanethiol, exhibiting roast coffee aroma in wines made from several *Vitis vinifera* grape varieties. J. Agric. Food Chem. 48, 1799–1802.

Verdú Jover, A.J., Montes, F.J.L., Fuentes, M.M.F., 2004. Measuring perceptions of quality in food products: The case of red wine. Food Qual. Pref. 15, 453–469.

Vidal, S., Cartalade, D., Souquet, J.M., Fulcrand, H., Cheynier, V., 2002. Changes in proanthocyanidin chain length in wine-like model solutions. J. Agric Food Chem. 50, 2261–2266.

Wansink, B., Payne, C.R., North, J., 2007. Fine as North Dakota wine: Sensory expectations and the intake of companion foods. Psychol. Behav. 90, 712–716.

Zarzo, M., 2011. Hedonic judgements of chemical compounds are correlated with molecular size. Sensors 11, 3667–3686.

9

Wine and Food Combination

INTRODUCTION

Food and wine pairing is usually considered the preserve of the sommelier and wine aficionado. Nonetheless, it behooves the wine professional to understand, as much as is scientifically possible, this association. Winemakers need no reminder that their wines should be compatible with, and preferably be ideally suited to, consumption with food. To be insouciant would be to their peril.

The view that wine is primarily a food beverage is traditional in many European cultures and has spread throughout much of the world (Unwin, 1992; Pettigrew and Charters, 2006). Wine, among alcohol-containing beverages, is uniquely associated with food and a positive social image (Lindeman and Lang, 1986, Unwin, 1992). Nonetheless, the profusion of essentially identical wines, combined with vintage variations, causes considerable consumer confusion. Thus, consumers are frequently searching for direction about what to buy and with what to pair it. The mental paralysis elicited on facing an overwhelming selection of any product is well known (Schwartz, 2004). Consternation and risk aversion (Roets et al., 2012) are recognized as major impediments to restaurant wine sales (see Wansink et al., 2006).

Food scientists have largely avoided the field of food and wine pairing. The topic is legitimately fraught with difficulties, as there are a myriad of interactions between foods (e.g., Ahn et al., 2011; Vilgis, 2013) and wines, not to speak of the complexities of perception. Combined, these make almost any data interpretation suspect. Whether development in molecular gastronomy (This, 2013) and gastrophysics (Vilgis, 2013) will help significantly is questionable.

Interest among consumers about food and wine pairing may be on the upswing, if the continuing profusion of books and articles on the subject is any indication. This may relate to enhanced concern about social correctness among young professionals, and how wine can donate an element of elegance and refinement to their lives. Nonetheless, it seems that many consumers are unaware that most wines go surprisingly well with most foods. If one likes both the wine and the food, their combination is unlikely to be a disaster; one is already primed to be pleased. The relation between food and wine seems equivalent to:

"When I'm not near the girl I love, I love the girl I'm near." from *Finian's Rainbow*

This should not be surprising. Most people make decisions about food (and presumably wine) based on a few, personally significant criteria (Scheibehenne et al., 2007).

The simplest guide to selection is to choose a wine equivalent in flavor intensity to that of the meal, respecting the importance of the occasion, and your mood. This usually equates to the more significant or celebratory the occasion, the more expensive the wine considered appropriate. Wines of higher price may not necessarily superior, but ideally should possess a more marked and distinctive flavor. Whether this is the "right" choice depends on the preferences and expectations of those attending the meal. Unfortunately, there is no linear correlation between wine price and its quality. If anything, the relationship is probably decreasingly logarithmic, i.e., the higher the price, the lower the sensory quality/price ratio.

However, for the aficionado, the goal is more than just compatibility – it is achieving a beatific duet. The perfect marriage of food and wine is often viewed as the quintessential vinous experience. Wine becomes a liquid salve for the soul. This affiliation, however, needs to be qualified by the realization that the best wines, and certainly the oldest, express their finest qualities when sampled alone (well-aged wines possess delicate flavors easily masked by

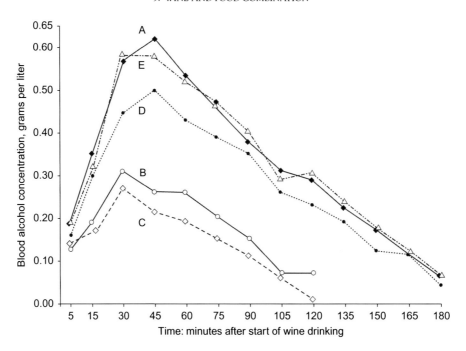

FIGURE 9.1 Blood alcohol concentrations after wine drinking in a single dose. (A) Fasting, (B) during a meal, (C) 2 hours after a meal, (D) 4 hours after a meal, (E) 6 hours after a meal. *From Serianni, E., Cannizzaro, M., Mariani, A., 1953. Blood alcohol concentrations resulting from wine drinking timed according to the dietary habits of Italians. Quart. J. Stud. Alcohol 14, 165–173, reproduced by permission.*

food flavors, and a fragrance that is often evanescent). Science cannot assist in selecting wine based on the context of the occasion, but it can begin to provide an explanation of some food and wine interactions.

However, in a search for that paradigm of synergies, the principal benefits of wine's association with food are often forgotten, that is, reducing the potential inebriating influence of alcohol. Consumption with food delays the movement of wine out of the stomach and into the small intestine (Franke et al., 2004), where most of the alcohol is absorbed and enters the blood system. By slowing and extending the period over which alcohol is absorbed, the liver is better able to limit any spike in blood alcohol content (BAC). Blood from the intestinal tract first passes through the liver, where most of the ethanol is metabolized, before circulating to the rest of the body. Although consumption with or without food marginally affects the timing of peak blood alcohol content (30 versus 45 min after consumption), it is only about 40 percent of what it would have been had wine been taken on an empty stomach (Fig. 9.1). Consumption with food also limits the cumulative alcohol content in the blood to about a third, especially when consumption occurs over the course of a meal (Serianni et al., 1953).

Because of the potential for a negative association between alcohol consumption and driving (Fig. 9.2), and the progressive lowering of permissible blood alcohol content, a few additional points on this issue seem warranted. Limiting consumption of wine during a meal to no more than ~250 mL (about one-third of a standard 750-mL bottle), the BAC is likely to remain below the legal driving limit in most jurisdictions. Nonetheless, this depends on a range of factors, including body weight, gender, amount of food consumed, the wine's alcohol content and type (e.g., sparkling) (Fig. 9.3), the period over which the wine is consumed, the interval between consumption and driving, and genetic factors. However, concentrating on the wine's sensory pleasures, rather than quantity, is likely to be a more effective and enjoyable means of assuring sobriety. Wine is a beverage to be savored, not drunk as if it were flavored water or a soft drink. However, for those interested in investigating the relationship between BAC and sobriety further, Wikipedia (http://en.wikipedia.org/wiki/Blood_alcohol_content) provides a fascinating range of data and references on the issue.

Compatibility

Most recommendations concerning marrying food and wine can be reduced to the common quip: white with white and red with red. The lighter, fruity/floral fragrances and more acidic nature of most white wines seem to match the milder-flavored attributes of fish and poultry. In contrast, the more intense flavors and tannic attributes

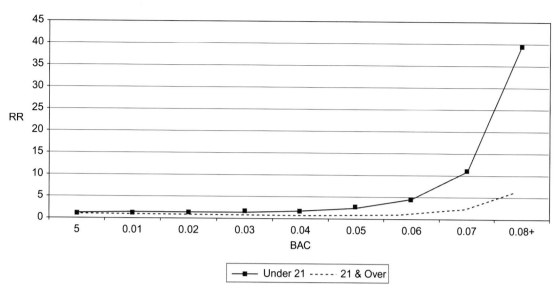

FIGURE 9.2 Relative crash risk as a function of blood alcohol content (BAC) with age (cubic model, BAC capped at 0.08+). *Reprinted from Peck, R. C., Gebers, M., Voas, R. B., Romano, E., 2008. The relationship between blood alcohol concentration (BAC), age, and crash risk. J. Safe. Res. 39, 311–319, with permission from Elsevier.*

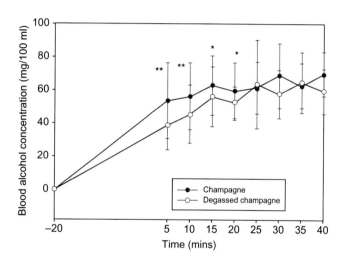

FIGURE 9.3 Mean blood alcohol concentration (BAC) time curves. Values are means ± SEM in samples taken 5–40 min after the end of a 20-min consumption with champagne and degassed champagne. Number of subjects = 6. *$p < 0.5$, **$p < 0.01$ (two tailed). Reprinted from Ridout, F., Gould, S., Nunes, C., Hindmarch, I., 2003. The effects of carbon dioxide in Champagne on psychometric performance and blood-alcohol concentration. Alcohol Alcoholism 38, 381–385. by permission of Oxford University Press.

of most red wines seem ideally suited to the more robust flavors of meats. However, in reality, it is often the method of food preparation and seasoning that are more critical in choosing a compatible wine. Fish boiled versus broiled are markedly different in flavor characteristics and intensity. Spices and condiments also can have marked effects on a food's flavor.

Highly specific wine–food suggestions often reflect no more than the personal preferences of the commentator, and usually are impossible to match for the reader. Possibly for that reason, rationales for the advice are rarely, if ever given. Although useful in a "paint-by-numbers" approach, without explanations, specific recommendations provide little basis on which to design future pairings. Suggestions noted in Harrington (2007) are a start, but do make the issue unduly complex. Few people have the detailed knowledge of food and wine flavor characteristics he expects. In addition, the sensory characteristics of wines often vary from producer to producer, from year to year, and with aging. Thus, without personal experience of a particular wine and recipe, it is often difficult to predict with certainty how well any combination will be expressed, assuming that any such blissful "marriages" do occur, are necessary, or will appeal to those at the table. Furthermore, the sensory characteristics of food and wine change in the mouth, and often throughout the meal. For examples of the dynamics of flavor release in the mouth during consumption see Cook et al. (2004), Linforth and Taylor (2006) and Muñoz-González et al. (2015).

Although consumers are unlikely to know sensory terminology, it would be considerate if food and wine writers helped their readers by adopting precise, definable language. It would also be wise to avoid ill-supported claims, such as acids "cut" through fats, or the saltiness of cheese "contrasts" the sweetness of dessert wines, etc. (e.g., Harrington, 2007, pp. 85, 172, 188). More useful would be a discussion of findings that have experimental backing, such as the mutual reduction of tannins and fats on each other's perception (Peyrot des Gachons et al., 2012). A thin coating of fatty acids might limit the access of tannins to trigeminal receptors (reducing the perception of astringency). However, oil does not appear to have such an effect. At least application of a thin layer of oil on the tongue did not affect the sweetness of a sucrose solution (Camacho et al., 2015b). Any gustatory effects depended on the oil suspension and the copresence of proteins and thickeners (Camacho et al., 2015a).

More regrettable than the simplistic comments that often pepper the popular literature are suggestions that smack of elitism, such as suggesting the inherent compatibility of simple foods with 'humble' Beaujolais. If popular or fruity wines are inherently plebeian, would its corollary, that expensive wines are aristocratic, automatically be true? Although extrinsic information about a wine often prejudices perception, tasting blind suggests expensive wines are no more necessarily appreciated than inexpensive wines. In addition, pejorative attitudes about inexpensive wines have, for far too long, given wine connoisseurship an aura of haughtiness that is as deplorable as it should be infelicitous.

Many popularly expressed opinions about pairings are more likely to reflect habituation than sensory logic. Reactions to gustatory sensations are hardwired, such that sweet, savory, salty, and fat perceptions are inherently liked whereas acidic, bitter, burning, and astringent sensations are instinctively disliked. In contrast, response to flavor is largely experience based. Palatability preferences begin in the womb (based primarily on what the mother eats). Subsequently, infants act like true scientists, tasting essentially everything within reach. However, their likes (and dislikes) soon become fixed and further experimentation limited by neophobia. Preferences tend to broaden later in childhood and into adulthood, possibly due to peer pressure, a newfound willingness to try, or incidental exposure.

Culture appears to be the primary factor directing regional food preferences (Chrea et al., 2004). Although unconfirmed, the same likely applies to wine preferences. The flexibility probably arises from our ancestors being omnivores, capable of living on almost anything that was nourishing and nontoxic. Food likes and dislikes are principally associated with oral-based sensations (Rozin and Vollmecke, 1986; Prescott, 2012). In contrast, innate olfactory revulsion seems to apply only to putrid odors. Other, apparent reflex dislikes to aromatic compounds seem to be responses to some negative association with the first (even if forgotten) experience with the odor.

In contrast, reflex responses to taste undoubtedly have evolutionary rationales. For example, the liking of sugar, creams, and fats probably relates to their caloric value. In contrast, aversion to bitter and irritant compounds involves a protective adaptation. Many wild plants contain alkaloids, saponins, and other bitter-tasting toxicants. One of the first benefits of cooking and crop domestication were, respectively, the inactivation and reduced production of plant toxins. In addition, the heating associated with cooking promoted the hydrolysis of proteins and starches, making both meat and tough grains more esculent. Heating also enhanced the flavor (sweetness) of vegetables, by releasing sugars from starch molecules. The aversive response to sourness, typically associated with spoilt food, is also likely an ancient, evolved, protective reflex.

Nonetheless, humans are, if nothing else, highly adaptive, and can develop acceptance, even passion, for sensations initially disliked and inherently unpleasant (see Moskowitz et al., 1975; Rozin and Schiller, 1999). For example, people appear to come to enjoy the irritants found in capsicum peppers and horseradish, because of, not in spite of, the pain induced. Thus, because a person may prefer milk in their tea, or sugar in their coffee, does not necessarily predict one's appreciation of, or aversion to, the bitter/astringent character of red wines.

Historically, diverse social, climatic, and geographic conditions have created a wide diversity of food preferences and prejudices. However, these norms can change with surprising speed. Examples include the recent and rapid acceptance of hot spices in Western cultures, and the acceptance of modern "junk foods" throughout much of the world. When it comes to food and wine association, availability and upbringing appear to be the primary defining factors, though conforming to social expectations is also very influential.

Despite the major impact of external factors, inherited predispositions can significantly mold individual preferences (see Fig. 3.6). Tasters of the bitter tastants phenylthiocarbamide (PTC) and 6-n-propylthiouracil (PROP) are more likely to reject bitter foods and beverages (Drewnowski et al., 2001), experience a heightened oral burn with capsaicin (Karrer and Bartoshuk, 1995), and be more sensitive to the irritation of high concentrations of ethanol than

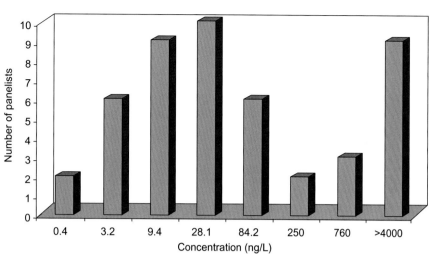

FIGURE 9.4 Distribution of best estimate threshold concentration (ng/L) for rotundone red wine. About 20–25% of panelists could not smell the compound at 4000 ng/L (the highest level tested). *From Wood, C., Siebert, T.E., Parker, M., Capone, D.L., Elsey, G.M., Pollnitz, A.P., et al., 2008. From wine to pepper: Rotundone, an obscure sesquiterpene, is a potent spicy aroma compound. J. Agric. Food Chem. 56, 3738–3744, reproduced by permission.*

nontasters. Other genetic factors affecting flavor sensitivity (and presumably food and wine preferences) include the relative number of fungiform papillae on the tongue and specific anosmias. Heritable variations in sensitivity to gustatory (Garcia-Bailo et al., 2009; Newcomb et al., 2012) and presumably aromatic compounds (Tempere et al., 2011, McRae et al., 2013) undoubtedly influence personal beverage and food selection, and may also play a role in the variation in food preferences among cultures. Fig. 9.4 illustrates an example of differences in individual sensitivity to a potent peppery aromatic compound, rotundone. Its threshold was assessed in Shiraz wines. Rotundone also occurs in black pepper, at much higher concentrations (in the range of 1200 ng/L), along with alkaloid piperine.

Habituation can involve not only psychological acceptance but also physiological adjustment. For example, the salinity of saliva adjusts to the level of salt intake (Christensen et al., 1986; Bertino et al., 1986). In the short term, though, salt addition enhances food flavor, whereas its absence is correlated with blandness. Physiological desensitization also follows repeat exposure to capsaicin (hot chilies), and is often associated with development of its appreciation.

Despite wine's centuries-long association with food, there seems little inherent logic to this association. The connection almost undoubtedly arose as a consequence of accessibility, safety (versus nonpotable water), and the subsequent establishment of cultural norms. A similar phenomenon may explain the atypical appreciation of sour and bitter tastes in the Karnataka region of India (Moskowitz et al., 1975), the love of chilies in Mexican, Korean, and several other East Asian foods, and the popularity of sweet-and-sour combinations in some Chinese regional cuisines. The appreciation of innately unpleasant tastes, such as chili peppers, horseradish, mustard, bitter chocolate, black coffee, burnt and acidic foods is an acquired taste. Inherently pleasurable food sensations often originate in the medial orbitofrontal cortex and the amygdala, whereas irritant foods activate the lateral orbitofrontal cortex (Pelchat et al., 2004). Appreciation/aversion can also be context sensitive (Hersleth et al., 2003). For example, both table and dessert wines may be savored, but not with the same foods or at the same time of day.

Most discussions of food and wine pairing concentrate on some illusory synergy of flavors. In reality, this is only one component of the subject. At one end of the spectrum are incompatible associations, by almost any standard, in which either (or both) the food and wine accentuate unpleasant aspects of one or both. For example, highly alcoholic wines often undesirably enhance the spiciness of a dish, the sweetness of a dessert augments the apparent sourness of a dry table wine, and a botrytized dessert wine is an abomination with bouillabaisse. Sometimes, the iron content of a wine can induce a metallic sensation by catalyzing lipid oxidation (Lawless et al., 2004). Most mismatches are almost self-evident and usually easily avoided. Other associations range from being neutral, to refreshing and harmonious, and, yes, occasionally rapturous.

It is probable that most food and wine combinations are perceived as comparatively neutral. This may be because people do not take the time to search for any, rather than there being an absence of any interaction. This certainly seems true in cultures where wine is the standard food beverage. Wine is often taken in tumblers and consumed without thought, simply as a mouth freshener during eating, except on special occasions, when better wines are served and relished.

Where a diversity of foods is being served simultaneously, such as a buffet or potluck, choice of a neutral to mild-flavored wine seems sensible. Why insult a fine wine with an eclectic mix? Most restaurant house wines are of the neutral genre. As such, they are unlikely to offend the sensibilities of those unfamiliar with wine, are relatively inexpensive, and readily available. The supposed natural partnership of local wines with regional dishes is likely founded more on ideas of promoting local produce, a reflection of habituation (or narcissism), or to represent ready availability or economic value. Any success chefs or winemakers have in marrying their respective recipes and wines is probably more fantasy than actuality. Even for connoisseurs, most wine is probably consumed more or less unconsciously. It is not often in our rushed lives that we have the liberty to savor our meals to the full. It is one of the sad realities of modern life. Even in a restaurant, most people are more interested in conversation than paying serious attention to the wine.

When the union of food and wine is simply compatible, the combination can be both enjoyable and fascinating, without being distracting. This is not negative, and can be an advantage. It does not demand extended attention. Here, the wine acts to mollify any less pleasant aspects of the food, and vice versa. By cleansing the palate, wine reestablishes sensitivity, so that the next morsel can be savored afresh. Ideally, the wine should supply an additional sensory element, enhancing the overall appreciation of the meal. In such instances, the wine acts almost like a condiment or garnish. It may be in this role that wine initially came to be viewed as pairing with food. However, even here, for this benefit, it is essential to actively concentrate on what is being detected each time the wine is sampled. In addition, wines, unbalanced by themselves, appear more harmonious. Many white wines are marginally too sour, and red wines overtly tannic. In white wine, the reaction between wine acids and food proteins mellows the perception of the wine. In addition, the coolness of white wines can dampen the "heat" of spicy foods. With red wines, balance may develop when tannins react with food proteins, preventing or reducing their ability to activate bitter receptors or disorganize epithelial cell membranes, eliciting astringency.

Occasionally, compatibility arises from both the wine and food possessing similar attributes. Examples might include the buttery character of a Chardonnay matching that of crab, the peppery aroma of Shiraz blending with that of pepper steak, the fruity flavor of a Riesling spätlese affiliating with a sweet, fruit-based sauce, or the nutty aspect of dry sherries harmonizing with a hazelnut cream soup. Conversely, similar chemistry may unfavorably heighten their perceived intensities. For example, the herbaceous aspect of most Sauvignon blanc and poorer Cabernet Sauvignon wines accentuate similar flavors in bell peppers or the vegetal aspect of green beans. However, usually, the basic characteristics of table wines (acidity, bitterness, astringency, and fruity to floral fragrances) are not found in food. Conversely, the fatty and meaty flavors of food, and adjoining vegetal, spice, and condiment elements find few equivalents in wines. A few wines (notably Gewürztraminer and Shiraz) possess intense ("spicy") attributes. Nevertheless, these are mild in comparison to what occurs in many recipes. The marked disparity between the sensory features of most wines and foods does severe damage to the oft-mentioned rapport suggested by food and wine critics.

In reality, compatibility seems to originate more from complementary differences than similarities. As noted above, the interaction between food proteins and a wine's acidic and phenolic constituents diminishes their perceptions, generating a mellowness the wine would not otherwise possess. Food lipoproteins (Katsuragi et al., 1995), long-chain fatty acids (Homma et al., 2012; Ogi et al., 2015), and sugars (Mennella et al., 2015) are also known to suppress the perception of bitterness. The active ion in salt (Na^+), found in most recipes and cheeses, is also an effective suppressor of bitter–astringent sensations (Breslin and Beauchamp, 1995). Salt-induced suppression of bitterness can also enhance the perception of sugars (Breslin and Beauchamp, 1997). As an additive to food, salt is well known as a flavor enhancer (Rabe et al., 2003; Mitchell et al., 2011), apparently disrupting weak associations between aromatic constituents and reducing solubility, promoting their liberation and retronasal detection. In addition, the hydration of sodium ions may decrease the "free water" and change solution polarity. Although the release of aromatic compounds is a partial function of salt content, part of the perceived enhancement may also reflect the pleasantness of detectable concentrations of salt (Bolhuis et al., 2016). Flavor is also well known to be enhanced by sweet (Hansson et al., 2001), fatty (Arancibia et al., 2011), umami (Linscott and Lim, 2016), and kokumi (Kuroda and Miyamura, 2015) tastants.

Possibly in contrast to most foods, cheeses appear to diminish rather than intensify the flavor of white (Nygren et al., 2003a) and red (Madrigal-Galan and Heymann, 2006) wines, and vice versa. Similarly, in a study of Shiraz–cheddar pairings, both the flavor and astringency of the wine were reduced, with the tannins being considered silkier (Bastian et al., 2010). This may relate to the precipitation of catechins by the salt in the cheese (Yan and Luo, 1999). In another study involving a diversity of cheeses and wines, there was marked specificity in which unions were considered optimal (Bastian et al., 2009). Comparing these results with other studies (King and Cliff, 2005; Harrington and Hammond, 2005) suggests that generalizations such as stronger-flavored (red) wines with stronger-flavored (drier) cheeses, and white wines with milder (softer) cheeses, as suggested by Immer (2002), are untenable. Thus, much of the perceived enhanced pleasure from pairing food or cheese with wine may be chimerical. Admittedly, most people would prefer to think in positive terms, i.e., flavor enhancement, rather than suppression of aversive attributes in either or both the food and wine. Such an interpretation lacks the comforting sentiment the public prefers, promotes wine sales in restaurants, and sells books on the subject.

Although current data suggest that what are considered desirable combinations are due to a mutual reduction in aversive attributes, other interactions seem at least theoretically possible. For example, identical or chemically related odorants in both could combine additively or synergistically, leading to an enhanced appreciation. Such is part of the gist of ideas behind molecular gastronomy (This, 2005, 2013; Ahn et al., 2011).

Typically, negative pairings are usually construed in terms of incompatible tastes, such as the acidity of table wines clashing with the sweetness of a dessert. However, less well-known consequences can occur, such a fishy smell associated with sampling some white wines with shellfish (Tamura et al., 2009). Another little-known instance may be the action of subthreshold concentrations of TCA (Takeuchi et al., 2013) suppressing olfaction, affecting not only the wine's fragrance but also flavor in general.

Usually, the sensory intensity of a selected wine should be less pronounced than that of the food, with the wine acting as a foil for food flavors. Nonetheless, with a superior wine, the food component is often selected to be mild flavored, to avoid its flavors conflicting with or masking those of the wine. For example, the delicateness of simply prepared chicken is often selected to act as a foil to allow the fascinating interplay of sensations found in the wine to express themselves (Plate 9.1). In contrast, ordinary wines may be best served with food possessing dominant flavors. Famous combinations may have their sensory values based on mollifying attributes of the food or wine, for example, the association of low-alcohol German Rieslings with hot or spicy foods – the coolness of the wine and its low alcohol and phenolic contents assuaging spice-induced oral burning. Conversely, wine may be used as a beverage condiment, acidity adding a zesty tang (like lemon juice), and phenolic constituents supplying complementary bitterness. Wine can, thus, provide an alternative to what is often supplied by seasoning.

When wine takes center stage, it receives considerable attention; in fact, scrutiny. This is especially so when an old wine of delicate flavor is served. In contrast, when a more intensely flavored fine wine is chosen, a more savory meal is probably preferable to favor mutual appreciation. What is preferable also depends considerably on those in attendance. Is a dynamic contrast of flavors desired, enhanced flavor expression preferred, or is the evolution of flavor harmony coveted?

In designing a specific food and wine pairing, much of the enjoyment often comes from planning and anticipation. However, it is often difficult for reality to match expectation. Truly memorable associations are usually unforeseen; therein lies their appeal. It is the unexpected transcendental experience that is so embracing and luminary. Once experienced, the elation is hard to reproduce, even with the same wine, meal, and surroundings. Personally, these unanticipated explosions of gustatory and olfactory ecstasy are at the heart of the "holy grail" of wine appreciation, forever haunting but irreproducible.

PLATE 9.1 Tasting of a series of red and other wines pared with a menu (chicken), so as to not conflict with the flavors of the wines. *Photo courtesy of R. Jackson.*

WINE SELECTION

As evident from Chapters 7 and 8, multiple factors affect wine quality and style. Nonetheless, in selecting a wine to go with a meal, the principal concerns should relate to how the wine's stylistic, varietal, and regional characteristics match with those of the food, and the preferences of the participants. It has been suggested that European wines are more subtle and, therefore, more suited to pairing with, and accentuating the characteristics of food, whereas New World wines are too fruit-forward to be food-friendly. Even if this were so, and that is debatable, is sensory nuance or distinctiveness more likely to refreshen the palate and intrigue the mind?

Another important factor relates to wine age. Young wines typically are more flavorful as well as varietally and stylistically distinctive. In addition, red wines tend to be more noticeably bitter and astringent when young. As wines mature, these attributes soften, becoming more subtle and mellow, the varietal character begins to fade, potentially being replaced by a more generic, aged bouquet. It is for these reasons that well-aged fine wines are usually reserved for enjoyment alone or after the meal, where their refinement can be more effectively appreciated.

The presence of an oak aspect can also significantly influence choice. With wines of little varietal flavor, oak character tends to mask (as well as the wood absorb) aspects of a mild wine's fragrance. With wines of more pronounced flavor, oak can ennoble their character with additional complexity. Whether this is appreciated depends on personal taste.

Regional differences affecting varietal character tend to be subtle and often somewhat nebulous. In regions with variable climates, vintage conditions are more likely to be obvious than differences due to provenance. Typically, extensive experience is necessary to permit confident detection of regional differences. Thus, unless one has some personal or cultural preference for particular countries or regions, provenance should be one of the least significant aspects determining wine selection.

For most people, price is a defining factor in choice. Unlike most products, price is not a good indicator of quality, or flavor. Within a country or region, price often can be of some predictive value, but between regions and countries, quality/price comparisons are valueless. For wine, historical repute and worldwide demand are often the prime drivers of price. The most economical way of assessing the characteristics and relative value of wines is to sample them blind with friends or a tasting society. But, in the end, personal appreciation should be the most important factor in any choice. Enjoyment should be the ultimate goal; the opinion of "authorities" should be immaterial.

HISTORICAL ORIGINS OF FOOD AND WINE ASSOCIATION

From a historical perspective, pairing a particular wine with a specific food is comparatively recent, possibly only going back several hundred years. Even the consumption of wine with food probably goes back little more than 3000 years, and may have become a cultural norm only with the ancient Greeks and Romans. Earlier, limited availability restricted wine's use to religious or ceremonial occasions, often involving only the priestly caste and the ruling elite. Beer was the alcoholic beverage of the masses (Hornsey, 2016), being simpler and more rapid to prepare. In addition, because the raw materials (barley or other grains) can be effectively stored dry and are, therefore, readily available when needed throughout the year, beer could be produced on demand (no need for extended storage). Wine production became a beverage staple only in regions where grapes grew indigenously, such as southern Europe and southwestern Eurasia, and later along river valleys leading into central Europe. Although made from a perishable, periodically available crop, wine's higher alcohol, low pH, and phenolic content gave it the potential for storage, and thus year-round supply. The antimicrobial effects of wine's alcohol, acidic, and phenolic contents made it safe to drink, which was especially valuable in sites where water supplies were often sources of food- and waterborne infections.

On the development of comparatively nonporous containers (amphoras with a pitch lining and oak cooperage), sealable with cork, wine could be stored in a drinkable state for about a year (longer if vitreous-lined amphoras were used) (Vandiver and Koehler, 1986). This set the stage for wine to become the standard food beverage in the ancient Greek and Roman worlds. That grapevines required little in the way of cultivation abetted this development. Vines could grow up trees, adjacent to field crops, or, with pruning, could be cultivated on poor soils, drier sites, or slopes unsuitable to food-crop production. Thus, grape culture did not necessarily compete with food production, any more than the cultivation of olives. What is unclear is whether southern European cuisines adapted to the phenolic and/or acidic content of the predominant beverage, or the populace simply became accustomed to the taste of wine. The answer may be a bit of both, but I suspect more likely the latter.

From the written record, wine's association with food was well established by ancient Greek and Roman times. Even then, it took on a degree of sophistication among the wealthy, with wines from particular regions and vintages being preferred, and occasionally lauded for their quality. However, striving for the perfect amalgam between wine and food appears to not have been a concern. Banquets of the wealthy were an eclectic mix of various items, all served simultaneously, and thus ill-designed for pairing particular wines with specific foods. Limited quantities of excellent wine may also have contributed to little inclination to pair individual wines with particular items. Most of the famous wines of antiquity seem to have been sweet, concentrated wines, inappropriate to modern tastes for associating with food.

The idea of matching a specific wine with particular items of food probably began in Renaissance Italy, when meals evolved into a series of separate courses we would recognize (Tannahill, 1973; Flandrin and Montanari, 1999), and possibly associated with serious (dissective) wine tasting (Eiximenis, 1384; Johnson, 1989). The Industrial Revolution subsequently favored the rise of a middle class, providing increased leisure time, and the disposable income for the development of an urbane clientele (Unwin, 1991). In addition, improved transportation provided the means for supplying the diversity of wines requisite for connoisseurship.

During the medieval period, winemaking making skill was relatively crude. Written accounts of the recommended use of sulfur dioxide begin only in the late 1400s. The first clear reference to its use is found in a report published in Rotenburg, Germany, 1487 (reproduced in Anonymous, 1986). Nonetheless, its use did not become common until the latter part of the 19th century. Wine storage was also "primitive." Wines, stored primarily in barrels, often spoiled by the following summer. To fill the shortfall, until the current year's wine was ready, makeshift wine (*verjus*) was produced from immature fruit during the summer. Because most wine did not travel well, especially under the archaic transport systems of the time, even the nobility would have had restricted access to the range of wine now considered necessary for refined food and wine harmonization.

Coinciding with the improving economy and sophistication of Western Europe was the disappearance of the morass of medieval cooking. Grand meals of the time often included a chaotic medley of simultaneously presented soup, meat, fish, poultry, and sweet dishes (Tannahill, 1973)—the original smorgasbord? Peasants had neither the time nor wherewithal to select a special wine to pair with their gruel. Without an orderly sequence of courses, there would have been little rational for pairing food with particular wines.

In addition, culinary practices, based on the theories of Hippocrates and Aristotle, were supplanted by those derived from thinkers such as Paracelsus, an early 16th-century Swiss physician (Laudan, 2000). Under the older nutritional concept, diet could affect health by balancing the four basic elements of life (heat, cold, wet, and dry). This theory was replaced by the view that proper nutrition involved three essential principles or elements. One elemental aspect, called "salt," gave food its taste (e.g., salt and flour). The second principle, designated "mercury," gave food its smell (e.g., wine and meat sauces). The final component, termed "sulfur," bound the first two elements together (e.g., oil and butter). Things have changed markedly for the better in the past few hundred years, thanks to a shift from the philosophical to scientific investigation of nutrition.

Nonetheless, dietary recommendations were one of the few pleasant remedies open to the physician in ancient times. Physicians were often employed by wealthy families for their restorative culinary advice. With a change in nutritional theory, heavily spiced dishes disappeared, sweetening was relegated to dessert (rather than added to almost every dish), and wines in their natural state replaced the almost universal use of hot, spiced, red wines (*hypocras*). That the goodness of a meal might be concentrated in "gases" (nourishing the brain) may have favored the early appreciation of sparkling wine (soon to be supplanted by celebration). The first mention of sparkling wine serendipitously coincided with a change in culinary views (late 17th century). Equally, the reputed salubrious benefits of distilled spirits (e.g., *eau de vie*) supplied a health rationale for adding brandy to wine. An additional benefit to the practice of adding distilled spirits to the low-acid wines in some southern European regions prevented their spoilage during transport to northern centers of commerce. The augmented alcohol content also supplied extra "warmth," a valued property on cold drafty nights in homes without central heating and efficient insulation. Developments in science also began to improve the stability and general quality of wine. Under such conditions, the stage was set for major refinements in culinary techniques and the development of cultured associations with wine.

Thus, the development of rational pairings of food and wine seems little more than three centuries old. It evolved in association with the development of French-modified Tuscan cuisine, itself derived and adopted initially by the Venetians from the Levant (Tannahill, 1973).

CONCEPT OF FLAVOR PRINCIPLES

In a study of world cuisines, Rozin (1982a) classified culinary styles according to their primary ingredients, cooking techniques, and unique flavorants, i.e., their flavor principles. Of these, the most distinctive was the use of flavorants. For example, particular cuisines are often distinguished by their use of one or more particular condiments or flavorants, such as soy sauce, sambal belacan, nam pla prik, kimchee, rice vinegar, coconut, ginger, garlic, herb blends, specific spices or curry blends, fermented black beans, tomato paste, chilies, or sauces often with an olive oil, butter, cream base, or sweet-and-sour flavorings. The relative importance of these flavorants to particular cuisines has been expounded on further by Rozin (1982b), Harrington (2005b), and Ahn et al. (2011).

The flavor intensity of some regional seasonings may seem to give the cuisine a monotonous character. However, the incredible variation in chili peppers, curry preparations, soy sauces, etc., can provide a rich diversity of sensory nuances to those habituated to their basic attributes. This is probably equivalent to the apparent similarity of wines to those unaccustomed to their consumption.

Especially intriguing is the appreciation of the burning sensation of items such as chilies and horseradish, the bitterness of coffee and tonic water, or the sourness of pickled foods. The rapid and widespread acceptance of intense flavors, initially perceived as painful or harsh, is in stark contrast to the slow spread of neutral-flavored foods, such as corn, potato, zucchini, cassava, and breadfruit. Intriguingly, Northern Europe and other northern climatic regions had, until recently, largely resisted the spread of chili pepper use in their cuisines (Andrews, 1985).

The active ingredients in some condiments have a numbing, desensitizing influence on trigeminal nerve receptors. This is clearly the case for capsaicin found in chili peppers. Habitual use probably leads to the degeneration of TRPV1 receptors in the mouth and throat, as it does in the skin (Nolano et al., 1999). The rate of recovery appears to be concentration and duration-of-exposure dependent (Karrer and Bartoshuk, 1991). How desensitization by capsaicin and other pungent flavorants (piperine in black pepper, eugenol in cloves, isothiocyanate in mustard and horseradish, and menthol in peppermint) influences wine perception appears not to have been studied. Laboratory studies suggest that the immediate effect of capsaicin is a moderate reduction in the sensitivity to sweet, bitter, and umami tastants (Simons et al., 2002; Green and Hayes, 2003), and possibly salty and acidic tastes (Gilmore and Green, 1993). These effects appear to be more marked in those unhabituated to capsaicin. However, this has not been consistently found (Lawless et al., 1985). The oft-noted suppression of taste and flavor may be due more to a sensory disruption, caused by the burning sensation, than an actual reduction in flavor detection. Thus, the problem of pairing hot spicy foods with wine may occur only with those unaccustomed to their presence on a daily basis. Traditional solutions for those unhabituated to hot spices have been to choose a simple white wine. It acts as a palate cleanser, while the wine's cool temperature may diminish the burning sensation (Babes et al., 2002). However, a cold temperature enhances the bite of carbon dioxide on the tongue, such as with sparkling wines (Amoore, 1977). Occasionally, more intensely flavored wines, such as those made from Gewürztraminer or Muscat varieties, have been suggested for pairing with spicy foods. Their more marked aromas might just partially mollify the sensory disruption caused by hot spices.

Because food preferences are so often culturally linked, value judgments must be viewed in relation to the norms on which they are based. For example, sweet-and-sour combinations are common in some Chinese and German dishes. However, a lemon syrup, instead of lemon juice, with fish would seem uncouth, and acidic wines with dessert are non sequiturs.

FOOD AND WINE PAIRING

Suggestions for food pairing vary from simplicity itself (essentially anything goes) to ludicrously specific (named producer and vintage). Neither extreme is particularly useful. But, on a fundamental level, problems can arise with any credenda. Preferences are too diverse to provide anything but views that are individually applicable. Furthermore, upbringing and experience markedly influence what is considered culturally or socially appropriate. Finally, and to be honest, other than for some professional chefs, sommeliers, and food/wine critics, few people seriously contemplate food and wine pairing. Nonetheless, discussion of the issue probably promotes more verbiage, and possibly wine sales, than almost any other wine-related issue, and may promote a broadening of wine's consumer base.

Despite extensive advertising, the majority of consumers select the house wine in restaurants, and purchase wines with which they are familiar and comfortable. However, the impression from the media would suggest that there is considerable demand for wine and food information. How else can one interpret the steady stream of books, magazines, and newspaper articles on the topic, as well as the modern cornucopia of wine blogs? If one can assume that these truly reflect demand, then most consumers want specific recommendations. However, explaining the underlying principles involved in suggestions is rarely provided; maybe because there are none. Certainly, it takes more mental effort to analyze potential combinations than the point-and-shoot approach provided by the media, thus their appeal. My feeling is that, like travel articles, reading is done for entertainment, there being little or no intention of following any of the specifics noted.

However, under special circumstances, wine consumers do genuinely desire good counsel. This may be relative to a gift for a loved one, a celebratory meal, the presence of someone important for dinner, or going out to a fine restaurant with a date or a client. If one or more of the participants are known to be a wine aficionado, then the host may be eager to honor their presence with an especially fine wine. Regrettably, wine choice can also be exploited to demonstrate the host's ability to purchase (or order) expensive wines. However, if those to be impressed have little knowledge of wine, the selection of an expensive wine may backfire. Being unaware of the price or prestige of the label, the wine may not please their taste, and those intended to be in awe may only wonder why the host did not offer (or order) a "good" wine. Ah, the problems of trying to inculcate importance.

Once, when my wife and I were at a fancy restaurant, celebrating the purchase of a condo, the maitre d', upon coming to realize my knowledge about wine, asked me if I would like to sample two expensive wines that had been rejected by young bucks out to impress their chicks. We were the recipient of two of the finest examples of a Barolo and an Amarone of my life. Thus, occasionally, if you are known to know about wine, you can benefit from the failing of the young and foolish.

Although detailed suggestions can have their place, in most situations wine should simply be viewed as a savory beverage to enjoy with a meal. Thus, its selection is too often given little more thought than the choice of vegetables. This is certainly not what artisanal wine producers want to read, but it is probably the attitude of the majority of wine purchasers. For all its prominence in the literature, prestigious wine is beyond the purchasing power of most people. Most consumers appear to focus on taste attributes, largely disregarding aroma (Bastian et al., 2005). This seems supported by cursory glances at restaurant clientele; nary a swirl or sniff, just down the hatch. What missed opportunities for sensory fulfillment.

For wine lovers, though, food and wine pairing is not only important, but also an essential element of a meal. As the expression goes, "A meal without wine is called breakfast." However, breakfast with champagne does raise "breaking fast" to an unaccustomed status.

Because expensive investigation of wine and food partnering is beyond the financial or time constraints of most of us, there is practical value in understanding at least the fundamentals of food and wine marriage. It is also important for chefs, sommeliers, and others in the hospitality industry to understand how clients' desires need not only be met, but also judiciously advanced.

The central theme in most food and wine pairing concepts is harmony between flavor intensities. At least, this is how it seems to normally be construed in Western cuisines (e.g., Paulsen et al., 2015). However, this may be another example of customary habits being rationalized, viewing one's own cultural percepts as correct. Thus, the logic of drinking a dry, alcoholic, acidic, tannic beverage with food may only reflect the historic coincidence of availability, nutritional value, and microbial safety. The role of habituation in directing food preferences has been repeatedly confirmed (e.g., Rozin, 1977; Blake, 2004).

Viewed traditionally, the concept of harmony is crystallized in the adage, "red with red, white with white." What red wines may lack in acidity, in comparison with white wines, is more than compensated by their phenolic content. Another element of food and wine harmony relates to color. White wines seem to look better when matched with pale-colored foods and sauces, just as red wines are visually more appealing with dark colored meats and sauces. Or is that just another example of custom dictating etiquette?

An alternative explanation of the "red with red" concept has been proposed by Ronca et al. (2003). It is based on the relative health benefits of consuming red wines with red meats. Red wines, with their higher phenolic content, have a greater propensity to bind metal ions (Brune et al., 1989) than their white equivalents. Because metal ions occur in higher concentrations in red meats, consumption with red wine could reduce the free concentration of these ions in the stomach, and their catalytic action in producing toxic free-oxygen radicals during digestion. In addition, red wine phenolics reduce the production of toxic lipid oxidation products (Gorelik et al., 2008; Fig. 9.5). These lipid byproducts can modify circulating low-density lipoproteins (LDLs), encouraging the formation of arterial plaque. Although some digestive enzymes are inactivated by tannins, this effect is probably diminished by their binding with food proteins. However, some monomeric phenolics can enhance the action of pepsin in simulated gastric digestion (Tagliazucchi et al., 2005). The intestinal flora slowly metabolize tannins to monomers, but the relative health significance of these changes is still speculative (e.g., Forester and Waterhouse, 2008; Nardini et al.,

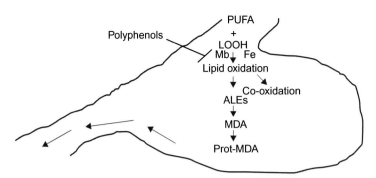

FIGURE 9.5   Limitation of food-lipid peroxidation (and cytotoxin production) in the stomach by the antioxidant action of polyphenols: PUFA, polyunsaturated fatty acids; *LOOH*, hydroperoxide; *Mb*, metmyoglobin; Fe, iron; ALEs, lipid oxidation end-products; *MDA*, malondialdehyde; *Prot-MDA*, protein-malondialdehyde). *Reprinted with permission from Kanner, J., Gorelik, S., Roman, S., Kohen, R., 2012. Protection by polyphenols of postprandial human plasma and low-density lipoprotein modification: The stomach as a bioreactor. J. Agric Food Chem. 60, 8790–8796. Copyright 2012, American Chemical Society.*

2009; Gross et al., 2010). In the mouth, phenolics have the potential to enhance the release of aromatics from food (Genovese et al., 2015). Finally, marinating in red wine reduces the production of toxic heterocyclic amines during frying meat (Viegas et al., 2012). Regrettably, marination can also degrade vitamins A, C, and E, carotenoids, and cholesterol (German, 1999). On the positive side, again, the essential absence of sulfur dioxide in red wines avoids any sulfite-induced destruction of thiamin (vitamin B1) (Skurray et al., 1986).

Inherently, the concept of flavor compatibility seems easy to comprehend. However, its precise application is anything but simple. There is no easy way of determining with precision the relative flavor intensity of either the food or the wine. Rietz (1961) attempted to quantify graphically the relative flavor intensities of different foods, sauces, condiments, etc. When items were selected from various columns, it was considered that menus could be constructed by combining foods of compatible flavor intensities, or to show a predesigned transition in flavor intensity throughout the meal. Although interesting, factors such as preparation technique (e.g., poaching, baking, broiling, grilling) modify not only the basic flavors of the food but also generate new flavorants. As a consequence, foods can vary markedly in their flavor complexity and intensity as well as texture. Serving temperature also significantly affects flavor intensity and characteristics. Each of these factors could independently influence the selection of an "ideal" wine. In spite of these constraints, modern attempts at designing meals on fundamental principles include Ahn et al. (2011), dealing only with recipe development, and Harrington (2005), with food and wine combination.

Establishing an estimate of wine flavor intensity is at least as complex as that for food. Fig. 9.6 indicates that considerable variation in flavor intensity is typical. For example, Gewürztraminer can vary from mildly flavored and slightly sweet, to bone dry with a robust fragrance. The former would go better with lightly braised chicken, while the latter with roast turkey and savory stuffing. The only way of assessing actual flavor intensity is by sampling. Tasting several potential wines in advance is, in most instances, as impractical as it would be uneconomical. Experience and personal preference are again the only certain guides. Although seraphic pairings are rare, it is encouraging to note that so are disasters.

Harmony is often considered essential to refined dining, but contrasting flavors can give a meal intrigue and verve. Typically, pizzazz in a meal is provided by spices, herbs, a salsa, sauce, or some other condiment. Occasionally, though, wine supplies the flavor zip desired. In this situation, the wine can act as a food flavorant, for example as in *coq au vin*. Conversely, a mild wine may soften the intense flavors of a dish. The combinations and permutations of food and wine pairings are almost endless, as are the associations of various ingredients in a recipe. The only real limitations are the imagination, preferences, and desires of the creator.

Initially, some renowned pairings appear to associate opposed tastes or flavors, for example, the combination of sweet, richly flavored wines (e.g., ports and sauternes) with salty, creamy, blue cheeses (e.g., Stilton and Roquefort). The success of this pairing may relate to their similar richness and the blend of salt and sweetness. Differences may also integrate, enhancing enjoyment. In Germany, the gamey character of well-aged venison is supposedly counterbalanced by the sweet, rich flavors of a mature Riesling auslese. In the Loire, the aggressive dry acidity of Sancerre is viewed as counterpoised when taken with the raciness of goat cheese. The sourness of Chablis is deemed to be offset by the sweet flavors of crab. Finally, there is the classic association of champagne and caviar. The wine's effervescence is considered to suppress the saltiness of the caviar, either by activating oral trigeminal (pain) receptors (Simons et al., 1999), or by removing sodium ions (by the scouring action of carbon dioxide). Or, is this a fabrication, designed to justify combining two products considered unique and prestigious at posh events? The ingenuity of

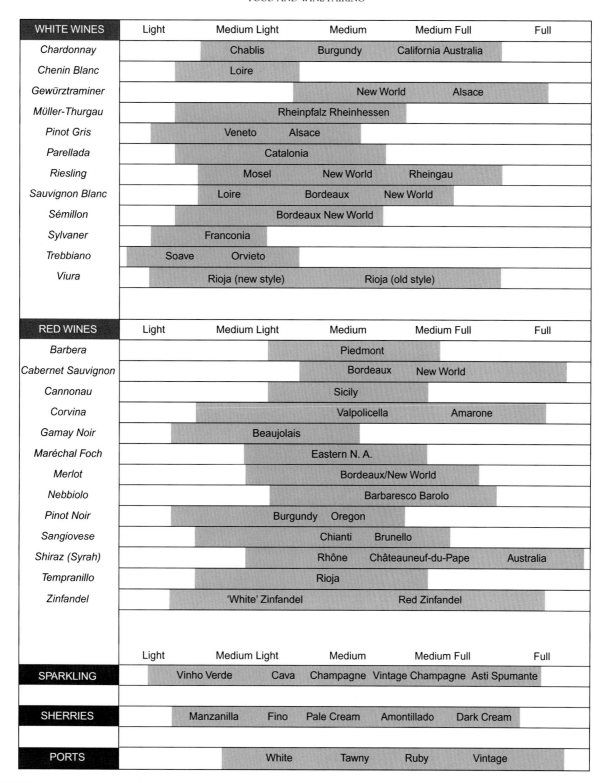

FIGURE 9.6 Relative flavor intensity of wines. *From Jackson, R.S., 2014. Wine Science: Principles, Practice, Perception, fourth ed., Elsevier Press, San Diego, CA (Jackson, 2014, reproduced by permission).*

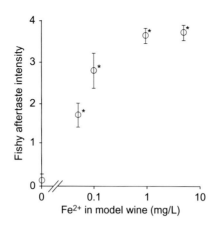

FIGURE 9.7   Effect of increasing ferrous ion concentration on the intensity of fishy aftertaste in a model wine–dried scallop pairing. Results are expressed as mean ± SEM (n = 7), *, $P < 0.0$, compared with the model wine without ferrous sulfate. *Reprinted with permission from Tamura, T., Taniguchi, K., Suzuki, Y., Okubo, T., Takata, R., Konno, T., 2009. Iron is an essential cause of fishy aftertaste formation in wine and seafood pairing. J. Agric. Food Chem. 57, 8550–8556. Copyright 2009, American Chemical Society reproduced with permission.*

some combinations may, in themselves, be enough to engender the perception of new and unexpected gastronomic delights. Novelty often has an appeal for the adventuresome.

Of the five principal taste sensations, wine possesses but three: sweet, sour, and bitter. As noted, most corporal food ingredients rarely exhibit either sour or bitter tastes, and except for dessert, rarely possess pronounced sweetness. Therefore, there is little apparent logic to wine's association with food. However, as previously noted, food can suppress the sour, bitter, and astringent aspects found in many table wines. Thus, often it is the food that may enhance the perception of wine, rather than the reverse. Nevertheless, the acidity of wine freshens the mouth, whereas moderate bitterness and astringency can enliven bland foods. Only rarely are there common elements. Examples are the nutty aspect of cream sherries pairing with a walnut dessert, the oaky character of a wine combining with the smoky flavors of meat roasted over charcoal, the joint presence of ethyl 3-mercaptopropionate, a compound found in both Muenster and Camembert cheeses (Sourabié et al., 2008) and considered to contribute to the flavor of Concord grapes (Kolor, 1983) and the toasty aspect of champagnes (Tominaga et al., 2003). Might chemical similarities help to explain why some wines go better (or worse) with particular cheeses?

Although most foods are not inimical to wine, several are, at least to Western sensibilities. For example, vinegar and vinegar-based condiments create an unpleasant harshness, and are likely to mask the wine's fragrance. Thus, wine is not taken with salads possessing a vinaigrette. Taste buds are also numbed by the burning sensation of chilies and most curries (at least to those unaccustomed to their presence). Highly salty foods and egg dishes are also considered to make poor wine companions. Nonetheless, the effervescence of champagne appears to make it particular fitting, being often served with caviar or eggs at elegant breakfasts. Such pairings have received some experimental support (Harrington and Hammond, 2009). Chocolate is another conundrum. In one experiment, appreciation of the pairing depended more on the liking of the beverage than the chocolate sample, with port generally preferred in such pairings (Donadini et al., 2012). Conversely, wine can be inimical to some foods, for example a wine's iron content generating a metallic perception (Fig. 4.22) or a fishy aftertaste (Fig. 9.7).

Although most commentators seem to hold as sacrosanct the view that wine directly enhances food appreciation, this is a shibboleth, the illusion of enhancement arising from the unrecognized reduction of unpleasant sensory attributes of one or both. Under laboratory conditions, mutual masking in flavorant mixtures is the norm (e.g., Brossard et al., 2007). For example, few individuals can correctly identify more than three constituents in complex, taste–odor combinations, and none consistently (Fig. 9.8). This may arise from competitive inhibition of receptor attachment sites and/or neural feedback suppression. The problem of identification is particularly noticeable with volatile compounds, where concentrations in the mouth are much lower than those in the headspace of a wine glass. Thus, it should not be surprising that food tends to diminish the detection of wine aromatics. These influences appear to be at least partially explained by dilution, reaction with food proteins, and potentially fats coating taste receptors (Kinsella, 1990). Fats and oils can also absorb nonpolar aromatics, reducing their volatility and passage into the nasal cavity (Buttery et al., 1973; Roberts et al., 2003; Hodgson et al., 2005; Bayarri et al., 2007).

Interactions between food and wine constituents probably explain the reduced acidity and bitterness of white wines in the presence of Hollandaise sauce (Nygren et al., 2001), and the general suppression of wine flavors by

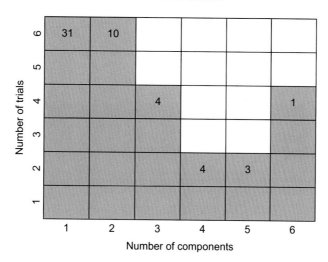

FIGURE 9.8 Identification of mixture components by individual subjects. Number of trials (*y*-axis) represents the number of presentations of a particular type of mixture, e.g., the four-component mixture, with each presentation involving a different composition. The *x*-axis shows the number of components present in a stimulus. Numbers within the matrix indicate the number of subjects who correctly identified all the components present in a solution, while the number of shaded boxes indicates the number of trials at which these particular subjects were successful. *Reproduced from Marshall, K., Laing, D. G., Jinks, A. J., Hutchinson, I., 2006. The capacity of humans to identify components in complex odor-taste mixtures. Chem. Senses 31, 539–545. by permission of Oxford University Press.*

cheese. Fig. 9.9 illustrates the effect of blue cheese on the "apple" flavor of several wines, and their effects on cheese saltiness. Influences on citrus, blackberry, oak, tropical, and dried fruit flavors, as well as taste (except for sweetness), were similar. The sodium content in cheese could also diminish the bitterness (Keast et al., 2001) and astringency (Yan and Luo, 1999) of wine tannins. The effects of wine on the flavor of Hollandaise sauce and cheese were less marked than the reverse. Those more limited effects may involve dilution effects, such as the bitter-tasting peptides in cheese (Roudot-Algaron et al., 1993; Karametsi et al., 2014).

Madrigal-Galan and Heymann (2006) have confirmed and extended the findings of Nygren to red wines, where cheese reduced the perception of bell pepper and oak flavors. In contrast, the buttery aspect of wine was enhanced by combination with Hollandaise sauce and cheese. These effects appear to be largely independent of the type of cheese. These findings lend support to the old maxim: "Sell wine over cheese, but buy it over water." In another study, white wines were perceived to be more balanced in flavor with a range of cheeses than were red or speciality wines (King and Cliff, 2005). Despite these generalities, marked variation in opinion was observed.

Although interesting, whether these studies have relevance to tasting under real-life (nonlaboratory) situations is unconfirmed. Contextual factors and prior knowledge are major factors influencing sensory perception and appreciation.

Certain aspects of food texture can also affect, or be affected by, flavor perception. For example, certain textural sensations are heightened in the presence of aromatic compounds (Bult et al., 2007). In contrast, flavor perception is often suppressed by thickening agents (Hollowood et al., 2002; Ferry et al., 2006). With starch, this may relate to hydroxyl groups interacting with polar aromatic compounds (Arvisenet et al., 2002), or their incorporation into amylose helices. Similar phenomena may involve other thickeners. The suppression of volatilization so induced probably explains the need for strong or concentrated seasonings in sauces.

Sensory perception, being a complex cerebral construct, is influenced by not only by the stimuli received, but also by how it is detected (ortho- or retronasally). Examples of the latter are the pronounced differences between the smell of Limberger cheese and durian fruit, and their in-mouth flavors. Similarly, wine sampled in the glass often appears considerably different when present in the mouth or taken with food. This is often explained as a function of reduced concentration reaching the nasal passages or masking by food flavorants. However, it may also relate to the anticipatory aspect of orthonasal olfaction, whereas retronasal olfaction is more directly associated with reward circuitry (Negoias et al., 2008). Each pathway involves separate neural routes as well as different neurotransmitters: opioid and GABA/benzodiazepine for liking, and dopamine for reward (Berridge, 1996). Complex cerebral interactions involve both the cognitive and emotional/memory centers of the brain (Small et al., 2005, Fig. 3.10). The interrelated nature of sensory perception helps explain why predicting how a wine will pair with food is so difficult.

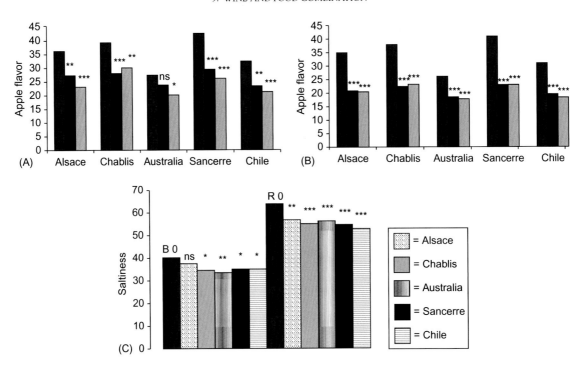

FIGURE 9.9   Influence of wine and cheese on flavor. Mean intensity of apple flavor of white wines (A) before and after tasting blue cheeses; (B) before and after mixed tasting with the cheeses. *Darker colored bars*, wine before cheese tasting; *black and paler colored bars*, wine after or with Bredsjö Blä and Roquefort cheeses. (C) Mean saltiness of the cheeses before and after sampling the same white wines. B, Bredsjö Blä; R, Roquefort; before sampling is represented by (0) and after wine sampling (see bar explanations). The wines were Riesling (Alsace), Chardonnay (Chablis and Australian), and Sauvignon (Sancerre and Chilean). Level of significance: $*p < .05$; $**p < .01$; $***p < .001$; ns, not significant. $p > .05$. *From Nygren, I.T., Gustafsson, I.-B., Johansson, L., 2003a. Effects of tasting technique – sequential tasting vs. mixed tasting – on perception of dry white wine and blue mould cheese. Food Service Technol. 3, 61–69; Nygren, I.T., Gustafsson, I.-B., Johansson, L., 2003b. Perceived flavour changes in blue mould cheese after tasting white wine. Food Service Technol. 3, 143–150. (Nygren et al., 2002, 2003a,b), respectively, reproduced by permission.*

However explained, the most memorable food and wine combinations relate to their aesthetically pleasing nature. Science is starting to unravel some of the entangled interactions, but consumers must still rely on developing their own set of sensory skills and wine experiences, or swallow the recommendations of taste gurus. Every culture has had its arbiters of good taste. If consumers discover that their own sensory preferences mimic those of some "authority," this is simpler, and their suggestions are likely to be of relevance.

USES IN FOOD PREPARATION

Basic Roles

At its least profound, wine can act as a savory palate cleanser. By rinsing food particles and substituting its own flavors, wine may minimize sensory fatigue. Thus, the appeal and freshness of food flavors are maintained throughout the meal. In its turn, food helps freshen the palate to receive the wine anew. Wine acids and phenolics also stimulate saliva production (Hyde and Pangborn, 1978), as does its alcohol content (Martin and Pangborn, 1971).

Wine can also be viewed as a foil for the meal, highlighting the central flavors of the food. Equally, food is often viewed as enhancing wine appreciation. Food modulates the typical taste imbalances of wines, i.e., their acidic, bitter, and astringent aspects. Recently, adjustment in wine production has been instituted in response to the increasing use of table wines as an aperitif, making them more balanced (less acidic, bitter, and astringent) and flavorful at bottling.

As a food beverage, though, adequate acidity freshens the mouth, as well as seemingly diminishing the oily or fattiness of some foods. However, there is no inherent physicochemical reason why acidity should reduce the perception of fats—acids being polar and fats being nonpolar—and correspondingly, should have no emulsifying effect. However, whether acid-induced activation of trigeminal receptors (Liu and Simon, 2000) might indirectly reduce the perception of fats is a possibility. Another unsupported contention is the enhanced flavor of meat by the bitter/astringency of red wines. Equally, does maturation in oak give white wines the flavor and bitterness sufficient to complement more tangy foods? As usual, there are more questions than answers.

Wine can have significant solubilizing action. Its acid and alcohol contents may help solubilize or volatilize food flavorants. In so doing, wine could accentuate food appreciation. Admittedly, if significant, this effect must act quickly. Food seldom remains in the mouth longer than it takes to masticate it sufficiently to permit swallowing. Conversely, the dilution of alcohol by food constituents could promote the liberation of wine aromatics (Fischer et al., 1996; Fig. 3.16). Due to time constraints associated with eating, if important, dilution is more likely to affect the wine's finish.

Wine also has several direct and indirect effects on food digestion. As noted, its phenolic and alcohol contents activate saliva production. Wine also promotes the release of both gastrin and gastric juices in the stomach. Wine constituents, notably succinic acid (Teyssen et al., 1999), malic acid (Liszt et al., 2012), and phenolics (Liszt et al., 2015) activate the release of gastric juices. Wine also significantly delays gastric emptying (Franke et al., 2004; Benini et al., 2003), possibly via stimulating bitter taste receptors in the stomach (Janssen et al., 2011). These influences both aid digestion (by extending acid hydrolysis) and enhance inactivation of foodborne pathogens. Delayed gastric emptying also retards the absorption of alcohol, giving the liver more time to metabolize ethanol, minimizing alcohol accumulation in the blood.

Despite these benefits, the reason most people take wine with their meals is simply because they enjoy, as well as are habituated to, the association. It tends to extend eating, permitting each morsel to be savored. Wine also encourages conversation and the social aspect of dining. Thus, wine raises a biologic need to one of life's more civilized delights.

Involvement in Food Preparation

Wine has had a long tradition of use in food preparation. Possibly the oldest example is as a marinade. Wine acids hydrolyze proteins, tenderizing meat. The acids also temporarily preserve the meat. Pickling items in wine vinegar is another example. Wine also has been employed to extract or mask the gamey flavor of wild meats. Wine used in this manner seldom adds significantly to the food's flavor since the marinade is usually discarded after use. Use as a marinade can also reduce the production of toxic heterocyclic amines during frying (Viegas et al., 2012). Wine is also occasionally used to marinate fruit in preparation for its use in a dessert. Here, the wine does double duty as both a marinade and a central flavorant in the recipe.

Another culinary wine use is as the liquid used in poaching, stewing, or braising. Fine wines are seldom used because cooking dramatically changes their flavor. Nevertheless, the quality of the wine should be adequate to not adversely affect the food's flavor. Shorter cooking times, or lower temperatures, result in more of the original attributes of the wine being retained. Thus, the significance of the color of the wine used depends primarily on whether the cooking time is short and/or its temperature low. Normally, prolonged cooking turns any wine brown, via the generation of Maillard and Strecker degradation products.

When one is poaching or braising, the wine is often reduced to make a sauce. Wine is also often added to deglaze a pan after frying to solubilize the *fond*, in preparation for making a sauce. Because deglazing exposes the wine to less heating, the sauce will possess more of the natural flavors and color of the wine. The more the original wine flavors are desired to be retained during cooking, the later the wine should be added. Correspondingly, this requires more care in selecting the wine.

Occasionally, dry wines can be used in dessert preparation, e.g., as a poaching fluid for firm fruit, be incorporated into a sherbet, or function as a blending medium for creamy custards. Otherwise, only sweet wines are compatible with dessert, notably fresh, fully mature, low-acid fruits such as sweet cherries, nectarines, plums, and apricots.

Because carbon dioxide escapes from wine even more rapidly during cooking than alcohol, there is little rationale for using sparkling wine in lieu of still wine. This is especially so with champagne. Such a waste would be almost a sacrilege.

Cooking speeds the loss of alcohol from wine, but its loss is slower than generally realized (Table 9.1). Thus, if wine is taken with a meal as well as added during cooking, the residual alcohol contents from both sources should be taken into account in guesstimating one's blood alcohol level and driving compentency.

TABLE 9.1 Comparison of various methods of food preparation on the loss of alcohol[a]

Preparation method	Alcohol remaining (%)
Flambée	75
Marinade (overnight)	70
Simmered (15 min)	45
Simmered (30 min)	35
Simmered (1 h)	25
Simmered (2 h)	10

From Jackson, 2000, reproduced by permission.
[a]*Data from Augustin et al., 1992.*

TYPES OF OCCASIONS

As the social circumstances of a dinner changes, so does wine presentation. Correspondingly, the quality or style of wine chosen should reflect the social conditions. Typically this will involve the selection of a fine (mature) wine to complement a special meal. Refined food preparation is likely to demand exceptional wines. As such, honored guests deserve the best one can offer (assuming they will appreciate your generosity). But, as noted above, old delicate wines are best reserved for after the meal, so that their finesse can express itself optimally. In contrast, fine wines would be ill-suited to summer relaxing by the pool.

Regrettably, the choices people make often reflect more habit than appropriateness. For example, the balance of a dry table wine is more in evidence combined with food than alone. This is even more so with red wines. Nevertheless, table wines are increasingly served in place of the more traditional sherries or ports as an aperitif. Consumer trends seem to change almost as fast as computer technology.

WINE PRESENTATION

Presentation Sequence

Grand meals, served with a multitude of wines, are now largely viewed as an anachronism. There is much greater awareness of the dangers of excessive food and wine consumption, both in terms of personal health as well as in one's ability to drive. In addition, the accumulation of alcohol, due to sampling multiple wines, progressively dulls the senses. If one cannot clearly remember the vinous pairings of the previous evening, what was their value? Nevertheless, a brief discussion of traditional views on presentation sequence is provided below.

Because sensory acuity tends to decline during a meal, the flavor intensity of a series of wines logically should increase throughout the meal. The choice of dry sherry as an aperitif is an exception. Their marked flavor stimulates the appetite and activates the secretion of digestive juices. More in keeping with incremental flavor augmentation is the choice of a light dry white or sparkling wine before dinner. They equally activate the release of digestive juices, and their subtle flavors do not blunt the appetite. Thus, their consumption can be closer to the meal. They also combine well with many hors d'oeuvres. A more intensely flavored white or red wine is typically served with the main course, depending on the character of the principal ingredients. With sumptuous meals, several table wines may be served consecutively. In this case, if the wines are of similar character, the youngest wine should be served first, reserving the finer, more mature vintage(s) for somewhat later. At the end of the meal, a full-flavored fortified wine, such as a port, palo cortado, or setubal may be served before or after dessert. Alternatively, a botrytized wine, such as an auslese or sauternes may be selected in lieu of, or in addition to, dessert.

Several serving traditions have developed over the years. One recommends that the finest wines should be presented last, the rationale being that they will appear all the better by the comparison. However, this does run the risk that satiety may numb the senses, reducing appreciation (or the ability to fully assess the wine's qualities). Another custom advises that the better the wine, the simpler the accompanying meal. The avoidance of strong food flavors allows the subtleties of the wine to be perceived more cleanly. Finally, convention states that guests should be informed of the vinous pleasures to come. This not only enhances anticipation, but also increases expectation. If this is the outcome, wonderful. Otherwise, it may produce expressed accolade, but unspoken disappointment–the greater the expectation, the greater must be the manifestation. This can be avoided by presenting the wine without acknowledgment of origin or price. If the guests recognize the wine's quality, in the absence of extrinsic factors, the estimation of your guest's connoisseurship, and your personal pleasure, will be much enhanced. If not, one must face the chagrin of having wasted your "treasure" on sensory simpletons. Although risky, the most memorable vinous experiences occur when a beatific wine is sampled without the expectation that heavenly delights are imminent.

My own preference is to reserve my best wines for myself and my spouse, or a select few, unbiased, open-minded friends. This will avoid any embarrassment of presenting a fine (or expensive) wine that fails to meet expectations. Conversely, if your anticipation of its quality is fulfilled, you will have more opportunity to enjoy the wine slowly and savor every nuance to the full with an appreciative audience. That could give rise to hope that you might be remembered, if for nothing else, for your generosity in presenting fine wines.

Cellaring

For many consumers, cellaring in the traditional sense is out of the question. A wine rack in an apartment or condo cannot be considered cellaring. In addition, for many wines there is no need, or opportunity, with most wines being consumed within hours or days of purchase. Occasionally, this might be justified on the contention that, without elaborate precautions, most wines will deteriorate quickly at modern room termperatures. If aging were to be considered be in the range of 20 or more years, this would have validity. However, only a few red and very sweet white wines are likely to retain much sensory appeal after half a century, except under optimal cool conditions. More relevant is whether the purchaser prefers the fresh, fruity, varietal fragrance of young wines, to the more subtle, delicate, aged bouquet of fully aged wines. Most modern wines are produced to be consumed with pleasure immediately upon, or up to several years, after purchase. Premium red wines are still produced to possess a higher tannic content, on the belief (probably true) that it enhances aging potential. It does protect some important flavorants against early oxidation. Correspondingly, premium wines require aging to develop their potential. Coincident with a 'softening' of the tannins is the progressive loss of fruity and varietal flavors, and their substitution with a wonderful, but distinctly different, subtle, aged bouquet.

The principle wines that definitely should not be stored for an extended period are *nouveau* wines. These often show obvious deterioration within several months of bottling. Neither are bag-in-box wines designed to be stored long term, nor do they age well in the container. Most wines sealed with synthetic corks should also be consumed within one, or at most, a few years. This situation may change as manufacturers solve problems associated with oxygen permeability. In contrast, modern screw caps appear as effective at aging premium wines as closures made from the best-quality cork (Hart and Kleinig, 2005). Aluminum caps also have the advantage that they will not crumble or split as cork on opening often does after many years in-bottle, a potential embarrassment and major annoyance at the dinner table. Avoiding this is probably one of the best reasons for opening older bottles closed with cork away from the table and out of sight.

In the absence of a belowground cellar, or a refrigerated wine cabinet, a dark, cool location is about as good as can commonly be expected. Temperatures below 20°C are preferable, with 10°C generally viewed as ideal for prolonged aging. However, because cool temperatures significantly retard aging, premium wines may take decades to reach maturity. Temperatures in the 15–20°C range are a convenient compromise between the desire to age and the impulse to sample. Another, complementary solution is to purchase several bottles, and sample them periodically over an extended period. Unless one takes excellent notes, or has an incredible memory, such an exercise is too often futile. Although moderate temperatures may achieve more rapid aging, extended storage at temperatures above 25°C is ill-advised. For example, fruity-smelling esters rapidly break down to their component (less flavorful) constituents at such temperatures, and compounds such as dimethyl sulfide can accumulate to undesirable levels (Marais, 1986).

The duration of aging required to reach some optimal plateau of quality is a contentious issue. Probably as good an indicator as any may be the price or prestige of the wine. Relatively inexpensive wines, at least from established and well-known regions, probably need little aging, though this does not necessarily imply that further aging will go unrewarded or is detrimental. Certainly, "jug" wines will unlikely benefit from aging. They are produced to have minimal character (to avoid offending the clientele for this genre of wine), and are unlikely to be sealed in a manner to avoid significant oxidation for more than several months. For many wines, the primary limiting factor relative to aging potential is how the bottle is sealed, and the antioxidant content of the wine (primarily tannins or sulfur dioxide). If a standard wine is adequately protected against oxidation, and kept relatively cool (15–20°C), it should retain much of its desirable character for at least several years.

PLATE 9.2   Example of a wine cellar designed with sections containing 12 bottles. *Photo courtesy of R. Jackson.*

For cork-closed bottles, cellaring bottles in a horizontal position is recommended. This avoids significant moisture loss from the stopper, retaining its resilience and maintaining a tight adherence to the bottle neck. Despite this general recommendation, some studies have shown that upright storage does not necessarily result in detectable wine oxidation within two (Lopes et al., 2006) or more years (Skouroumounis et al., 2005). Storage position is of little or no importance for bottles closed with screw caps, except for convenience.

The humidity level of most storage areas is rarely an issue, except in earthen cellars where the humidity may be high. In such situations, mold growth can damage the appearance of the label and eventually make it illegible. The scenic photos of old bottles festooned with cobwebs in dark cellars are without labels, and for good reason.

An incredible diversity of holding racks and construction materials have been used to store wine, e.g., clay tiles, brick, stone, metal, plastic, and wood. Each has its benefits and drawbacks. Wood is possibly the most common construction material. It possesses high strength/weight, is easily worked, joined, and repaired, and can donate an aesthetically warm ambience. This can be further enhanced by the periodic application of beeswax. The example shown in Plate 9.2 was designed to lessen the aspect of the rack and accentuate the visual impact of the bottles. The openness also facilitates identification of the wine. Designing each section to potentially contain 12 bottles of any wine is standard practice. The 8 × 8-cm slots accept most 750-mL bottles.

Glasses

Glasses have been discussed in both Chapters 5 and 6. In the home and restaurant, the desires and needs are usually different from those in the sensory lab or during training/tasting sessions. Nonetheless, the ISO wine-tasting glass is fully appropriate in both home and restaurant situations. Its one major drawback is its fragile crystal nature. Thicker, sturdier, regular glass versions are more practical for everyday use (Plate 5.13, right). They can also be purchased in more generous volumes, possibly more appropriate for the dinner table. Regrettably, any of these are rarely readily available in most retail stores. Restaurant supply stores are usually the best source, but ISO glasses can also be purchased through several Internet sites. Certainly the more common bulbous-shaped glasses will do, but lack finesse for the same or higher price. What they often do have in their favor is impressive size, too often equated with elegance (or is that excess).

At the upper end of the market are crystal glasses produced by firms such as Reidel, Waterford, Bohemia, Luigi Bormioli, and Ravenscroft. They produce a wide range of wine glasses. Glass shape and size do affect a wine fragrance, by changing the dynamics of how aromatics escape and become concentrated in the bowl. However, there is no published evidence that particular shapes enhance the appreciation of specific wines. Nonetheless, this does not negate the psychological impact of presenting separate wines in different voluminous glasses over the course of an elegant dinner.

Serving Temperature

The range of temperatures typically recommended for various types of wine were noted earlier (Chapters 5 and 6). These preferences probably reflect the effects of temperature on gustatory sensations in the mouth and the volatilization of wine aromatics. However, habituation cannot also be ruled out as a major factor (see Zellner et al., 1988). The wine's temperature also has direct effects on wine appreciation in its traditional association with food. The standard cool temperature of white wines adds to its fresh, cleansing influence in the mouth. This is especially appreciated with hot, spicy foods. Even red wines at room temperature will often provide some refreshing contrast with the food's temperature.

Breathing and Aeration

Breathing was discussed in Chapter 6. Correspondingly, little further needs to be noted here, other than to reiterate that it is unnecessary. The current aerator fad is both silly and unnecessary. Enhancing the release of aromatics from the wine is far more rapid and effective when the wine is swirled in the glass. There are also no known undesirable reactions that occur sufficiently rapidly to be of sensory significance within the duration of sampling a wine. It took several days for 20-mL samples of wine, in open-topped 60-mL-capacity bottles, to show significant reductions in the concentration of aromatics (Roussis et al., 2009). This is about 50 times the surface area/volume ratio of a newly opened 750-mL bottle of wine, and occurred over days, not minutes. What can be useful, though, is focusing on changes in the wine's fragrance while the wine is in the glass, especially for finer quality wines. Occasionally the wine shows a progressive increase and transformation in fragrance, what is termed development. To miss this experience, by attempting to rush to some hypothetical optimum moment, would prevent detecting one of the more fascinating sensory delights a wine can express.

Wine Preservation After Opening

Usually, no preservation is required, most bottles being emptied by the end of the meal. Where a partially full bottle remains, there is the proverbial issue of conservation. The common, but inadequate solution is to reseal the bottle and store it in the refrigerator. While the cool temperature does delay deterioration, it also increases oxygen uptake by the wine (oxygen solubility increases relative to lowered temperature). Usually there is noticeable character loss within a day, and the wine may be unrecognizable (but still drinkable) within several days.

An inexpensive means of reducing the degree of oxidation is to use one of the commercially available vacuum pumps. Although partially effective in that role, it is inadequate in preserving the wine's fragrance. During the interval between opening and resealing, the wine can absorb oxygen. This initiates the oxidation of aromatics. In addition, hand pumps do not fully evacuate the air, leaving oxygen that can be absorbed by the wine. Finally, and more importantly, the partial vacuum encourages the escape of aromatic compounds into the headspace of the bottle (frequently this is more than half the bottle's volume). When the wine is subsequently poured, these aromatics remain in the bottle, eventually escaping into the air. The result is a loss of character due the escape of volatiles into the headspace gases above the wine and some minor oxidation. The loss of volatiles into the headspace of the bottle can be detected by comparing the wine's fragrance in a glass with that remaining in the empty bottle (smelling at the neck of the bottle).

An inelegant but more effective solution is to pour the portion of the bottle one expects not to consume into one or more clean, small, screwcap bottles immediately upon opening. Later is better than not at all, but not as preferable as upon opening the bottle. If the bottle(s) chosen are filled to the rim, this should essentially prevent any oxidative deterioration. Also, there will be no irreversible loss of aromatics into the headspace, since there is essentially none. Additional protection could be provided by flushing the empty bottles with carbon dioxide, nitrogen, or argon, but this is unlikely to be worth the effort or expense.

For the oenophile who is preserving samples for more than a day or two, the former technique is cumbersome. A commercial solution may be a product such as the Pek system (Plate 9.3). It preserves wine by displacing air that enters the bottle after opening and pouring. Argon gas is injected from a cylinder in the cap of the device. Because argon is heavier than air, it effectively displaces air (and its associated oxygen). The internal light and transparent front panel permit the bottle label to remain in prominent display. Refrigerated units can also maintain the wine at an appropriate serving temperature. Although applicable for home use, they are ill-designed for restaurant application.

For aficionados or commercial establishments who wish to serve wine by the glass from a series of bottles, several producers (e.g., Cruvinet, Eurocave, WineKeeper) produce a variety of refrigerated units with this option (Plate 5.11). Units dispensing four bottles (two white, two red) are appropriate for home use. Larger units are typically

PLATE 9.3   Pek Supremo wine-preservation system. *Photo courtesy of Pek Preservation Systems, Windsor, CA.*

PLATE 9.4   WineKeeper individual dispensing units. *Photo courtesy of WineKeeper, Santa Barbara, CA.*

used in restaurants. Some, such as Enomatic, can be supplied with a card system that allows customers to serve themselves. Samples are dispensed through individual spigots in refrigerated units (Plate 5.12), or may be attached to individual bottles where appearance is unimportant (Plate 9.4). Compressed nitrogen or argon gas are typically used to both dispense the wine and fill the volume voided by the poured wine.

It is essential to periodically cleanse and disinfect the units. Even traces of oxygen left in the headspace after inserting the replacement stopper can permit the growth of spoilage bacteria, notably acetic acid bacteria. Bacterial multiplication in the spigot is even more of a problem. Any residual wine in the spigot absorbs oxygen, favoring bacterial growth and the conversion of ethanol to acetic acid. Without frequent wine dispensing (essentially daily), vinegarization of the residual wine in the spigot can taint the first sample poured. These problems are avoided in the dispenser produced by FreshTech Inc. (Plates 9.5 and 9.6). It has a system where the cork is removed and replaced with a dispensing stopper in a nitrogen-filled environment. The shooter is also flushed with nitrogen after dispensing a wine sample. The system is gravity fed, with nitrogen (produced by its own generator) replacing the dispensed wine at atmospheric pressure. Small systems are available with self-contained nitrogen or argon gas cylinders under the Vino-Barista label. Many of these wine-dispenser systems come in a range of sizes or can be designed to suit the specifications of the client. However, if sales are sufficient, such precautions are unnecessary, as in some wine bars (Plate 9.7).

All dispenser systems are designed to minimize or avoid wine oxidation prior to dispensing into a glass. However, there have been few studies on what actually occurs to residual wine in a bottle, after opening, with no attempt at preservation. Left in the bottle, wine begins to show distinguishable sensory changes in character within

PLATE 9.5   Basic WHYNOT wine-preservation system. *Photo courtesy of Fresh Tech Inc., Tokyo, Japan.*

PLATE 9.6   On design installed WHYNOT unit. *Photo courtesy of Fresh Tech Inc., Tokyo, Japan.*

PLATE 9.7   Illustration of a wine bar that when sufficiently popular does not need a wine-dispenser system other than servers. *Photo courtesy of Gordon's Wine Bar, London, UK, photo taken by Simon Burgess.*

about 8–12 h. On a longer term, Poulton (1970) found that oxygen consumption showed a first-order reaction rate, becoming complete within 10–12 days in white wine. For example, if half a bottle were poured, the headspace in the bottle would contain about 120 mg oxygen, which is more than sufficient to initiate noticeable oxidation. Aromatic deterioration has been correlated with a decline in the presence of fruit esters and terpenoid aromatics (Roussis et al., 2005). Because the decline was reduced (delayed) in the presence of some phenolics and antioxidants (Roussis et al., 2007), it was concluded that the aromatic deterioration was at least partially due to oxidation. It may also result from the progressive volatilization and escape of aromatics from the wine. Contrary to common belief, short-term deterioration has not been associated with the accumulation of acetaldehyde (Escudero et al., 2002; Silva Ferreira et al., 2003), produced indirectly as a consequence of phenol oxidation. This probably results from the rapid formation of nonvolatile complexes between acetaldehyde and various wine constituents. This would effectively prevent the development of an oxidized (aldehyde) odor in the wine. Extensive contact with oxygen is required for the development of an oxidized (sherry-like) acetaldehyde taint.

The results noted above have all involved exposure to air for periods considerably longer than typical during a tasting. However, Russell et al. (2005) did investigate exposure periods lasting 0, 5, 15, and 30 min. Tasters detected no sensory differences in a Merlot over the test period, although there were discernable changes in the concentration of some phenolics. The only wine aromatic known to degrade quickly on exposure to air is hydrogen sulfide, the source of a rotten egg odor. Nonetheless, there is ample evidence that wine fragrance changes qualitatively after pouring. In the glass, finer wines often show development, an interesting aromatic transformation over the duration of a tasting. Like the opening of a flower, these changes are highly appreciated (until the fragrance begins to fade). It is interpreted as a sign of quality. Part of this phenomenon undoubtedly involves modifications to the dynamic (partition coefficient) equilibrium between dissolved and weakly-bound aromatic in the wine, and their presence in the headspace gases above the wine. This process is greatly facilitated by swirling. Swirling also increases the rate at which oxygen dissolves in the wine. However, what, if any, effect this might have on a wine's aroma, within the timeframe of a tasting, remains speculative.

Label Removal

The human brain is remarkably skilled at retaining information, but few of us are endowed with the ability to remember even a fraction of the impressions the thousands of wines we may taste over our lifetime. One solution is to record impressions in a book dedicated to the purpose, such as a wine log. Although writing the name, producer, and vintage are sufficient to denote the wine's origin, most people prefer to add the label.

For labels attached with old, water-based glues, soaking in warm water for an hour is more than sufficient for label removal. When the label had a waxy, plastic, or aluminum foil coating, several days were required for water to seep in from along the edges, and the label float off. For some other glues, use of hot water, strong detergent, or the addition of ammonia was necessary. Occasionally, rubbing alcohol was effective. Once free, rubbing the back of the label removed any remaining glue and the label was ready for pressing. However, with the increased use of heat-activated or other non-water-soluble glues, these treatments are ineffective. The only method that is occasionally effective is a soak in hot water, followed by the use of a single-sided razor blade (e.g., GEM or PAL blades). With slow, meticulous slicing action, the label can usually be extricated. However, a far simpler alternative is to print a copy of the label from the producer's website, or use a smart phone or camera to take a picture of the label. Using a digital wine log on a tablet or computer avoids having to print the label.

For actual labels, place the label between two sheets of heavy felt paper or several pieces of newsprint. Once damp-dry, place the label between dry sheets of felt paper or newsprint. It can be sandwiched between sheets of corrugated cardboard, in-between sheets of plywood (8 × 11 in. is typical). Heavy weights are applied (or the whole apparatus is tied tightly with straps, as in a plant press). The latter will be very familiar to anyone having taken a course in plant taxonomy. Within several hours the label will be dry and perfectly flat. If there are wrinkles or creases, these can usually be easily removed with pressure applied with a cold iron. For labels removed by slicing, place the label first on an equivalently sized piece of cling wrap or wax paper. This will prevent the label from sticking to the newsprint or felt paper during pressing and drying. Otherwise, the procedure of drying and pressing is identical to other labels. So prepared, labels make an accurate and attractive record of the wine's origin.

Postscript

The combination of wine with food can be conceived in a wide diversity of ways. It can be viewed in terms of simplicity (personal preference), following "expert" opinion, as a complement or contrast to the food, as a palate

PLATE 9.8 Illustration of the warm ambiance for tasting wine possible in a wine bar. *Photo courtesy of Gordon's Wine Bar, London, photo taken by Simon Burgess.*

cleanser or condiment, as a component in food preparation or cooking, as a predinner appetizer or dessert substitute, or consumed partially for its health benefits. In all, though, it is couched in some manner in terms of what Prescott (2015) considers "units of pleasure that influence our motivation to consume."

Finally, there is the question of whether there ever is the "right" wine. As Andrea Immer (2002) points out, the right wine is often a matter of convenience, or what you happen to want at the time, in other words, *de gustibus non est disputandum*. Nonetheless, half the joy of preparing a special meal may come from selecting the wine(s) that will grace the table. For the guests, too, much pleasure can be derived from contemplating the vinous gems to be offered. When one is appreciating wine outside the lab, psychological factors that may enhance enjoyment definitely have their place. The pleasing ambiance of a cosy wine bar (Plate 9.8), or posh restaurant, candlelight, and delicate background music is well known to soothe one into a positive frame of mind, conducive to appreciation, and far different than the perception one would get with the same wine, were one hurried (harried) at home. Even using wine to impress some business associate can occasionally be justified, although it seems a shame if one needs label prestige rather than sensory quality to impress.

But, when all is said and done, the old Spanish proverb seems to sum it up perfectly:

For wine to taste like wine, it should be drunk with a friend.

Suggested Readings

Berridge, K.C., 1996. Food reward: Brain substrates of wanting and liking. Neurosci. Biobehav. Rev. 20, 1–25.
Blake, A.A., 2004. Flavor perception and the learning of food preferences. In: Taylor, A.J., Roberts, D.D. (Eds.), Flavour Perception, Blackwell Publ. Ltd, Oxford, UK, pp. 172–202.
Charters, S., 2006. Wine and Society: The Cultural and Social context of a Drink. Elsevier Butterworth-Heinemann, Burlington, MA.
Freedman, P. (Ed.), 2006. Food: The History of Taste, Universities of California Press, Berkeley, CA.
Kiple, K.F., Ornelas, K.C., 2000. The Cambridge World History of Food. Cambridge University Press, Cambridge.
Laudan, R., 2000. Birth of the modern diet. Sci. Am. 283 (2), 76–81.
McGee, H., 2004. On Food and Cooking—The Science and Lore of the Kitchen, 2nd ed. Scribners, New York.
Prescott, J., 2012. Taste Matters: Why We Like the Food We Do. Reaktion Books, London, UK.
Rietz, C.A., 1961. A Guide to the Selection, Combination and Cooking of Food, Vol. 1 and 2. AVI, Westport, CT.
Risbo, J., Mouritsen, O. (2013). The emerging science of gastrophysics. Special January issue of Flavour.
Rozin, E., 1983. Ethnic Cuisine: The Flavor-Principle Cookbook. Stephen Green, Brattleboro, VI.
Salles, C., 2006. Odour-taste interactions in flavour perception. In: Voilley, A., Etiévant, P. (Eds.), Flavour in Food, Woodhouse Publ. Inc, Cambridge, UK, pp. 345–368.
Shepherd, G.M., 2013. Neurogastronomy: How the Brain Creates Flavor and Why it Matters. Columbia University Press, New York, NY.
Spence, C., Wang, Q.J., 2015. Wine and music (III): So what if music influences the taste of the wine. Flavour 4, 36 (15 pp).
Tannahill, R., 1973. Food in History. Stein & Day, New York.

Web Sites

The science of taste. KQED Quest. https://www.youtube.com/watch?v=0HxAB54wlig
Food, the Brain and Us: Exploring our historical, cultural and sensory perceptions of food https://www.youtube.com/watch?v=ygdbRCdsM6g

References

Ahn, Y.-Y., Ahnert, S.E., Bagrow, J.P., Barabási, A.-L., 2011. Flavor network and the principles of food pairing. Sci. Reports 1 (196), 1–7. http://dx.doi.org/10.1038/srep00196.

Amoore, J.E., 1977. Specific anosmia and the concept of primary odors. Chem. Senses Flavor. 2, 267–281.

Andrews, J., 1985. *Peppers.* University of Texas Press, Austin.

Anonymous, 1986. The history of wine: sulfurous acid–used in wineries for 500 years. German Wine Rev. 2, 16–18.

Arancibia, C., Jublot, L., Costell, E., Bauarri, S., 2011. Flavor release and sensory characteristics of o/w emulsions. Influence of composition, microstructure and rheological behavior. Food Res. Int. 44, 1623–1641.

Arvisenet, G., Voilley, A., Cayot, N., 2002. Retention of aroma compounds in starch matrices: Competitions between aroma compounds toward amylose and amylopectin. J. Agric. Food Chem. 50, 7345–7349.

Augustin, J., Augustin, E., Cutrufelli, R.L., Hagen, S.R., Teitzel, C., 1992. Alcohol retention in food preparation. J. Am. Diet. Assoc. 92, 486–488.

Babes, A., Amuzescu, B., Krause, U., Scholz, A., Flonta, M.-L., Reid, G., 2002. Cooling inhibits capsaicin-induced currents in cultured rat dorsal root ganglion neurones. Neurosci. Lett 317, 131–134.

Bastian, S., Bruwer, J., Alant, K., Li, E., 2005. Wine consumers and makers: Are they speaking the same language? Aust. NZ Grapegrower Winemaker 496, 80–84.

Bastian, S.E.P., Payne, C.M., Perrenoud, B., Joselyne, V.L., Johnson, T.E., 2009. Comparisons between Australian consumers' and industry experts' perceptions of ideal wine and cheese combinations. Aust. J. Grape Wine Res. 15, 175–184.

Bastian, S.E.P., Collins, C., Johnson, T.E., 2010. Understanding consumer preferences for Shiraz wine and Cheddar cheese pairings. Food Qual. Pref. 21, 668–678.

Bayarri, S., Taylor, A., Hort, J., 2007. The role of fat in flavor perception: Effect of partition and viscosity in model emulsions. J. Agric. Food Chem. 54, 8862–8868.

Benini, L., Salandini, L., Rigon, G., Tacchella, N., Brighenti, F., Vantini, I., 2003. Effect of red wine, minor constituents, and alcohol on the gastric emptying and the metabolic effects of a solid digestible meal. Gut 52 (Suppl. 1), pA79–pA80.

Berridge, K.C., 1996. Food reward: Brain substrates of wanting and liking. Neurosci. Biobehav. Rev. 20, 1–25.

Bertino, M., Beauchamp, G.K., Engelman, K., 1986. Increasing dietary salt alters salt taste preference. Physiol. Behav. 38, 203–213.

Blake, A.A., 2004. Flavor perception and the learning of food preferences. In: Taylor, A.J., Roberts, D.D. (Eds.), Flavour Perception. Blackwell Publ. Ltd, Oxford, UK, pp. 172–202.

Bolhuis, D.P., Newman, L.P., Keast, R.S.J., 2016. Effects of salt and fat combinations on taste preference and perception. Chem. Senses 41, 189–195.

Breslin, P.A.S., Beauchamp, G.K., 1995. Suppression of bitterness by sodium: Variation among bitter taste stimuli. Chem. Senses 20, 609–623.

Breslin, P.A.S., Beauchamp, G.K., 1997. Salt enhances flavour by suppressing bitterness. Nature 387, 563.

Brossard, C., Rousseau, F., Dumont, J.-P., 2007. Perceptual interactions between characteristic notes smelled above aqueous solutions of odorant mixtures. Chem. Senses 32, 319–327.

Brune, M., Rossander, L., Hallberg, L., 1989. Iron absorption and phenolic compounds: Importance of different phenolic structures. Eur. J. Clin. Nutr. 43, 547–558.

Bult, J.H.F., de Wijk, R.A., Hummel, T., 2007. Investigations on multimodal sensory integration: Texture, taste, and ortho- and retronasal olfactory stimuli in concert. Neurosci. Lett. 411, 6–10.

Buttery, R.G., Guadagni, D.G., Ling, L.C., 1973. Flavor compounds: Volatilities in vegetable oil and oil-water mixtures. Estimation of odor thresholds. J. Agric. Food Chem. 21, 198–201.

Camacho, S., den Hollander, E., ven de Velde, F., Stieger, M., 2015a. Properties of oil/water emulsions affecting the deposition, clearance and after-feel sensory perception of oral coatings. J. Agric. Food Chem. 63, 2145–2153.

Camacho, S., van Eck, A., van de Velde, F., Stieger, M., 2015b. Formation dynamics of oral oil coatings and their effect on subsequent sweetness perception of liquid stimuli. J. Agric. Food Chem. 63, 8025–8030.

Christensen, C.M., Bertino, M., Beauchamp, G.K., Navazesh, M., Engelman, K., 1986. The influence of moderate reduction in dietary sodium on human salivary sodium concentration. Archs. Oral Biol. 31, 825–828.

Chrea, C., Valentin, D., Sulmont-Rossé, C., Mai, H.L., Nguyen, D.H., Adbi, H., 2004. Culture and odor categorization: Agreement between cultures depends upon the odors. Food Qual. Pref. 15, 669–679.

Cook, D.J., Davidson, J.M., Linforth, R.S.T., Taylor, A.J., 2004. Measuring the sensory impact of flavor mixtures using controlled delivery. In: Deibler, K.D., Delwiche, J. (Eds.), Handbook of Flavor Characterization. Marcel Dekker, Inc, New York, NY, pp. 135–149.

Donadini, G., Fumi, M.D., Lambi, M., 2012. The hedonic response to chocolate and beverage pairing: A preliminary study. Food Res. Int. 48, 703–711.

Drewnowski, A., Henderson, S.A., Barratt-Fornell, A., 2001. Genetic taste markers and food preferences. Drug Metab. Disposit. 29, 535–538.

Eiximenis, F. (1384). Terc del Crestis (Tome 3, Rules and Regulation for Drinking Wine). [cited in Johnson (1989) p. 127.]

Escudero, A., Asensio, E., Cacho, J., Ferreira, V., 2002. Sensory and chemical changes of young white wines stored under oxygen. An assessment of the role played by aldehydes and some other important odorants. Food Chem. 77, 325–331.

Ferry, A.-L., Hort, J., Mitchell, J.R., Cook, D.J., Lagarrigue, S., Valles Pamies, B., 2006. Viscosity and flavor perception: Why is starch different from hydrocolloids? Food Hydrocolloids 20, 855–862.

Fischer, C., Fischer, U., Jakob, L., 1996. Impact of matrix variables, ethanol, sugar, glycerol, pH and temperature on the partition coefficients of aroma compounds in wine and their kinetics of volatization. In: Henick-Kling, T., Wolf, T.E., Harkness, E.M. (Eds.), Proc. 4th Int. Symp. Cool Climate Vitic. Enol., Rochester, NY, July 16–20, 1996. NY State Agricultural Experimental Station, Geneva, New York. pp. VII 42–46.

Flandrin, J.-L., Montanari, M. (Eds.), *(Sonnefeld, A., English edition)* 1999. Food. A Culinary History. Penguin, Harmondsworth, England.

Forester, S.C., Waterhouse, A.L., 2008. Identification of Cabernet Sauvignon anthocyanin gut microflora metabolites. J. Agric. Food Chem. 56, 9299–9304.

Franke, A., Teyssen, S., Harder, H., Singer, M.V., 2004. Effects of ethanol and some alcoholic beverages on gastric emptying in humans. Scand. J. Gastroenterol. 39, 638–645.

Garcia-Bailo, B., Toguri, C., Eny, K.M., El-Sohemy, A., 2009. Genetic variation in taste and its influence on food selection. OMICS 13, 69–80.

Genovese, A., Caporaso, N., De Luca, L., Paduano, A., Sacchi, R., 2015. Influence of olive oil phenolic compounds on headspace aroma release by interaction with whey proteins. J. Agric. Food Chem. 63, 3838–3850.

German, J.B., 1999. Food processing and lipid oxidation. Adv. Exp. Med. Biol. 459, 23–50.

Gilmore, M.M., Green, B.G., 1993. Sensory irritation and taste produced by NaCl and citric acid: Effects of capsaicin desensitization. Chem. Senses 18, 257–272.

Gorelik, S., Ligumsky, M., Kohen, R., Kanner, Joseph, 2008. The stomach as a "bioreactor": When red meat meets red wine. J. Agric. Food Chem. 56, 5002–5007.

Green, B.G., Hayes, J.E., 2003. Capsaicin as a probe for the relationship between bitter taste and chemesthesis. Physiol. Behav 79, 811–821.

Gross, G., Jacobs, D.,M., Peters, S., Possemiers, S., van Duynhoven, J., Vaughan, E.E., et al., 2010. In vitro bioconversion of polyphenols from black tea and red wine/grape juice by human intestinal microbiota displays strong interindividual variability. J. Agric. Food Chem. 58, 10236–10246.

Hansson, A., Andersson, J., Leufvén, A., 2001. The effect of sugars and pectin on flavour release from a soft drink-related model system. Food Chem. 71, 363–368.

Harrington, R.J., 2005. The wine and food pairing process: Using culinary and sensory perspectives. J. Culin. Sci. Technol. 4, 101–112.

Harrington, R.J., 2005b. Defining gastronomic identity: The impact of environment and culture on prevailing components, texture and flavors in wine and food. J. Culin. Sci. Technol. 4, 129–152.

Harrington, R.J., 2007. Food and Wine Pairing: A Sensory Experience. John Wiley & Sons, Inc, Hoboken, NJ.

Harrington, R.J., Hammond, R., 2005. The direct effects of wine and cheese characteristics on perceived match. J. Foodser. Bus. Res. 8, 37–54.

Harrington, R.J., Hammond, R., 2009. The impact of wine effervescence levels on perceived palatability with salty and bitter foods. J. Foodser. Bus. Res. 12, 234–246.

Hart, A., Kleinig, A., 2005. The role of oxygen in the aging of bottled wine. Wine Press Club of New South Wales, Rozelle, Australia, pp. 1–14.

Hersleth, M., Mevik, B.-H., Naes, T., Guinard, J.-X., 2003. Effect of contextual factors on liking for wine – use of robust design methodology. Food Qual. Pref. 14, 615–622.

Hodgson, M.D., Langridge, J.P., Linforth, R.S.T., Taylor, A.J., 2005. Aroma release and delivery following the consumption of beverages. J. Agric. Food Chem. 53, 1700–1706.

Hollowood, T.A., Linforth, R.S.T., Taylor, A.J., 2002. The effect of viscosity on the perception of flavour. Chem. Senses 27, 583–591.

Homma, R., Yamashita, H., Funaki, J., Ueda, R., Sakurai, T., Ishimaru, Y., et al., 2012. Identification of bitterness-masking compounds from cheese. J. Agric. Food Chem. 60, 4492–4499.

Hornsey, I.S., 2016. Beer: History and types. In: Caballero, B., Finglas, P., Toldrá, F. (Eds.), The Encyclopedia of Food and Health. Academic Press, London, UK, pp. 345–354.

Hyde, R.J., Pangborn, R.M., 1978. Parotid salivation in response to tasting wine. Am. J. Enol. Vitic. 29, 87–91.

Immer, A., 2002. Great Tastes Made Simple: Extraordinary Food and Wine Pairing for Every Palate. Broadway Books, New York, NY.

Jackson, R.S., 2014. Wine Science: Principles, Practice, Perception, 4th ed. Elsevier Press, San Diego, CA.

Janssen, S., Laermans, J., Verhulst, P.-J., Thijs, T., Tack, J., Depoortere, I., 2011. Bitter taste receptors and α-gustducin regulate the secretion of ghrelin with functional effects on food intake and gastric emptying. PNAS 108, 2094–2099.

Johnson, H., 1989. Vintage: The Story of Wine. Simon & Schuster, New York, NY.

Kanner, J., Gorelik, S., Roman, S., Kohen, R., 2012. Protection by polyphenols of postprandial human plasma and low-density lipoprotein modification: The stomach as a bioreactor. J. Agric Food Chem. 60, 8790–8796.

Karametsi, K., Kokkinidou, S., Ronningen, I., Peterson, D.G., 2014. Identification of bitter peptides in aged cheddar cheese. J. Agric. Food Chem. 62, 8034–8341.

Karrer, T., Bartoshuk, L., 1991. Capsaicin desensitization and recovery on the human tongue. Physiol. Behav. 49, 757–764.

Karrer, T., Bartoshuk, L., 1995. Effects of capsaicin desensitization on taste in humans. Physiol. Behav. 57, 421–429.

Katsuragi, Y., Sugiura, Y., Lee, C., Otsuji, K., Kurihara, K., 1995. Selective inhibition of bitter taste of various drugs by lipoprotein. Pharm. Res. 12, 658–662.

Keast, R.S.J., Breslin, P.A.S., Beauchamp, G.K., 2001. Suppression of bitterness using sodium salts. CHIMIA Int. J. Chem. 55, 441–447.

King, M., Cliff, M., 2005. Evaluation of ideal wine and cheese pairs using a deviation-from-ideal scale with food and wine experts. J. Food Qual. 28, 245–256.

Kinsella, J.E., 1990. Flavour perception and binding. Intern. News Fats Oils Related Materials: Inform. 1, 215–226.

Kolor, M.K., 1983. Identification of an important new flavor compound in Concord grape, ethyl 3–mercaptopropionate. J. Agric. Food Chem. 31, 1125–1127.

Kuroda, M., Miyamura, N., 2015. Mechanism of the perception of "*kokumi*" substances and the sensory characteristics of the "*kokumi*" peptide, γ-Glu-Val-Gly. Flavour 4, 11 (3 pp).

Laudan, R., 2000. Birth of the modern diet. Sci. Am. 283 (2), 76–81.

Lawless, H., Rozin, P., Shenker, J., 1985. Effects of oral capsaicin on gustatory, olfactory and irritant sensations and flavor identification in humans who regularly or rarely consume chili pepper. Chem. Senses 10, 579–589.

Lawless, H.T., Schlake, S., Smythe, J., Lim, J., Yang, H., Chapman, K., et al., 2004. Metallic taste and retronasal smell. Chem. Senses 29, 25–33.

Lindeman, R., Lang, A.R., 1986. Anticipated effects of alcohol consumption as a function of beverage type: A cross-cultural replication. Intl. J. Psychol. 21, 671–678.

Linforth, R., Taylor, A., 2006. The process of flavour release. In: Voilley, A., Etiévant, P. (Eds.), Flavour in Food. Woodhouse Publ. Inc, Cambridge, UK, pp. 287–307.

Linscott, T.D., Lim, J., 2016. Retronasal odor enhancement by salty and umami tastes. Food Qual. Pref. 48, 1–10.

Liszt, K.I., Walker, J., Somoza, V., 2012. Identification of organic acids in wine that stimulate mechanisms of gastric acid secretion. J. Agric. Food Chem. 60, 7022–7030.

Liszt, K.I., Eder, R., Wendelin, S., Somoza, V., 2015. Identification of catechin, syringic acid, and procyanidin B2 in wine as stimulants of gastric acid secretion. J. Agric. Food Chem. 63, 7775–7783.

Liu, L., Simon, S.A., 2000. Capsaicin, acid and heat-evoked currents in rat trigeminal ganglion neurons: Relationship to functional VR1 receptors. Physiol. Behav. 69, 363–378.

Lopes, P., Saucier, C., Teissedre, P.L., Glories, Y., 2006. Impact of storage position on oxygen ingress through different closures into wine bottles. J. Agric. Food Chem. 54, 6741–6746.

Madrigal-Galan, B., Heymann, H., 2006. Sensory effects of consuming cheese prior to evaluating red wine flavor. Am. J. Enol. Vitic. 57, 12–22.2.

Marais, J., 1986. Effect of storage time and temperature of the volatile composition and quality of South African *Vitis vinifera* L. cv. Colombar wines. In: Charalambous, G. (Ed.), The shelf life of foods and beverages. Elsevier, Amsterdam, pp. 169–185.

Martin, S., Pangborn, R.M., 1971. Human parotid secretion in response to ethyl alcohol. J. Dental Res. 50, 485–490.

McRae, J.F., Jaeger, S.R., Bava, C.M., Beresford, M.K., Hunter, D., Jia, Y., et al., 2013. Identification of regions associated with variation in sensitivity to food-related odors in the human genome. Curr. Biol. 26, 1596–1600.

Mennella, J.A., Reed, D.R., Mathew, P.S., Roberts, K.M., Mansfield, C.J., 2015. "A spoonful of sugar helps the medicine go down": Bitter masking by sucrose among children and adults. Chem. Senses 40, 17–25.

Mitchell, M., Brunton, N.P., Wilkenson, M.C., 2011. Impact of salt reduction on the instrumental and sensory flavor profile of vegetable soup. Food Res. Int. 44, 1036–1043.

Moskowitz, H.W., Kumaraiah, V., Sharma, K.N., Jacobs, H.L., Sharma, S.D., 1975. Cross-cultural differences in simple taste preferences. Science 190, 1217–1218.

Muñoz-González, C., Sémon, E., Martín-Álvarez, P.J., Guichard, E., Moreno-Arribas, M.V., Geron, G., et al., 2015. Wine matrix composition affects temporal aroma release as measured by proton transfer reaction – time-of-flight – mass spectrometry. Aust. J. Grape Wine Res. 21, 367–375.

Nardini, M., Forte, M., Vrouvsek, U., Mattivi, F., Viola, R., Scaccini, C., 2009. White wine phenolics are absorbed and extensively metabolized in humans. J. Agric. Food Chem. 57, 2711–2718.

Negoias, S., Visschers, R., Boelrijk, A., Hummel, T., 2008. New ways to understand aroma perception. Food Chem. 108, 1247–1254.

Newcomb, R.D., Xia, M.B., Reed, D.R., 2012. Heritable differences in chemosensory ability among humans. Flavor 1, 9 (9 pp).

Nolano, M., Simone, D.A., Wendelschafer-Crabb, G., Johnson, T., Hazen, E., Kennedy, W.R., 1999. Topical capsaicin in humans: Parallel loss of epidermal nerve fibers and pain sensation. Pain 81, 135–145.

Nygren, I.T., Gustafsson, I.B., Haglund, A., Johansson, L., Noble, A.C., 2001. Flavor changes produced by wine and food interactions: Chardonnay wine and Hollandaise sauce. J. Sens. Stud. 16, 461–470.

Nygren, I.T., Gustafsson, I.-B., Johansson, L., 2002. Perceived flavour changes in white wine after tasting blue mould cheese. Food Service Technol. 2, 163–171.

Nygren, I.T., Gustafsson, I.-B., Johansson, L., 2003a. Effects of tasting technique – sequential tasting vs. mixed tasting – on perception of dry white wine and blue mould cheese. Food Service Technol. 3, 61–69.

Nygren, I.T., Gustafsson, I.-B., Johansson, L., 2003b. Perceived flavour changes in blue mould cheese after tasting white wine. Food Service Technol. 3, 143–150.

Ogi, K., Yamashita, H., Terada, T., Homma, R., Shimizu-Ibuka, A., Yoshimura, E., et al., 2015. Long-chain fatty acids elicit a bitterness-masking effect on quinine and other nitrogenous bitter substances by formation of insoluble binary complexes. J. Agric. Food Chem. 63, 8493–8500.

Paulsen, M.T., Rognsa, G.H., Hersleth, M., 2015. Consumer perception of food–beverage pairings: The influence of unity in variety and balance. Int. J. Gastron. Food Sci. 2, 83–92.

Peck, R.C., Gebers, M., Voas, R.B., Romano, E., 2008. The relationship between blood alcohol concentration (BAC), age, and crash risk. J. Safe. Res. 39, 311–319.

Pelchat, M.L., Johnson, A., Chan, R., Valdez, J., Ragland, J.D., 2004. Images of desire: Food-craving activation during fMRI. Neuroimage 23, 1486–1493.

Pettigrew, S., Charters, S., 2006. Consumers' expectations of food and alcohol pairing. Br. Food J. 108, 169–180.

Peyrot des Gachons, C., Mura, E., Speziale, C., Faveau, C.J., Dubreuil, G.F., Breslin, P.A.S., 2012. Opponency of astringent and fat sensations. Curr. Biol. 22, R829–R830.

Poulton, J.R.S., 1970. Chemical protection wine against oxidation. Die Wynboer 466, 22–23.

Prescott, J., 2012. Taste Matters: Why We Like the Food We Do. Reaktion Books, London, UK.

Prescott, J., 2015. Flavours: The pleasure principle. Flavor 4, 15. (3 pp).

Rabe, S., Kring, U., Berger, R.G., 2003. Initial dynamic flavour release from sodium chloride solutions. Eur. Food Res. Technol. 218, 32–39.

Ridout, F., Gould, S., Nunes, C., Hindmarch, I., 2003. The effects of carbon dioxide in Champagne on psychometric performance and blood-alcohol concentration. Alcohol Alcoholism 38, 381–385.

Rietz, C.A., 1961. A Guide to the Selection, Combination and Cooking of Food, Vols. 1 and 2. AVI, Westport, CT.

Roberts, D.D., Pollien, P., Antille, N., Lindinger, C., Yeretzian, C., 2003. Comparison of nosespace, headspace, and sensory intensity ratings for the evaluation of flavor absorption by fat. J. Agric. Food Chem. 51, 3636–3642.

Roets, A., Schwartz, B., Guan, Y., 2012. The tyrany of choice: A cross-cultural investigation of mazimizing-satisficing effects on well-being. Judgem. Decis. Making 7, 689–704.

Ronca, G., Palmieri, L., Maltinti, S., Tagliazucchi, D., Conte, A., 2003. Relationship between iron and protein content of dishes and polyphenol content in accompanying wines. Drugs Exp. Clin. Res. 29, 271–286.

Roudot-Algaron, F., Le Bars, D., Einhorn, J., Adda, J., Gripon, J.C., 1993. Flavour constituents of aqueous fraction extracted from Comte cheese by liquid carbon dioxide. J. Food Sci. 58, 1005–1009.

Roussis, I.G., Lambropoulos, I., Papadopoulou, D., 2005. Inhibition of the decline of volatile esters and terpenols during oxidative storage of Muscat-white and Xinomavro-red wine by caffeic acid and N-acetyl-cysteine. Food Chem. 93, 485–492.

Roussis, I.G., Lambropoulos, I., Tzimas, P., 2007. Protection of volatiles in a wine with low sulfur dioxide by caffeic acid or glutathione. Am. J. Enol. Vitic. 58, 274–278.

Roussis, I.G., Papadopoulou, D., Sakarellos-Daitsiotis, M., 2009. Protective effect of thiols on wine aroma volatiles. Open Food Sci. J. 3, 98–102.

Rozin, E. (1982a). The structure of cuisine. In: The Psychobiology of Human Food Selection, (L.M. Barker, ed.), pp. 189–203. (L.M. Barker, Ed). AVI Publ. Co., AVI, Westport, CT.

Rozin, P., 1977. The use of characteristic flavourings in human culinary practice. In: Apt, C.M. (Ed.), Flavor: Its Chemical Behavioural and Commercial Aspects. Westview Press, Boulder, CO, pp. 101–127.

Rozin, P., 1982b. Human food selection: The interaction of biology, culture and individual experience. In: Barker, L.M. (Ed.), The Psychophysiology of Human Food Selection. AVI Publ. Co, Westport, CT, pp. 225–254.

Rozin, P., Schiller, C., 1999. The nature and acquisition of a preference for chili pepper by humans. Motiv. Emotion 4, 77–101.

Rozin, P., Vollmecke, T.A., 1986. Food likes and dislikes. Annu. Rev. Nutr. 6, 433–456.

Russell, K., Zivanovic, S., Morris, W.C., Penfield, M., Weiss, J., 2005. The effect of glass shape on the concentration of polyphenolic compounds and perception of Merlot wine. J. Food Qual. 28, 377–385.

Scheibehenne, B., Miesler, L., Todd, P.M., 2007. Fast and frugal food choices: Uncovering individual decision heuristics. Appetite 49, 578–589.

Schwartz, B., 2004. The tyranny of choice. Sci. Am. 290 (4), 71–75.

Serianni, E., Cannizzaro, M., Mariani, A., 1953. Blood alcohol concentrations resulting from wine drinking timed according to the dietary habits of Italians. Quart. J. Stud. Alcohol 14, 165–173.

Silva Ferreira, A.C., Hogg, T., Guedes de Pinho, P., 2003. Identification of key odorants related to the typical aroma of oxidation-spoiled white wines. J. Agric. Food Chem. 51, 1377–1381.

Simons, C.T., Dessirier, J.-M., Carstens, M.I., O'Mahony, M., Carstens, E., 1999. Neurobiological and psychophysical mechanisms underlying the oral sensation produced by carbonated water. J. Neurosci. 19, 8134–8144.

Simons, C.T., O'Mahony, M., Carstens, E., 2002. Taste suppression following lingual capsaicin pre-treatment in humans. Chem. Senses 27, 353–365.

Skouroumounis, G.K., Kwiatkowski, M.J., Francis, I.L., Oakey, H., Capone, D.L., Duncan, B., et al., 2005. The impact of closure type and storage conditions on the composition, colour and flavour properties of a Riesling and a wooded Chardonnay wine during five years' storage. Aust. J. Grape Wine Res. 11, 369–384.

Skurray, G.R., Perkes, J.M., Duff, J., 1986. Effect of marinading with wine, sodium bicarbonate or soy sauce on the thiamin content of beef. J. Food Sci. 51, 1059–1060.

Small, D.M., Gerber, J.C., Mak, Y.E., Hummel, T., 2005. Differential neural responses evoked by orthonasal versus retronasal odorant perception in humans. Neuron 47, 593–605.

Sourabié, A.M., Spinnler, H.-R., Bonnarme, P., Sainte-Eve, A., Landaud, S., 2008. Identification of a powerful aroma compound in Munster and Camembert cheeses: Ethyl 31mercaptopropionate. J. Agric. Food Chem. 56, 4674–4680.

Tagliazucchi, D., Verzelloni, E., Conte, A., 2005. Effect of some phenolic compounds and beverages on pepsin activity during simulated gastric digestion. J. Agric. Food Chem. 53, 8706–8713.

Takeuchi, H., Kato, H., Kurahashi, T., 2013. 2,4,6-Trichloroanisole is a potent suppressor of olfactory signal transduction. PNAS 110, 16235–16240.

Tamura, T., Taniguchi, K., Suzuki, Y., Okubo, T., Takata, R., Konno, T., 2009. Iron is an essential cause of fishy aftertaste formation in wine and seafood pairing. J. Agric. Food Chem. 57, 8550–8556.

Tannahill, R., 1973. Food in History. Stein & Day, New York.

Tempere, S., Cuzange, E., Malik, J., Cougeant, J.C., de Revel, G., Sicard, G., 2011. The training level of experts influences their detection thresholds for key wine compounds. Chem. Percept. 4, 99–115.

Teyssen, S., González-Calero, G., Schimiczek, M., Singer, M.V., 1999. Maleic acid and succinic acid in fermented alcoholic beverages are the stimulants of gastric acid secretion. J. Clin. Invest. 103, 707–713.

This, H., 2005. Molecular gastronomy. Nature Materials 4 (1), 5–7.

This, H., 2013. Molecular gastronomy is a scientific discipline, and note by note cuisine is the next culinary trend. Flavour J. 2 (1), 1–8.

Tominaga, T., Guimbertau, G., Dubourdieu, D., 2003. Role of certain volatile thiols in the bouquet of aged champagne wines. J. Agric. Food Chem. 51, 1016–1020.

Unwin, T., 1991. Wine and the Vine, An Historical Geography of Viticulture and the Wine Trade. Routledge, London.

Unwin, T., 1992. Images of alcohol: Perceptions and the influence of advertising. J. Wine Res. 3, 205–233.

Vandiver, P., Koehler, C.G., 1986. Structure, processing, properties, and style of Corinthian amphoras In: Kingery, W.D. (Ed.), Ceramics and Civilization, Vol. 2. Technology and Style American Ceramics Society, Columbus, OH, pp. 173–215.

Viegas, O., Amaro, L.F., Ferreira, I.M., Pinho, O., 2012. Inhibitory effect of antioxidant-rich marinades on the formation of heterocyclic aromatic amines in pan-fried beef. J. Agric. Food Chem. 60, 6235–6240.

Vilgis, T.A., 2013. Texture, taste and aroma: Multi-scale materials and the gastrophysics of food. Flavour J. 2 (17), 1–5.

Wansink, B., Cordua, G., Blair, E., Payne, C., Geiger, S., 2006. Wine promotions in restaurants. Cornell Hotel Restaurant Admin. Quart. 47, 327–336.

Wood, C., Siebert, T.E., Parker, M., Capone, D.L., Elsey, G.M., Pollnitz, A.P., et al., 2008. From wine to pepper: Rotundone, an obscure sesquiterpene, is a potent spicy aroma compound. J. Agric. Food Chem. 56, 3738–3744.

Yan, M., Luo, M.Z., 1999. Influences of salt-out on extraction of catechin. China Tea 21, 29.

Zellner, D.A., Stewart, W.F., Rozin, P., Brown, J.M., 1988. Effect of temperature and expectations on liking for beverages. Physiol. Behav. 44, 61–68.

Glossary

Acidity the concentration of nonvolatile organic acids in must or wine, or the perception of acids in the mouth.

After-Taste the lingering taste perception in the mouth after wine has been swallowed.

After-Smell the fragrance that lingers in the mouth after swallowing wine.

Aging changes in wine chemistry that occur after bottling; occasionally includes maturation.

Amygdala the part of the brain most associated with the perception of strong emotions, but also with decision making and memory processing.

Anthocyanin flavonoid pigments that generate the red to purple color of red grapes and wine (see Table 6.6).

Aroma the fragrant perception that is derived from aromatic grape constituents, usually a desired attribute, but in some instances and to some individuals, a negative quality, e.g., the foxy attribute of some *V. labrusca* cultivars.

Aromatic a relatively lipid-soluble compound sufficiently volatile to stimulate the olfactory receptors in the nose.

Baking the heating used in processing wines such as madeira to obtain their distinctive bouquet.

Bouquet the fragrant sensation in wine derived from aromatics produced during fermentation, maturation, or aging.

Breathing a term that refers to either the exposure of wine to air shortly following the opening or the decanting of bottled wine.

Browning an undesired increase in the yellow-brownish cast of a wine; primarily considered to be due to the oxidation of phenolic compounds.

Carbonic Maceration the intracellular fermentation of grape cells that may precede yeast fermentation; used in the production of Beaujolais-like wines.

Caudalie the unit (seconds) of flavor duration (finish) in the mouth after swallowing or expectorating a wine.

Chemesthesis sensations that arise from receptors of the trigeminal nerve, such as astringency, touch, pain, heat, cold, and prickling.

Color Density the sum of the absorbency of a wine, typically measured for red wines at 420 and 520 nm ($E_{420} + E_{520}$).

Color Stability the long-term retention of a wine's red color; favored by low pH, oxygen exclusion, and anthocyanin polymerization with tannins.

Congener compounds that influence the sensory quality of related substances; usually refers to alcohols other than ethanol in wines or distilled beverages.

Dry having no perceived sweetness.

Dumping the tendency of tasters to include relevance of an important attribute they detect, for which there is no term, to another seemingly related term.

Fatty Acid a long, straight hydrocarbon possessing a carbonyl (acid) group at one end.

Flavor the integrated percept of taste, touch, and odor of food and beverages; often influenced by color and sound.

Fortification the addition of wine spirits to arrest fermentation, increase alcohol content, or influence the course of wine development.

Fragrance the aromatic aspect of wine.

Fusel Oil an organic compound containing more than two carbon atoms and an alcohol group; volatile forms often possess a pronounced fusel or petroleum odor (see Higher Alcohol).

Halo Effect the tendency of one assessment to affect a second assessment.

Headspace the volume of gas left in a container after filling and attaching the closure.

Herbaceous describing an odor induced by the presence of above-threshold levels of several hexanols and hexanals, or certain pyrazines.

Higher Alcohol alcohols with more than two carbon atoms that correspondingly have higher molecular weights and boiling points (also called fusel oils); produced in small amounts in wines due to yeast metabolism but often occurring at recognizable concentrations only in distilled products such as wine spirits and brandy.

Hippocampus the part of the brain that seems most central to the retention of short-term memory and its subsequent processing into long-term memory.

Lees sediment that forms during and after fermentation; it includes material such as dead and dying yeasts and bacteria, grape cell remains, seeds, tartrate salts, and precipitated tannins.

Maillard Product the product of nonenzymatic reactions between reducing sugars and amine compounds (i.e., amino acids and proteins), which produce polymeric brown pigments and caramel-like aromatics.

Malolactic Fermentation the decarboxylation of malic to lactic acid by several lactic acid bacteria and some yeast strains; a biologic form of wine deacidification.

Micelle aggregation of molecules in a solution, usually water, that typically form microspheres; the more water-soluble portion faces out into the water whereas the less water-soluble portion is positioned inward; micelles may contain higher concentrations of additional compounds inside the sphere than is found in the suspending solution.

Mineral Taste a retronasal odor, thought to be derived from the oxidation of certain fatty acids in the mouth.

Minerality a general term used variously by popular wine writers, presumably to describe a nebulous sensation that may incorporate a range of taste and flavor sensations.

Mouth-feel the sensation produced by compounds dissolved or suspended in the saliva that activate trigeminal receptors in the mouth, producing the various perceptions of astringency, burning, pain, prickling, temperature, touch, viscosity, etc.

Mouthspace the gaseous (air) phase above food and liquid in the mouth (analogous to headspace).

Oak Lactones a pair of optical isomers found in oak that contribute to the characteristic flavor of wine matured in oak cooperage.

Off-odor a fragrant or pungent compound that is generally considered undesirable.

Off-taste an atypical or imbalance in taste sensations considered undesirable.

Olfaction the sensation produced by volatile compounds carried by inspiration or expiration to the olfactory patches in the nose and able to reach and stimulate receptive neurons in the olfactory epithelium.

Olfactory Bulb the region of the brain just above the olfactory receptor regions of the nose; the site in which sensations are collected before being sent to the piriform cortex.

Orbitofrontal Complex the region of the brain where various sensory inputs (olfaction, taste, mouth-feel, vision, sound) are integrated into the percept termed flavor.

Oxidation a reaction in which a compound loses an electron (or hydrogen atom) and becomes oxidized; although molecular oxygen is the principal initiator of most oxidation–reduction reactions in wine, it is oxygen's more reactive radicals, such as peroxide, that directly oxidize most wine constituents; in a more restricted sense, oxidation may be used to refer to the browning of white and red wines, the development of a pungent, cooked vegetable off-odor in bottled wines, the development of a distinct aldehyde odor in sherries, the development of "bottle sickness," or the fragrance loss after bottle opening.

Partition Coefficient an indicator of the relative solubility of a compound in a solution of immiscible solvents (such as oil and water) or between a liquid and the gaseous (air) phase above it.

Pétillance oral sensation generated by a slight amount of carbon dioxide (about 200 kPa—2 atm).

Phenolic Compounds compounds containing one or more benzene-ring structures and at least one hydroxyl (OH^-) group.

Piriform Cortex the region of the brain to which most impulses from the olfactory bulb go; the site in which odor quality is interpreted and odor memories appear to be stored.

Quality the property of wine showing marked aromatic and flavor complexity, subtlety, harmony, and development, associated with a distinct aroma and aged bouquet; in aromatic compounds, quality signifies the subjective similarity to a known flavor or aroma, for example apple-like (in contrast to the intensity of the sensation).

Residual Sugar the sugar content that remains in wine after fermentation is complete; in a dry wine this primarily involves the nonfermentable sugars arabinose and rhamnose.

Sequence Error where the sequence of samples can influence perception, e.g., a strongly astringent sample can enhance the apparent astringency of the next sample.

Taste the sensation produced by substances dissolved or mixed in the saliva that activate receptors in the mouth (mostly tongue); includes bitter, salty, sour, sweet, and umami perceptions.

Tannins polymeric phenolic compounds that can tan (precipitate proteins); in wine they contribute to bitter and astringent sensations, promote color stability, and are potent antioxidants.

Tears the droplets that slide down the sides of a swirled glass of wine; they form as alcohol evaporates and the increased surface tension of the film pulls the fluid on the glass together.

Terroir the combined influences of vineyard atmospheric, soil, and cultural conditions on vine growth and fruit ripening; the term is often misused in an attempt to justify the supposedly unique quality of wines from certain vineyard sites.

Trigeminal Nerve the fifth cranial nerve, three branches of which carry impulses from the nose and mouth; it gives rise to the sensations of astringency, heat, body, prickling, and pain in wine.

Typicity the term used to describe the unique characteristics of wines from a particular region.

Viscosity the perception of the resistance of wine to flow; a smooth, velvety mouth-feel.

Volatile in wine, it refers to the escape of aromatic compounds from wine into the air; it is affected by the partition coefficient of the compound, and how it is affected by other wine constituents, the surface area/volume contact, the formation of micelles, and the formation of reversible complexes with other wine constituents (matrix).

Volatile Acids organic acids that can be readily removed by steam distillation, almost exclusively acetic acid in wine.

Wine Spirit distilled wine used to fortify wines such as sherry and port; it may be highly rectified to produce a neutral-flavored source of high-strength alcohol.

TASTING TERM GLOSSARY

Appearance any visual wine perception.

Clarity–degree of brilliance (absence of haze-causing colloids or particulate matter); can vary from *clear* to *dull* to *cloudy*.

Color–presence of perceptible amounts of yellow, red, or brown pigments in solution.

Sparkle–chains of carbon dioxide bubbles rising in the wine; *still* refers to their absence, *pearl* refers to slight effervescence, and *sparkling* refers to marked, prolonged effervescence.

Tears–formation of droplets of wine on the sides of the glass after swirling the wine in a glass; *arches* refers to the overall appearance of the formation of tears and as they begin to descend.

Balance the perception of harmony, notably between the sweet, sour, bitter, and astringent sensation in the mouth, but clearly influenced by the intensity of the aromatic sensation of the wine; one of the most highly regarded of wine attributes.

Fragrance olfactory perceptions that may come from sniffing the wine (orthonasal) or vapors reaching the nasal passages via the mouth (retronasal).

Aroma–the fragrance derived from grapes; typically resembling some complex of *fruity, floral, spicy, herbaceous,* or other aromatic attributes (see Fig. 1.3).

Bouquet–the fragrance derived either from alcoholic fermentation (e.g., *fruity, yeasty*), processing (e.g., *buttery, nutty, oaky, madeirized*), or aging (e.g., *oxidized, leathery, cigar-box*).

Complexity–a quantitative/qualitative term referring to the perceptible presence of many aromatic compounds, combining to generate pleasure; a highly desirable attribute.

Development–the change in the aromatic quality during the period the wine is sampled; a highly regarded wine attribute.

Duration–the length of time the wine maintains its distinctive character, before becoming just vinous (generically wine-like); long duration is a highly regarded attribute.

Expression–the relative evolution of the fragrance; *closed-in,* if not apparent in a young wine; *opening,* progressive increase in aromatic intensity; *faded* when absent in an old wine; *well developed* when amply present.

Off-odors–the detectable presence of olfactory compounds considered unacceptable or atypical (see Fig. 1.4 for specific examples).

Quality for specific odors, it refers to descriptive terms applied to the odor; for wine, it refers to the ranking wine relative to some standard (personal, varietal, regional, stylistic, etc.).

Artistic–features such a complexity, subtlety, dynamism, development, duration, harmony (balance) uniqueness, and memorableness that distinguish the flavor of the wine.

Regional–the flavor attributes that are thought to characterize the wines from a particular region.

Stylistic–presence of a fragrance typical of a particular wine-making style (i.e., *carbonic maceration, recioto, botrytized, flor, baked*).

Varietal–presence of an aroma distinctive to as single, or a group of related, grape cultivars (see Tables 7.2 and 7.3).

Taste oral perceptions derived from the taste buds.

Sweet–the perception of sweetness; a complex response to compounds such as sugars, glycerol and ethanol, as influenced by sensations to the acidic and phenolic compounds in the wine; *cloying* refers to an intense, unpleasant sensation of sweetness; the opposite is *dry.*

Acidity–a sour perception derived as a complex response to organic acids, wine pH, and as affected by other sapid substances, notably sugars, ethanol, and phenolic compounds; *flat* refers to the absence of sufficient acidity, the opposite of *acidic; tart* usually denotes an appropriate, pleasant acidic perception.

Bitter–a perception induced primarily by the presence of small molecular weight phenolic compounds that is influenced marginally by the presence of sugars, ethanol, and acids.

Mouth-Feel perceptions derived from trigeminal receptors in the mouth.

Alcoholic–a negative expression indicating the excessive presence of alcohol, relative to other sensory attributes.

Astringency–a set of tactile sensations including dryness, puckeriness, and dust-in-the-mouth perceptions; provoked principally by the polyphenolic content of wine, but also induced by acids; *smooth* implies a positive response to astringency; *rough* refers to excessive astringency.

Body–a term of imprecise meaning, generally referring to the summary perception of "weight" or "richness" in the mouth; a tactile sensation induced primarily by the presence of alcohol, but clearly influenced by the presence of sugars, glycerol (in high concentration), and phenolics; *full-bodied* is a positive perception of weight in the mouth; *watery* is the negative perception of the absence of sufficient body.

Burning–an intense sensation of heat that can be generated either by high alcohol or very high sugar contents.

Heat–a perception of warmth generated by the presence of ethanol.

Pain–a sharp sensation occasionally induced by excessive tannin contents, or high carbon dioxide contents under cold conditions.

Prickling–the pleasant sensation of pain induced by the bursting of bubbles of carbon dioxide on the tongue.

Finish the perceptions that linger in the mouth after the wine has been swallowed or expectorated; when measured (in seconds), each unit is termed a *caudalie.*

After-smell–the flavor aspect of finish; usually a highly regarded attribute.

Aftertaste–the taste–mouth-feel aspects of finish.

Index

Note: Page numbers followed by "*f*" and "*t*" refer to figures and tables, respectively.

Printed in the United States
By Bookmasters